Chitosan

Scrivener Publishing
100 Cummings Center, Suite 541J
Beverly, MA 01915-6106

Publishers at Scrivener
Martin Scrivener (martin@scrivenerpublishing.com)
Phillip Carmical (pcarmical@scrivenerpublishing.com)

Chitosan

Derivatives, Composites and Applications

Edited by

Shakeel Ahmed and Saiqa Ikram

Scrivener Publishing

WILEY

This edition first published 2017 by John Wiley & Sons, Inc., 111 River Street, Hoboken, NJ 07030, USA and Scrivener Publishing LLC, 100 Cummings Center, Suite 541J, Beverly, MA 01915, USA

For more information about Scrivener publications please visit www.scrivenerpublishing.com.

Wiley Global Headquarters
111 River Street, Hoboken, NJ 07030, USA

For details of our global editorial offices, customer services, and more information about Wiley products visit us at www.wiley.com.

Library of Congress Cataloging-in-Publication Data
Names: Ahmed, Shakeel, 1989- editor. | Ikram, Saiqa, editor.
Title: Chitosan : derivatives, composites and applications/edited by Shakeel Ahmed and Saiqa Ikram.
Description: Hoboken, NJ, USA : Wiley, 2017. | Includes bibliographical references and index. |
Identifiers: LCCN 2017023345 (print) | LCCN 2017024016 (ebook) | ISBN 9781119364818 (epub) | ISBN 9781119364801 (pdf) | ISBN 9781119363507 (cloth)
Subjects: LCSH: Chitosan--Biotechnology. | Chitosan--Industrial applications.
Classification: LCC TP248.65.C55 (ebook) | LCC TP248.65.C55 C547 2017 (print) | DDC 660.6--dc23
LC record available at https://lccn.loc.gov/2017023345

Cover image: Pixabay.Com
Cover design by: Russell Richardson

Set in size of 11pt and Minion Pro by Exeter Premedia Services Private Ltd., Chennai, India

10 9 8 7 6 5 4 3 2 1

Contents

Section III Chitosan in Biomedical Applications

Section IV Chitosan in Agriculture and Water Treatment

Preface

Many books are devoted to the description of polymers that are not of biopolymers or the polymers which are originating from the natural resources; the techniques for their mass/industrial production, their art, value, and their distribution reflecting the wealth of the information existing in the field of science and engineering to therapeutics. On the other hand, there is no book exclusively devoted to the fascinating topic of biopolymers especially to "Chitosan" and its unique properties and applications. A few books contain scattered chapters and details on chitosan emphasizing the possibilities of locating fragments of information elsewhere. However, again there is no book that is solely devoted to chitosan and its versatile applications. In the meantime, the expanded applications of chitosan are on the rise in many fields, making a book that covers both past and novel applications of this "Biopolymer for the Future" as well as its properties and ways in which to manipulate them, crucial.

The main aim in compiling this book is to present, in easy-to-follow sequence, a description of chitosan production methods and of techniques which can be used to estimate and modify its physical and chemical properties. This book offers a full description not only of the traditional and recent developments following applications of chitosan in the fields of biotechnology, environmental studies, food, medicine, water treatments, drug delivery including all therapeutically usages, but also topics which have never been covered in the literature, making it of utmost importance to industry and academia.

We hope that this book will assist all levels of readers. It is dedicated not only to academic community but also to the researchers and industrialist who will find this book to be a source of knowledge as well as a launching pad for novel ideas and inventions. In particular, this book is expected to be of interest to the people involved in formulations, water-treating methods, food scientists, technologists, industrial chemists and engineers. Potential readers also include both professional and dedicated non-professional environmentalists, agriculturists, and those working on the development of novel chitosan-based matrices and their applications.

Finally, it is expected that this book will find a prominent place in the traditional universities and research institutions libraries where chemistry, biotechnology, medicine as well as environmental studies, and other practical and theoretical mechanized topics are taught, studied, and implemented.

Shakeel Ahmed and Saiqa Ikram
Delhi, India
June 2017

Section I

PRODUCTION AND DERIVATIVES OF CHITOSAN

1

Chitin and Chitosan: History, Composition and Properties

Annu[1*], Shakeel Ahmed[1,2] and Saiqa Ikram[1*]

[1]Bio/Polymers Research Laboratory, Department of Chemistry, Jamia Millia Islamia (A Central University), New Delhi, India
[2]Department of Textile Technology, Indian Institute of Technology Delhi, New Delhi, India

Abstract

Chitin and chitosan are most abundant naturally occurring polymers, ranked second after cellulose. Chitin is generally extracted from natural sources viz. terrestrial organisms, marine organisms, microorganisms like fungi and enzymatically from crustaceans shell waste materials. On the other hand, chitosan is obtained from the deacetylation of the former, chitin.

Nowadays, chitin and chitosan are commercially manufactured from biowastes obtained from aquatic organisms. But due to the seasonal and capricious availability of raw materials, terrestrial crustaceans and mushrooms are the alternative source for their production. Structurally, chitin and chitosan are *N*-acetyl-D-glucosamine units and D-glucosamine units, respectively, with only difference in hydroxyl group of cellulose. Both the biopolymers are biodegradable and possess many beneficial properties such as biocompatibility, antimicrobian, hemostatic, anti-inflammatory, antioxidant, mucoadhesion, analgesic, non-toxicity, adsorption enhancing, antihypertensive, anticholesterolemic, anticancer, and antidiabetic. Due to having such respectable properties chitin, chitosan, and their corresponding derivatives are greatly exploiting by the scientists and also getting tremendously better results in medical and engineering fields.

***Keywords*:** Chitin, chitosan, history, structure, properties, solubility, viscosity, biomedical, anticancer

**Corresponding authors*: annuchem92@gmail.com; sikram@gmail.com

Shakeel Ahmed and Saiqa Ikram (eds.) Chitosan, (3–24) © 2017 Scrivener Publishing LLC

1.1 Chitin

1.1.1 History

French scientist Prof. Henri Braconnot for the very first time discovered chitin in 1811 in mushroom. After that Odier found the same compound in the cuticles of insects in 1823 and called it "Chitin" from the Greek word meaning tunic, covering or envelope [1].

This was how it begins the research in a new direction of polymers present in our nature. Gradually as the time passed away in 1859, Prof. C. Rouget coined another noval biopolymer, actually derived from previous chitin, and it was known as "Chitosan." In 1878, Ledderhose revealed that the chitin consists of glucosamine and acetic acid. Thereafter, in 1930s and 1940s, both former and latter attract considerable attention as evidenced by about 50 patents. Chitin is the most abundant naturally occurring polymer, ranked second after cellulose and also most abundant naturally occurring polysaccharide possesses amino acid and sugars. Chemically, chitin is composed of *N*-acetyl-D-glucosamine units with β-(1-4) glycosidic linkage [2].

1.1.2 Sources of Chitin

Chitin is generally produced from natural sources viz. terrestrial organisms, marine organisms, microorganisms like fungi and enzymatically from crustaceans shell waste materials. On the other hand, chitosan is obtained from the deacetylation of the former, chitin. Nowadays, chitin and chitosan are commercially manufactured from biowastes obtained from aquatic organisms. But due to the seasonal and capricious availability of raw materials terrestrial crustaceans and mushrooms are the alternative source for their production [2]. The industrial manufacturing of synthetic polymers was restricted by the time because of the inadequate facilities as well as the cutthroat competition in synthetic polymers. Therefore, again the shellfish shells, crustaceans and shrimps revitalized the interest in late 1970s (Global industry analysis since 2004). Various sources for extraction and production of chitin can be categorized as follows:

i. Terrestrial organisms
ii. Marine or Aquatic organisms
iii. Microorganisms (e.g., Fungi)

1.1.2.1 Terrestrial Organisms

From commercial point of view, these organisms are mainly used for the extraction of chitin, due to their easy availability and processability.

Table 1.1 Percentage of chitin produced from different sources [3].

Source	% Chitin produced
Cockroach	30–37% (abdomen, legs, and head) 25–29% (genitalia, antennae, thorax, and cerci) 19% (fore and hind wings)
Extatosoma tiaratum	24% (exuviae) 3.4% (eggs) 3.8% (eggshells)
Sipyloidea sipylus	14%
Honey bee	23–32%
Silkworms	20%

Terrestrial species generally includes crustaceans such as *Porcellio scaber*, *Armadillidium vulgare;* arthropods, nematodes, insects, silkworms, mosquitoes, honeybee, *Sipyloidea sipylus*, *Drosophila melanogaster*, *Extatosoma tiaratum*, and many more [3]. As the composition of these organisms is quite different, there is a variation in the contribution to the percentage of chitin produced as depicted in Table 1.1.

1.1.2.2 *Marine or Aquatic Organisms*

Chitin produced from aquatic species includes diatoms, algae, crabs, shrimps, lobster, squids, and krill. The hazardous waste materials generated from head, thorax, shells, and claws of shellfish are utilized as raw materials for processing, containing 15–40% chitin, 20–40% proteins, and 20–50% $CaCO_3$. The waste streams of molluscs and crustaceans are the main source of chitin. Also they constitute a rich source of proteins, flavor compounds, and various pigments and hence are of great attention for different research fields and industries as well. Actually, if they are disposed of in the open environment of the water bodies, such as sea or river, then they are problematic due to the higher biochemical oxygen demand and eutrophication [4].

1.1.2.3 *Microorganisms (e.g., Fungi or Mushroom)*

Chitin can be obtained from microorganisms either by fermentative methods or biotechnological methods. Utilization of various microbes makes it easier for industries to produce chitin widely and hence microbes are considered as the economic source of extracting chitin as well as

Table 1.2 Examples of sources of extraction of chitin [5].

Terrestrial	Marine or aquatic	Microorganisms
Porcellio scaber	*Euphausia* sp. (Krill)	*Mucor rouxii*
Armadillidium vulgare	*Paralithodes* sp. (King crab)	*Mucor racemosus*
Sipyloidea sipylus	*Chionoecetes* sp. and *Carcinus* sp. (Crab)	*Cunninghamella Echinulate*
Drosophila melanogaster	*Callinectes* sp. (Blue crab)	*Aspergillus niger*
Extatosoma tiaratum	*Loliginidae* sp. (Squid)	*Rhizopus oryzae*
Bombyx mori	*Pandalus* sp. (Shrimp)	*Mucor circinelloides*
Apis mellifera	*Nephrops* sp. (Lobster)	*Pleurotus ostreatus*
Periplaneta americana	*Lepas* sp. (Goose Barnacle) *Penaeus* sp. (Prawn)	*Absidia. Coerulea*
	Crassostrea sp. (Oyster)	*Lentinus edodes*
	Mytilus sp.	*Auricularia auricula-judae*

chitosan. Microbial world mainly includes fungi (cell wall, mycelia, septa) molds, chrysophyte algae, yeasts, prosthecate bacteria, spores of streptomycete and ciliates. Except Oomycetes, remaining Ascomycetes, Basidiomycetes, Deuteromycetes, and Zygomycetes consists of 2–65% chitin/chitosan. Especially, Mucoralean strains viz. *Syncephalastrum racemosum* and *Cunninghamella echinulata* cell wall exhibited maximum chitin/chitosan yield of 7% per mycelia dry weight under optimum conditions [5–7]. Some of the examples of extraction of chitin from terrestrial, marine and microorganisms are listed in Table 1.2.

1.1.3 Extraction of Chitin

Chitin can be extracted from insect cuticle, tracheae and peritrophic matrix [25], shellfish waste such as shrimps, crabs, krill, lobster, fishes and microorganisms such as fungi or mushroom mycelia and some bacteria as discussed above. Generally, extraction of chitin involves the following steps:

i. Demineralization
ii. Deproteination
iii. Decolorization

There may be difference in the sequence of these steps [3, 8]. For instance, Kumari *et al.* described the extraction of chitin from fish scales of *Labeo rohita* but after demineralization they performed decolorization followed by deproteination [9].

1.1.3.1 Demineralization

It can be performed by using strong acids such as HCl, H_2SO_4, HNO_3 and weak acids such as CH_3COOH and HCOOH. But generally HCl seems to be best one. Previous researchers showed that the concentration of HCl was about 1N or 2N for 0.3–96 h from 25–100 °C (for insect cuticles). But from last decade, it has been found that only 0.25M to 2M concentration of HCl for 1–36 or 48 h is sufficient at temperature 0–100 °C for just 15 min [8]. This step particularly performed to remove the minerals, especially $CaCO_3$.

1.1.3.2 Deproteination

It is generally performed under alkaline medium of 0.75N–2.5N or 1M NaOH aqueous solution for 2–42 or 1–72 h at temperature 40 or 65–100 °C in crustaceans and marine shrimp shells, respectively. NaOH, KOH, $NaHCO_3$, Na_2CO_3, K_2CO_3, $Ca(OH)_2$, $NaHSO_3$, and Na_2SO_3 are the reagent that can be used for deproteination of the crustaceans. Due to the adverse effect of these perilous chemicals on environment, nowadays, proteolytic enzymes such as trypsin, pepsin, or papain have been used to raise the efficiency of deproteination [10, 11].

1.1.3.2.1 Enzymatic Deproteination

Proteolytic enzymes can be extracted from various sources such as *Bacillus mojavensis* A21, *B. subtilis* A26, *B. licheniformis* NH1, *B. licheniformis* MP1, *Vibrio metschnikovii* J1, alkaline protease from Sardinelle (*Sardinella aurita*), Goby (*Zosterisessor ophiocephalus*), *Aspergillus clavatus* ES1, and Grey triggerfish (*Balistes capriscus).* One unit of protease activity can be defined as the amount of enzyme required to liberate 1 μg of tyrosine per minute [12]. Degree of Deproteination can be calculated as follows:

$$\% \ DDP = \frac{[(Pi \times S) - (Pf \times R)]}{Pi \times S} \times 100 \quad (1.1)$$

where *Pi* and *Pf* are the concentrations of protein at initial and final hydrolysis; while *S* and *R* are the mass of original sample and hydrolysed residue, respectively [12].

1.1.3.3 Decolorization

In order to remove pigmentation and melanin, the mixture has been treated either with hydrogen peroxide or potassium permanganate solution [3].

According to Mohammed *et al.* approximately 35% of chitin (dry weight) were obtained from prawn shell after deproteination, decolorization, and demineralization [13]. Chitin can be extracted either via using biotechnological method or via chemical method as depicted in Figures 1.1 and 1.2.

1.1.4 Structure and Composition

Chitin is a linear copolymer of β-(1-4)-linked 2-amino-2-deoxy-D-glucan and 2-acetamido-2-deoxy-D-glucan. Structurally, chitin is composed of β-(1-4)-linked *D*-glucosamine units. Unlike cellulose, the hydroxyl group of cellulose is replaced by *N*-acetyl group ($-NHCOCH_3$) as shown in Figure 1.3. Chitin has been existed mainly in three solid state polymorphic forms viz. α, β, and γ. Out of these allomorphs, α-chitin is the most abundant one

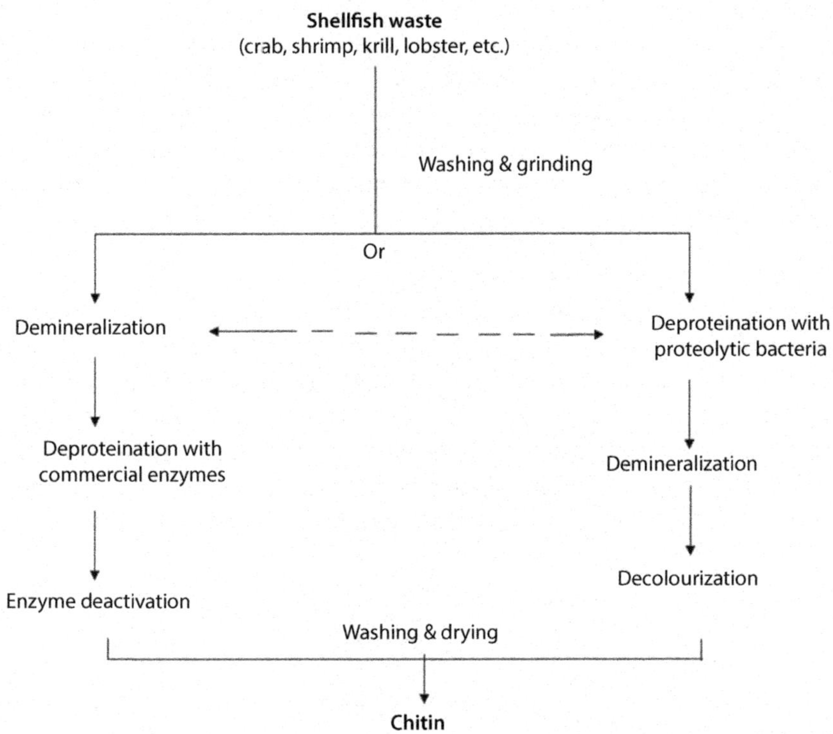

Figure 1.1 Extraction of chitin.

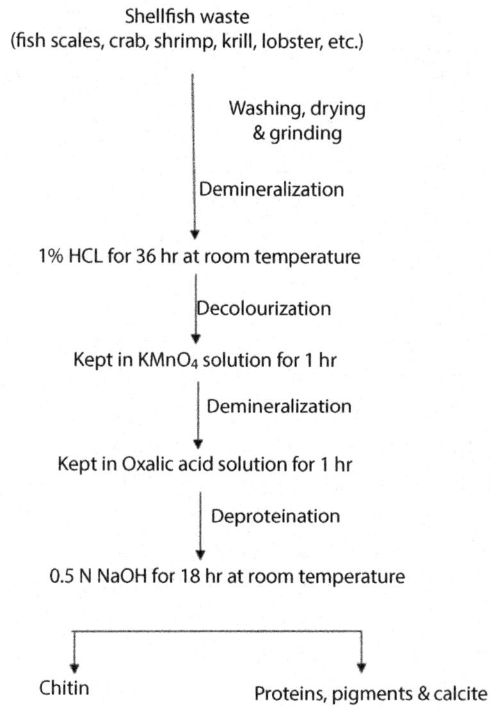

Figure 1.2 Extraction of chitin by alkaline solution.

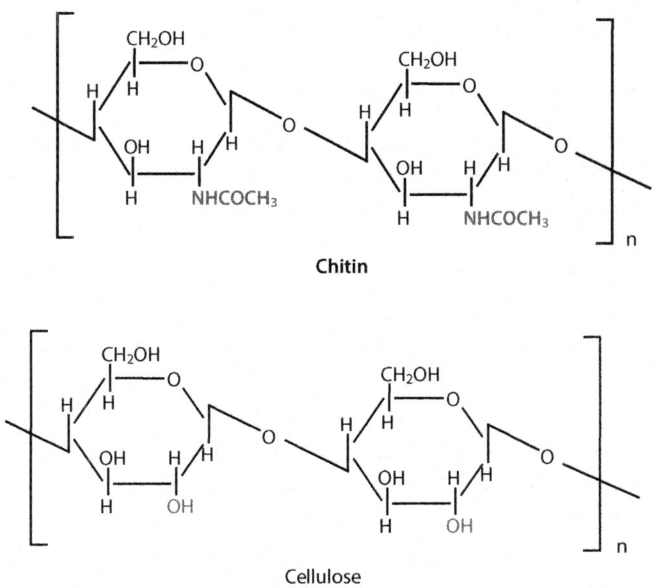

Figure 1.3 Structural comparison of chitin and cellulose.

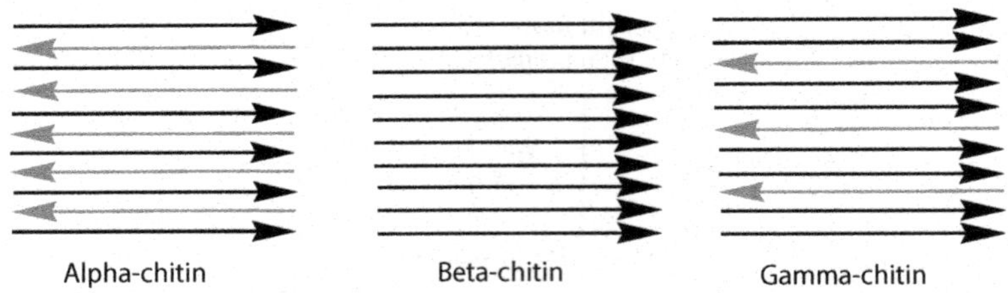

Figure 1.4 Antiparallel and parallel arrangements of different allomorphs of chitin.

followed by β and then rarest one is γ-chitin. The major source of α-form of chitin is generally shrimps, insect cuticle, crab, krill, lobster, cell wall of yeast and Zygomycetes. The abundancy of α-chitin favors the significant quality of chitin as high crystallinity and purity due to the absence of calcium carbonate, proteins, and pigments. Instead β-chitin is found in connotation with proteins in squid pens while γ-chitin is found in cuttlefish stomach lining. X-Ray diffraction revealed that the inner ring present in α-form of chitin is unaffected from hydration while the inner ring of β-chitin is sensitive to hydration. Moreover crystallographically, α-chitin exhibits two antiparallel molecules per unit cell, whereas β-chitin exhibits one parallel arrangement as depicted in Figure 1.4. As far as similarity is concerned, both the allomorphs have same moiety of *N*-acetylglycosamine [14–17].

1.1.5 Properties of Chitin

1.1.5.1 Solubility, Reactivity, and Viscosity

Like cellulose, chitin as biopolymer is highly crystalline materials having specific solvent behavior. As chitin is β-(1,4)-linked *N*-acetyl-D-glucosamine therefore, its solubility and reactivity are highly influenced by $-NH_2$ and –OH functional groups. Being a typical natural polymer obtained from different sources, chitin possesses excessive hydrogen bonding which in turn degrade it before melting and thus making it essential to be dissolve in suitable solvent. Chitin is generally hydrophobic in nature and hence insoluble in water and organic solvents at room temperature but soluble in hexafluoroisopropanol, chloroalcohols, and hexafluoroacetone in association with mineral acid aqueous solution and dimethylacetamide having 5% lithium chloride [18]. In other words, water is a thermodynamically poor solvent for uncharged chitin unit due to strong interactions. Thus, charge on the polymeric chain can make the chitin soluble by releasing the attractive forces. It has also been reported that small amount of chitin can be dissolved in 1% acidic solution at 121 °C for 20 min and on reacting it with HCl and NaOH, decrement in crystallinity index [19, 20]. The hydrolysis of chitin with concentrated acids under drastic conditions produces relatively pure *D*-glucosamine. Additionally, the intrinsic viscosity of chitin can be considered as a function of demineralization time. Temperature is also an important criterion for purity of chitin which if carried out at high temperature produces better results as compared to the low temperature and obtains stable viscosities with time as well. Nitrogen content available in chitin depends on the extent of degree of deacetylation and is found to be 5–8%. However, the nitrogen content available in chitosan is present in the form of primary aliphatic amino groups and hence undergo important amine *N*-acylation and Schiff reactions [18].

1.1.5.2 Miscellaneous Properties

Chitin has also been known for its various properties such as water-binding and fat-binding capacity. Knorr revealed from his studies that 0.5–2.0% of microcrystalline chitin on treating with wheat flour bread or with potato protein white bread, the water binding, fat binding capacity, and emulsifying ability were enhanced from 170–315% (w/w) which is better than microcrystalline cellulose. Generally, chitin cannot be able to yield emulsions but the microcrystalline chitin showed improved emulsifying properties which increases on increasing its concentration (0.12–0.8 g/100 ml water) and 65% water addition is suitable for manufacturing the loaf of bread of chitin [21]. Chitin exhibits almost similar properties as chitosan including biodegradability, biocompatibility, nontoxicity, antimicrobial, anti-inflammatory, anticancer, antioxidant, etc. [22].

1.2 Chitosan

1.2.1 History

In 1859, C. Rouget published his findings that "modified chitin" could be prepared by treating chitin with boiling, concentration solutions of potassium hydroxide in water and he observed that chitin can be manipulated via chemical- and temperature-dependent treatment. Chitin was renamed as "Chitosan" by Hoppe-Seiler in 1894, pronounced as "kite-O-san." Till 1920s, chitosan had been studied by the researchers as a subject and it seeks its importance in different fields and different ways of its extraction. In 1930s chitosan derived from chitin source such as shrimp shells, crabs, lobster, krill, and mushroom had been confirmed by Rammelberg. Till 1950s, some advance diffraction techniques had already been developed such as X-Ray diffraction, which were the most reliable one and it was therefore being proved that the cell wall of fungi, mycelia, and septa consists of chitin and chitosan. In 1960s, chitosan was studied as a hemostatic agent for its ability to bind with red blood cells, for water treatments and detoxification. Recently, it is considered as "fat magnet" because it acts as a fat inhibitor and hence beneficial dietary supplement for weight [23, 24].

1.2.2 Sources and Extraction

Chitosan is found in similar sources as of chitin. Similarly, terrestrial organisms include silkworm, insects, honeybee, arthropods, and nematodes. The exoskeleton of crustaceans, shrimp shells, crab, krill, lobster, etc. has been successfully utilized in the extraction of chitosan. On the other hand, mushrooms; cell wall of fungi, mycelia, and septa; spores of Zygomycetes and Ascomycetes are good source of chitosan besides Basidiomycetes. Generally, chitosan is extracted from the deacetylation of the chitin. As already explained above, chitin can be extracted from different sources, mainly crustaceans and shellfish waste via demineralization, deproteination, and decolorization. After the decolorization of the shellfish (or any other source), deacetylation can be performed in order to obtain the desired product, chitosan. The deacetylation is a chemical process and can be achieved by different methods. The methods involved either chemical or enzymatic method. The degree of deacetylation of chitosan obtained from insect were found to be 70–95% [3, 25]. The flow chart representation of extraction of chitosan has been depicted in Figure 1.5.

1.2.2.1 Deacetylation of Chitin by Chemical Method

Previously, chitosan had been extracted chemically by means of fungi cell wall using alkali and acid treatments. In that process, cell wall constituents mainly proteins, lipids, and chitosan were first treated with 2–4% NaOH for 15–120 min at 100 °C and then the chitosan containing material (alkali insoluble) of cell wall with 2–10% CH_3COOH solution for 1–24 h at 25–95 °C. The components of cell wall which neither dissolve in alkali nor in acid medium are called alkali-acid insoluble material and the acetic acid soluble material is referred to "fungal chitosan" [26, 27].

Chemically, deacetylation can be achieved by treating the washed and purified chitin with 25% and 50% NaOH by 1:5 (w/v) at 80 and 100 °C for 5 and 10 h, respectively. Thereafter, deacetylation the product is washed with deionized water till neutral pH and placed under vacuum oven at 60 °C for getting dried weight chitosan product [13]. Zamani *et al.* extracted chitosan

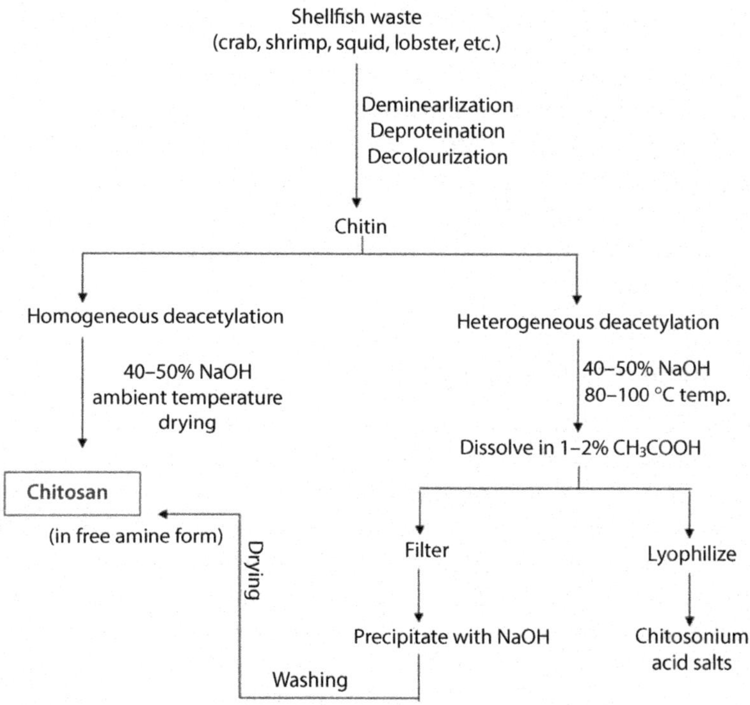

Figure 1.5 Extraction of chitosan.

from Zygomycetes fungus *Rhizomucor pusillus* cell wall by using sulphuric acid as a medium to dissolve it. He revealed from his studies that 45.3% of chitosan can be extracted from the fungi cell wall when treated with 1% H_2SO_4 for 20 min. at 121 °C followed by NaOH alkali solution. Contrary, he concluded that the acetic acid soluble material did not contain chitosan while the alkali-acid insoluble material contained chitosan and phosphate was the major source of impurity of acetic acid soluble material. Moreover, chitosan present in 8% of biomass and 45.3% in alkali insoluble material [28].

1.2.2.2 Deacetylation of Chitin by Enzymatic Method

In order to achieve chitosan, the decolorized and purified chitins attained after demineralization and deproteination can be reacted with NaOH. The concentration of NaOH could be 12.5M at 140 °C for 4 h. The ratio of the mixture of chitin to NaOH can be around 1:10 (w/v). The chitosan thus obtained is completely water soluble in acidic conditions. The residues are washed by distilled water and after 12 h drying in incubator at 50 °C, dry weight chitosan is obtained.

1.2.2.3 Deacetylation of Chitin by Microwave Assisted Method

Recently, it has been found that the three extraction steps of chitin can also be achieved by microwave assisted mechanism as a time saving as well as eco-friendly method. Knidri *et al.* successfully produced chitosan after deacetylation with the help of microwave irradiation, with degree of deacetylation of 82.73% within just 24 min. which is quite good as compared to the conventional method where it took 81.5% degree of deacetylation in much longer time of 5–10 h [29].

1.2.3 Structure and Composition

Chitosan has similar structure like cellulose and chitin, the only difference is the functional group at C-2 position. Cellulose and chitin possess hydroxyl group (–OH) and *N*-acetylamine group ($-NHCOCH_3$), respectively, on the other hand, chitosan consists of amino group ($-NH_2$) at C-2 position. Chitosan is a linear-chain polysaccharide which consists of *N*-acetyl-2-amino-2-deoxy-D-glucopyranose (acetylated unit) and 2-amino 2-deoxy-D-glucopyranose (deacetylated unit), where the repeating units are linked by β-(1 → 4)-glycosidic bonds as shown in Figure 1.6 [30]. Formation of chitosan from deacetylation of chitin can be confirmed by the Fourier transform infrared spectroscopy (FT-IR) due to the two amide bands I and II at 1655 cm^{-1} and 1583 cm^{-1}, respectively. Higher intensity of band I and lower intensity of band II indicate efficient deacetylation and formation of $-NH_2$ group [31]. On the basis of the molecular weight or density of the polymeric chain, chitosan can be divided into two types viz. low density chitosan (LDC) and high density chitosan (HDC).

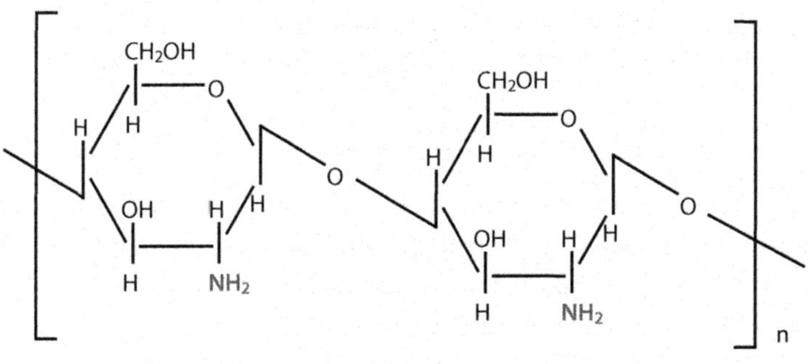

Figure 1.6 Structure of chitosan.

1.2.4 Properties

Chitosan obtained from chitin after various procuring steps can have different conditions such as temperature, concentration, time, and deacetylation which can affect the physical, chemical, or biological properties of the product.

1.2.4.1 Physical Properties

1.2.4.1.1 Viscosity
Generally, viscosity of the chitosan solution is affected by many factors such as temperature, pH, concentration, molecular weight, degree of deacetylation, and method of extraction. Starting with temperature, if the temperature of the chitosan solution increases, its viscosity will decrease. Reverse is observed in case of concentration where there is an increase in viscosity as the concentration increases. In acidic medium, chitosan acts as an excellent viscosity enhancer. Type of acid used as a solvent to dissolve the chitosan is the criteria of pH of the solution which in turn affects the viscosity of the solution. The intrinsic viscosity of chitosan plays a vital role in its storage and stability. Basically, intrinsic viscosity deals with the ability of the solution to become viscous with suitable solvent and temperature and is directly proportional to the polymer average molecular weight. The intrinsic viscosity can be evaluated by using Mark-Houwink equation:

$$[\eta] = k \times M_w^{\alpha} \tag{1.2}$$

where, η is intrinsic viscosity of the chitosan solution, k and α are polymer conformation constants, and M_w is the viscosity average molecular weight. Compact spherical structure of chitosan is defined by $\alpha = 0$, random coil by $\alpha = 0.5–0.8$, and rigid coil by $\alpha = 1.8$. Moreover, viscosity of chitosan also influences biomedical properties such as wound dressing and biodegradation as well [17, 32, 33].

1.2.4.1.2 Molecular Weight
Molecular weight of chitosan greatly influences the physicochemical properties of the biopolymer. Generally molecular weight of chitosan can deal with the average of all the molecules present in the sample and can be evaluated with the help of advanced techniques such as light scattering, osmometry, NMR, viscometry, gel permeation chromatography, and size exclusion chromatography. These techniques applied on chitosan give varied results and hence comparison of finally achieved polymeric material from different manufacturers is tedious and complicated. Low and high molecular weight chitosan shows different effect on their physicochemical

properties such as viscosity, hydrophilicity, moisture content, thermal properties, and stability. As chitosan polymeric chains of shorter length are of low molecular weight therefore indicates less interaction with hot sulfuric acid and hence lower degree of hydrolysis in comparison with longer ones. Additionally, the low molecular weight chitosan has ability to penetrate inside the bacterial cell thereby inhibits the RNA transcription leading to the death of the cell [34, 35]. Furthermore, the degree of deacetylation decreases the molecular weight of the chitosan [36]. Polydispersity, ratio of molecular weight and average molecular weight, indicates the uniformity and functionality of the polymer which is considered as respectable if occur in between 0.85 and 1.15 because of the good polymer homogeneity at this value. Also, the high molecular weight chitosan is considered as more stable. Various factors viz. thermal stability, temperature, pH, and mechanical shearing of chitosan affect the molecular weight of chitosan and are responsible for the variation in polydispersity index [32].

1.2.4.1.3 Degree of Deacetylation

Degree of deacetylation is an extremely important property of chitosan as it influences almost all the other properties to a significant extent. Ample of studies and different methods of production of chitosan revealed that most of the properties including physical properties such as viscosity, thermal, swelling, stability and chemical properties such as solubility, pH of the solution and reactivity are greatly influenced by the degree of deacetylation. It has been generally found that the temperature and time can modify the characteristic pattern of the deacetylation of chitosan which leads to alterations in its physicochemical as well as biological behavior. The high positive charge density on the polymeric chain leads to a high degree of deacetylation of around 97.5% which in turn responsible for good antimicrobial activity as compared to mild or moderate degree of deacetylation of about 83.7% [16]. Actually, the degree of deacetylation is the ratio of glucosamine to the *N*-acetylated glucosamine units. Commercially, the appropriate degree of deacetylation of chitosan should be 75–98% for biomedical purpose, as manufactured by pharmaceutical industries. Since, higher the degree of deacetylation, higher is its purity hence is of great interest in studying their degradation behavior. This is due to the fact that chitosan with high degree of deacetylation does not induce inflammation because of its lower affinity toward enzyme and therefore, exhibit slower rate of enzymatic degradation as compared to the lower (chitosan with lower degree of deacetylation) one [32]. Additionally, increase in concentration of the alkaline solution can increase the degree of deacetylation of the polymeric chain with increasing temperature and reaction time [36, 37].

1.2.4.1.4 Stability or Polymer Degradation

Stability is also an important property of chitosan biopolymer. The degradation of the polymeric chain of chitosan defines its stability under various conditions. Generally, during acid hydrolysis at low pH, the polymeric chain of chitosan cleaves and thus degrades the polymer. The cleavage mechanism involves depolymerisation followed by deacetylation by the splitting of β-1,4-glycosidic bonds followed by *N*-acetylglucosamine linkage, respectively. This depolymerisation leads to the generation of free radical species in the solution which in turn induces oxidation reactions. On the other hand, increased deacetylation results in decrease in molecular weight. After splitting, there has been observed strong intermolecular attractions among them as interchain cross-linking which can change its structure and structure-based properties. The degradation of chitosan mainly depends on degree of deacetylation, molecular weight, temperature, moisture content, polydispersity, and purity level. Besides this, *in vivo* degradation of chitosan can be performed by several enzymes such as lysozyme, a protease found in mammalian cells, generating oligosaccharides which are nontoxic and can be introduced into glycoproteins and glycosaminoglycans. Amid them, *in vitro* degradation of chitosan can be carried out under controlled conditions by various chemical reactions such as oxidation and enzymatic hydrolysis in order to obtain low molecular weight chitosan [32].

1.2.4.1.5 Thermal Properties

Elevated temperatures can alter the physicochemical properties of the chitosan solution such as solubility, viscosity, structural changes, etc. Generally, heating of polymeric solution causes the degradation and loss of stability. This degradation due to excessive heating at different rate is referred to thermal degradation and can be measured by thermogravimetric analysis (TGA) of the solution. This degradation can be performed in three steps:

i. At temperature range 30–110 °C, evaporation takes place in order to remove moisture from the polymeric solution.
ii. At temperature range 180–340 °C, decomposition takes place.
iii. At temperature 470 °C, subsequent weight loss of chitosan has been observed.

However, it has been found that the physicochemical properties of chitosan are unaffected below glass transition temperature. Also, elevated glass transition temperature leads to an increase in molecular weight of chitosan. As far as biomedical application is concerned, the temperature of chitosan should not exceed above 100 °C because excessive and uncontrolled heating may cause discoloration as well as depolymerisation which in turn alter its rheological properties [32].

1.2.4.1.6 Hygroscopicity and Swelling Ability

Chitosan is a hygroscopic pseudo-plastic natural polymer. It can form hydrogen bond with the functional group present in the polymeric chain, i.e., $-NH_2$ and –OH with the O-atom and H-atom of water molecule, respectively. The rate of water take up ability can be determined by inherent moisture already present in the sample and the conditions in which it has been stored. The water content present in dry chitosan is found to be increased on decreasing the degree of deacetylation. The swelling property of chitosan decreases with an increase in the concentration of cross-linking agent [17]. Apart from this, viscosity, compressibility, and flow properties have also been affected by the absorbed water content and cause a little decrease in tensile strength. However, 6% (w/w) water content can enhance the binding ability due to weak hydrogen bonding interactions. It has also been reported that the storage of chitosan for long duration can increase the moisture content of chitosan but it reduces its ability to further binding to water as well as increase its degradation time [38]. This leads to the pronounced disintegration of chitosan by hydrolysis reaction. In order to measure the water content present in chitosan, a modest and fast technique, known as loss on drying technique can be used in which chitosan sample is weighed, heated, and then again weighed after cooling. Besides, for long duration storage of chitosan, swelling index test has also been utilized. The important point to be kept in mind is that the selection of type of medium erstwhile to the water studies. This is because of the varying effect of ionic strength, solubility, swelling behavior, and viscosity on chitosan. The swelling ratio can be evaluated by using the following formula after getting constant value of weighed sample at a predefined duration, usually at 37 °C temperature. The formula is:

$$[SR] = \frac{W_f - W_i}{W_i} \tag{1.3}$$

where, SR is the swelling ratio, W_i represents the initial weight, and W_f is the final weight of dosage form before and after the swelling, respectively [32].

1.2.4.2 Chemical Properties

1.2.4.2.1 Solubility

Chitosan is generally soluble in some organic as well as inorganic acids having pH < 6.0 of the solution and form a non-Newtonian, thin shearing fluid. Usually, the organic acids display solubility includes methanoic acid (HCOOH), acetic acid (CH_3COOH), hydrochloric acid (1% HCl), dilute nitric acid (HNO_3), lactic acid, etc. Out of these 0.2–100% aqueous methanoic acid considered as the best solvent to dissolve chitosan. But the most frequently used solvent is 1% CH_3COOH at pH 4.0. It should be note here that at high temperature concentrated CH_3COOH can cause depolymerisation of chitosan. Chitosan usually insoluble in phosphoric and sulphuric acid but at elevated temperatures around 95–121 °C both low as well as medium molecular weight chitosan can be dissolved. Chitosan can form the water soluble salts such as pyruvate, malate, lactate, malonate, ascorbate, acetate, tartarate, glyoxylate, and glycolate. The formation of water soluble salts takes place by the neutralization of acids viz. HCl, HCOOH, CH_3COOH, and lactic acid. On the other hand, chitosan is insoluble in water and aqueous bases. This is due to the fact that the $-NH_2$ group of chitosan cannot be protonated in basic or neutral medium but at low pH due to the electrostatic repulsion the free $-NH_2$ group become protonated and hence soluble and allowing polymer solvation. Additionally, chitosan is also insoluble in organic nonpolar solvents such as dimethylformamide and dimethylsulphoxide but substantively soluble in acidified polyol. Solubility of chitosan solution mainly depends on the degree of deacetylation, method of extraction, time, temperature, concentrations, and molecular weight [17, 28, 32].

1.2.4.2.2 pH of Chitosan Solution

The dilute acidic chitosan solutions are found to have pH less than 6.0. Since chitosan itself known to be a strong base due to the presence of free $-NH_2$ groups throughout the chain having pKa value of 6.3, hence pH of the solution can alter the properties and charge on chitosan. At low pH, the free $-NH_2$ groups get protonated and acquire positive charge which lead chitosan a cationic polyelectrolyte and hence water soluble. Besides, it can form quaternary salts with nitrogen at low pH values. Amid them, at high pH, i.e, >6.0, again the amino groups become deprotonated thus made chitosan insoluble. The pKa value of chitosan also greatly depends on degree of deacetylation and method of extraction. Furthermore, chitosan has gelation ability to form gels with anionic hydrocolloids because of its acidic pH and hence can be utilized in slow release drug delivery [39].

1.2.4.2.3 Reactivity

Chitosan is a linear chain biopolymer having $-NH_2$ and -OH functional group as a reactive site of its backbone. The extent of these functional groups, especially, amino group are protonated, that gives rise to more reactive the chitosan is. More the protonated $-NH_2$ group available in the polymeric chain more is its reactivity and hence ability to bind with toxic metal ions and also form chelates with transition metal ions. This chelating ability of chitosan exemplifies its importance in water treatment and air purification. The nitrogen content available in chitosan is present in the form of primary aliphatic amino groups and hence undergo important amine *N*-acylation and Schiff reactions. Besides, chitosan can react with aldehydes and ketones to yield aldimines and ketimines, respectively, at room temperature. Additionally, proteic and nonproteic amino groups containing glucans can also be obtained by reacting chitosan with ketoacids followed by reaction with sodium borohydride, e.g., *N*-Carboxybenzyl chitosans (nonproteic glucan) obtained from *o*- and *p*-phthalaldehydic acids. Furthermore, on treating chitosan with simple aldehydes, hydrogenation takes place producing *N*-alkyl chitosan. Since, presence of hindered bulky group tends to decrease the strength of hydrogen bond of chitosan, this leads to enhance its swelling ability instead of being hydrophobic and hence can be utilized in film formation [18]. Amid them, the cationic nature of chitosan is being high positive charge density, greatly affect the reactivity, solubility, adsorption, and biodegradability of chitosan [17].

1.2.4.3 Miscellaneous Properties

Besides above physical and chemical properties, chitosan possesses many biological and miscellaneous properties. These includes [17, 33, 40]:

- Non-toxicity
- Gelation ability
- Biodegradability
- Biocompatibility
- Chelating ability
- Antimicrobial activity
- Haemostaticity
- Fungistaticity
- Spermicidal activity
- Anticancer activity
- Anti-inflammatory
- Mucoadhesion ability
- Wound healing ability
- Bone regeneration ability
- Drug releasing ability
- Immunoadjuvant ability

Being a natural polysaccharide, chitosan has been illustrated with many beneficial properties. The high positive charge density on the polymeric chain leads to a high degree of deacetylation of around 97.5% which in turn responsible for good antimicrobial activity as compared to mild or moderate degree of deacetylation of about 83.7%. Actually, the polycationic nature of chitosan tends to interact it with the negative charge density or anionic components of the bacterial cell and thus alters the function as well as permeability of the cell thereby causing cell death because of the rupture and leakage of intercellular components, sometimes by inhibiting RNA transcription. That's why chitosan exhibits remarkable antibacterial or antimicrobial activity. This is the most accepted mechanism of its antibacterial activity as put forth by many researchers but the actual mechanism is still unknown. Chitosan also acquired excellent mucoadhesion property as it possesses free –OH and $-NH_2$ groups which permit the polysaccharide chain to interact with mucin via electrostatically as well as via hydrogen bonding [32].

1.3 Conclusion

Chitin and chitosan are not unknown to the researchers and scientists of polymeric fields. And nowadays, these are not only confined to polymeric fields, instead being applied in electronics and pharmaceutical to biomedical fields also. Both chitin and chitosan are the naturally occurring polysaccharide with *N*-acetyl-D-glucosamine long chain polymeric unit having β-(1,4) linkage and glycosidic linkage. Chitin can be extracted from terrestrial (silkworm or honeybee), marine (crab, krill, lobster, squids), microorganisms (fungi), and enzymes (protease or lipase). In the extraction of chitin from these sources, there are three major steps termed as: demineralization, deproteination, and decolorisation. Chitosan is extracted from the same sources as of chitin and can be obtained after deacetylation of chitin. Various properties of chitin and chitosan such as purity, solubility, pH, viscosity, hygroscopicity, thermal, swelling, and reactivity have been greatly influenced by each other, especially the molecular weight and degree of deacetylation. Degree of deacetylation plays an important role in altering almost all the properties of chitosan and hence the results too. Viewing biomedical aspects of chitin as well as chitosan, it has been observed that marine originated more deacetylated products gives better results in terms of its properties, as compared to less deacetylated products.

References

1. History of Chitin. http://www.primex.is/quality-and-environment/history-of-chitin/, Accessed Feb 26, 2017.

2. Kim, S.K. *Chitin, Chitosan, Oligosaccharides and Their Derivatives: Biological Activities and Applications*; Taylor & Francis, 2010.
3. Nwe, N., Furuike, T., Tamura, H. Chitin and Chitosan from Terrestrial Organisms. In: *Chitin, Chitosan, Oligosaccharides and Their Derivatives*, S.K. Kim, Ed., pp. 3–10, CRC Press, 2010.
4. Brück, W., Slater, J., Carney, B. Chitin and Chitosan from Marine Organisms. In: *Chitin, Chitosan, Oligosaccharides and Their Derivatives*, S.K. Kim, Ed., pp. 11–23, CRC Press, 2010.
5. Knezevic-Jugovic, Z., Petronijevic, Z.;,Smelcerovic, A. Chitin and Chitosan from Microorganisms. In: *Chitin, Chitosan, Oligosaccharides and Their Derivatives*, S.K. Kim, Ed., pp. 25–36, CRC Press, 2010.
6. Tan, S.C., Tan, T.K., Wong, S.M., Khor, E. The Chitosan Yield of Zygomycetes at Their Optimum Harvesting Time. *Carbohydr. Polym.*, 30, 239–242, 1996.
7. Amorim, R.V.S., Ledingham, W.M., Kennedy, J.F., Campos-Takaki, G.M. Chitosan from Syncephalastrum Racemosum Using Sugar Cane Substrates as Inexpensive Carbon Sources. *Food Biotechnol.*, 20, 43–53, 2006.
8. Roberts, G.A.F. *Thirty Years of Progress in Chitin and Chitosan, XIII*, 7–15, 2008.
9. Kumari, S., Rath, P.K. Extraction and Characterization of Chitin and Chitosan from (Labeo Rohit) Fish Scales. *Procedia Mater. Sci.*, 6, 482–489, 2014.
10. Zuber, M., Zia, K.M., Barikani, M. Chitin and Chitosan Based Blends, Composites and Nanocomposites. In: *Advances in Natural Polymers: Composites and Nanocomposites*, S. Thomas, P.M. Visakh, A.P. Mathew, Eds., pp. 55–119, Springer Berlin Heidelberg: Berlin, Heidelberg, 2013.
11. Percot, A., Viton, C., Domard, A. Optimization of Chitin Extraction from Shrimp Shells. *Biomacromolecules*, 4, 12–18, 2003.
12. Younes, I., Hajji, S., Frachet, V., Rinaudo, M., Jellouli, K., Nasri, M. Chitin Extraction from Shrimp Shell Using Enzymatic Treatment. Antitumor, Antioxidant and Antimicrobial Activities of Chitosan. *Int. J. Biol. Macromol.*, 69, 489–498, 2014.
13. Mohammed, M.H., Williams, P.A., Tverezovskaya, O. Extraction of Chitin from Prawn Shells and Conversion to Low Molecular Mass Chitosan. *Food Hydrocoll.*, 31, 166–171, 2013.
14. Rudall, K.M. Chitin and Its Association with Other Molecules. *J. Polym. Sci. Part C Polym. Symp.*, 28, 83–102, 1969.
15. Acosta, N., Jiménez, C., Borau, V., Heras, A. Extraction and Characterization of Chitin from Crustaceans. *Biomass Bioenerg.*, 5, 145–153, 1993.
16. Rinaudo, M. Chitin and Chitosan: Properties and Applications. *Prog. Polym. Sci.*, 603–632, 2006.
17. Ahmed, S., Ikram, S. Chitosan & Its Derivatives : A Review In Recent Innovations, *Int. J. Pharma. Sci. Res.*, 6, 2015.
18. Kumar, M.N.V.R. A Review of Chitin and Chitosan Applications. *React. Funct. Polym.*, 46, 1–27, 2000.
19. Philippova, O.E.; Korchagina, E. V. Chitosan and Its Hydrophobic Derivatives: Preparation and Aggregation in Dilute Aqueous Solutions. *Polym. Sci. Ser. A*, 54, 552–572, 2012.

20. Pillai C.K.S, Paul W. Sharma, C.P. Chitin and Chitosan Polymers: Chemistry, Solubility and Fiber Formation. *Prog. Polym. Sci.*, 641–678, 2009.
21. Knorr, D. Functional Properties of Chitin and Chitosan. *J. Food Sci.*, 47, 593–595, 1982.
22. Park, B.K.; Kim, M.-M. Applications of Chitin and Its Derivatives in Biological Medicine. *Int. J. Mol. Sci.*, 11, 5152–5164, 2010.
23. The History of Chitosan. http://www.streetdirectory.com/etoday/the-history-of-chitosan-euolup.html, Accessed Jan 23, 2017.
24. Chitosan: Knowing the History. http://chitosan.in-honolulu.com/, Accessed Jan 23, 2017.
25. Nemtsev, S. V, Zueva, O.Y., Khismatullin, M.R., Albulov, A.I., Varlamov, V.P. Isolation of Chitin and Chitosan from Honeybees. *Appl. Biochem. Microbiol.*, 40, 39–43, 2004.
26. Niederhofer, A., Müller, B.W. A Method for Direct Preparation of Chitosan with Low Molecular Weight from Fungi. *Eur. J. Pharm. Biopharm.*, 57, 101–105, 2004.
27. Synowiecki, J., Al-Khateeb, N.A.A.Q. Mycelia of Mucor Rouxii as a Source of Chitin and Chitosan. *Food Chem.*, 60, 605–610, 1997.
28. Zamani, A., Edebo, L., Sjöström, B., Taherzadeh, M.J. Extraction and Precipitation of Chitosan from Cell Wall of Zygomycetes Fungi by Dilute Sulfuric Acid. *Biomacromolecules*, 3786–3790, 2007.
29. Knidri, H. El, Khalfaouy, R. El, Laajeb, A., Addaou, A., Lahsini, A. Eco-Friendly Extraction and Characterization of Chitin and Chitosan from the Shrimp Shell Waste via Microwave Irradiation. *Process Saf. Environ. Prot.*, 104, *Part*, 395–405, 2016.
30. Annu, Ahmed, S., Ikram, S. Perspectives of Chitosan and Alginate Membranes for Biomedical Applications. In: *Natural Polymers: Derivatives, Blends and Composites*, Ahmed, S. and I.S., Eds.; pp. 157–162, Nove Science Publishers, USA, 2017.
31. Moussout, H., Ahlafi, H., Aazza, M., Bourakhouadar, M. Kinetics and Mechanism of the Thermal Degradation of Biopolymers Chitin and Chitosan Using Thermogravimetric Analysis. *Polym. Degrad. Stab.*, 130, 1–9, 2016.
32. Szymańska, E., Winnicka, K. Stability of Chitosan—A Challenge for Pharmaceutical and Biomedical Applications. *Mar. Drugs*, 13, 1819–1846, 2015.
33. Ahmed, S., Ikram, S. Chitosan Based Scaffolds and Their Applications in Wound Healing. *Achiev. Life Sci.*, 10, 27–37, 2016.
34. Klaykruayat, B., Siralertmukul, K., Srikulkit, K. Chemical Modification of Chitosan with Cationic Hyperbranched Dendritic Polyamidoamine and Its Antimicrobial Activity on Cotton Fabric. *Carbohydr. Polym.*, 80, 197–207, 2010.
35. Islam, S., Bhuiyan, M.A.R., Islam, M.N. Chitin and Chitosan: Structure, Properties and Applications in Biomedical Engineering. *J. Polym. Environ.*, 1–13, 2016.
36. Tsaih, M.L., Chen, R.H. The Effect of Reaction Time and Temperature during Heterogenous Alkali Deacetylation on Degree of Deacetylation and

Molecular Weight of Resulting Chitosan. *J. Appl. Polym. Sci.*, 88, 2917–2923, 2003.
37. Yaghobi, N., Hormozi, F. Multistage Deacetylation of Chitin: Kinetics Study. *Carbohydr. Polym.*, 81, 892–896, 2010.
38. No, H.K., Prinyawiwatkul, W. Stability of Chitosan Powder during Long-Term Storage at Room Temperature. *J. Agric. Food Chem.*, 57, 8434–8438, 2009.
39. Pillai, C.K.S., Paul, W., Sharma, C.P. Chitin and Chitosan Polymers: Chemistry, Solubility and Fiber Formation. *Prog. Polym. Sci.*, 34, 641–678, 2009.
40. Ahmed, S., Ahmad, M., Ikram, S. Chitosan: A Natural Antimicrobial Agent-A Review, *J. Applicable Chem.*, 3, 493–503, 2014.

2

Nitrogenated Polysaccharides - Chitin and Chitosan, Characterization and Application

Michael Ioelovich

Designer Energy Ltd, Rehovot, Israel

Abstract

In this chapter preparation, characterization and applications of chitin and chitosan have been described and discussed. These nitrogenated polysaccharides occupy second place prevalence in the nature after cellulose. Diverse physical, physicochemical, and chemical methods are applied to characterize structure and properties of these biopolymers. Being nanostructured, chitin and chitosan can be isolated from natural sources in a form of nanoscale particles, fibrils, and filaments having unique features such as small dimensions, variety shapes, enhanced specific surface, high sorption and absorption ability, and other specific characteristic in combination with biocompatibility, biodegradability, and complex of unique therapeutic properties. In addition, these biopolymers serve as a basis for the production of some therapeutically active substances. Therefore, chitin and chitosan find a great commercial interest in biomedicine, pharmaceutics, cosmetics, personal care, and some other areas. Small particles and fibrils made of the biopolymers can be used as fillers for biocomposites and thickeners imparting to liquid systems an increased viscosity and gel consistence. The nanocarriers of nitrogenated polysaccharides can bind various therapeutic active substances, which expand application areas of these biopolymers. Antibacterial nonwoven materials made of nanofilaments have an increased absorption ability and accelerate healing process. Some other application areas of the nitrogenated polysaccharides are also described.

Keywords: Chitin, chitosan, features, nanomaterials, preparation, characterization, applications

Corresponding author: bd895892@zahav.net.il

Shakeel Ahmed and Saiqa Ikram (eds.) Chitosan, (25–70) © 2017 Scrivener Publishing LLC

2.1 Introduction

As is known, chitin is one from the most widespread biopolymers that occupies second place prevalence in the nature after cellulose. Resources of this biopolymer are estimated in 100 billion tons [1]. In the nature, chitin is found in shells of mollusks, shrimps and lobsters, pens and beaks of cephalopods, exoskeletons of arthropods, crustaceans and insects, cell walls and membranes of some fungi and microorganisms, etc. Chitin implements a skeleton function in lower eukaryotes similar to that of cellulose in plants. Commercially, chitin can be isolated from shells of crabs, shrimps and lobsters by acidic treatment to remove of calcium carbonate followed by alkaline extraction of proteins and bleaching [2]. Annual production volume of chitin in the world is estimated at about 80–100 thousand metric tons.

Macromolecules of this nitrogenated polysaccharide consist of 1,4-β-*N*-acetylglucosamine (more precise 1,4-β-*N*-acetyl-2-aminodeoxyglucose) units [2, 3]. Chitin can be regarded as a nitrogenated derivative of cellulose, where the hydroxyl group at C2 atoms in each repeat units is replaced with an acetylamino group. Inter and intramolecular hydrogen bonds impart to chitin chains an increased stiffness. Molecular weight (MW) of chitin samples varies in the range of 100–1000 kDa. Similar to cellulose, chitin is a linear semicrystalline biopolymer [4, 5]. The linear macromolecules joined by hydrogen bonds form a supramolecular structure of the polymer that consists of nanoscale fibrils [6, 7]. Each fibril is built of ordered crystallites and low-ordered noncrystalline (amorphous) domains statistically alternating along the fibril.

Structural studies showed that crystallites of chitin can be in three allomorph forms: α, β, and γ [8-11]. The most abundant α-form is present in chitin samples isolated from sea animals such as crabs, shrimps, lobsters, krill, etc., insects, fungi, and some microorganisms. The rare β-allomorph is found in the chitin of squid pens, tube-worms and some others sources. The γ-form of chitin can be isolated from some beetles [12]. The β- and γ-forms of chitin are instable and can undergo an intra-crystalline swelling; moreover, after some treatments these forms of chitin are transformed irreversibly into more stable α-polymorph [10, 11].

Chitin is also feedstock for production of another known nitrogenated polysaccharide – chitosan, by means of alkaline deacetylation process. Global market of chitosan in 2015 was around 20–25 thousand metric tons. The Asia-Pacific region has the leading chitosan market with a share of 55%, while the US represents the second biggest market for chitosan, with an estimated share of 25%.

Native chitosan occurs in cell walls of some fungi, for example, *Mucoraceae* [3]. Chitosan is a linear semicrystalline nitrogenated cellulose derivative composed of 1,4-β-2-glucosamine units. Degree of deacetylation (DD) of chitosan samples isolated from chitin ranges from 60% to 100%. MW of chitosan samples can be in the range 5–500 kDa. Unlike cellulose and chitin, chitosan dissolves in an aqueous medium having a weak acidic pH value. After neutralization, acidic solutions of chitosan turn into hydrogels [13].

Both chitin and chitosan can be isolated from natural sources in a form of nanoscale particles, fibrils and filaments having unique features, such as small dimensions, variety shapes, enhanced specific surface, high sorption and absorption ability, and other specific characteristic in combination with biocompatibility, biodegradability, and complex of unique therapeutic properties - antibacterial, analgesic, fungistatic, haemostatic, etc. In addition, these biopolymers serve as a basis for the production of diverse therapeutically active substances, for example, glucosamine, *N*-acetylglucosamine, nitrogenated oligosaccharides, etc. Therefore, chitin and chitosan can be applicable in biomedicine, pharmaceutics, cosmetics, personal care, and some other areas.

Small particles and fibrils made of the biopolymers can be used as fillers for biocomposites and thickeners imparting to liquid systems an increased viscosity and gel consistence. The solid and gel-like nanocarriers of the nitrogenated polysaccharides can bind various therapeutic active substances, which expand application areas of these biopolymers. Antibacterial nonwoven materials made of nanofilaments have an increased absorption ability and accelerate healing process. Multilayer dressings containing the nitrogenated polysaccharides are used for active treatment of injuries, wounds, and burns. Moreover, both biopolymers can find application in some other areas such as production of membranes, adhesives, coatings, sorbents, cleansing additives, fining agents, etc.

The purpose of this papers was to describe preparation methods, structure, properties and main applications of chitin and chitosan, and their based materials, including nanoscale particles, fibrils, and filaments.

2.2 Extraction of Nitrogenated Polysaccharides from Natural Sources

Chitin can be extracted from broad variety of natural sources such as aquatic animals (crustaceans, arthropods, cephalopods, mollusks, shrimps, lobsters, etc.), insects (scorpions, spiders, beetles, ants, etc.), microorganisms (yeast, microalgae, etc.), fungi and some other sources [14]. The chitin sources can also contain proteins, lipids, minerals, mainly calcium carbonate, and some other admixtures [15]. For example, nanofibrils of chitin in sea animals are strongly bound with proteins, and this organic complex is surrounded by a mineral layer [16, 17]. The different sources of chitin have different structure and chemical composition. Content of chitin in shells of various crustaceans can be from 14% to 70%, in shells of insects from 18% to 65% and in cell walls of microorganisms from 20% to 45% [2, 17, 18]. The average chitin content in shrimp shells was 18–20% of the dry weight of the shells. Furthermore, shell of crustaceans contains 10–50% of proteins and 20–60% of minerals.

Since natural sources contain proteins, minerals, lipids, and pigments, these admixtures should be removed to isolate pure chitin. Extraction procedure of chitin usually involves steps of demineralization, deproteinization, and decoloration [19, 20]. The chosen raw-material is preliminary purified by washing with water and organic solvents, followed by drying and pulverization. The purified powder is treated with 0.2–2 M hydrochloric acid at various temperatures to remove mineral substances. The demineralized material is deproteinized by extraction with 1–2 M sodium hydroxide at elevated temperatures. An alternative method of deproteinization was treatment with proteolytic enzymes [21, 22]. However this method is slow and expensive, therefore it cannot be applied in industry. Decoloration of the chitin is carried out by bleaching agents such as sodium hypochlorite, hydrogen peroxide, etc. [23]. Since tissues of fungi, insects, and microorganisms contain small content of minerals, the demineralization stage for isolation of chitin from these sources can be omitted.

The extracted chitin serves as a feedstock for production of chitosan [2, 20]. To isolate chitosan, a starting chitin material is soaked in 4–5 M alkali at room temperature and then treated with 10–20 M sodium hydroxide at elevated temperature, usually at 100 °C. The one-stage procedure also can be used for this purpose, when the feedstock is treated with 10–20 M sodium hydroxide at 100–150 °C. After separation from alkali solution, the obtained chitosan was washed with water several times to neutral pH and dried to constant weight. The conditions of chitosan isolation influence DD of this biopolymer. Rise of temperature and increase duration of the alkaline treatment contribute to higher DD-value of obtained chitosan.

2.3 Research Methods of Nitrogenated Polysaccharides

To characterize structure and properties of nitrogenated polysaccharides diverse physical, physicochemical, chemical, and biological methods are used such as X-ray scattering, NMR, spectroscopy, electron and atomic force microscopy, calorimetry, sorption, solubility, viscometry, chemical assays, etc.

2.3.1 X-Ray Diffraction

X-ray methods are widely used for structural investigations of nitrogenated polysaccharides of various origins and allomorphs. Wide angle X-ray diffraction (XRD) is designed to study the positions, intensities, and widths of the diffraction peaks in order to determine the types and content of allomorphs, interplanar distances, crystallinity, sizes of crystallites, distortion degree of the lattice, etc. For this purpose standard X-ray diffractometers can be applied. A typical diffractometer consists of source of radiation, monochromator to choose the wavelength, slits to adjust the shape of the beam, goniometer, and detector.

The dry samples of the equal mass are pressed into tablets that are used for XRD experiments. Three to five X-ray diffractograms of the sample, crystalline standard (e.g., glucosamine) and amorphous standard (e.g., amorphized biopolymer) are recorded in the range of 2θ diffraction angle from 5° to 50° at least, using the CuK_α-radiation (λ = 0.15418 nm) and reflection mode. The incoherent diffuse background caused by air, apparatus, Compton scattering, thermal agitation of atoms and molecules, as well as structural defects and distortions should be subtracted from the diffractograms, and then intensities of the peaks should be corrected. Further, the total intensity at the angle 2θ is divided into the correction coefficient, $K(\theta)$, which includes the Lorentz, polarization factor and initial intensity of the X-ray beam (J): $K(\theta) = J\,(1 + \cos^2 2\theta)/(\sin^2\theta \cos^2\theta)$.

Finally, the corrected diffractograms are obtained.

For example, X-ray diffractogram of the α-chitin sample isolated from crab shells has sharp diffraction peaks at 9.2°–9.6° from (020) planes and at 19.0°–19.6° from (110) planes, middle intensity peaks from planes of (021) and (013) in crystalline unit cells, as well as some weak peaks (Figure 2.1(a)). On the other hand, X-ray diffractogram of α-chitosan sample (DD = 85%) produced from α-chitin shows visible diffraction peaks at 10°–11° from (100) and (010) planes and at 20°–21° from (020) planes of crystalline unit cells together with some weak peaks (Figure 2.1(b)). The diffractograms of nitrogenated polysaccharides can be used for determination of various structural characteristics of the samples, as follows:

(a) Determination of positions of diffraction peaks and calculation of interplanar distances by means of Bragg's equation: $d = \lambda/2\sin\theta_{hkl}$ (See Table 2.1).

Determination of positions of the peaks and calculation of interplanar distances create a possibility to establish the type of crystal structure and calculate parameters of crystalline unit cell of the biopolymer.

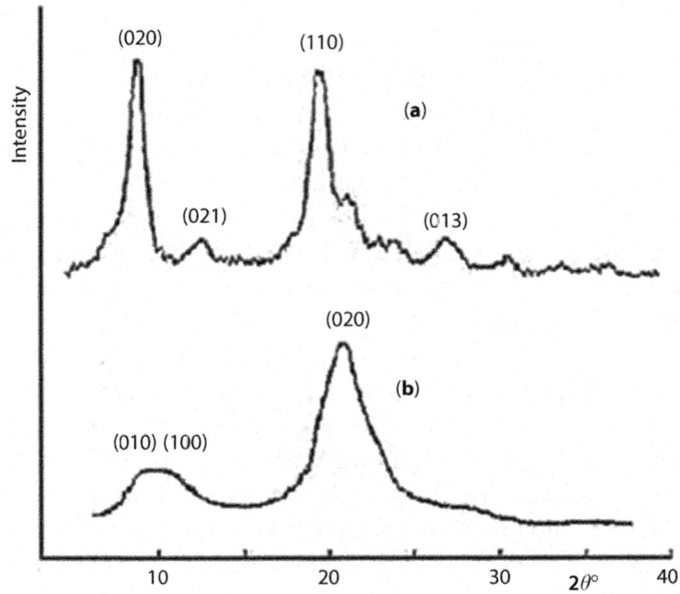

Figure 2.1 X-ray diffractogram of α-chitin (a) and α-chitosan (b).

Table 2.1 Positions of the main diffraction peaks and interplanar distances for α-chitin and α-chitosan.

Biopolymer	Index	Position of peaks, $2\theta^{\circ}_{hkl}$	*d*, nm	Intensity of peaks
α-chitin	**(020)**	**9.4**	**0.943**	**H**
	(021)	12.6	0.700	M
	(110)	**19.3**	**0.460**	**H**
	(101)	21.0	0.424	W
	(130)	23.6	0.376	W
	(013)	26.5	0.336	M
α-chitosan	(010)	10.4	0.850	M
	(100)	10.8	0.815	M
	(020)	**20.8**	**0.426**	**H**
	(210)	24.0	0.370	W
	(211)	25.6	0.348	W

*Note: H – high intensity; M – middle intensity; W – weak intensity.

(b) Calculation of lateral sizes of nanocrystallites

Knowledge of sizes of crystallites is very important because it allows clarify the supramolecular structure of biopolymers. WAXS method is considered as a direct, nondestructive and convenient method for determining the sizes of the crystallites. The common X-ray method for measurement of average size of small crystallites is based on the equation of Scherrer:

$$D_{sh} = K\lambda/(B\cos\theta_{hkl})$$

where B is width of the peak at the half of maximum height (in radians); θ_{hkl} is diffraction angle of the peak; shape factor K is usually taken close to 1.

Measurement of the peak width can be exemplified by (020)-peak of chitosan sample (Figure 2.2).

His equation Scherrer was developed in 1918 in order to calculate the size of nanocrystallites from measuring the width at half maximum of peaks [24]. Despite the past 96 years from its creation, this equation is widely used also nowadays, although it has considerable limitations, as follows:

- Equation of Scherrer cannot be used if the crystallite size exceeds 100 nm.
- Width of the peaks depends not only on the size of the crystallite but also on other factors, such as instrumental effects and paracrystalline distortions of crystalline lattice caused, for example, by dislocations, twinning, internal stresses, surfaces, boundaries of crystallites, etc. [25].

Currently, the instrumental effect only is taken into consideration, while other significant factors are ignored, which leads to clearly understated sizes

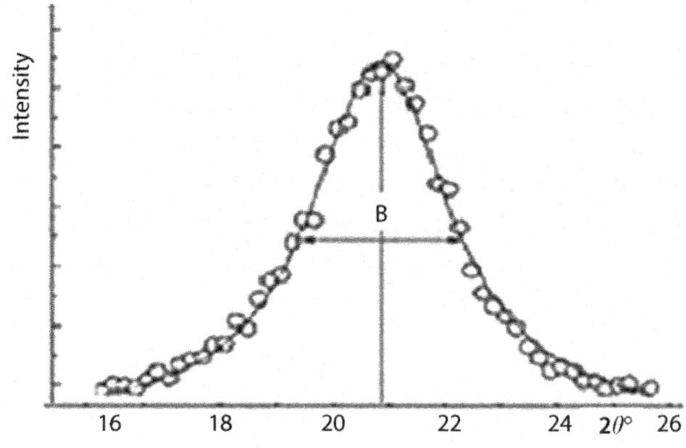

Figure 2.2 Measurement of width of crystalline peak.

of the crystallites. To determine the actual sizes of crystallites, an improved WAXS method should be used taking into consideration the contribution both of instrumental factor (***b***) and also of paracrystalline distortions (Δ) in the experimental width (B) of the peak:

$$B_o = (B^2 - \boldsymbol{b}^2 - \Delta^2)^{1/2}$$

where B_o is corrected width of the peak at the half of maximum height; the instrumental factor, ***b***, can be measured using a crystalline standard.

Theoretical contribution of the paracrystalline distortions in the width of the peak is:

$$\Delta = 2\delta\,\lambda/(d\cos\theta_{hkl})$$

where δ is distortion degree; d is inter-plane distance in the direction perpendicular to (hkl) planes of crystallites [26–28].

Thus, the updated equation for calculating the actual size of crystallites, D_{cr}, has been derived:

$$D_{cr} = [(B^2 - b^2)(\cos\theta_{hkl}/\lambda)^2 - (2\delta/d)^2]^{-1/2}$$

For example, the lateral size of nanocrystallites in the direction perpendicular to (110)-planes of chitin or (020)-planes of chitosan can be calculated by the equation:

$$D_{cr} = [(B^2 - \mathrm{b}^2)(\cos\theta_{200}/\lambda)^2 - \mathrm{C}]^{-1/2}$$

where $C = (2\delta/d)^2 = 0.016\ \mathrm{nm}^{-2}$ is a correction coefficient of lattice distortions.

To determine the sizes of crystallites, X-ray diffractograms are recorded by reflection step-scan mode using 0.01° step, and with the accumulation of pulses at the each step. To separate the crystalline peak, the incoherent background and scattering from noncrystalline domains should be subtracted. Then, shape of the peak is corrected using the coefficient, $K(\theta)$. Finally, the actual size (D_{cr}) of crystallites and standard deviation (SD) is calculated as follows:

$$D_{cr} = (\Sigma D_i)/n$$

$$SD = \pm\sqrt{\Sigma(D_i - D_{cr})^2/n}$$

where D_i is the actual size obtained from each diffractogram; $n = 3–5$ is number of the recorded diffractograms for the same sample.

Comparative results of calculation of crystallite sizes of nitrogenated polysaccharides sample using the Scherrer equation and improved equations reveal that the actual lateral size of crystallites is 20–30% greater than the estimated Scherrer's size. Consequently, the improved WAXS method should be used in order to obtain reliable results about the actual sizes of crystallites.

(c) Determination of crystallinity

The XRD is the most widespread technique allowing determination the crystallinity of biopolymers. However, the determination of the actual crystallinity degree is laborious and complicated procedure, and therefore most researchers are limited by a more simple way – estimation of index of crystallinity, CrI. This index shows a comparative crystallinity only, that is, which of the samples has greater crystallinity and which less crystallinity, but it does not indicate the true degree of crystallinity, that is, weight part of the crystalline fraction in the sample.

The index of crystallinity of nitrogenated polysaccharides may be estimated by various calculation methods. The most widespread is a method based on measuring the heights of crystalline peak and amorphous scattering:

$$\mathrm{CrI} = (I_o - I_{am})/I_o,$$

where I_o is total height of crystalline peak and I_{am} is height of amorphous scattering.

To calculate CrI of chitin samples Focher *et al.* [29] proposed to measure the total height of (110)-peak (I_o) at $2\theta = 19.3°$ and height of amorphous scattering (I_{am}) at $2\theta = 12.6°$. For chitosan samples, I_o was measured at $2\theta = 20°$, while I_{am} was measured at $2\theta = 12°$ [30] or at $2\theta = 16°$ [31]. Zhang *et al.* [32] used also the total height of peak (I_o) at 2θ of 9°–10° and height of amorphous scattering (I_{am}) at $2\theta = 12.6°$ to calculate CrI of chitin and chitosan samples. It should be noted that heights of crystalline peaks and amorphous scattering are not proportional to content of these phases in the sample, and therefore the height method does not characterize the actual degree of crystallinity [28].

Another method of crystallinity estimation was based on the dividing of total area of X-ray diffractogram by coherent crystalline, diffuse amorphous and background areas [33, 34]. The problem is that the dividing procedure was performed arbitrarily, without reliable justification; therefore, also this method gives comparative index, instead of degree of crystallinity. For example, according to method of Ruland, a part of incoherent background scattering caused by structural defects of crystalline lattice and thermal oscillations should be added to the amorphous scattering. However, these lattice distortions and oscillations are not the part of amorphous phase, but belong rather to the crystalline phase. Furthermore, this method requires the use of disorder parameter (*D*) – correction coefficient providing the convergence of ratio:

$$\mathrm{CrI} = F_c/(D\,F_0)$$

where F_c is total area of all crystalline peaks, whereas F_0 is total area of X-ray scattering for the whole sample. As a result, the Ruland's method gives conditional index of crystallinity only.

Calculation of various indexes of crystallinity for chitin samples showed that depending on the calculation method a different value of CrI, from 0.6 to 0.9, can be obtained [35, 36]. Crystallinity index of chitosan samples calculated by different methods can also vary within a wide range [31, 37–39]. Thus, results about the crystallinity of chitin and chitosan samples based on estimation of CrI are doubtful.

The determination of actual degree of crystallinity is a very important challenge to establish the real structure of biopolymers, predict their properties and study the relationships between structure and properties of polymer materials. It should be noted that the term "actual degree of crystallinity" means the mass fraction of the whole crystalline phase including various lattice defects in the sample. To determine the actual degree of crystallinity, the quantitative X-ray phase analysis should be performed, which requires compliance of certain conditions, and namely:

- Firstly, to prevent the texture factor tablets of powdered samples should rotate during recording the diffractograms.
- Secondly, the incoherent background scattering (bg) caused by air, apparatus, Compton factor and some other factors should be subtracted.
- Thirdly, the experimental diffractogram should be corrected using factors of Lorentz, polarization, stability of X-ray beam, etc.
- Fourthly, the scattering areas related to crystalline (cr) and amorphous (am) domains should be separated from the corrected diffractogram (see, e.g., Figures 2.3 and 2.4).

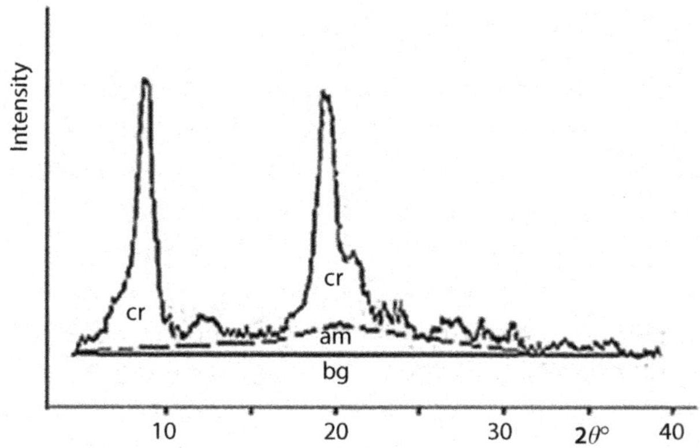

Figure 2.3 Illustration of method for determining the crystallinity degree of α-chitin sample.

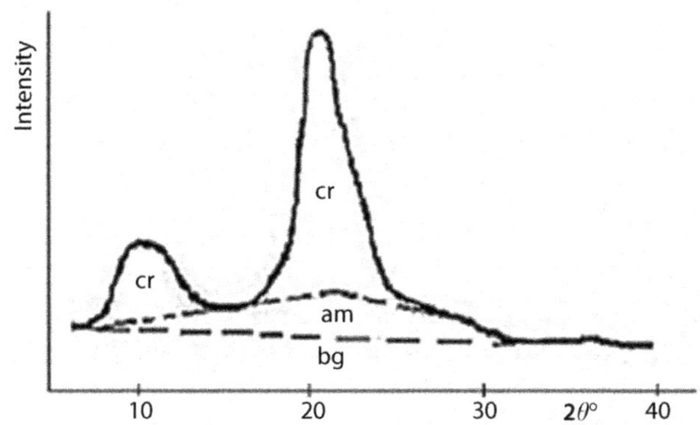

Figure 2.4 Illustration of method for determining the crystallinity degree of α-chitosan sample.

Finally, the integrated intensities (areas) related to crystalline and amorphous (noncrystalline) scatterings are determined to calculate the actual degree of crystallinity (*X*):

$$X = \int I_{cr}\, d\theta / \int I_{o}\, d\theta$$

The standard deviation of crystallinity also can be calculated.

Application of this method to nitrogenated polysaccharides showed that actual crystallinity degree of chitin extracted from crab shell was 0.70–0.74, and of chitosan 0.55–0.60 [40]. Moreover, the obtained values of actual crystallinity degree were significantly lower than values of crystallinity index calculated according to procedure described in [29, 30].

2.3.2 Nuclear Magnetic Resonance

Two versions of NMR, ^{1}H and ^{13}C, are widely used to characterize the nitrogenated polysaccharides, mainly to determine the degree of acetylation (DA) or deacetylation (DD) [41–43]. NMR ^{1}H is a conventional method for studying the dissolved chitosan. On the other hand, solid state ^{13}C and ^{15}N NMR are applied to study of poorly soluble chitin samples [44, 45].

To implement ^{1}H NMR, a dilute solution (cca. 0.5%) of chitosan in 1% DCl was used [46, 47]. Spectra were recorded at a constant temperature in the range of chemical shift (δ) between 1 and 6 ppm. The ^{1}H NMR spectrum of chitosan solution contains signals at 2.0–2.2 ppm from protons of methyl groups, 3.2–3.3 ppm from H2 protons in complete deacetylated units, 3.6–4.1 ppm from H2, H3, H4, H5, and H6 protons in the acetylated sample, as well as 4.8–5.0 ppm from H1 protons (Figure 2.5). The DA of chitosan is estimated by equation [47]:

$$\text{DA}, \% = 200\ (F_a/F_o)$$

Accordingly, the DD can be calculated, as follows:

$$\text{DD}, \% = 100 - \text{DA}$$

where F_a is an area of the signal from protons of methyl groups at 2.0–2.2 ppm, whereas F_o is a sum of the areas of all signals in the range of 3.1–4.2 ppm.

Solid state, high-resolution cross-polarization magic-angle spinning ^{13}C NMR technique has been reported to characterize solid chitin samples [45]. Typical spectrum of chitin sample recorded by solid state CP/MAS

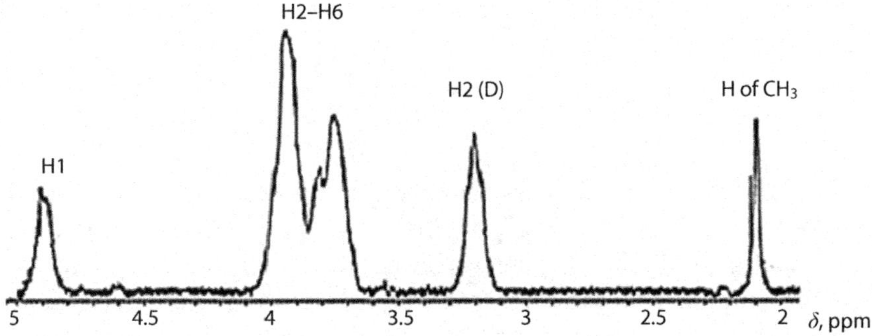

Figure 2.5 Illustration of ^{1}H NMR spectrum of chitosan solution.

^{13}C NMR spectrometer has eight peaks (Figure 2.6), corresponding to signals from C1 to C8 atoms of N-acetylglucosamine units (Figure 2.7).

The DA of chitin is estimated by equation:

$$DA, \% = 100\, k\, (F_{C8}/F_{C1})$$

where k is calibration coefficient; F_{C8} and F_{C1} is area of the signal from C8 and C1, respectively.

2.3.3 FTIR Spectroscopy

The FTIR technique is used to analyze the functional groups present in nitrogenated polysaccharides. Measurements are carried out by IR translucence of tablets containing a low amount of the powdered biopolymer blended with potassium bromide. Study of infrared spectra of various chitin samples showed specific bands at 1550–1560 cm^{-1} corresponding to

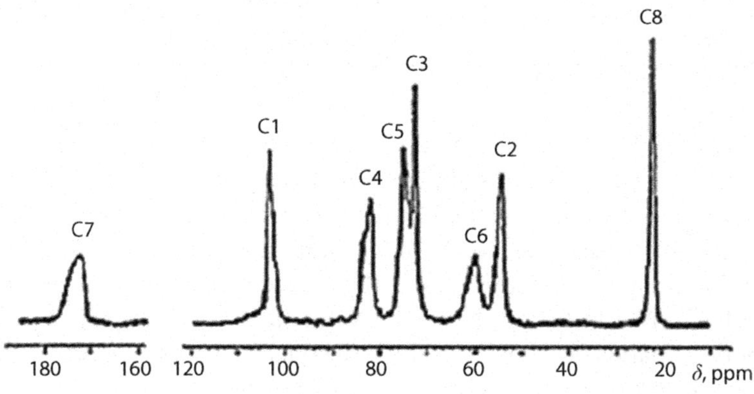

Figure 2.6 Illustration of solid state CP/MAS ^{13}C NMR spectrum of chitin.

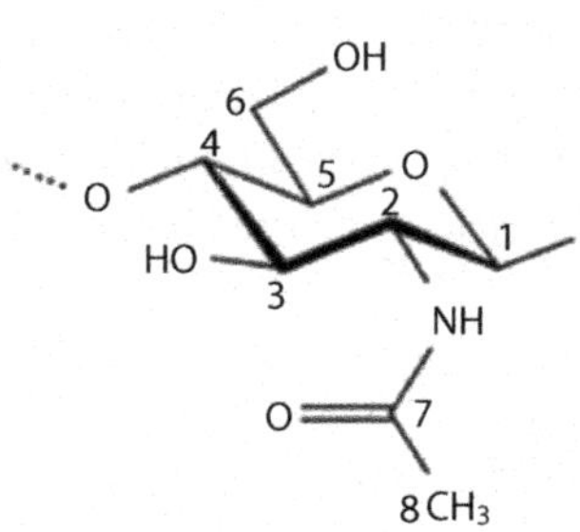

Figure 2.7 N-acetylglucosamine unit of chitin.

Table 2.2 Characteristics IR-bands of chitin samples [23, 48–54].

Frequency (cm^{-1})	Assignment
850–900	CH & COC stretching
1000–1050	COC stretching
1100–1150	Asymmetric ring stretching
1300–1320	Amide III band
1400–1420	CH2 bending
1550–1560	Amide II band
1620–1630	Amide I band
1640–1660	Amide I band
2850–2900	CH stretching
2930–2950	CH_2/CH_3 stretching
3250–3270	NH stretching
3400–3450	OH stretching

bending of N-H groups of amide II, as well as at 1630–1660 cm^{-1} attributed to stretching vibrations of the C=O of acetyl groups and C–N groups of amide I [23, 48–54]. Samples of α-chitin have two components of amide I band, at 1660 cm^{-1} and 1630 cm^{-1}, whereas β-chitin samples have one amide I band at around 1630 cm^{-1}. The amide bands I and II are used to characterize *N*-acetylation degree of chitin samples. Reducing the intensity of these bands demonstrates the deacetylation process of chitin. Other characteristic bands of chitin are presented in Table 2.2.

For samples of chitosan containing residual *N*-acetylamino groups (DD < 100%) the band of amide I at 1640–1660 cm^{-1} was observed [54]. Other characteristic bands of chit are presented in Table 2.3.

To estimate the DA, the height ratio of amide II band (at around 1550 cm^{-1}) to CH-stretching band (at around 2880 cm^{-1}) or to OH-stretching band at around 3450 cm^{-1}) relatively to baselines is used [59, 60]:

$$DA = 100\, k\, H_{1550}/H_{2880}$$

$$DA = 100\, k\, H_{1550}/H_{3450}$$

where k is calibration coefficient obtained by independent method.

Table 2.3 Characteristics IR-bands of chitosan samples [54–60].

Frequency (cm^{-1})	Assignment
1030–1070	COC stretching
1150–1160	Asymmetric ring stretching
1310–1330	Amide III band
1420–1430	CH_2 bending
1540–1560	Amide II band
1580–1600	Amine groups bending
1640–1660	Amide I band
2850–2900	CH stretching
3250–3300	NH stretching
3400–3500	OH stretching

2.3.4 Calorimetry

Various calorimetry methods are used to study thermodynamic and physicochemical properties of biopolymers, such as heat of combustion, enthalpy of polymer formation, enthalpy of interaction between polymers and liquids, specific heat capacity, phase and isophase transitions, etc. These studies are conducted by a variety types of calorimeters: adiabatic, isothermal, microcalorimeters, etc. [61–65].

Adiabatic calorimetry is the most widespread method for determination the specific heat capacity (C_p) under standard conditions, as well as temperature dependence of C_p to calculate thermodynamic functions, and study of phase and isophase transitions in biopolymers [63, 64].

To calculate enthalpy of formation, a combustion enthalpy of the sample should be determined experimentally using an isothermal bomb calorimeter [63, 65–67]. The combustion process of one unit of the nitrogenated polysaccharides can be written, as follows.

For chitosan:

$$C_6H_{11}O_4N(s) + 6.75O_2\,(g) \rightarrow 6CO_2\,(g) + 5.5H_2O\,(l) + 0.5N_2\,(g)$$

For chitin:

$$C_8H_{13}O_5N(s) + 8.75O_2\,(g) \rightarrow 8CO_2\,(g) + 6.5H_2O\,(l) + 0.5N_2\,(g)$$

Then, standard enthalpy of formation ($\Delta_f H°$) of the biopolymers is calculated.

For chitosan:

$$\Delta_f H° = 6\Delta_f H° (CO_2, g) + 5.5\Delta_f H° (H_2O, l) - \Delta_c H°$$

For chitin:

$$\Delta_f H° = 8\Delta_f H° (CO_2, g) + 6.5\Delta_f H° (H_2O, l) - \Delta_c H°$$

where $\Delta_f H° (CO_2, g) = -393.51$ (kJ/mol) is standard enthalpy of formation of gaseous carbon dioxide, and $\Delta_f H°(H_2O, l) = -285.83$ (kJ/mol) is standard enthalpy of formation of liquid water; while $\Delta_c H°$ is standard combustion enthalpy of nitrogenated polysaccharides, which is in the range from −3000 to −4000 (kJ/mol unit) [66]. Accordingly, the standard enthalpy of formation of the nitrogenated polysaccharides can be in the range of −700 to −1200 (kJ/mol unit) [63, 66]. For comparison, $\Delta_c H°$ value for various cellulose samples is −2800 to −2850 (kJ/mol unit), while $\Delta_f H°$ value ranges from −940 to −990 (kJ/mol unit) [67].

Measurement of enthalpy of wetting ($\Delta_w H°$) is used to determine the degree of crystallinity and amorphicity of biopolymers [68, 69]. Calorimetric determination of the structural characteristics is based on the fact that crystalline domains are inaccessible for water, whereas molecules of this polar liquid interacts selectively only with amorphous domains (AD) of biopolymers. Thus, content of amorphous domains or amorphicity degree (Y) and crystallinity degree (X) of the sample can be calculated by the equations:

$$Y = \Delta_w H° / \Delta_w H°_a$$

$$X = 1 - (\Delta_w H° / \Delta_w H°_a)$$

where $\Delta_w H°$ and $\Delta_w H°_a$ denote the enthalpy of wetting of semicrystalline and completely amorphous sample, respectively.

2.3.5 Sorption

Sorption and desorption of gases and vapors are widely used for investigation of structural characteristics of various hydrophilic polysaccharides and their derivatives. The precise experiments were carried out at 25 °C on a vacuum Mac-Ben apparatus having helical spring quartz scales. Prior to starting of the experiments the samples were dried at 100–110 °C up to constant weight and additionally degassed in the sorption device.

Sorption is usually described through isotherms having various shapes that express the relative amount of sorbate in the sorbent (*A*, *g/g*) as a function of relative vapor pressure ($\phi = p/p_o$) at a constant temperature [70]. Isotherms of type I exhibit a saturation plateau due to adsorption of an inert sorbate on the surface of pores. In this case the Langmuir equation is used for linearization of the isotherms and for calculations of monolayer adsorption value (A_m) and specific surface area (S_{sp}) of the sorbent.

Commonly occurring case involves the multilayer adsorption of inert sorbate onto the surface of mesoporous sorbents. Adsorption isotherms of this type are sigmoidal and related to type II or IV. The linearization of such sigmoidal isotherms is usually performed by means of the BET equation in order to calculate values of A_m and S_{sp}, as well as an energetic constant, *C*. Specific surface value of the sample is calculated using the BET equation [70, 71]:

$$\frac{p}{a(p_o - p)} = \frac{1}{Ca_m} + \frac{(C-1)p}{Ca_m p_o}$$

After the adsorption value of monolayer, a_m, was measured, the specific surface of the cellulose sample can be calculated by the equation:

$$S = a_m w N_A$$

where *w* is cross-sectional area occupied by one molecule of sorbate; N_A is Avogadro number.

If φ is low and the **C**-constant is high, the BET equation turns into the Langmuir equation.

Sorption isotherms of such active polar vapor as water by hydrophilic polysaccharides and their nitrogenated derivatives also have sigmoidal shape, similar to multilayer adsorption isotherms. However, in this case the sorption mechanism is not an adsorption of the sorbate on the surface of pores, but absorption of the water molecules into noncrystalline amorphous domains of the biopolymers. This process is accompanied by swelling, plasticization, and other structural changes of the hydrophilic polymers. High values of "imaginary" surface area of the samples obtained by sorption of water vapors and calculated by BET-equation do not reflect an actual surface area. In reality, it is a result of interaction between active molecules and polar groups of individual units inside the amorphous fractions of hydrophilic biopolymers. Water absorption into amorphous domains of cellulose and its nitrogenated derivatives occurs in a manner similar to the absorption of water molecules by amorphous starches. To describe an isotherm of water vapor absorption into amorphous domains of hydrophilic biopolymers a special thermodynamic equation has been proposed [40, 72]:

$$A = A_{m,a} Y/(1 - K \ln\phi)$$

Or in a linear form

$$A^{-1} = (A_{m,a} Y)^{-1} - K (A_{m,a} Y)^{-1} \ln\phi$$

where Y is actual degree of amorphicity; $A_{m,a}$ is maximum equilibrium absorption value of water molecules into amorphous polymer ($Y = 1, X = 0$) at $\phi = 1$; K is coefficient.

At constant relative pressure of water vapor, $\phi = 0.5$, degree of amorphicity of nitrogenated polysaccharides can be found from sorption experiments as follows:

$$Y = nA_{0.5}$$

Accordingly the degree of crystallinity $X = 1-nA_{0.5}$
where $A_{0.5}$ is sorption value at $\phi = 0.5$; coefficient $n = 7.15$ for chitin samples with DA $\geq$ 85%, whereas for chitosan samples with DD $\geq$ 85% $n = 5.65$ [40].

2.3.6 Electron Microscopy

Two main methods of the electron microscopy are widespread for the study of biopolymers, and namely, transmission electron microscopy and scanning electron microscopy [73]. A transmission electron microscope (TEM) is an apparatus, in which a beam of electrons pass through an ultra-thin specimen (thickness $\leq$ 100 nm) [74]. An image is formed by the interaction of the electrons with its substance of the specimen; the image is magnified and focused onto an imaging device, such as a fluorescent screen or photographic film, or to be detected by a sensor such as a CCD camera. The electron beam generated by an electron gun is focused by condensers with creating of a small electronic spot. A drop of dilute aqueous dispersion ($\leq$0.1% solids) is deposited on the copper grid (300–400 mesh) coated with an ultrathin carbon or formvar film and air-dried; then a drop of the solution of phosphotungstic acid or uranyl acetate is added to the dry sample for staining and air-dried before observation in TEM. The grid together with a sample is placed in a microscope and evacuated. To obtain an electron image the specimen is placed in a magnetic field of the objective lens with high optical power that determines the maximum resolution of TEM. After passing through the object, the electrons pass through an aperture, and then they are focused in the object plane of the projection lens that forms an image on the imaging device. The magnification of a TEM image can reaches about 10^6, and resolution 0.2 nm.

Field emission gun scanning electron microscope (FEG-SEM) produces images of an object by scanning it with a focused beam of electrons (electron probe) [75]. Thin electron probe is generated by an electron gun and is focused by electron condenser lenses into a very fine focal spot sized up to 1 nm. Scanning coils deflect the probe in two mutually perpendicular directions, scanning the surface of the specimen, like scanning electron beam screen of a cathode-ray in TV tube. The electron probe interact with atoms of the specimen producing low-energy secondary electrons, which are collected by a detector and give information about the size, topography of surface and composition of the specimen. A drop of the dilute dispersion of the solid object is deposited on the mirror surface of a substrate (mica or glass) and dried. After drying the specimen is subjected to deposition of thin conductive layer of gold to remove the charge and the screening of the incident beam on the accumulated charge. The coated specimen is placed in a microscope for investigation. The tests can carry out in high vacuum, and at a wide temperature range. The magnification of FEG-SEM image can reaches about 10^5, and resolution 0.5 nm.

2.3.7 Atomic Force Microscopy

Atomic force microscope (AFM) consists of a cantilever with a sharp tip at its end that is used to scan the specimen surface [76]. The cantilever is typically silicon or silicon nitride with a tip radius of curvature on the order of nanometers. When the tip is brought into proximity of a sample surface, forces between the tip and the sample lead to a deflection of the cantilever. Typically, the deflection is measured using a laser spot reflected from the top surface of the cantilever into an array of photodiodes. Other methods that are used include optical interferometry, capacitive sensing, or piezoresistive AFM cantilevers.

AFM can operate in several modes, depending on the material and application. In general, the possible display modes are divided into static (also called contact) modes and various dynamic (noncontact and semicontact) modes, where the cantilever vibrates. When operating in contact mode, AFM is an analogue of the profilometer. Cantilever tip is in direct contact with surface of the sample, and the scanning is carried out usually in a constant force mode. However, this mode is not suitable for study of organic materials having a low strength. To investigate the nanoscale organic object, the contact mode is limited by use a small force, while the noncontact or semicontact (tapping) modes are preferable.

When operating in noncontact mode, the tip of the cantilever does not contact the sample surface. The cantilever is oscillated at its resonant frequency (frequency modulation) or just above (amplitude modulation) where the amplitude of oscillation is typically a few nanometers (<10 nm). The force exerted from the surface of the specimen causes a shift of the amplitude-frequency and phase-frequency characteristics of the probe. Feedback system usually supports constant oscillation amplitude of the probe, and the change in frequency and phase at each point is recorded. Similarly to noncontact mode, in tapping (semicontact) mode the cantilever is driven to oscillate up and down at near its resonance frequency by a piezoelectric element. However, the amplitude of this oscillation is greater than 10 nm, typically 100–200 nm. The servo adjusts the height to maintain a set cantilever oscillation amplitude as the cantilever is scanned over the sample. A tapping AFM image is therefore produced by imaging the force of the intermittent contacts of the tip with the sample surface. The vertical resolution of AFM can be up to 0.1 nm, but the lateral resolution is much lower due to the convolution.

2.3.8 Laser Light Scattering

The technique of laser light scattering (LLS) is based on the principle that particles passing through a laser beam will scatter light at an angle that is related to their size. As the particle size decreases, the observed scattering angle increases logarithmically. The observed scattering intensity is also dependent on particle sizes and diminishes with decrease of cross-sectional area of the particle. Thus, small particles scatter the light at wider angles but with low intensity, whereas large particles scatter at narrow angles with high intensity.

A typical device of laser light scattering consists of:

- Laser to provide a source of intense coherent light of fixed wavelength;
- Sample feeding system to ensure that the dispersion can pass through the laser beam as a homogeneous stream of particles in a reproducible state of dispersion;

- Series of detectors which are used to measure the light pattern produced over a wide range of angles.

Modern LLS-instruments (e.g., Mastersizer-3000 of Malvern Co.) measure the scattering angle in the range from 0.02° to 135°. A logarithmic detector sequence is used, where the detectors are grouped closely together at small angles and more widely spaced at wide angles, yields the optimum sensitivity. The LLS-instrument uses two light colors – blue for determining sizes of nanoparticles and the red for determination the sizes of microparticles. The diluted dispersion of particles after incubation at a constant temperature and sonication passes through the transparent slit illuminated by laser beams. Detectors entrap the diffracted light that is converted into electrical signals and processed by the computer. Laser light diffraction reports about sizes that relate to the particle volume or number.

The final result can be differential distribution curves of particle sizes by volume or number in the range from 10 nm to hundreds microns. For spherical particles the result is a hydrodynamic diameter of the particles. In the case of elongated rod-like particles, the LLS-instrument shows predominantly a longitudinal size, since the intensity of the scattered light in the longitudinal direction is much higher than in the lateral direction of the particles.

2.3.9 Thermoanalytical Methods

Differential scanning calorimetry (DSC) is a thermoanalytical technique, in which the difference in the amount of heat energy between sample and etalon is measured as a function of temperature [63, 77]. Both the sample and etalon are maintained throughout the experiment at the same temperature. Generally, the temperature program for a DSC analysis is designed such that the sample holder temperature increases linearly as a function of time. The etalon sample should have a well-defined heat capacity over the range of temperatures to be scanned.

Differential thermal analysis (DTA) is a technique, in which sample and etalon are heated identically, whereas the exothermic or endothermic processes cause a difference in temperature between both [78]. Thus, DTA measures the difference in temperature between the sample and the etalon, when they receive the same amount of heat. Both DSC and DTA methods are intended to measure exothermic or endothermic processes such as melting, crystallization, oxidation, cross-linking, decomposition, etc. These methods may also be used to observe more subtle physical changes, such as glass transitions.

Thermogravimetric analysis (TGA) is a method of thermal analysis in which changes in weight of sample are measured as a function of increasing temperature [79]. TGA can provide information about various physical phenomena, such as vaporization, sublimation, sorption, and desorption. Likewise, TGA can provide information about chemical phenomena including chemisorption, oxidation, reduction, dehydration, decomposition, etc.

The application of various thermoanalytical methods to nitrogenated polysaccharides allows to study the processes of sorption and desorption, oxidation and thermal decomposition, phase and isophase transitions, etc.

2.3.10 Viscometry

This method is used for determination of average degree of polymerization (DP) and MW of nitrogenated polysaccharides in dilute solutions (£1 mg/ml). Samples of chitin can be dissolved only in certain solvents such as 2.77 M NaOH, *N,N*-dimethylacetamide (DMAA) containing 5% of LiCl and some others, which are used to estimate the MW of the biopolymer [80, 81]. To determine MW of chitosan, samples of this biopolymer are dissolved in dilute acetic acid (Ac) containing a low amount of salt (NaCl, KCl, etc.) [82, 83].

Viscosity measurements were performed in capillary viscometers at 25 °C to measure the intrinsic viscosity $[\eta]$ of the solutions. The MW of biopolymers is calculated by the equation of Mark-Houwink-Sakurada:

$$[\eta] = K\,(\mathrm{MW})^n$$

Coefficients *K* and *n* of this equation are shown in Table 2.4.

2.3.11 Composition Assay Methods

Isolated samples of nitrogenated polysaccharides can contain some admixtures such as residual lipids minerals, proteins, etc. To estimate the content

Table 2.4 Coefficients in equation of Mark-Houwink-Sakurada.

Biopolymer	Solvent	*K*, ml/g	*n*	References
Chitin	2.77 M NaOH	0.1	0.68	[80]
Chitin	DMAA/LiCl	$2.4 \cdot 10^{-3}$	0.69	[81]
Chitosan	0.1 M Ac/0.2 M NaCl	$1.81 \cdot 10^{-3}$	0.93	[82,83]

of these admixtures, special analytical methods are used. Residual lipids in biopolymers were determined by extraction of sample with mixture of ethanol and benzol in accordance with standard procedure T 204 of TAPPI. Content of mineral admixtures (ash) was determined by combustion of the sample (2–10 g) in an oven at 550 °C [84].

Residual proteins in chitin samples were determined by weight loss after extraction of dry sample with 1M NaOH at ambient temperature for 20 h [85]. To determine residual proteins in chitosan its sample was dissolved in 0.1 M acetic acid to final concentration 10 mg/ml and after addition of Bradford reagent the optical density was measured by a spectrometer at 595 nm [31]. Content of residual proteins is calculated from a calibration graph using solutions of bovine albumin in 0.1 M acetic acid.

Elemental analysis of biopolymers is carried out using standard elemental analyzers in order to determine the content of carbon, nitrogen, and hydrogen, as well as to calculate DA or DD [86]. Since the fully deacetylated chitosan contains 8.696% nitrogen and fully acetylated chitin 6.897% nitrogen, the DA of the sample can be calculated by equations:

$$DA, \% = 55.586\,(8.696 - N)$$

Or more precise

$$DA, \% = (58.34\,C/N) - 300$$

where C and N is percentage of carbon and nitrogen in the analyzed sample.

Before elemental analysis the polymer samples should be completely deproteinized.

Several other methods (^{15}N NMR, X-ray photoelectron spectroscopy, UV-Vis-spectroscopy, mass spectrometry, HPLC, etc.) are also applied for characterization of nitrogenated polysaccharides, but to a lesser extent.

2.4 Characterization of Nitrogenated Polysaccharides

2.4.1 Chitin

Chitin is the most widespread nitrogen-containing polysaccharide in nature, which is present in exoskeletons and shells of some aquatic animals and insects, as well as in cell walls of some fungi and microorganisms [87]. Chitin is regarded as a nitrogenated derivative of cellulose, where the hydroxyl group at C2 atoms in each repeat units is replaced with an acetylamino group. Macromolecules of this natural polymer contain up to 10000–15000 units of 1,4-β-N-acetylglucosamine [88]. Similar to cellulose, chitin is a linear semicrystalline biopolymer. The linear macromolecules of this biopolymer joined by hydrogen bonds form the nanoscale fibrils. Each such nanofibril is built of ordered crystallites and low-ordered amorphous domains statistically alternating along the fibril [88, 89].

Chitin occurs in nature in the form of nanofibrils with lateral size of 3–5 nm, embedded in an amorphous matrix composed of proteins, glycoproteins, saccharides and some other components [90, 91]. In aquatic animals and insects, nanofibrils of chitin are bound with proteins, while in fungal cell wall nanofibrils of this biopolymer can be associated also with glucans, mannans and other saccharides [91–94]. Nanofibril-matrix complex in some natural sources (e.g., sea animals) is surrounded by a mineral layer containing mainly calcium carbonate.

During isolation process of chitin from natural sources, the surface of nanofibrils is released from the matrix components. As a result, the direct contact between the nanofibrils becomes possible, which causes the lateral aggregation of the nanofibrils via cocrystallization of adjacent nanocrystallites followed by increase in their size and forming of nanofibrillar bundles called microfibrils. So, structural investigations have shown that after isolation of chitin from natural sources the lateral sizes of nanocrystallites increased from 3–5 nm to 8–10 nm, probably due to cocrystallization process [38, 63, 95], whereas average diameter of the formed nanofibrillar bundles reached 25–90 nm [88, 96, 97].

Content of crystalline fraction in various chitin samples was estimated by XRD method using so called index of crystallinity (CrI). However, it was found that depending on the method of calculation CrI for fungi chitin varied from 0.43 to 0.80, for insect chitin from 0.54 to 0.89, of shrimp chitin from 0.64 to 0.89, while CrI of crab chitin ranged from 0.67 to 0.90 [20, 34–36, 97, 98]. On the other hand, actual degree of crystallinity (X) of the crab chitin was 0.73, and this reliable value has been confirmed by independent sorption and calorimetric methods [99]. Chitins isolated from insects and sea animals have very close crystallinity [100]. On the other hand, crystallinity of fungal chitins was lesser than of chitins of sea animals [39, 69]. DA of commercial chitin samples can be of usually 80–95%. Decreasing of DA or nitrogen content causes reduction in crystallinity of the sample [32, 101].

Crystalline domains of chitin can be in α-, β-, or γ-allomorph [3, 8, 9]. The most abundant α-allomorph of crystalline chitin is present in majority of aquatic animals, insects, fungi and microorganisms. The rarer β-form of crystalline chitin is found in the squid pens, diatoms spines and tubes of *Pogonophora*, whereas the γ-form can be isolated from lining of the *Loligo* stomach and cocoons of the beetles *Ptinus tectus* or *Rhynchaenus fangi*.

The β-and γ-forms of chitin are instable and after some treatments they can be irreversibly transformed into more stable α-allomorph [10, 11].

Structural studies have shown that α-allomorph has orthorhombic crystalline unit cell (Figure 2.8) with a space group $P2_12_12_1$ and antiparallel arrangement of macromolecular chains [3, 9, 12, 51, 102]. In contrast to α-form, the β-allomorph of crystalline chitin has monoclinic unit cell having a space group $P2_1$ and parallel arrangement of macromolecular chains [3, 8, 12, 51, 103]. Structure of γ-form is still not clear; it is only know that this form of crystalline chitin has a mixed arrangement of the macromolecular chains.

As a result of numerous structural studies, the parameters of crystalline unit cells for α – and β-allomorphs of chitin have been established [3, 8, 9, 12, 51, 102–104]. As can be seen from Table 2.5, at the transition from orthorhombic α-chitin to monoclinic β-chitin and further to monoclinic crystalline cellulose, the cell parameter c remains unchanged, parameter a increases, whereas parameter b decreases.

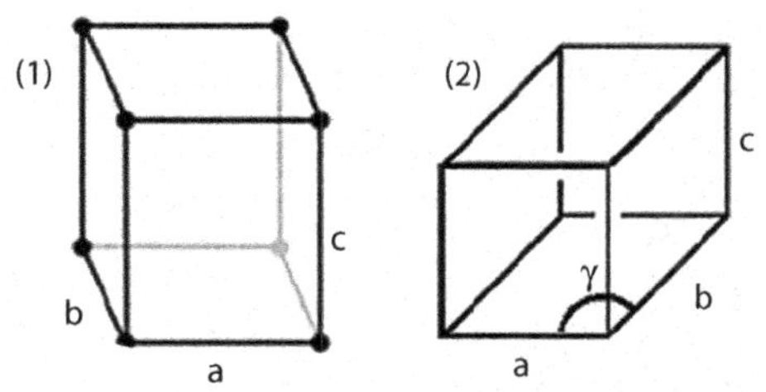

Figure 2.8 Image of orthorhombic (1) and monoclinic (2) crystalline unit cells.

Table 2.5 Characteristics of crystalline unit cells of chitins, chitosan and natural cellulose $CI_β$.

Characteristics	α-chitin [104]	β-chitin [8]	α-chitosan [116]	$CI_β$ [28]
a, nm	0.474	0.485	0.807	0.790
b, nm	1.886	0.926	0.844	0.820
c, nm	1.032	1.038	1.034	1.034
γ°	90	97.5	90	97
V_c, cm³/mole	139	140	106	100
$ρ_c$, g/cm³	1.46	1.46	1.52	1.62

Note: V_c is molar volume and $ρ_c$ is specific gravity of crystallites.

Although the α – and β-chitins have a close packing density, these two allomorphs differ greatly from each other. Firstly, due to different types of crystalline unit cells, α – and β-chitins show the different XRD patterns and interplanar distances (Figure 2.9, Table 2.6). Furthermore, the crystallinity of α-chitin samples is higher than samples of β-chitin [49].

Secondly, α-chitin forms a developed system of strong intra and intermolecular H-bonds along a-, b – and c-axes of the unit cell, which makes the crystalline structure stable and inaccessible for molecules of water and other polar liquids [3]. Unlike orthorhombic α-chitin, monoclinic β-chitin has strong H-bonds within (010) layers only, while between these layers H-bonding is absent, resulting in the interlayer swelling of β-chitin

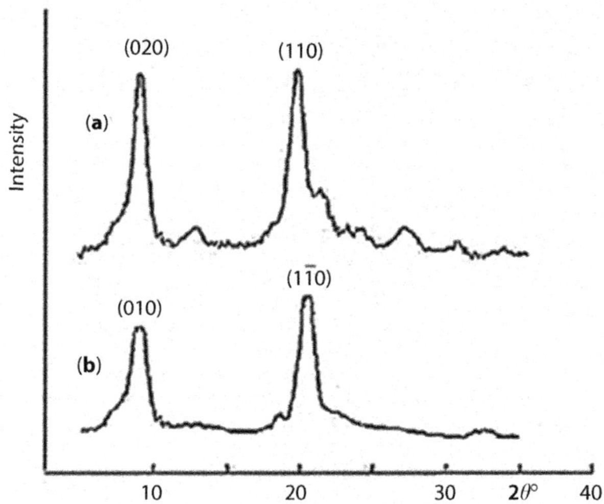

Figure 2.9 X-ray diffractograms of (a) α – and (b) β-chitins.

Table 2.6 X-ray diffraction patterns of α – and β-chitins.

α-chitin			**β-chitin**		
Index	**2θ°$_{hkl}$**	***d*, nm**	**Index**	**2θ°$_{hkl}$**	***d*, nm**
(020)	**9.4**	**0.943**	**(010)**	9.6	0.918
(021)	12.6	0.700	(011)	12.8	0.688
(002)	17.5	0.505	(002)	17.1	0.519
(110)	**19.3**	**0.460**	**$(1\bar{1}0)$**	19.8	0.45
(013)	26.5	0.336	(110)	22	0.405

crystallites [3, 8, 11, 105, 106]. This phenomenon is a specific feature of layered monoclinic crystalline structure of β-chitin and cellulose CI_{β} [107].

Penetration of molecules of polar liquids into crystallites of β-chitin leads to increase of the distance between (010)-planes proportional to volume (*V*) of the guest molecule, from water to *n*-octanol (Figure 2.10). Due to labile and accessible structure β-chitin is deacetylated also easier than stable α-chitin.

Additional differences are observed in the study of α – and β-chitin by methods of FTIR and NMR. IR-spectrum of α-chitin has two bands of amide I, at 1660 and 1630 cm^{-1}, while β-chitin gives one amide I band at around 1630 cm^{-1} [54]. The solid state ^{13}C NMR reveals that α-chitin has a double signal at 73 ppm and 75 ppm, while β-chitin gives single resonance at 74 ppm [12, 108].

The nanoarchitecture of chitin promotes the release of free nanofibrillar bundles and nanoparticles. These nanoscale objects are a subject of continuing research and have a great commercial interest in terms of manufacturing and application of new nanoproducts. To isolate nanofibrillar bundles (NFBs) from α-chitin with diameter of 10–20 nm, the starting material was dispersed in water by high-speed blender at solids concentration of 0.2–0.4 wt.% and pH value of 3–4 to form a slurry, which further was disintegrated by high-pressure homogenizer to obtain a transparent gel [109–111]. The partially carboxylated NFBs with lateral size of 20–50 nm have been isolated from α- and β-chitins using the pretreatment with hypochlorite/TEMPO system followed by disintegration in an aqueous medium [111, 112]. A more productive technology is the use of high-pressure homogenizers or similar industrial equipment for large-scale production of chitin NFB.

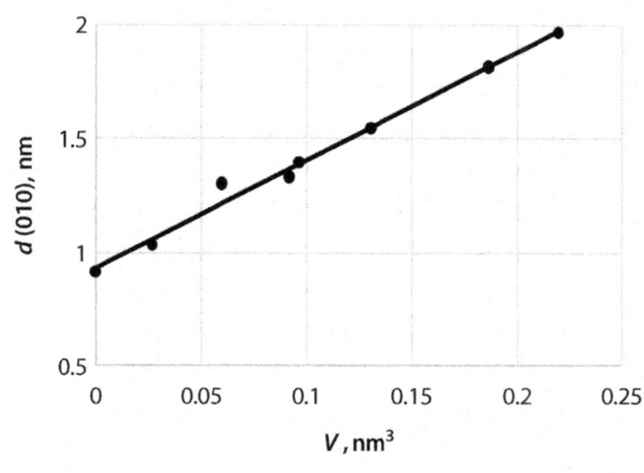

Figure 2.10 Dependence of interplanar distance, *d* (010), of β-chitin on volume of quest molecules.

Nanocrystalline particles of chitin were prepared mainly by acid hydrolysis of the isolated biopolymer with boiling 3M HCl followed by ultrasonic or mechanical disintegration in an aqueous medium having a pH value of 3–4 [113, 114]. The obtained individual nanocrystallites or nanowhiskers of α-chitin were rod-like with average diameter about 8 nm and length of 50–200 nm. However, also aggregates of individual nanocrystallites with sizes of 40–80 nm × 300–500 nm were present. Another method for isolation of rod-like nanocrystallites with a width of 8 nm was implemented by selective oxidation of starting chitin samples with hypochlorite/TEMPO system followed by disintegration in an aqueous medium [35]. Dilute aqueous dispersions of nanocrystalline particles were phase stable due to charged surface of the particles containing the carboxylic groups.

The methods for isolation of free nanocrystallites are based on the nanofibrillar structure of chitin, where each nanofibril contains ordered nanocrystallites and disordered amorphous domains statistically alternating along the fibril. The nanocrystallites with three-dimensional order are strong and inaccessible constituents. As against, the disordered amorphous domains are weak and accessible places of the nanofibril. Thereby, cleavage of glycosidic bonds under the action of acids, oxidants, and some other chemicals occurs selectively in the amorphous domains of nanofibrils, while the nanocrystallites remain intact and can be released after vigorous ultrasonic or mechanical disintegration of the cleaved fibrils in aqueous medium.

2.4.2 Chitosan

Chitosan is an amino-polysaccharide containing amino groups at C2 of glucosamine units of macromolecules. This biopolymer is produced mainly from sources of α-chitin by deacetylation of refined feedstock with a hot 10–20 M sodium hydroxide [2, 20]. Therefore the obtained product can be called as α-chitin. The conditions of chitosan preparation affect the DD and MW of this biopolymer. Commercially available chitosan can have DD in the range 70–98% and MW from 4 kD to 500 kD [54, 115].

Like chitin, chitosan is a semicrystalline linear biopolymer comprising nanofibrils which consist of nanocrystallites and amorphous domains. However during the extraction of chitosan from natural sources, a partial decrystallization occurs resulting in reduced crystallinity and size of crystallites. For example, actual degree of crystallinity of the α-chitosan samples was $X = 0.57–0.58$, that is, significantly lower than crystallinity degree of chitin samples, $X = 0.67–0.72$ [40, 69, 99]. The lateral sizes of crystallites of α-chitosan are 4–5 nm, unlike isolated samples of α-chitin containing larger crystallites with lateral sizes of 8–10 nm [38, 63, 95]. Furthermore, after dissolution and regeneration of chitosan from solutions its crystallinity and size of crystallites decreases [69].

X-ray diffractograms of α-chitosan samples show visible diffraction peaks at 10°–11° from (100) and (010) planes and at 20°–21° from (020) planes of crystalline unit cells together with some weak peaks (Figure 2.2). With increasing the DA of chitosan these X-reflections are shifted toward lower 2θ-angles, whereas intensity of the reflections increases [32, 54]. Like the α-chitin, α-chitosan has orthorhombic crystalline unit cell with space group $P2_12_12_1$ and antiparallel arrangement of macromolecular chains [3]. Parameters of the unit cell of anhydrous α-chitosan were estimated, and namely: a = 0.807 nm, b = 0.844, and c =1.034 nm [116, 117]. However, some other parameters also were proposed [54].

The calculations showed that in the sequence: chitin – chitosan – cellulose, the increasing in the packing density of crystallites is observed (Table 2.5) as a result of substitution of large acetylamino groups having Van-der-Waals volume $w = 0.052$ nm^3 of chitin onto smaller amino groups ($w = 0.012$ nm^3) of chitosan, and further onto very small hydroxyl groups ($w = 0.010$ nm^3) of cellulose.

FTIR-spectroscopic studies have shown that absorbance in the area of amide I (1640–1660 cm^{-1}) and II (1540–1560 cm^{-1}) bands of chitosan having high DD is relatively low, but it increases with increasing of the acetylation degree [53, 54]. For chitosan with high DD additional band at 1590 cm^{-1} also appears, which can be related to bending vibrations of amine groups [54].

According to solid state CP/MAS ^{13}C NMR spectroscopy, weak intensity of signal at 20–30 ppm from carbon atoms of methyl groups and/or weak intensity of signal at 170–180 ppm from carbon atoms of carbonyl groups evidence about low content of acetyl groups in chitosan [118].

A distinctive feature of chitosan is easy dissolution in dilute acids and subsequent regeneration from the solutions after neutralization and addition of some polyanionic salts. This feature can be used to obtain chitosan nanoparticles (CsNP) [119]. The known preparation method of CsNP is based on the electrostatic interaction between positively charged amine groups of chitosan and negatively charged group of polyanions, for example, tripolyphosphate (TPP). Chitosan was dissolved in dilute acetic acid, and then solution of TPP was added with vigorous stirring to neutral pH value up to obtaining an opalescent dispersion of CsNP, which were separated by centrifugation [120]. At optimal conditions (low concentration, high stirring speed, etc.), the spherical gel-like CsNP can be isolated with a diameter of 40–60 nm.

The nanofibrillated chitosan is prepared by mechanical or ultrasound disintegration in an aqueous medium. Along with nanoparticles and nanofibrils also nanofilaments may be manufactured from chitosan solutions using an electrospinning technique. In this technique a high electrostatic voltage is imposed on a drop of biopolymer solution at the end of a capillary, which causes distortion of the drop shape with forming of a cone [121]. When the electrostatic force overcomes the solution surface tension, a stable liquid jet is ejected from the cone tip. The solvent is removed, leaving the ultrafine filaments that are collected on an electrically grounded target. The apparatus for the wet electrospinning comprises a high voltage source, pump, syringe, and water bath collector [122]. When the fine jet of the chitosan solution in dilute acetic or trifluoroacetic acid emerging from the syringe falls into the water bath having basic pH, the solvent dissolves, while the biopolymer regenerates with formation of wet nanofilaments, which then are dried. The resulting nanofilaments of chitosan had a diameter of 60–100 nm [122–124].

2.5 Properties of Nitrogenated Polysaccharides

Among the chemical properties a hydrolysis of nitrogen-containing biopolymers has of particular importance. The hydrolysis is used for depolymerization and production of nanoparticles, oligomeric and monomeric products. To decrease MW or DP the hydrolysis of initial polysaccharide is carried out with dilute mineral acids at room or elevated temperatures. For example, after hydrolysis of chitosan sample with 0.36 M H_2SO_4 at 120 °C for 15 min, the decrease of MW from 1388 kDa to 66 kDa was observed [125]. Sample of chitin having MW of 1200 kDa after hydrolysis with 3M HCl at 25 °C for 24 h acquired the MW of 200 kDa only [91]. Nanocrystalline particles of chitin can be prepared by hydrolysis of this biopolymer with boiling 3M HCl for 90 min followed by disintegration in an aqueous medium [113, 114].

As is known main part of the extracted chitin in the world is used for production glucosamine (GLA), which is used for treatment of arthritis [91]. To produce the GLA the refined chitin was hydrolyzed with concentrated acid, usually 10–12 M HCl, at 80–90 °C for 60–90 min while stirring [126, 127]. The obtained dark solution was decolorized, filtered, evaporated in vacuum at 45 °C to increase the concentration of product and remove the excess of acid. The concentrated solution was kept with cooling to crystallize the GLA-HCl salt. The crystals were resuspended in water and then recrystallized by ethanol. Finally, the GLA-HCl salt was recovered, washed with ethanol and dried at 50 °C. Concentrated solutions of sulfuric acid is also used in production of GLA. The hydrolysis reaction of chitin proceeds in two main stages: the first stage is a fast cleavage of glycosidic linkages and release of monomeric *N*-acetylglucosamine, while the second stage is a slower process of deacetylation of *N*-acetylglucosamine [91]. Laboratory method for production of GLA involves enzymatic hydrolysis of dissolved chitosan with endo-acting α-amylase and exo-acting glucoamylase [128].

Nitrogenated polysaccharides can be chemically modified to synthesize various derivatives using reactions of oxidation, esterification, and etherification of hydroxyl groups, as well as specific reactions involving amino groups of chitosan [1, 3, 129–131].

The nanostructure of the chitin and chitosan promotes the release of free nanoparticles, nanofibrils, and nanofilaments having unique features such as small dimensions, variety shapes, enhanced specific surface, high sorption ability, and other specific characteristic in combination with biocompatibility, biodegradability, and complex of unique therapeutic properties. These nanoscale substances reinforce the biocomposites and thicken the liquid systems.

The most pronounced physicochemical feature of nitrogenated polysaccharides is their hydrophility. Therefore, a great attention is paid to the interaction of hydrophilic biopolymers with water [72, 132]. Theoretical isotherms sorption of water vapor by completely amorphous chitin (DA = 100%) and chitosan (DA = 0%) can be calculated using the method of additive functions of Van Krevelen [133]. As can be seen from Figure 2.11, the isotherms have sigmoidal shape similar to the isotherm of amorphous cellulose [72]. Comparison of sorption isotherms for amorphous biopolymers showed that the sorption ability of chitosan containing hydrophilic amino groups at C2 was higher than of chitin containing hydrophobic

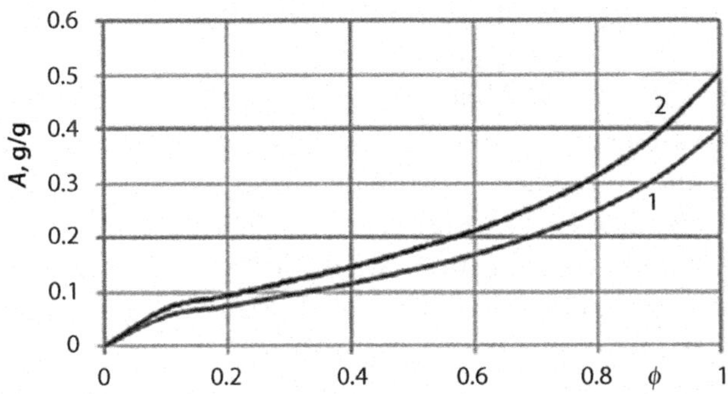

Figure 2.11 Theoretical isotherms sorption of water vapor by completely amorphous chitin (1) and chitosan (2) at 25 °C.

Table 2.7 Value of $A_{m,a}$ as function of DA.

DA, %	$A_{m,a}$ (g/g)
0	0.505
30	0.467
50	0.450
70	0.425
100	0.399

acetyl groups. Maximum equilibrium water sorption value at relative vapor pressure φ =1 for amorphous chitosan with DA = 100% is $A_{m,a}$ = 0.505 (g/g), while for amorphous chitin with DA = 0% $A_{m,a}$ = 0.399 (g/g).

Values of maximum equilibrium water sorption ($A_{m,a}$) at 25 °C for the amorphous biopolymers with various degrees of acetylation (DA) are shown in Table 2.7. Thereby, increasing of the DA of amorphous biopolymers reduces the maximum sorption value, $A_{m,a}$

In the case when the biopolymers are semicrystalline, along with DA also degree of crystallinity (*X*) or amorphicity (*Y* = 1–*X*) must be taken into account, because increasing the degree of crystallinity reduces the sorption of water vapor. The isotherms sorption of water vapor for semicrystalline nitrogenated polysaccharides can be calculated by the equation [40]:

$$A = A_{m,a}(1 - \mathrm{X})/(1 - K\ln\phi)$$

or

$$A = A_{m,a}Y/(1 - K\ln\phi)$$

where coefficient $K = 2.7$, while value of $A_{m,a}$ can be found from Table 2.7.

Using these equations, the degree of crystallinity of the biopolymer sample with known DA can be estimated from the sorption experiment.

Standard enthalpy of wetting ($\Delta_w H°_a$) of the dry amorphous biopolymers correlates with the maximum water sorption value ($A_{m,a}$), and namely, for chitosan with DA = 0% $\Delta_w H°_a = -171$ kJ/kg, which is more exothermic than $\Delta_w H°_a = -134$ kJ/kg of chitin with DA = 100% [99]. Increasing the DA and/or degree of crystallinity of biopolymers decreases the exothermic value of the wetting enthalpy ($\Delta_w H°$). Furthermore, studies have shown that the specific enthalpy of interaction ($h = \Delta_w H°/A$) for chitin and chitosan samples with various X and DA, as well as for other polysaccharides (cellulose, starch, etc.) is almost the same: $h = -335$ kJ/kg liquid. This makes it possible to calculate the maximum equilibrium sorption, if the wetting enthalpy is known, and vice versa.

Due to the presence of basic amino groups with a pKa value of 6.3, chitosan possesses a number of specific properties [134, 135]. At acidic pH, the amino groups are protonated and acquire a positive charge; whereby chitosan is converted into a soluble cationic polyelectrolyte. The dissolving mechanism includes repulsion of oppositely charged groups, disruption of hydrogen bonds, collapse of crystallites, hydration of polar groups, and finally diffusion of hydrated macromolecules in a solvent (water).

On the other hand, the deprotonation of amino groups of the biopolymer leads to a loss in solubility, association of macromolecules and formation of a hydrogel. For example, when the protonated groups of chitosan are neutralized by negatively charged anions of sodium triphosphate or sodium dodecyl sulfate, the gel formation is observed [136]. The cationic chitosan polyelectrolyte can form gel-like complexes with anionic polyelectrolytes, such as pectin, xanthan, carrageenan, alginate, polyacrylic acid, carboxymethylcellulose, some proteins, etc. [137, 138]. Using a special technique, gels of chitosan or its complexes can be prepared in the shape of nanoparticles [120].

The amino groups also possess the chelating functions, and therefore chitosan can form chelates with multivalent cations such as Cu (II), Hg (II), Zn (II), Cd (II), Co (II), Fe (III), etc. [139, 140]. This feature can be used to purify water from heavy metals.

Unlike chitosan, highly acetylated chitin contains almost no free amino groups and therefore does not have such specific features as solubility in weakly acidic aqueous medium, formation of complexes with negatively charged anions and polyanions, chelation of heavy metals cations, etc.

Chitin can be dissolved only in a few solvents such as 3–4 M sodium hydroxide, *N,N*-dimethylacetamide-LiCl system, *N*-methylmorpholine-*N*-oxide, and some ionic liquids such as 1-butyl-3-methylimidazolium acetate [141].

Study of thermo-physical properties revealed that basic glass temperature (T_g) of chitosan and chitin is 140–160 °C [63], which are smaller than T_g of cellulose, 220–230 °C [142]. Heating of the biopolymers to elevated temperatures causes their thermal decomposition. Maximum weight loss due to thermal decomposition is observed at 370–380 °C for α-chitin and at 300–310 °C for β-chitin and chitosan [20, 143]. Thus, thermal stability of highly crystalline α-chitin is higher than of low crystalline β-chitin and chitosan. Standard enthalpy of combustion for chitosan samples can be from −3000 kJ/mol to

−3200 kJ/mol unit, while exothermic value of combustion enthalpy for chitin samples is higher, in the range from −3900 kJ/mol to −4200 kJ/mol unit [63, 66]. This is due to the fact that chitosan contains a lesser amount of energy atoms C and H (total 17 moles) than chitin (total 21 moles).

To measure the Young's axial elastic moduli for crystallites of the biopolymers, the XRD technique was applied [144]. The results showed that the elastic modulus (E) was 41 GPa for crystallites of α-chitin, 65 GPa for crystallites of chitosan and about 140 GPa for crystallites of cellulose I, which correlates with the packing density of crystallites of these biopolymers (see Table 2.5). Of course, physico-mechanical properties of semicrystalline films were significantly lower than crystallites. So, the cast film of common chitosan possessed very small Young's modulus about 0.6 GPa only and tensile strength (TS) about 36 MPa, while the film made of nanofibrillated chitosan had mechanical characteristics typical for liquid crystalline polymers: E about 2.2 GPa and TS about 100 MPa [145, 146].

2.6 Applications

As is known about 65% of the world's chitin resources is used for production of such valuable substance as glucosamine, which is a dietary supplement and effective remedy for the treatment of arthritis [91]. In most of European counties, glucosamine sulfate has been approved as a medical drug. Furthermore, about a quarter of all the resources of chitin is used for production of chitosan. A smaller part of chitin is intended for the production of *N*-acetylglucosamine and oligosaccharides.

Main applications of nanoparticles and nanofibrils of chitin are focused on environmentally friendly nanocomposite materials with improved compatibility and mechanical properties [113, 114, 147]. For example, introduction of 3% of chitin nanofibrils in polyurethane latex causes a twofold increase in the Young modulus compared to net polymer. The strong films made of chitin nanofibrils and acrylic or methacrylic resins were transparent to 40% chitin content [148].

In view of biodegradability, nontoxicity, physiological inertness, antibacterial properties, and affinity chitin has found applications in biomedicine and food industry. Positive cytocompatible effect was found after introduction of chitin nanoparticles in some biomedical materials, such as hydrogels, scaffolds, and wound dressing [114]. Study of electrospun nanofilaments produced from chitosan solutions containing chitin nanoparticles indicated a good compatibility between the nanoparticles and matrix, which led to improved morphology and thermo-mechanical properties of the electrospun nanofibers. Nanoparticles of chitin are effective in stabilizing of oil-in-water emulsions used in cosmetics [149]. Chitin with immobilized amylases and invertases has applications in the food industry for clarification of fruit juices and milk processing [150].

Another nitrogenated polysaccharide – chitosan, received a great interest in biomedicine, pharmaceutics, cosmetics, and food industry because of its nontoxicity, biocompatibility, and biodegradability, as well as complex of therapeutic properties such as antioxidative, antimicrobial, analgesic, fungistatic, haemostatic, and some others [2, 151–154]. Nanoparticles of chitosan can serve as carriers for drug delivery systems in order to prolong duration of drug activity and improve therapeutic efficiency [155–157]. For example, chitosan nanobiocarrier binds and transports such acidic therapeutically active substances as ketoprofen, diclofenac, gadopentetic acid, aspartic and glutamic amino acids, etc., according to scheme:

$$\text{Ch-NH}_2 + \text{HOOC-R} \rightarrow \text{Ch-NH}_3^{+\,-}\text{OOC-R}$$

The chitosan nanocarrier can attach also cations of transition metals by means of chelating mechanism, for example,

$$2\text{Ch-NH}_2 + \text{Me}^{+2} \rightarrow \text{Ch-NH}_2 \ldots \text{Me}^{+2} \ldots \text{H}_2\text{N-Ch}$$

The formed nanocomplex can be used in biomedicine. As it is reported, the complex of nanochitosan and gadolinium-157 has a great potential in cancer therapy [154].

Nonwoven biomaterials made of electrospun chitosan nanofilaments with a high absorption ability are applicable in biomedicine, personal care, and cosmetics. Novel smart dressings contain two-three functional layers, which impart to these dressings the therapeutic, inadherent, absorbing, cleansing, breathing, occlusive, and other desired properties required for the fast and effective treatment [158–160]. The therapeutic layer of the dressings can contain, for example, the hydrogel of chitosan or its composite with others biopolymers, which successfully cleans the wounds and burns, decreases temperature, and reduces pains. Furthermore, the hydrogel may incorporate special therapeutically active substances (e.g., proteolytic enzymes, antibiotics, etc.), which accelerate the healing process. Chitosan-based bandages having excellent bactericide and hemostatic properties are used by the US army and the UK military.

Chitosan serves also as an active ingredient of shampoos, conditioners, and colorants. Being positive charged, chitosan macromolecules bind to the negative charged hair surface. As a result, chitosan forms on the hair surface a transparent and elastic layer, which imparts to hair strength, smoothness, and softness [154]. The addition of chitosan in composition of toothpaste prevents formation of odontolith and tooth decay. Furthermore, this biopolymer can be used in compositions of creams, lotions, lipstick, and other cosmetic remedies.

Chitosan has been used in multiple nutritional supplement products due to its ability to bind fat [2]. Besides chitosan possesses a food preservative properties.

In agriculture chitosan has been used as spray for foliage to induce disease resistance and increase quality and production of ornamental plants [161]. Chitosan was successfully applied also as protective coating for such crops as rice, palm, corn, cassava and many tropical fruits. The role of chitosan coating in plant protection is probably the result of its barrier function along with fungistatic properties.

Chitosan is involved in papermaking for surface coating and binding of cellulose fibers [154]. Paper coated with chitosan layer has smooth surface and increased wet strength. Chitosan serves as high-quality binder in production of toilet and wrapping paper.

Some other application areas of nitrogenated polysaccharides also are discussed [2, 3, 114, 154].

2.7 Conclusion

The widely spread nitrogenated polysaccharides – chitin and chitosan contain nanoscale fibrils which composed of ordered nanocrystallites and disordered amorphous nanodomains. The nanostructure of these biopolymers promotes the release of free nanoparticles, nanofibrils, and nanofilaments having unique features such as small dimensions, variety shapes, enhanced specific surface, high sorption ability, and other specific characteristic in combination with biocompatibility, biodegradability, and complex of unique therapeutic properties: antioxidative, antibacterial, analgesic, fungistatic, haemostatic, etc. Diverse methods, such as XRD, NMR, IR and UV-spectroscopy, electron and atomic force microscopy, sorption calorimetry, thermal analysis and some others, are applied to characterize parameters of molecular structure (MW, DP, DA, or DD) and supramolecular structure (crystallinity, type of crystalline allomorph, size of structural constituents, etc.), as well as various chemical, physico-chemical and physico-mechanical properties of the nitrogen-containing biopolymers.

Both nitrogenated polysaccharides, chitin and chitosan, find a great commercial interest in biomedicine, pharmaceutics, cosmetics, personal care, agriculture, and some other areas. Small particles and fibrils of the biopolymers can be used as fillers for biocomposites and thickeners imparting to liquid systems an increased viscosity and gel consistence. The nanocarriers bind various therapeutic active substances, which expand application areas of these nitrogenated biopolymers. Antibacterial nonwoven materials made of nanofilaments have an increased absorption ability and accelerate healing process. Further progress in the nanotechnology and development of nanomaterials will expand the application areas of both biopolymers, chitin and chitosan.

References

1. Roberts, G., *Chitin chemistry*, MacMillan, 1998.
2. Aranaz, I., Mengibar, M., Harris, R., Panos, I., Miralles, B., Acosta, N., Functional characterization of chitin and chitosan. *Curr. Chem. Biol.,* 3, 203, 2009.
3. Rinaudo M., Chitin and chitosan: properties and applications. *Progr. Polym. Sci.*, 31, 603, 2006.
4. Blackwell J., *Biopolymers*, Academic Press, 1973.
5. Watthanaphanit, A., Rujiravanit, R., *Handbook of carbohydrate polymers: development, properties and applications*, Nova Publication, 2010.
6. Chen, P.Y., Lin, A.Y.M., McKittrick, J., Meyers, M.A., Structure and mechanical properties of crab exoskeletons. *Acta Biomater.,* 4, 587, 2008.
7. Revol, J.F., Marchessault, R.H., *In vitro* chiral nematic ordering of chitin crystallites. *Int. J. Biol. Macromol.*, 15, 329, 1993.
8. Gardner, K.H., Blackwell, J., Refinement of the structure of β-chitin. *Biopolymers*, 14, 1581, 1975.
9. Rudall, K.M., Kenchington, W., The α-chitin system. *Biol. Rev.,* 48, 597, 1973.
10. Saito, Y., Putaux, J.-L., Okano, T., Gaill, F., Chanzy, H., Structural aspects of the swelling of β-chitin in HCl and its conversion into α-chitin. *Macromolecul.,* 30, 3867, 1997.
11. Saito, Y., Okano, T., Gaill, F., Chanzy, H., Putaux, J.-L., Structural data on the intra-crystalline swelling of β-chitin. *Int. J. Biol. Macromol.,* 28, 81, 2000.
12. Jang, M.K., Kong, B.-G., Jeong, Y.-I., Lee, C.H., Nah, J.-W., Physicochemical characterization of α-chitin, β-chitin, and γ-chitin separated from natural resources. *J. Polym. Sci.*, Part A, 42, 3423, 2004.
13. Moussaoui, Y., Mnasri, N., Elaloui, E., Ben Salem, R., Lagerge, S., De Menorval, L.C., Preparation of chitosan gel. *EPJ Web Conf.,* 29, 34, 2012.
14. Mathur, N.K., Narang, C.K., Chitin and chitosan, versatile polysaccharides from marine animals. *J. Chem. Educ.*, 67, 938, 1990.
15. Gooday, G.W., Jeuniaux, C., Muzzarelli, R., *Chitin in nature and technology,* Plenum Press, 1986.

16. Hackman R.H., The occurrence of complexes in which chitin and protein are covalently linked. *Aust. J. Biol. Sci.,* 13, 568, 1960.
17. Diaz-Rojas, E.I., Arguelles-Monal, W.M., Higuera-Ciapara, I., Hernandez, J., Lizardi-Mendoza, J., Goycoolea, F.M., Determination of chitin and protein contents during the isolation of chitin from shrimp waste. *Macromol. Biosci.,* 6, 340, 2006.
18. Arbia, W., Arbia, L., Adour, L., Amrane, A., Chitin extraction from crustacean shells using biological methods – a review. *Food Technol. Biotechnol.,* 51, 12, 2013.
19. Percot, A., Viton, C., Domard, A, Optimization of chitin extraction from shrimp shells. *Biomacromolecules,* 4, 12, 2003.
20. Zelencova, L, Erdogan, S, Baran, T, Kaya, M., Chitin extraction and chitosan production from Chilopoda (Scolopendra cingulata) with identification of physicochemical properties. *J. Polym. Sci., Part A.*, 57, 437, 2015.
21. Younes, I., Ghorbel-Bellaaj, O., Nasri, R., Chaabouni, M., Rinaudo, M., Chitin and chitosan preparation from shrimp shells using optimized enzymatic deproteinization. Process Biochem., 47, 2032, 2012.
22. Gortari, M.C., Hours, R.A., Biotechnological processes for chitin recovery out of crustacean waste: a mini-review. *Electron. J. Biotechnol.,* 16, 14, 2013.
23. Acosta, N, Jimenez, C, Borau, V, Heras, A., Extraction and characterization of chitin from crustaceans. *Biomass Bioenerg.,* 5, 145, 1993.
24. Scherrer, P., Bestimmung der Grösse und der inneren Struktur von Kolloidteilchen mittels Röntgenstrahlen. *Nachrichten Gesellsch. Wissensch.* 2, 98, 1918.
25. Singh, K., *Advanced X-ray techniques in research and industries.* IOS Press Inc, 2005.
26. Ioelovich, M., For supramolecular structure of native and isolated celluloses. *Acta Polymerica,* 43, 110, 1992.
27. Dinilchenko, S., Kukharenko, O., Moseke, C., Protsenko, I., Sukhodub, L., Sulkio-Cleff, B., Determination of bone mineral crystallite size and lattice strain from diffraction line broadening. *Res. Technol. Crystal.,* 37, 1234, 2002.
28. Ioelovich, M., *Cellulose: nanostructured natural polymer,* LAP, 2014.
29. Fosher, B., Beltranme, P.L., Naggi, A., Torri, G., Alkaline deacetylation of chitin enhanced by flesh treatment: reaction kinetics and structural modification. *Carbohydr. Polym.,* 12, 405, 1990.
30. Cruz, M.C.P., Ravagnani, S.P., Brogna, F.M.S., Campana, S.P., Trivino, G.C., Lisboa, A.C.L., Mei, L.H.I., Evaluation of the diffusion coefficient for controlled release of oxytetracycline from alginate/chitosan/poly(ethylene glycol) microbeads in simulated gastrointestinal environments. *Biotech. Appl. Biochem.,* 40, 243, 2004.
31. Yuan, Y., Chesnutt, B.M., Haggard, W.O., Bumgardner, J.D., Deacetylation of chitosan: material characterization and *in vitro* evaluation via albumin adsorption and pre-osteoblastic cell cultures. *Materials,* 4, 1399, 2011.

32. Zhang, Y., Xye, C., Xye, Y., Gao, R., Zhang, X., Determination of degree of deacetylation of chitin and chitosan by X-ray powder diffraction. *Carbohydr. Res.*, 340, 1914, 2005.
33. Abdou, E.S., Nagy, K.S.A., Elsabee, M.Z., Extraction and characterization of chitin and chitosan from local sources. *Biores Technol.*, 99, 1359, 2008.
34. Mogilevskaya, E.L., Akopova, T.A., Zelenetskiy, A.N., Ozerin, A.N., The crystal structure of chitin and chitosan. *J. Polym. Sci., Part A*, 48, 116, 2006.
35. Fan, Y., Saito, T., Isogai, A., Chitin nanocrystals prepared by TEMPO-mediated oxidation of alpha chitin. *Biomacromolecules*, 9, 192, 2008.
36. Fan, Y., Saito, T., Nano-fibrillation of chitins by TEMPO-mediated oxidation or protonation of amino groups. *Funct Mater.*, 29, 19, 2009.
37. Chebotok, E.N., Novikov, V.Y., Konovalova, I.N., Effect of crystallinity of chitin and chitosan of kinetics of alkaline deacetylation. *J. Appl. Chem.*, 80, 1724, 2007.
38. Vihoreva, G.A., Gorbacheva, I.N., Galbraich, L.S., Chemical modification of polysaccharides of hydrobiotes. *Chem Fibers.*, 5, 36, 1994.
39. Maryin, A.P., Feophylova, E.P., Genin, Y.V., Effect of crystallinity on sorption and thermal properties of chitosan. *J. Polym. Sci., Part B*, 24, 658, 1982.
40. Ioelovich, M., Crystallinity and hydrophility of chitin and chitosan. *Res. Rev. J. Chem.*, 3, 7, 2014.
41. Kasaai, M.R., Determination of the degree of N-acetylation for chitin and chitosan by various NMR spectroscopy techniques: a review. *Carbohydr. Polym.*, 79, 4801, 2010.
42. Czechowska-Biskup, R., Jarosińska, D., Rokita, B., Ulański, P., Rosiak, J.M., Determination of degree of deacetylation of chitosan – comparison of methods. *Prog. Chem. Appl. Chitin. Deriv.*, 17, 5, 2012.
43. Lavertu, M., Xia, Z., Serreqi, A., Berrada, M., Rodrrigues, A., Wang, D., Buschmann, M., Gupta, A., A validated H^1 NMR method for the determination of the degree of deacetylation of chitosan. *J. Pharm. Biomed. Anal.*, 32, 1149, 2003.
44. Heux, L., Brugnerotto, J., Desbrières, J., Versali, M.-F., Rinaudo, M., Solid state NMR for determination of degree of acetylation of chitin and chitosan. *Biomacromolecules*, 1, 746, 2000.
45. Duarte, M.L., Ferreira, M.C., Marvao, M.R., Rocha, J., Determination of the degree of acetylation of chitin materials by C-13 CP/MAS NMR spectroscopy. *Int. J. Biolog. Macromolec.*, 28, 359, 2001.
46. Kasaai, M., A review of several reported procedures to determine the degree of N-acetylation for chitin and chitosan using infrared spectroscopy. *Carbohydr. Polym.*, 71, 497, 2008.
47. De Alvarenga, E.S., De Oliveira, C.P., Bellato, C.R., An approach to understanding the deacetylation degree of chitosan. *Carbohydr. Polym.*, 80, 1155, 2010.
48. Paulino, A.T., Simionato, J.I., Garcia, J.C., Nozaki, J., Characterization of chitosan and chitin produced from silkworm chrysalides. *Carbohydr. Polym.*, 64, 98, 2006.

49. Al Sagheer, F.A., Al Sughayer, M.A., Muslim, S., Elsabee, M.Z., Extraction and characterization of chitin and chitosan from marine sources in Arabian Gulf. *Carbohydr. Polym.*, 77, 410, 2009.
50. Prabu, K., Natarajan, N., Isolation and FTIR spectroscopy characterization of chitin from local sources. *Adv. Appl. Sci. Res.*, 3, 1870, 2012.
51. Rumengan, I.F.M., Suryanto, E., Modaso, R., Wullur, S., Tallei, T.E., Limbong, D., Structural characteristics of chitin and chitosan isolated from the biomass of cultivated rotifer, *Brachionus rotundiformis. Int. J. Fisher. Aquat. Sci.*, 3, 12, 2014.
52. Kaya, M., Halici, M.G., Duman, F., Erdogan, S., Baran, T., Characterization of α-chitin extracted from a lichenised fungus species *Xanthoria parietina. Nat. Prod. Res.*, 29, 1280, 2015.
53. Negrea, P., Caunii, A., Sarac, I., Butnariu, M., The study of infrared spectrum of chitin and chitosan extract as potential sources of biomass. *Digest. J. Nanomater. Biostruct.*, 10, 1129, 2015.
54. Kumirska, I., Czerwicka, M., Kaczyński, Z., Bychowska, A., Brzozowski, K., Thöming, J., Stepnowski, P., Application of spectroscopic methods for structural analysis of chitin and chitosan. *Mar. Drugs*, 8, 1567, 2010.
55. Queiroz, M.F., Melo, K.R.T, Sabry, D.A., Sassaki, G.L., Rocha, H.A.O., Does the use of chitosan contribute to oxalate kidney stone formation. *Mar. Drugs.*, 13, 141, 2015.
56. Lim, S.H., Hudson, S.M., Synthesis and antimicrobial activity of a water-soluble chitosan derivative with a fiber-reactive group. *Carbohydr. Res.*, 339, 313, 2004.
57. Vino, A.B., Ramasamy, P., Shanmugam, V., Shanmugam, A., Extraction, characterization and *in vitro* antioxidative potential of chitosan and sulfated chitosan from cuttlebone of *Sepia Aculeata Orbigny. Asian. Pac. J. Trop. Biomed.*, 2, 334, 2012.
58. Song, C., Yu, H., Zhang, M., Yang, Y., Zhang, G., Physicochemical properties and antioxidant activity of chitosan from the blowfly *Chrysomya megacephala larvae. Int. J. Biol. Macromol.*, 60, 347, 2013.
59. Baxter, A., Dillon, M., Taylor, K.D., Roberts, G.A.F., Improved method for IR determination of the degree of N-acetylation of chitosan. *Int. J. Biol. Macromol.*, 14, 166, 1992.
60. Hussain, R., Iman, M., Maji, T.K., Determination of degree of deacetylation of chitosan and their effect on the release behavior of essential oil from chitosan and chitosan-gelatin complex microcapsules. *Int. J. Adv. Eng. Appl.*, 2, 4, 2013.
61. Calvet, E., Prat, H., *Recent progress in microcalorimetry*, Pergamon Press, 1963.
62. Wadső, I., Isothermal microcalorimetry near ambient temperature: an overview and discussion. *Thermochim. Acta.*, 294, 1, 1997.
63. Uryash,V.F., Kokurina, N.Y., Larina, V.N., Varlamov, P.V., Ilyina, A.V., Grishatova, N.V., Gruzdeva, A.E., Effect of acid hydrolysis on specific

heat capacity and physical translations of chitin and chitosan. *News of NN University,* 3, 98, 2007.

64. Blokhin, A,V., Olga V. Voitkevich, O.V., Kabo G.J., http://pubs.acs.org/doi/abs/10.1021/je200270t - cor1Paulechka, Y.U., Shishonok M.V., Kabo, A.G., Simirsky, V.V., Thermodynamic properties of plant biomass components: heat capacity, combustion energy and gasification equilibria of cellulose. *J. Chem. Eng. Data*, 56, 3523, 2011.
65. Kabo, G.J., Voitkevich, O.V., Blokhin AV., Kohut S.V., Stepurko, E.N., Paulechka, Y.U., Thermodynamic properties of starch and glucose. *J. Chem. Thermodyn.*, 59, 87, 2013.
66. Gartman O.R., *Thermodynamics of chitin and chitosan*, Dissertation, University Press, 1998.
67. Ioelovich M., Study of thermodynamic stability of various allomorphs of cellulose. *J. Basic Appl. Res. Int.*, 16, 96, 2016.
68. Ioelovich, M., Physicochemical methods for determination of cellulose crystallinity. *Chem. Xpress,* 9, 245, 2016.
69. Prusov, A.N., Prusova, S.M., Radugin, M.V., Zakharov A.G., Interrelation between the crystallinity of polysaccharides and water absorption. *J. Phys. Chem.*, 88, 813, 2014.
70. Gregg, S.I., Sing K.S.W., *Adsorption, surface area and porosity*, Academic Press, 1982.
71. Brunauer, S., Emmett P.H., Teller, E., Adsorption of gases in multimolecular layers. *J. Amer. Chem. Soc.*, 60, 309, 1938.
72. Ioelovich, M., Leykin A. Study of sorption properties of cellulose and its derivatives. *Bioresources,* 6, 178, 2011.
73. Egerton, R., *Physical principles of electron microscopy*, Springer-Verlag, 2005.
74. Hall, J.L., Hawes, C.R., *Electron microscopy of plant cells*, Academic Press, 1991.
75. Fultz, B., Howe, J., *Transmission electron microscopy and diffractometry of materials,* Springer-Verlag, 2007.
76. Binnig, G., Quate, C.F., Gerber, C., Atomic force microscope. Phys. Rev. Lett., 56, 930, 1986.
77. Wunderlich, B., Baur, H., Heat capacities of linear high polymers. *Adv. Polym. Sci.*, 7, 151, 1970.
78. Bhadeshia, H., *Thermal analyses techniques - differential thermal analysis*, Mater. Sci. Press, 2002.
79. Coats, A.W., Redfern, J.P., Thermogravimetric analysis: a review. Analyst., 88, 906, 1963.
80. Einbu, A., Naess, S.N., Elgsaeter A., Vårum, K.M., Solution properties of chitin in alkali. *Biomacromolecules,* 5, 2048, 2004.
81. Terbojevich, M., Carraro, C., Cosani, A., Marsano E., Solution studies of the chitin in lithium chloride-N,N-dimethylacetamide system. *Carbohyd. Res.*, 180, 73, 1988.
82. Roberts. G.A., Domszy, J.G., Determination of the viscometric constants for chitosan. *Int. J. Biol. Macromol.*, 4, 374, 1982.

83. Kasaai, M.R., Calculation of Mark-Houwink-Sakurada (MHS) equation viscometric constants for chitosan in any solvent–temperature system using experimental reported viscometric constants data. *Carbohyd. Polym.*, 68, 477, 2007.
84. Jiang, T.D., *Chitosan*, Chem. Ind. Press, 2001.
85. Toan, N.V., Production of chitin and chitosan from partially autolyzed shrimp shell materials. *Open Biomater. J.*, 1, 2009.
86. Dos Santos, Z.M., Caroni, A.L.P.F., Pereira, M.R., Da Silva, D.R., Fonseca, J.L.C., Determination of deacetylation degree of chitosan: a comparison between conductometric titration and CHN elemental analysis. *Carbohyd. Res.*, 344, 2591, 2009.
87. Mathur, N.K., Narang, C.K., Chitin and chitosan, versatile polysaccharides from marine animals. *J. Chem. Educ.*, 67, 938, 1990.
88. Galbraih, L.S., *Chitin and chitosan: structure, properties and application*, Chemistry, 2001.
89. Ito, R., Matsuo, Y., *Handbook of carbohydrate polymers: development, properties and applications*, Nova Publishers, 2010.
90. Gupta, N.S., *Chitin: formation and diagenesis*, Springer, 2011.
91. Einbu, A., *Characterization of chitin and a study of its acid-catalyzed hydrolysis*, NTNU, 2007.
92. Giraud-Guille, M.M., Plywood structures in nature. *Curr. Opin. Solid State Mater. Sci.*, 3, 221, 1998.
93. Kozloff, E., *Invertebrates*, Saunder College Publishing, 1990.
94. Roberts, G.A.F., *Chitin chemistry*, Macmillan Press, 1992.
95. Lucia, L.A., Rojas, O., *The nanoscience and technology of renewable biomaterials*, Wiley, 2009.
96. Dobrovolskaya, I.P., Kasatkin I.A., Yudin V.E., Ivankova E.M., Elokhovsky V.Yu., Supramolecular structure of chitin nanofibrils. *J. Polym. Sci., Part A*, 57, 52, 2015.
97. Kaya, M., Lelešius, E., Nagrockaitė, R., Sargin, I., Arslan, G., Mol, A., Differentiations of chitin content and surface morphologies of chitins extracted from male and female grasshopper species. *PLoS One.*, 10, 1, 2015.
98. Liu, S., Sun, J., Yu, L., Zhang, C., Bi, J., Zhu, F., Qu, M., Jiang, C., Yang, Q., Extraction and characterization of chitin from the beetle *Holotrichia parallela Motschulsky. Molecules,* 17, 4604, 2012.
99. Ioelovich, M., Structure and physicochemical properties of nitrogenated derivatives of cellulose. *Amer. J. Biosci.*, 2, 6, 2014.
100. Zhang, M., Haga, A., Sekiguchi, H., Hirano, S., Structure of insect chitin isolated from beetle larva cuticle and silkworm (bombyx mori) pupa exuvia. *Int. J. Biol. Macrom.*, 27, 99, 2000.
101. Ifuku, S., Nomura, R., Morimoto, M., Saimoto, H., Preparation of chitin nanofibers from mushrooms. *Materials,* 4, 1417, 2011.
102. Cárdenas, G., Carbrera, G., Taboada, E., Miranda, S.P., Chitin characterization by SEM, FTIR, XRD and 13C cross-polarization/mass angle spinning NMR. *J. Appl. Polym. Sci.*, 93, 1876, 2004.

103. Kim, S.S., Kim, S.H., Lee, Y.M., Preparation, characterization and properties of β-chitin and N-acetylated β-chitin. *J. Polym. Sci. Pol. Phys.*, 34, 2367, 1996.
104. Minke, R., Blackwell, J., The structure of a-chitin. *J. Mol. Biol.*, 120, 167, 1978.
105. Noishiki, Y., Nishiyama, Y., Wada, M., Okada, S., Kuga, S., Inclusion complexes of b-chitin an aliphatic amines. *Biomacromolecules,* 4, 944, 2003.
106. Noishiki, Y., Kuga, S., Wada, M., Hori, K., Nishiyama, Y., Guest selectivity in complexation of b-chitin. *Macromolec.,* 37, 6839, 2004.
107. Goihman, A.S., Solomko, V.P., *High-molecular inclusion complexes*, Science, 1982.
108. Hajji, S., Younes, I., Ghorbel-Bellaaj, O., Hajji, R., Rinaudo, M., Nasri, M., Jellouli, K., Structural differences between chitin and chitosan extracted from three different marine sources. See comment in PubMed Commons below*Int. J. Biol. Macromol.,* 65, 298, 2014.
109. Ifuku, S., Nogi, M., Yoshioka, M., Morimoto, M., Yano, H., Saimoto, H., Fibrillation of dried chitin into 10-20 nm nanofibrils by a simple grinding method under acidic conditions. *Carbohydr. Polym.,* 81, 134, 2010.
110. Shams, I., Yano, H., Doubly curved nanofibers-reinforced optically transparent composites. *Sci. Reports,* 5, 1, 2015.
111. Fan, Y., Fan Y., *[C]Studies on preparation and structural characterization of nano-dispersed chitins*, University of Tokyo, 2009.
112. Fan, Y., Saito, T., Isogai, A., TEMPO-mediated oxidation of β-chitin to prepare individual nanofibrils. *Carbohydr. Polym.,* 77, 832, 2009.
113. Morin, A., Dufresne, A., Nanocomposites of chitin whiskers from *Riftia* tubes and poly(caprolactone). *Macromolec.,* 35, 2190, 2002.
114. Mincea, M., Negrulescu, A., Ostafe, V., Preparation, modification, and applications of chitin nanowhiskers: a review. *Rev. Adv. Mater. Sci.,* 30, 225, 2012.
115. Tajik, H., Moradi, M., Rohani, S.M.R., Erfani, A.M., Jalali, F.S.S., Preparation of chitosan from brine shrimp (*Artemia urmiana*) cyst shells and effects of different chemical processing sequences on the physicochemical and functional properties of the product. *Molecules,* 13, 1263, 2008.
116. Cartier, N., Domard, A., Chanzy, H., Single crystals of chitosan. *Int. J. Biol. Macromol.,* 12, 289, 1990.
117. Mazeau, K., William, W.T., Chanzy, H., Molecular and crystal structure of a high-temperature polymorph of chitosan from electron diffraction data. *Macromolecules,* 27, 7606, 1994.
118. Raymond, L., Morin, F.G., Marchessault, R.H., Degree of deacetylation of chitosan using conductometric titration and solid-state NMR. *Carbohydr. Res.,* 246, 331, 1993.
119. Calvo, P., Remunan-Lopez, C., Vila-Jato, J.L., Alonso, M.J., Novel hydrophilic chitosan-polyethylene oxide nanoparticles as protein carrier. *J. Appl. Polym. Sci.,* 63, 125, 1997.
120. Tang, Z.-X., Qian, J.-Q., Shi, L.-E., Preparation of chitosan nanoparticles as carrier for immobilized enzyme. *Appl. Biochem. Biotech.,* 136, 77, 2007.

121. Shin, Y.M., Hohman, M.M., Brenner, M.P., Rutledge, G.C., Experimental characterization of electrospinning: the electrically forced jet and instabilities. *Polymers,* 42, 9955, 2001.
122. Zhao, L.-M., Shi, L-E., Zhang, Z.-L., Chen, J.-M., Shi, D.-D., Yang, J., Preparation and application of chitosan nanoparticles and nanofibers. *Brazil. J. Chem. Eng.,* 28, 353, 2011.
123. Ohkawa, K., Cha, D.I., Kim, H., Nishida, A., Yamamoto, H., Electrospinning of chitosan. *Macromol. Rapid Commun.,* 25, 1600, 2004.
124. Schiffman, J.D., Schauer, C.L., One-step electrospinning of cross-linked chitosan fibers. *Biomacromolecules,* 8, 2665, 2007.
125. Zamani, A., Taherzadeh, M.J., Production of low molecular weight chitisan by hot sulfuric acid hydrolysis. Bioresouces, 5, 1554, 2010.
126. Li, J.-L., Zhou, Y., Zhang, M., Zhang, H.-B., Optimization of preparation process of glucosamine hydrochloride. *Anhui Med. Pharm. J.,* 11, 11, 2007.
127. Benavente, M., Arias1, S., Moreno, L., Martínez, J., Production of glucosamine hydrochloride from crustacean shell. *J. Pharma. Pharmacol.,* 3, 20, 2015.
128. Pan, S.-K., Wu, S.-J., Kim, J.-M., Preparation of glucosamine by hydrolysis of chitosan with commercial α-amylase and glucoamylase. *J. Zhejiang Univ., Part B,* 12, 931, 2011.
129. Roberts, G.A.F., *Chitin chemistry,* MacMillan Press, 1992.
130. Morimoto, M., Saimoto, H., Shigemasa, Y., Control of functions of chitin and chitosan by chemical modifications. *Trends Glycosci. Glycotech.,* 14, 205, 2002.
131. Philippova, O.E., Korchagina E.V., Chitosan and its hydrophobic derivatives: preparation and aggregation in dilute aqueous solutions. *J. Polym. Sci., Part A,* 54, 552, 2012.
132. Mucha, M,. Ludwiczak, S., *Water sorption by biodegradable chitosan/polylactide composites,* Polish Chitin Soc., 2007.
133. Van Krevelen, D.W., *Properties of polymers and correlation with chemical structure,* Elsevier, 1972.
134. Yi, H., Wu, L.-Q., Bentley, W.E., Ghodssi, R., Rubloff, G.W., Culver, J.N., Biofabrication with chitosan. *Biomacromolecules,* 6, 2881, 2005.
135. Pillai, C.K.S., Paul, W., Sharma, C.P., Chitin and chitosan polymers: chemistry, solubility and fiber formation. *Prog. Polym. Sci.,* 34, 641, 2009.
136. Muzzarelli, R.A.A., Muzzarelli, C., *Chitosan in pharmacy and chemistry,* ATEC, 2002.
137. Muzzarelli, R.A.A., *Natural chelating polymers: alginic acid, chitin and chitosan,* Pergamon Press, 1973.
138. Kurita, K., Chemistry and application of chitin and chitosan. *Polym. Degrad. Stabil.,* 59, 117, 1995.
139. Kurita, K., Sannan, T., Iwakura, Y., Studies on chitin: binding of metal cations. *J. Appl. Polym. Sci.,* 23, 511, 1979.
140. Rhazi, M., Desbrieres, J., Tolaimate, A., Rinaudo, M., Vottero, P., Alagui, A., Influence of the nature of the metal ions on the complexation with chitosan: application to the treatment of liquid waste. *Eur. Polym. J.,* 38, 1523, 2002.

141. Kadokawa, J., Takegawa, A., Mine, S., Prasad, K., Preparation of chitin nanowhiskers using an ionic liquid and their composite materials with polyvinyl alcohol. *Carbohydr. Polym.*, 84, 1408, 2011.
142. Ioelovich, M., Isophase and phase transitions of cellulose – a short review. *SITA*, 18, 14, 2016.
143. Kumari, S., Rath, P., Kumar, A.H.A., Chitosan from shrimp shell (*Crangon crangon*) and fish scales (*Labeorohita*): extraction and characterization. *Afr. J. Biotech.*, 15, 1258, 2016.
144. Nishino, T., Matsui, R., Nakamae, K., Elastic modulus of the crystalline regions of chitin and chitosan. *J. Polym. Sci., Part B.* 37, 1191, 1999.
145. Liu, D.G., Chang, P.R., Chen, M.D., Wu, Q.L., Chitosan colloidal suspension composed of mechanically disassembled nanofibers. *J. Coll. Interface Sci.*, 354, 637, 2010.
146. Liu, D.G., Wu, Q.L., Chang, P.R., Gao, G.Z., Self-assembled liquid crystal film from mechanically defibrillated chitosan nanofibers. *Carbohydr. Polym.*, 84, 686, 2011.
147. Muzzarelli, R.A.A., Biomedical exploitation of chitin and chitosan via mechano-chemical disassembly, electrospinning, dissolution in imidazolium ionic liquids, and supercritical drying. *Mar. Drugs*, 9, 1510, 2011.
148. Shams, M.I., Ifuku, S., Nogi, M., Oku, T., Yano, H., Fabrication of optically transparent chitin nanocomposites. *Appl. Phys.*, Part A, 102, 325, 2011.
149. Tzoumaki, M.V., Moschakis, T., Kiosseoglou, V., Biliaderis, C.G., Oil-in-water emulsions stabilized by chitin nanocrystal particles. *Food Hydrocol.*, 25, 1521, 2011.
150. Krajewska B. Application of chitin – and chitosan-based materials for enzyme immobilizations: a review. *Enz. Microbiol. Technol.*, 35, 126, 2004.
151. Koide, S.S., Chitin-chitosan: properties, benefits and risks. *Nut. Res.*, 18, 1091, 1998.
152. Kumar, M.N.V.R., A review of chitin and chitosan applications. *React. Funct. Polym.*, 46, 1, 2000.
153. Kumar, M.N.V.R., Muzarelli, R.A.A., Muzarelli, C., Sashiwa, H., Domb, A.J., Chitosan chemistry and pharmaceutical perspectives. *Chem. Rev.*, 104, 6017, 2004.
154. Dutta, P.K., Dutta, J., Tripathi, V.S., Chitin and chitosan: chemistry, properties and applications. *J. Sci. Ind. Res.*, 63, 20, 2004.
155. Bodek, K.H., Study of rheological properties of microcrystalline chitosan hydrogels used as drug carriers. *Polimery*, 45, 818, 2000.
156. Bodek, K.H., Study on the stability of microcrystalline chitosan systems with selected non-steroidal anti-inflammatory drugs. *Polimery*, 49, 29, 2004.
157. Masarudin, M.J., Cutts, S.M., Evison, B.J., Phillips, D.R., Pigram, P.J., Factors determining the stability, size distribution, and cellular accumulation of small, monodisperse chitosan nanoparticles as candidate vectors for anticancer drug delivery: application to the passive encapsulation of 14-C-doxorubicin. *Nanotech. Sci. Appl.*, 8, 67, 2015.

158. Belozky, S., Breitman, R., *Wounds and dressings*, DDB Publication, 2000.
159. Moon, C.H., Crabtree, T.G., New wound dressing techniques to accelerate healing. *Curr. Treat. Options Infect. Dis.*, 5, 251, 2003.
160. Grumezescu, A.M., *Fabrication and self-assembly of nanobiomaterials: applications of nanobiomaterials*, Elsevier, 2016.
161. Limpanavech, P., Chaiyasuta, S., Vongpromek, R., Pichyangkura, R., Khunwasi, C., Chadchawan, S., Lotrakul, P., Bunjongrat, R., Chaidee, A., Bangyeekhun, T., Chitosan effects on floral production, gene expression, and natomical changes in the Dendrobium orchid. *Sci. Horticult.*, 116, 65, 2008.

3

Enzymes for Production of Chitin, Chitosan, and Chitooligosaccharide and Determination of Activities of Enzymes Using Chitinous Substrates

Sai Pyae Sone Aung[1,2*], Wint Le Nu[1], Khin Chaw Win[1,2], Htun Inzali[1,2], Hla Ngwe[2] and Nitar Nwe[1*]

[1]*Ecological Laboratory, Advancing Life and Regenerating Motherland, Yangon, Union of Myanmar*
[2]*Department of Chemistry, University of Yangon, Yangon, Myanmar*

Abstract
Conventionally, commercial chitin, chitosan, and chitooligosaccharide (CTO) are produced from the crustacean shells by mean of chemical methods. Chemical waste from this chemical process results in environmental pollutions and the waste treatment cost is high. Therefore, many researches have been done to produce chitin, chitosan, and CTO using enzymatic methods which are less environmental pollution and produce better-quality products. Many different microorganisms such as bacteria and fungi have abilities to produce protease, chitin deacetylase, chitinase, and chitosanase. Selected microorganisms are cultivated under optimal conditions to produce enzymes. The enzymes are then purified and used for production of chitin, chitosan, and CTO. This review paper describes about fermentation processes for production of enzymes, methods for purification of enzyme, determination of enzyme activity, storage conditions of enzymes, and treatments of enzyme with various chitinous substrates to produce chitin, chitosan, and CTO.

Keywords: Chitin, chitinase, chitin deacetylase, chitosanase, chitosan chitooligosaccharides, protease

**Corresponding authors*: saipyaesoneaung19@gmail.com; nitarnwe@gmail.com

Shakeel Ahmed and Saiqa Ikram (eds.) Chitosan, (71–88) © 2017 Scrivener Publishing LLC

3.1 Introduction

Chitin is a polymer of *N*-acetyl-D-glucosamine (GlcNAc) units linked together by β-(1-4) glycosidic bonds. Figure 3.1(a) shows the structure of chitin with 100% degree of acetylation. Chitosan is a polymer of D-glucosamine (GlcN) units linked together by β-(1-4) glycosidic bonds. Figure 3.1(b) shows the structure of chitosan with 100% degree of deacetylation. In nature, chitin and chitosan do not occur as polymer with single repeating unit. Hence, chitin and chitosan are nitrogen containing polysaccharide which are copolymer of (GlcNAc) and (GlcN) units linked by β-(1-4) glycosidic bonds and are defined depending on the percentage of GlcN group in the polymer. Figure 3.1(c) shows the structure of chitin and chitosan with various degree of acetylation [1, 2].

Chitin and chitosan are found in aquatic organisms, crustaceans, insects, terrestrial organisms, and some microorganisms. Chitosan has been obtained by the deacetylation of chitin which is the hydrolysis of the acetamido group of *N*-acetylglucosamine units [3–6]. Chitin and chitosan have been produced from shell of shrimps and crabs, bone plates of squids, silkworm pupae, honey bees, mushroom, cell wall of fungi, etc., in laboratory-scale and from biowastes of aquatic organisms (e.g., shells of shrimps and crabs and bone plates of squids) in commercial scale [7, 8]. Chitosan from cell wall of Zygomycetes (as chitosan-glucan complex) has uniform properties and specific characteristics compared to chitosan from deacetylation of crustacean chitin [9–11].

Chitin, chitosan, and chitooligosaccharides (CTO) can be used in varies sectors such in food industries, medical sector, agriculture sector, etc., and so their productions get the great interest [1, 12, 13]. For large-scale production of chitin, chitosan, and CTO, the chemical methods are mostly used and fermentation methods are less used due to the ecological feasibility. In chemical methods, weak acids and strong bases were used in the production process which resulting in contamination of environment. Therefore, the production of chitin, chitosan, and CTO by enzymatic hydrolysis is preferred for the bioactive materials and the high-quality products and then this method is eco-friendly [14–17]. However, the higher production cost, low yields of product, and the time-consuming during production process are the drawbacks of enzymatic methods [14, 18].

To be cost effective and environmental friendly, the chemical and enzymatic methods should be combined [19]. In addition, genetic transformation of fungi is also developed to obtain higher yield of enzyme to produce chitinous compounds [9].

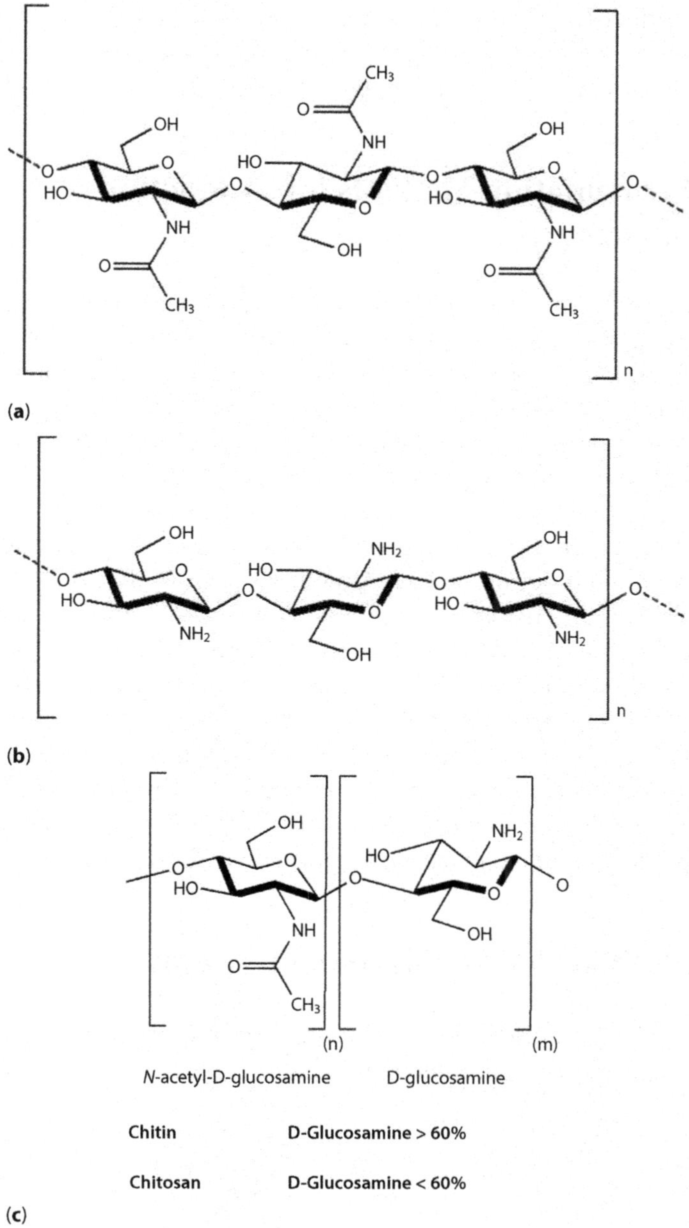

Figure 3.1 (a) The structure of chitin with 100% degree of acetylation; (b) The structure of chitosan with 100% degree of deacetylation; (c) Structure of chitin and chitosan with various degree of acetylation [1].

This chapter highlights the production of enzyme using fermentation methods and purification methods, as well as focus on the enzymatic production (biotechnological production) of chitin, chitosan, and CTO.

3.2 Fermentation Methods for Production of Enzymes

Generally, different enzymes are needed for demineralization, deproteination, and deacetylation steps (decolorization and removing of lipid are sometime needed) to produce shells of crustaceans to chitin and then chitin to chitosan. Chitooligosaccharides are produced from chitin and chitosan by hydrolysis as showed in Figure 3.2 [1, 14]. Different microorganisms including bacteria and fungi have been use to produce enzymes [protease, chitin deacetylase (CDA), chitinase, and chitosanase] for the production of chitin, chitosan, and chitooligosaccharides from shells of crustaceans by fermentation methods or enzymatic methods [15, 16, 20, 21].

Submerged fermentations (shake flask, batch, and continuous fermentation) are commonly used for production of enzyme from microorganisms in laboratory-scale [22]. For the fermentation to be effective, different factors are required such as – fermentation types, medium compositions, size of inoculum, pH, temperature, agitation, aeration, and growth time [9, 14]. Many microorganisms and their optimal conditions for the production of enzymes have been investigated. Some microorganisms and their fermentation methods for production of their respective enzymes are shown in Table 3.1.

3.3 Methods for Purification of Enzymes

Several methods are used to purify enzyme from the culture media. General methods and steps are summarized in Figure 3.3.

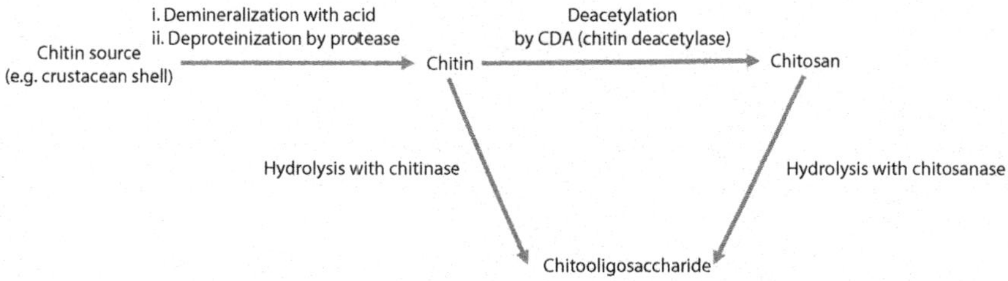

Figure 3.2 Steps for production of chitin, chitosan, and chitooligosaccharide.

Table 3.1 Fermentation methods and optimal conditions for the productions of chitinase, chitosanase, chitin deacetylase and protease.

Microorganisms	Fermentation method	Medium	Inoculum size	Conditions				Enzyme	Enzyme activities
				Temperature (°C)	rpm	pH	Time (hour)		
Aeromonas hydrophila HS4 [23]	Shake flask	Luria bertaini broth	–	37	120	8	24–48	Extracellular chitinase	93.27 U/ml
A. punctate HS6 [23]	Shake flask	Luria bertaini broth	–	37	120	7	48	Extracellular chitinase	82.36 U/ml
Pantoea dispersa [24]	Shake flask	Chitin containing medium	1 × 10^8 CFU/ml	30±2	180	7.2	144	Chitinase I, II, III	–
Aeromonas sp. [25]	Semisolid	2% Colloidal chitin containing medium	0.5 ml of bacterial suspension solution	20	–	6	192	Chitinase	~0.0084 μmol/ml/h
A. hydrophila [25]	Semisolid	2% Colloidal chitin containing medium	0.5 ml of bacterial suspension solution	40	–	6	192	Chitinase	~0.0088 μmol/ml/h
A. salmonicida [25]	Semisolid	1.5% Colloidal chitin containing medium	0.5 ml of bacterial suspension solution	30	–	6	192	Chitinase	0.051 μmol/ml/h
Massilia timonae [26]	Shake flask	1% Colloidal chitin containing medium	–	25–30	150	6.5	36 or 96	Chitinase	10.1 U/ml
Bacillus sp. Hu1 [27]	-	Chitin containing medium	–	45	180	7	72	Extracellular chitinase	–

Table 3.1 Cont.

Microorganisms	Fermentation method	Medium	Inoculum size	Conditions				Enzyme	Enzyme activities
				Temperature (°C)	rpm	pH	Time (hour)		
Microbispora sp. V2 [28]	Shake flask	colloidal chitin broth	–	40	100	7	96	Extracellular Chitinase (chitobiosidase)	~ 3.8 mU/ml
Pseudomonas aeruginosa K-187 [29]	Shake flask	SCSP containing medium	–	45	–	9	72	Chitinase I & II	0.5 u/ml
Streptomyces zaomyceticus [6]	Batch	Chitosan minimal salt	–	32	–	7.2	72	Chitosanase	–
		Glucose minimal salt	–	–	–	–	–	–	–
Gongronella sp. JG [30]	–	0.5% colloidal chitosan containing medium	5 ml of homogenized mycelia	30	180	6	72	Chitosanase	800 U/L
Bacillus cereus TKU031 [31]	Shake flask	Squid pen powder containing medium	–	30	150	7	96	Chitosanase	0.039 U/ml
Bacillus sp. strain KCTC 0377BP [32]	Shake flask	Minimal salt chitosan broth (2% soluble starch, 2.5% polypeptone, and 0.005% silicon oil)	–	30	100	6.8	–	Endo-chitosanase	45.8–100 U/ml
Aeromonas sp. HG08 [33]	–	1% powder chitosan containing medium	–	30	150	6.5	–	Chitosanase	–
S. cyaneogriseus [34]	–	Colloid chitosan containing medium	1 ml spore suspension	30	150	5	120	Chitosanase	~ 19 U/ml

Scopulariopsis brevicaulis [35]	Shake flask	Minimal synthetic medium containing 2% 3,6-di- *O*-carboxymethylchitin	–	29	200	7	96	Extracellular CDA	36 ± 0.4 U/ml
Cunninghamella bertholletiae IFM 46114 [3]	Shake flask	YPD medium	10^5 spores/ml	28	150	4.5	120	Intracellular CDA	0.075 U/mg
Colletotrichum lindemuthianum (ATCC 56676) [36]	Shake flask	Malt, yeast, and glucose containing medium	–	22	100	5.8	432	Extracellular CDA	–
B. amyloliquefaciens [37]	–	Basal medium containing starch	–	37	–	6	–	Endo-CDA	17.84 Eu/ml
Penicillium oxalicum SAE_M-51 [38]	Solid state	Mustard oil cake as solid support	–	30	–	9	144	CDA	–
Paenibacillus woosongensis TKB2 [39]	Shake flask	Chicken feather containing medium	–	30	–	8.5	48	Protease	–
Erwinia chrysanthemi [40]	–	High-calcium *Xanthomonas campestris* medium	–	30	180	7	12	Protease	22.4 U/ml
B. subtilis Y-108 [15]	Shake flask	Minimal synthetic medium containing SCSP	–	30	180	6	72	Protease	20.2 U/ml
A. hydrophila [41]	Shake flask	Minimal synthetic medium containing 7% ATSS	1×10^7 cells/ml	30±0.5	150	8	144	Protease	534.53 U/ml

SCSP, shrimp and crap shell powder; ATSS, acid treated shrimp shell (demimeralized).

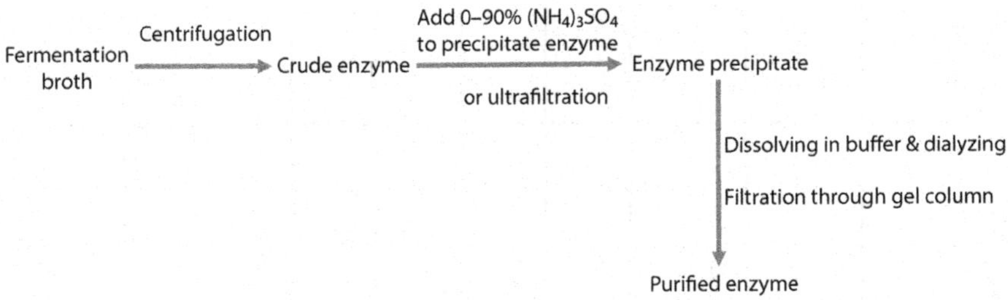

Figure 3.3 General method for the purification of crude enzyme.

The crude extraction is done by the centrifugation or filtration. The purifications of enzymes from their crude enzymes are briefly shown in Table 3.2. Generally, most of the steps are carried out at 4 °C.

3.4 Storage Conditions of Enzyme

The purified enzyme should be stored at appropriate conditions to retain the original structure for an extended period of time. Shelf-life of enzymes can vary from days to years depending on the natures of enzymes and the storage conditions. Generally, the enzyme should be stored at ≤4 °C in autoclave glassware or plastic containers to prevent spoilage conditions.

For short storage time (about 24 h) of enzyme, it can be stored at 4 °C. Lyophilized powder of enzyme can be stored at −70 °C for more than a year. However, the enzyme can be damage during lyophilization process. Moreover, stabilizing agent such as 5–50% glycerol can be added to make enzyme stable.

3.5 Commercial Enzymes

Protease, chitinase, and chitosanase have been produced commercially [28, 32, 39]. For the commercial chitinase and chitosanase, it is stable between pH 3.0–10.0 and temperature of 25–35 °C with corresponding products. However, it is recommended to store below 4 °C (https://www.alibaba.com/).

3.6 Determinations of Enzyme Activities Using Chitinous Materials

The enzymes obtained from different organisms have their respective optimal conditions and it is shown in Table 3.3. Generally, the enzyme activity

Table 3.2 Purifications of chitinase, chitosanase, chitin deacetylase, and protease.

Enzymes	Purification methods
Extracellular chitinase [27]	1. Collect crude extract from fermentation broth 2. Add 80% $(NH_4)_2SO_4$ for precipitation 3. Dissolve and dialysis in sodium acetate buffer (Conc. 50mM, pH 4.2) 4. Separate in DEAD-Sepharose fast flow column 5. Add 80% $(NH_4)_2SO_4$ to precipitate enzyme 6. Dissolve and dialysis in Tris-HCl buffer (Conc. 50mM, pH 7.2) 7. Separate in Sephacryl S-400 high resolution column
Extracellular chitinase [28]	1. Collect crude extract from fermentation broth 2. Add 80% $(NH_4)_2SO_4$ for precipitation 3. Dissolve and dialysis in citrate phosphate buffer (Conc. 50mM, pH 6) 4. Separate in Sephadex G-100 column
Extracellular chitinase [29]	1. Collect crude extract from fermentation broth 2. Add 80% $(NH_4)_2SO_4$ for precipitation 3. Dissolve and dialysis in sodium phosphate buffer (Conc. 50mM, pH 6) 4. Separate in DEAD-Sepharose CL-6B column 5. Add 80% $(NH_4)_2SO_4$ to precipitate enzyme 6. Dissolve and dialysis in phosphate buffer (Conc. 50mM, pH 6) 7. Separate in Econo-pac q column (Bio-Rad)
Intracellular chitinase [42]	1. Collect crude extract from mycelia 2. Separate in bacitracin-CNBr–Sepharose 4B column (Flow rate in bacitracin-CNBr–Sepharose is 14 $cm^3/cm^2/h$. The elute was obtained in 2.5 ml fractions) 3. Separate in Sephadex G-100 column (Flow rate in Sephadex G-100 is 8 $cm^3/cm^2/h$. The elute was obtained in 4 ml fractions) [Phosphate buffer (Conc. 0.2M, pH 7.2) is used to equilibrated each column.]
Intracellular chitinase [43]	1. Collect crude extract from mycelia 2. Separate in regenerated-chitin column (Flow rate in regenerated-chitin column is 50 cm^3/h) 3. Neutralize with 1M Tris 4. Dialysis in Tris-HCl containing 0.1% Triton X-100 (Conc. 20 mM, pH 8.0)

Table 3.2 Cont.

Enzymes	Purification methods
Intracellular chitinase [43]	5. Separate in DEAD-Sephadex A50 column (Flow rate in DEAD-Sephadex A50 is 50 cm^3/h) 6. Dialysis in sodium phosphate containing 0.1% Triton X-100 (Conc. 1 mM, pH 6.8) 7. Separate in hydroxyapatite column 8. Dialysis in octyl-D-glucoside (Conc. 40 mM)
Chitosanase [30]	1. Collect crude extract from fermentation broth 2. Ultrafiltration 3. Dialysis in sodium acetate (Conc. 20mM, pH 5) 4. Separate in CM-Sepharose fast flow column 5. Separate in Sephacryl S200 gel filtration column 6. Dialysis in sodium acetate (Conc. 20mM, pH 5) 7. Separate in SP-Sepharose fast flow column
Chitosanase [31]	1. Collect crude extract from fermentation broth 2. Add $(NH_4)_2SO_4$ for precipitation 3. Dissolve and dialysis in sodium phosphate buffer (Conc. 50mM, pH 7) 4. Separate in DEAD-Sepharose CL-6B column 5. Separate in Macro-prep DEAD column
Endo-chitosanase [32]	1. Collect crude extract from fermentation broth 2. Concentrate with polyethylene glycol (PEG) 3. Separate in CM-Toyopearl 650 column 4. Separate in Superose 12HR column FPLC (This is the laboratory-scale production and has 1700 U/mg enzyme activities)
	1. Collect crude extract from fermentation broth 2. Ultrafiltration 3. Separate in CM-Sepharose CL-6B 4. Lyophilization (This is the industry-scale production and has 1000 U/mg enzyme activities)
Chitosanase [33]	1. Collect crude extract from fermentation broth 2. Add 0–70% $(NH_4)_2SO_4$ for precipitation 3. Dissolve in Tris-HCl buffer (Conc. 50mM, pH 7.8) 4. Separate in Q-Sepharose fast flow column (Flow rate in Q-Sepharose is 0.4 ml/min.) 5. Separate in Sephadex G-75 (Flow rate in Sephadex G-75 is 9 ml/h)

Table 3.2 Cont.

Enzymes	Purification methods
Chitosanase [34]	1. Collect crude extract from fermentation broth 2. Add 80% $(NH_4)_2SO_4$ for precipitation 3. Dialysis and dissolve in citrate-phosphate buffer (Conc. 0.1M, pH 5) 4. Separate in Sephadex G_{200}column (The elute was obtained in 5 ml fractions)
Chitosanase [44]	1. Collect crude extract from fermentation broth 2. Add 50–90% $(NH_4)_2SO_4$ for precipitation 3. Separate in Mono-S cation exchanger (0.06–0.08 M KCl) 4. Separate in Superose gel column
CDA [45]	1. Collect crude extract from fermentation broth 2. Ultrafiltration 3. Dialysis in Bis-Tris-HCl buffer (Conc. 20 mM, pH 5.8) 4. Separate in Q-Sepharose fast flow column (Flow rate in Q-Sepharose is 300 ml/h) 5. Ultrafiltration 6. Separate in Sephacrysl S300 HR column 7. Dialysis in sodium formate buffer (Conc. 50 mM, pH 4.0) 8. Separate in Mono S column (Flow rate in Mono S is 30 ml/h)
Intracellular CDA [46]	1. Collect crude extract from mycelia 2. Treat to 50 °C (15min) and cool in ice 3. Separate in immunoadsorbent (Flow rate is 35 ml/h) 4. Adjusted to pH 7 5. Dialysis in Tris-HCl buffer (Conc. 25mM, pH 7.4) 6. Ultrafiltration
Extracellular CDA [38]	1. Collect crude extract from fermentation broth 2. Ultrafiltration 3. Separation in CM-Sepharose column (Flow rate in CM-Sepharose is 1 ml/min) 4. Separation in DEAE-Sepharose column
Extracellular CDA [36]	1. Collect crude extract from fermentation broth 2. Add 80% $(NH_4)_2SO_4$ for precipitation 3. Dissolve and dialysis in sodium tetraborate/HCl buffer (Conc 20mM, pH 8.5) 4. Saturated with 25% $(NH_4)_2SO_4$

Table 3.2 Cont.

Enzymes	Purification methods
Extracellular CDA [36]	5. Separate in Butyl-Toyopearl (hydrophobicinteraction column) (Flow rate in Butyl-Toyopearl is 3 ml/min) 6. Dialysis in sodium tetraborate/HCl buffer (Conc. 1mM, pH 8.5) 7. Addition of Triethanolamine/HCI buffer (Conc. 200 mM, pH 7.5) 8. Separate in Q-Sepharose fast flow column (anion exchange column) (Flow rate in Q-Sepharose is 2 ml/min) 9. Dialysis in Triethanolamine/HCI buffer (Conc. 200 mM, pH 7.5) 10. Separate in Resource-Q (anion exchange column) (Flow rate in Resource-Q is 1 ml/min)
Protease [15]	1. Collect crude extract from fermentation broth 2. Add $(NH_4)_2SO_4$ for precipitation 3. Separate in DEAD-Sepharose CL-6B column 4. Separate in Sephacryl S-200
Protease [47]	1. Collect crude extract from fermentation broth 2. Filter with Whitman No.1 filter paper 3. Add $(NH_4)_2SO_4$ for precipitation 4. Dissolve and dialysis in phosphate buffer (pH 7) 5. Separate in Sephadex G-200 column
Extracellular protease [48]	1. Collect crude extract from fermentation broth 2. Filter with 0.8 mm 3. Add 30–80% $(NH_4)_2SO_4$ for precipitation 4. Dissolve and dialysis in Tris buffer (Conc. 20mM, pH 8) 5. Separate in DEAD-Sepharose fast flow column

is measured by the amount of reducing sugar release from the substrate [23]. For chitinase and chitosanase, one unit of enzyme activity is the amount of enzyme that produces 1 μmol of reducing sugar in 1 min under reacting conditions [6, 25, 31]. For chitin deacetylase, it is measured on the amount of acetate production in one minute [37]. However, the definition on enzyme activities can be different depends on the substrates used and reaction conditions.

Table 3.3 Optimal conditions for enzyme activities.

Types of enzyme	Substrates	Conditions				Products and enzyme activities
		Temperature (°C)	rpm	pH	Time (min)	
Chitinase [24]	Acid Swollen chitin	50	–	5	10	GlcNAc & CTO
Chitinase [26]	Colloidal chitin	50	125	5	30	CTO
Chitinase [27]	1% (w/v) Colloidal chitin	60	–	6.5	45	GlcNAc & CTO
Chitinase [28]	[pNP-$(GlcNAc)_2$]	60	–	3	–	*p*-nitrophenol
Chitinase I [29]	Colloidal chitin	50	–	8	10	CTO
Chitinase II [29]	Colloidal chitin	40	–	7	10	CTO
Chitinase A [49]	*N*-acetylated chitosan	70	–	5	30	CTO
Chitosanase [30]	Colloidal chitosan	55	–	5.5	30	GlcN & CTO
Chitosanase[31]	Water-soluble chitosan (60% DDA)	55	–	5	60	CTO
Chitosanase [32]	chitosan	40	–	5	1440 (24h)	GlcN & CTO
Chitosanase [33]	Chitosan (100% DDA)	55	–	6.2	120	CTO 31 U/ml
Chitosanase [34]	Colloidal chitosan & KNO_3	50	–	5	–	GlcN & CTO
Chitosanase A [44]	Chitosan solution	45	–	4.5–6.5	30	CTO
CDA [3]	*O*-hydroxyethylated chitin (Glycol chitin)	50	–	4.5	-	Chitosan 6.5 U/L

(*Continued*)

Table 3.3 Cont.

Types of enzyme	Substrates	Conditions				Products and enzyme activities
		Temperature (°C)	rpm	pH	Time (min)	
CDA [45]	Glycol chitin	50	–	8.5	15	Chitosan
CDA [37]	Hexa-*N*-acetylchitohexaose	50	–	4.5	15	Chitosan
CDA [46]	Glycol chitin	50	–	4.5	10	Chitosan
	Chitohexaose	50		4.5	10	Chitosan
CDA [38]	Ethylene glycol chitin	50	–	9	1440 (24h)	Chitosan
CDA [36]	Partially *N*-deacetylated water-soluble chitin	60	–	11–12	–	Chitosan
Protease [15]	Casein	50	–	8	30	Remove protein

[pNP-$(GlcNAc)_2$], *p*-nitrophenyl-β-D-*N,N′*-diacetylchitobiose; CDA, chitin deacetylase; CTO, chitooligosaccharides; DDA, degree of deacetylation; GlcN, D-glucosamine; GlcNAc, *N*-acetyl-D-glucosamine.

3.7 Conclusion

Chitin, chitosan, and CTO have been produced commercially all over the world using enzymatic or fermentation method. The enzymes are mainly produced from bacterial and fungi in laboratory and industrial scale. Nowadays, commercial enzymes of protease, chitinase, and chitosanase have been produced. These enzymes are not only used in production of chitin, chitosan, and CTO but also used in other applications.

Acknowledgement

The authors would like to thank to members of Advancing Life and Regenerating Motherland for their encouragement.

References

1. Nwe, N., Furuike, T., Tamura, H., Chitosan from aquatic and terrestrial organisms and microorganisms: production, properties and applications, In; *Biopolymers*, B.M. Johnson, Z.E. Berkel, Eds., pp. 29–50, Nova Science Publishers, USA, 2011.
2. Nwe, N., Furuike, T., Tamura, H., The mechanical and biological properties of chitosan scaffolds for tissue regeneration templates are significantly enhanced by chitosan from *Gongronella butleri*. *Materials*, 2, 374–398, 2009.
3. Amorim, R.V.S., Ledingham, W.M., Fukushima, K., Campos-Takaki, G.M., Screening of chitin deacetylase from Mucoralean strains (Zygomycetes) and its relationship to cell growth rate. *J. Ind. Microbiol. Biotechnol.*, 32, 19–23, 2005.
4. Nwe, N., Furuike, T., Osaka, I., Fujimori, H., Kawasaki, H., Akarawa, R., Tokura, S., Stevens W. F., Kurozumi, S., Takamori, Y., Fukuda, M., Tamura, H., Laboratory scale production of ^{13}C labeled chitosan by fungi *Absidia coerulea* and *Gongronella butleri* grown in solid substrate and submerged fermentation. *Carbohydr. Polym.*, 84, 743–750, 2011.
5. Nwe, N., Furuike, T., Tamura, H., Production, properties and applications of fungal cell wall polysaccharides: chitosan and glucan., *Adv. Polym. Sci.*, 244, 187–208, 2011.
6. Sinha, S., Chand, S., Tripathi. P., Production, purification and characterization of a new chitosanase enzyme and improvement of chitosan pentamer and hexamer yield in an enzyme membrane reactor. *Biocatal. Biotransformation*, 32(4), 208–213, 2014.
7. Nwe, N., *Biopolymers based micro- and nano- materials*, pp. 8-9, ASME Press, 2015.

8. Nwe, N., Furuike, T., Tamura, H., Isolation and characterization of chitin and chitosan from marine origin, In: *Advances in food and nutrition research, marine carbohydrates: fundamentals and applications part A*, S. Kim, Ed., pp. 1–15, Academic Press, London, 2014.
9. Knezevic-Jugovic, Z., Petronijevic, Z., Smelcerovic, A., Chitin and chitosan from microorganisms, In: *Chintin, chitosan, oligosaccharides and their derivatives: biological activities and applications*, S. Kim, Ed., pp. 25–36, CRC Press, New York, 2010.
10. Nwe, N., Furuike, T., Tokura, S., Tamura, H., Characterization of chitosan and chitosan-glucan complex extracted from the cell wall of fungus *Gongronella butleri* USDB 0201 by enzymatic method. *Enzyme Microb. Technol.*, 42, 242–251, 2008.
11. Tai, C., Li, S., Xu, Q., Ying, H., Huang, H., Ouyang, P., Chitosan production from hemicellulose hydrolysate of corn straw: impact of degradation products on *Rhizopus oryzae* growth and chitosan fermentation. *Lett. Appl. Microbiol.*, 51, 278–284, 2010.
12. Pal, J., Verma, H.O., Munka, V.K., Maurya, S.K., Roy, D., Kumar, J., Biological method of chitin extraction from shrimpwaste an eco-friendly low cost technology and itsadvanced application. *Int. J. Fish. Aquac. Stud.*, 1(6), 104–107, 2014.
13. Younes, I., Rinaudo, M., Chitin and chitosan preparation from marine sources. Structure, properties and applications. *Mar. Drugs*, 13, 1133–1174, 2015.
14. Jo, G., Park, R., Jung, W., Enzymatic production of chitin from crustacean shell waste, In: *Chintin, chitosan, oligosaccharides and their derivatives: biological activities and applications*, S. Kim, Ed., pp. 37–45, CRC Press, New York, 2010.
15. Yang, J., Shih, I., Tzeng, Y., Wang, S., Production and purification of protease from a *Bacillus subtilis* that can deproteinize crustacean wastes. *Enzyme Microb. Technol.*, 26, 406–413, 2000.
16. Zhang, Y., Zhou, Z., Liu, Y., Cao, Y., He, S., Huo, F., Qin, C., Yao, B., Ringø, E., High-yield production of a chitinase from *Aeromonas veronii* B565 as a potential feed supplement for warm-water aquaculture. *Appl. Microbiol. Biotechnol.*, 98, 1651–1662, 2014.
17. Zhao, Y., Ju, W., Park, R., Enzymatic modifications of chitin and chitosan, In: *Chintin, chitosan, oligosaccharides and their derivatives: biological activities and applications*, S. Kim, Ed., pp. 185–192, CRC Press, New York, 2010.
18. Kim, S., Je, J., Continuous production of chitooligosaccharides by enzymatic hydrolysis, In: *Chintin, chitosan, oligosaccharides and their derivatives: biological activities and applications*, S. Kim, Ed., pp. 47–51, CRC Press, New York, 2010.
19. Brück W.M., Slater J.W., Carney, B.F., Chitin and chitosan from marine organisms, In: *Chintin, chitosan, oligosaccharides and their derivatives: biological activities and applications*, S. Kim, Ed., pp. 11–23, CRC Press, New York, 2010.

20. Aida, F.M., Taghreed, A.S., Production, optimization, characterization and antifungal activity of chitinase produced by *Aspergillus terrus*. *Afr. J. Biotechnol.*, 13(14), 1567–1578, 2014.
21. Zhao, Y., Park, R., Muzzarelli, R.A.A., Chitin deacetylases: properties and applications. *Mar. Drugs.*, 8, 24–46, 2010.
22. Roopavathi, A.S., Vigneshwari, R., Jayapradha, R., Chitinase: production and applications. *J. Chem. Pharm. Res.*, 7(5), 924–931, 2015.
23. Kuddus, S.M., Ahmad, R.I.Z., Isolation of novel chitinolytic bacteria and production optimization of extracellular chitinase. *J. Genet. Eng. Biotechnol.*, 11, 39–46, 2013.
24. Gohel, V., Naseby, D.C., Thermalstabilizaton of chitinolytic enzymes of *Pantoea dispersa*. *Biochem. Eng. J.*, 35, 150–157, 2007.
25. Brzezinska, M.S., Donderski, W., Occurrence and activity of the chitinolytic bacteria of *Aeromonas* genus. *Pol. J. Environ. Stud.*, 10, 27–31, 2001.
26. Faramarzi, M.A., Fazeli, M., Yazdi, M.T., Adrangi, S., Ahmadi, K.J.A., Tasharrofi. N., Mohseni F.A., Optimization of cultural conditions for production chitinase by a soil isolate of *Massilia timonae*. *Biotechnol.*, 8(1), 93–99, 2009.
27. Dai, D., Hu, W., Huang, G., Wei Li, W., Purification and characterization of a novel extracellular chitinase from thermophilic *Bacillus* sp. Hu1. *Arf. J. Biotechnol.*,10(13), 2476–2485, 2011.
28. Nawani, N.N., Kapadnis, P.B., Das A.D., Rao A.S., Mahajan, S.K., Purification and characterization of a thermophilic and acidophilic chitinase from *Microbispora* sp. V2. *J. Appl. Microbiol.*, 93, 965–975, 2002.
29. Wang, S., Chang, W., Purification and characterization of two bifunctional chitinases/lysozymes extracellularly produced by *Pseudomonas aeruginosa* K-187 in a shrimp and crab shell powder medium. *Appl. Environ. Microbiol.*, 63, 380–386, 1997.
30. Zhou, W., Yuan, H., Wang. J., Yao J., Production, purification and characterization of chitosanase produced by *Gongronella* sp. JG. *Lett. Appl. Microbiol.*, 46, 49–54, 2008.
31. Wang, C., Su, J., Liang, T., Nguyen, A.D., Wang, S., Production, purification and characterisation of a chitosanase from *Bacillus cereus*. *Res. Chem. Intermed.*, 40, 2237–2248, 2014.
32. Choi, Y.J., Kim, E.J., Piao. Z., Yun. Y.C., Shin. Y.C., Purification and characterization of chitosanase from *Bacillus* sp. strain KCTC 0377BP and its application for the production of chitosan oligosaccharides. *Appl. Environ. Microbiol.*, 70, 4522–4531, 2004.
33. Sun, Y., Zhang, J., Wang, S., Purification and characterization of the chitosanase from *Aeromonas* sp. HG08. *Arf. J. Biotechnol.*, 8(12), 2830–2834, 2009.
34. El-Sherbiny, E.A., Purification and characterization of chitosanase enzyme from *Streptomyces cyaneogriseus*. *Asian J. Biol. Sci.*, 4, 15–24, 2011.
35. Cai, J., Li, J., Wang, C., Lin, J., Hu, Y., Yang, J., Du, Y., Zheng, H., Parametric optimization of extracellular chitin deacetylase production by *Scopulariopsis brevicaulis*. *J. Biocatal. Biotransformation*, 2(1), 1–5, 2013.

36. Tokuyasu, K., Ohnishi-Kameyama, M., Hayashi, K., Purification and characterization of extracellular chitin deacetylase from *Colletotrichum lindemuthianum. Biosci. Biotech. Biochem.*, 60(10), 1598–1603, 1996.
37. Zhou, G., Zhang, H., He, Y., He, L.,Identification of a chitin deacetylase producing bacteria isolated from soil and its fermentation optimization. *Afr. J. Microbiol. Res.*, 4(23), 2597–25603, 2010.
38. Pareek, N., Vivekanand, V., Saroj, S., Sharma, A.K., Singh, R.P., Purification and characterization of chitin deacetylase from *Penicillium oxalicum* SAE_M-51. *Carbohydr. Polym.*, 87, 1091–1097, 2012.
39. Paul, T., Halder, S.K., Das, A., Ghosh, K., Mandal, A., Payra, P., Barman, P., Mohapatra, P.K.D., Pati, B.R., and Mondal, K.C., Production of chitin and bioactive materials from black tiger shrimp (*Penaeus monodon*) shell waste by the treatment of bacterial protease cocktail. *3 Biotech.*, 5, 483–493, 2015.
40. Giyose, N.Y., Mazomba N.T., Mabinya L.V., Evaluation of proteases produced by *Erwinia chrysanthemi* for the deproteinization of crustacean waste in a chitin production process. *Arf. J. Biotechnol.*, 9(5), 707–711, 2010.
41. Cheong, J.Y., Azway, A.A.N., Rusea, G., Noormasshela, U.A., Shaziera, A.G.N., Azleen A.A., Muskhazli, M., The availability ofastaxanthin from shrimp shell wastes through microbial fermentations, *Aeromonas hydrophila* and cell disruptions. *Int. J. Agric. Biol.*, 16, 277-284, 2014.
42. Struszczyk, K., Szczęsna-Antczak, M., Walczak, M., Antczak, T., Isolation and purification of intracellular chitosanolytic enzymes of *Mucor circinelloides. PCACD*, 13, 107–116, 2008.
43. Takaya, N., Yamazaki, D., Horiuchi, H., Ohta, A., Takagi, M., Intracellular chitinase gene from *Rhizopus oligosporus*: molecular cloning and characterization. *Microbiology*, 144, 2647–2654, 1998.
44. Pelletier, A., Sygusch, J., Purification and characterization of three chitosanase activities from *Bacillus megaterium* P1. *Appl. Environ. Microbiol.*, 56, 844–848, 1990.
45. Tsigos, I., Bouriotis, V., Purification and characterization of chitin deacetylase from *Colletotrichum lindemuthianum. J. Biol. Chem.*, 270, 26286–26291, 1995.
46. Martinou, A., Kafetzopoulos, D., Bouriotis, V., Isolation of chitin deacetylase from Mucor rouxii by immunoaffinity chromatography. *J. Chromatogr.*, 644, 35–41, 1993.
47. El-Safey, E.M., Abdul-Raouf, U.M., Production, purification and characterization of protease enzyme from *Bacillus subtilis*. International conferences for development and the environment in the Arab world, Assiut University, Egypt, March 23–25, 2004.
48. Ghafoor, A., Hasnain, S., Purification and characterization of an extracellular protease from *Bacillus subtilis* EAG-2 strain isolation from ornamental plant nursery. *Pol. J. Microbiol.*, 59, 107–112, 2010.
49. Ueda, M., Chitinolytic enzymes from the moderately thermophilic bacterium *Ralstonia* sp. A-471: characterization and application, In: *Chintin, chitosan, oligosaccharides and their derivatives: biological activities and applications*, S. Kim, Ed., pp. 371–376, CRC Press, New York, 2010.

4

Production of Chitin, Chitosan, and Chitooligosaccharide from Shrimp and Crab Shells Using Green Technology and Applications of Their Composite Materials

Wint Le Nu[1*], Khin Chaw Win[1,2], Sai Pyae Sone Aung[1,2], Htun Inzali[1,2], Tetsuya Furuike[3], Hiroshi Tamura[3*] and Nitar Nwe[1*]

[1]Ecological Laboratory, Advancing Life and Regenerating Motherland, Yangon, Union of Myanmar
[2]Department of Chemistry, University of Yangon, Yangon, Myanmar
[3]Faculty of Chemistry, Materials and Bioengineering, Kansai University, Osaka, Japan

Abstract

Chitin, chitosan, and chitooligosaccharides have been produced from biowaste obtained from seafood processing industries such as shrimp and crab shells using various methods such as chemical, enzymatical, and biological methods. Along with increased demands on environmental friendly society and rapid development of fermentation technology, more eco-friendly processes using enzymatic and microbiological methods for producing of chitin, chitosan, and chitooligosaccharide have attracted great interests. Chitin from crustacean shells has been produced using microbiological method, in which deproteination takes place by function of proteases obtained from microorganisms and demineralization by acid produced from microorganisms during fermentation. The enzymatic deacetylation of various chitin preparations has been investigated using the fungal chitin deacetylase (CDA) under solid condition and produced chitosan. An enzymatic process to produce oligosaccharides from chitosan using chitinases or chitosanases from several microorganisms has been tried. The resulted chitooligosaccharide

Corresponding authors: nitarnwe@gmail.com; tamura@kansai-u.ac.jp; wintlenu@gmail.com

Shakeel Ahmed and Saiqa Ikram (eds.) Chitosan, (89–114) © 2017 Scrivener Publishing LLC

is widely applied in agriculture due to their positive effects on plant growth and development. This chapter reviews the production of chitin, chitosan, and chitooligosaccharide from shrimp and crab shells using green technology and applications of their composite materials.

Keywords: Agriculture, chitin, chitooligosaccharides, chitosan, crab shells, enzyme, fermentation technology, green technology, microorganisms, shrimp shells

4.1 Introduction

Nowadays, environmental pollution problems are being facing all over the world. Therefore, environmental protection regulations have become stricter. In the environmental pollution, the basic options for minimization of solid wastes are to "reduce," "reuse," and "recycle" of the wastes. To follow this, appropriate technology should be applied to prevent decay and to convert the biomaterial into valuable products. Each year, large amount of solid waste is generated from shrimp processing depending upon shrimp species, types of processed products such as heads, shells, and tails and the seasonal variations. These biowastes contain valuable components such as chitin, protein, and pigments [1, 2]. These protein and pigments have been proven to be an excellent animal feed supplement isolated from shrimp biowaste [3].

Chitin, one of the most abundant renewable biopolymers on the earth, is a versatile, environmentally friendly modern material which composed of β-1-4-*N*-acetylglucosamine and glucosamine [4–6]. Nowadays, chitin from shrimp and crab shells has been prepared using chemical, enzymatic and microbiological methods [7]. In particular, chemical method is used for production of chitin using shrimp and crab shells involves hazardous chemicals such as alkalis (usually 4% NaOH) for deproteinization and dilute acids (4% HCl) for demineralization, respectively [5]. The traditional chemical method creates a disposable problem to environment due to release of large amounts of toxic chemical waste. Moreover, the chemical treatment methods seriously pollute the ecological environment, damage human health for worker and very high cost for waste treatment [7, 8]. Therefore, a less expensive and environmentally friendly method for chitin extraction is needed. An alternative method for production of chitin is fermentation method using proteolytic enzymes from different microorganisms to remove the proteins and minerals contents [4, 9]. Most of researchers have reported on treatment of shrimp and crab biowastes using fermentation method under various [10].

It has been reported that fermentation carried out with microorganisms include *Lactobacillus plantarum, Bacillus subtilis, Lactobacillus helveticus, Pseudomonas aeruginosa, Lactobacillus paracasei, Lecanicillium fungicola,*

Penicillium chrysogenum, Pseudomonas maltophilia, Pediococcus acidilactici [5, 6] together with shrimp shells produced lactic acid and proteases. In which, lactic acid solubilized calcium carbonate and others minerals from the exoskeletal shells [4] and protease digested protein from shells to water-soluble protein hydrolysates [4]. Furthermore, the efficiency of fermentation using microorganisms depends on the quantity of inoculums, the glucose concentration, the initial pH, and pH during culture and fermentation time [5].

Chitosan is a natural, safe, and cheap biopolymer produced from chitin with the process of chemical or enzymatic chitin deacetylation [11–13]. The deacetylated form of chitin is chitosan (i.e., higher percentage of glucosamine in polymer chain) and is highly soluble in dilute organic acid such as acetic acid, lactic acid, inorganic acid, etc. [14, 15]. It has attracted great interest due to its biocompatibility, high charge density, nontoxicity, and mucoadhesion [6].

Chitooligosaccharides (COS), bioactive compound, is prepared from chitosan by several methods, including enzymatic degradation and hydrolysis of chitosan through oxidation, UV, ultrasonication, and irradiation. Above all, the most frequently used process for preparation of COS is acid hydrolysis (HCl, nitrous acid, phosphoric acid, hydrogen fluoride, HNO_2) of chitosan [16]. To produce chitooligosaccharides, chemical, and enzymatic processes have been used in various industries. "Acid hydrolysis produces a large amount of short-chain oligosaccharides, including a monomeric unit which requires an additional process for the fractionation of longer oligosaccharides, resulting in low yields of oligosaccharides" [17]. Alternatively, enzymatic hydrolysis of chitosan is regarded to be a potential method for produce oligosaccharides using chitosanases that can be instead of the chemical method in future [17]. Several microorganisms including bacteria and fungi were reported to efficiently produce chitosanases to degrade chitosan to glucosamine oligomers [17, 18].

Chitin, chitosan, and chitooligosaccharides are widely applied in agriculture due to defensive mechanism and due to its positive effects on plants growth and development. Chitosan have antifungal, antibacterial, and antivirus activities and being applied as biofungicide for plants [19]. Along with, chitooligosaccharide enhancing crop production and quality of agricultural products [19] can be used with various ways as seed, leaf, fruit and vegetable coating, seedling root dipping, soil enrichment as fertilizer and in controlled agrochemical release, stimulate the immunity of plants, stimulate plant growth and supplement into plant tissue media [19, 20]. Moreover, chitin and chitosan have numerous applications including medical, therapies, cosmetics, paper, pulp and textile industries, biotechnology, and food applications [3, 21]. In addition, chitooligosaccharides are used as food additives, pharmaceuticals, and elicitors in plant cell culture [22].

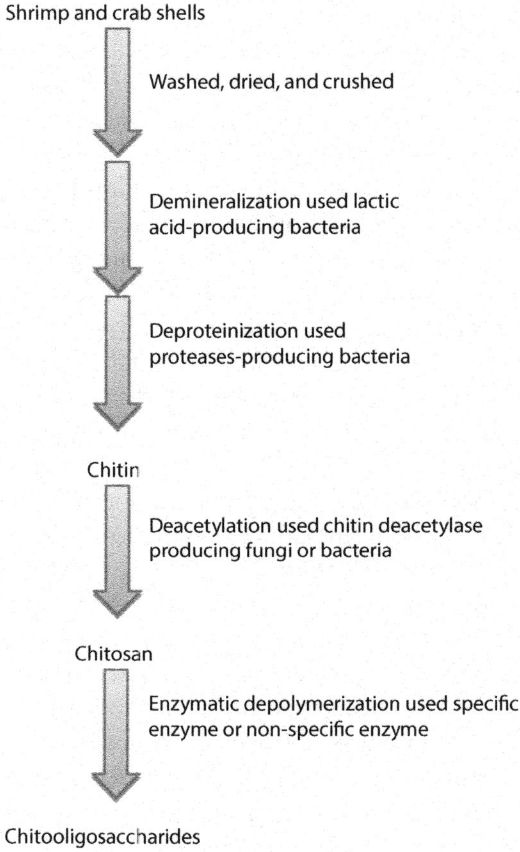

Figure 4.1 Flow chart for production of chitin, chitosan, and chitooligosaccharides by green technology [23].

This chapter describes the production of chitin, chitosan, and chitooligosaccharides from shrimp and crab shells using green technology and applications of their composite materials. Figure 4.1 shows the step by step production of chitin, chitosan, and chitooligosaccharides by green technology from shrimp and crab shells.

4.2 Microorganisms for Production of Chitin and Chitosan Using Green Technology

The shrimp and crab shells have been used to produce chitin using chemical and biological method in which there are advantages and disadvantages coupling in both methods. Chemical method involves deproteinization (DP) and demineralization (DM) using weak bases (NaOH) and acids (HCl), respectively [3]. This method used for industrial scale production of chitin and it takes short processing time but the used of these chemicals can seriously pollute the ecological environment, produces abundant waste, and is harmful to human health [3]. Furthermore, after acidic and alkali treatment the removed proteins and minerals cannot be used as human and animal food supplements [23]. Along with increased demands on environment-friendly society and to avoid acidic and alkali treatments, biologically treatment offers an alternative way to extract chitin from crustacean shells [23].

Since the early days of microbiology, large numbers of microorganisms were isolated from natural sources. These microorganisms are used not only for scientific research but also for industrial fermentation [24]. Microorganisms can be isolated from soil sample, natural probiotic (milk curd), yogurt, etc. [5, 25]. Most of isolated microorganisms are stored in culture collections such as World Federation for culture collections (WFCC), American type culture collection (ATCC), European collection of authenticated cell culture (ECACC), Persian type culture collection (PTCC), Korean collection type cultures (KCTC), Institute (CFTRI, Mysore) culture collection, and the strain collection of the Thailand institute of scientific and technological research (TISTR) [6, 10, 26–28]. Isolated microorganisms or microorganisms obtained from culture collection for production of chitin and chitosan from shrimp and crab shells were grown and maintained on agar slant and plate in which medium compositions and cultivation conditions depend on selected strain. Some microorganisms such as *Lactobacillus plantarum, Lactobacillus paracasei, Serratia marcescens, Pediococcus acidolactici, Pseudomonas aeruginosa, Bacillus pumilus, Paenibacillus woosongensis, Aspergillus niger, Pentosus,* and *Bacillus thuringiensis* are used for biological extraction of chitin from shrimp and crab shells [6].

4.3 Production of Chitin Using Microorganisms

Fermentation processes utilize microorganisms to convert solid or liquid substrates into various products. Mostly useful fermentation maybe classified as either solid-state or submerged cultures. In solid state fermentation (SSF), there are two types: solid state fermentation and solid substrate fermentation. In solid state fermentation, the microorganisms grow on and inside a moist solid with little or no free water, although capillary water may be present. In solid substrate fermentation, the solid matrix acts as carbon/energy source and solid support but in solid state fermentation, the solid matrix acts as a solid support only. In submerged fermentation (SMF), there are three types: batch fermentation, fed-batch fermentation, and continuous fermentation that may use a dissolved substrate [21]. Fermentation may require only a single species of microorganisms to effect the desired chemical change and it has been envisaged as one of the most eco-friendly, safe, technologically flexible, and economically viable alternative methods [29, 30]. The biological process for production of chitin from shrimp and crab shells involved demineralization by lactic acid – producing bacteria and deproteinization by protease-producing microorganisms [3, 23]. Most of researchers have reported fermentation conditions for production of chitin from shrimp and crab shells (Table 4.1).

4.4 Production of Chitosan from Chitin Using Chitin Deacetylase from Microorganisms

According to literature survey, the production of chitosan from chitin using microorganisms is not carried out yet. Chitosan has been produced by deacetylation of chitin by enzymatic methods. However the *N*-deacetylation is almost never complete [23]. Table 4.2 shows the production of chitosan by chitin deacetylase (CDA) treatment.

Win *et al.*, reported that shrimp chitin was studied as substrate for CDA from *Absidia coerulea* [31]. The chitin was treated physically (heated and sonicated) and chemically (substituted by succinic anhydride or heated with phosphoric acid and 2-propanol) in order to obtain a better accessibility of acetyl groups in chitin chain for deacetylation. In the kinetic study of CDA from *Absidia* with partially deacetylated chitin (PDC) (58% DD) as substrate, the formation of acetate occurred initially at a high speed but reached a plateau value after about 1 h. Aye *et al.*, (2006) [32] reported that the enzymatic deacetylation of various chitin preparations (natural chitin, partially deacetylated chitin and superfine chitin) was investigated using the fungal CDA isolated from *Rhizopus oryzae* in which the specific extracellular enzyme activity after solid state fermentation was 10 times higher than that after submerged fermentation. The fermentation fluid of *Rhizopus oryzae* grown on soybean residue under solid substrate fermentation had a much higher enzyme activity on all three chitin substrates. Activity of CDA (91 AU/100 ml) measured with reprecipitated chitin as substrate was higher than that measured with PDC as substrate (61 AU/100 ml). CDA from *Rhizopus*

Table 4.1 Producing of chitin from shrimp and crab shells using various microorganisms under different conditions deproteination (DP), demineralization (DM).

Raw materials	Microorganisms	Conditions				% DP	% DM	Remark
		Temp °C	Time (day)	Mixing (rpm)	pH			
Shrimp shell [6]	*Lactobacillus plantarum*	30	6	180	–	45	54	Batch culture
Crab shell [25]	*P. Aeruginosa F722*	30	7	180	–	63	92	–
Shrimp heads and shells (with tail) [2]	*Lactobacillus plantarum*	37	7 days	–	–	84.4	82	–
	Lactococcussp							
Red crab shell (*Chionoecetesjaponicus*) [22]	L. paracasei subsp. tolerans KCTC-3074(L. 3074)	30	7 days	180 rpm	–	52.6	97.2	–
	S. marcescens FS-3 (FS-3)							
Shrimp waste [27]	*L. plantarum*541 (non-amylolitic strain)	30	1	–	–	59.8	81.4	Stirring with glass rod at 1 h interval
	*L. plantarum*A6 (amylolytic strain)					52.2	65.5	
Shrimp shell [3]	B. pumilus A1	35	6 days	150 rpm	5	94	88	Similar to commercial α-chitin
Shrimp biowaste [11]	*Pediococcus acidolactici* CFR2182	37±1	72 h	–	4.3	97.9	72.5	–
Shrimp shell [4]	*Pseudomonas aeruginosa*	37	6 days	100 rpm	–	74.76	78.46	–
	Serratia marcescens					–	31.62	
	Bacillus pumilus					–	–	

(*Continued*)

Table 4.1 Cont.

Raw materials	Microorganisms	Conditions				% DP	% DM	Remark
		Temp °C	Time (day)	Mixing (rpm)	pH			
Shrimp waste (*Penaeussemisulcatus*) [28]	*Lactobacillus plantarum*	60	7 days	–	–	–	–	–
Black tiger shrimp (*Penaeusmondon*) shell [7]	*Paenibacillus woosongensis* TKB2	50	4 days	100 rpm	8	80	–	–
Shrimp waste (*Penaeus* spp.) [49]	*Lactobacillus* spp. strain B2	30	6 days	–	–	85.0	87.6	–
Shrimp waste (*Penaeusmonodon*) [5]	Natural probiotic (milk curd)	–	72 h	–	–	89	69	5.65% chitin recovery
Shrimp shell (*Metapenaeopsisdobsoni*) [9]	*B. subtilis*	–	15 days	–	–	–	–	84% protein and 72% minerals were removed
Shrimp shell (*Pandalus borealis*) [8]	*Aspergillusniger*	21.8	72 h	–	8.64	–	–	Solid state deproteinization
Shrimp shells (*Litopenaeusvannamei*) [50]	*Lactobacillus, Pentosus* L7and *Bacillus thuringiensis* SA	37	72	200 rpm	–	–	–	–

Table 4.2 Production of chitosan from chitin using deacetylase

			Conditions for production of enzyme from microorganisms treatment conditions						CDA treatment conditions for production of chitosan from CT				
Raw materials	**Modified form of chitin**	**Source of enzyme**	**Medium**	**Time (day)**	**Fermentation type**	**pH**	**rpm**	**Temp**	**pH**	**rpm**	**Temp**	**Time**	**Enzyme activity**
Chitin from Shrimp shell [32]	Natural chitin (10%DD)	Fungus (*Rhilzopusoryzae*)	Soybean residue	5 or 15	Solid state	–	–	–	8.5	155	45 °C	1 h	18.4 ± 0.5 (AU/100 ml)
	partially deacetylated chitin (58% DD)												91.4 ± 3.0 (AU/100 ml)
	Superfine chitin												60.9 ± 0.5 (AU/100 ml)
	Natural chitin (10% DD)	Fungus (*Absidiacoerulea*)	Liquid growth (2% glucose, 2% yeast extract, 0.1%K_2HPO_4, 0.1% NaCl, 0.5%$MgSO_4.7H_2O$, and 0.01% $CaCl_2.2H_2O$)	5 days	Liquid state	6.5	150	30 °C	5.8	155	45 °C	1 h	1.9 ± 0.4 (AU/100 ml)
	partially deacetylated chitin (58%DD)												3.5 ± 0.7 (AU/100 ml)
	Superfine chitin												33.6 ± 1.0 (AU/100 ml)
Chitin from Shrimp shells [31]	partially deacetylated chitin (58% DD)	*Absidia coerulea*	2% glucose, 2% yeast extract, 0.1%K_2HPO_4, 0.1% NaCl, 0.5% $MgSO_4.7H_2O$, and 0.01% $CaCl_2.2H_2O$	3 days	Liquid state	4.5	150	30 °C	5.8	–	–	–	45.9 U/100 ml

Abbreviations: CT – chitin; DD – degree of deacetylatio.

was a good candidate for chitin modification and its production by solid state fermentation seemed to be appropriate. Chitin treated by the enzyme behaved as a chitosan particle in its protein binding behavior [32]. In 2011, Suresh *et al.*, studied the production of extracellular chitin deacetylase by *Colletotrichum lindemuthianum* using chitin from shrimp shell waste (CSSW) as a substrate under solid substrate fermentation. Along with chitin deacetylase, *C. lindemuthianum* produced 0.38±0.04 U/gIDS of β-*N*-acetylhexosaminidase at 96 h and 0.49±0.05 U/g IDS of endo-chitinase at 264 h in CSSW medium [33].

4.5 Production of Crude Chitooligosaccharide from Shrimp and Crab Shells Using Fermentation Technology

Chitooligosaccharide (COS) is readily soluble in water due to their shorter chain lengths and free amino acid groups in D-glucosamine (GlcN) units [34]. Conversion of chitosan into chitosan oligosaccharide (COS) has been done either traditionally by chemical digestion with strong acids or enzymatic hydrolysis [35–37]. Among these two methods chemical hydrolysis is used more commonly in the industrial scale production, on the other hand, many problems existing in this method, such as production of a large amount of short-chain oligosaccharides produced, low yields of oligosaccharides, high cost of separation, and environmental pollution [34, 35]. According to these disadvantages, enzymatic hydrolysis has become more popular in recent years because of environmental compatibility, low cost and reproducibility [35].

Chitosanase has been found in various microorganisms, including various bacterial species and fungi. For the enzymatic/fermented production of chitosan oligosaccharides from chitosan, chitosanase-producing bacteria has been grown in various media containing chitosan, chitin, or shrimp head powder as the substrate or carbon source [16, 35, 38, 39]. "Microbial chitosanase shows excellent performances in COS production, they are expensive to be utilized in large-scale industrial applications, because enzyme hydrolysis requires multisteps, particularly, enzyme preparation and purification." [36] Table 4.3 shows production of chitosanase from various bacteria strains under different fermentation conditions. Enzyme purification steps could not be necessary for production of chitooligosaccharides in agriculture sector.

Table 4.3 Production of chitosanase from various bacteria strains under different fermentation conditions.

Strain	Fermentation			Inducer/carbon source	Enzyme activity
	Method	Media	Condition		
Bacillus sp. [38]	Liquid fermentation	0.1% KH_2PO_4, 0.05% $MgSO_4.7H_2O$, 50 mM Sodium Phosphate buffer	pH 6.0, 30 °C	3% w/v chitin	–
Bacillus cereus D-11 [18]	Liquid fermentation	1% yeast extract and 1% NaCl	30 °C, 3 days	0.7% colloidal chitosan	4.85 U/ml
Acinetobacter sp. Strain CHB101 from soil sample [51]	Liquid fermentation	M9 synthetic medium	25 °C, 7 days, 150 rpm	0.25% chitosan	No activity
Paenibacillusehimensis from soil sample [36]	Submerged fermentation	peptone 6.0g/L, yeast extract 6.0/L, glucose 1.0g/L, K_2HPO_4 1.0g/L, $MgSO_4.7H_2O$ 0.5 g/L	pH 7.0,36 °C, 120 rpm	0.2% (w/v) of soluble chitosan	500 U/L
B. cereus TKU027 [35]	Liquid fermentation	0.1% K_2HPO_4, 0.05% $MgSO_4{\cdot}7H_2O$	pH 7, 37 °C, 2 days	1% shrimp head powder	
Bacillus sp. strain KCTC 0377BP from soil sample [17]	Liquid fermentation	0.5% yeast extract, 0.2% K_2HPO_4, 0.1% H_2PO_4, 0.07% $MgSO_4{\cdot}7H_2O$, 0.05% NaCl, 0.05% KCl, 0.01% $CaCl_2$	pH 6.8, 30 °C, 5 days, 180 rpm	0.5%(w/v) colloidal chitosan	1.2 U/ml
Bacillus sp. KFB-C108 [53]	Liquid fermentation	0.15% K_2HP0_4, 0.05% KH_2P0_4, 0.01% $MgS0_4{\cdot}$ $7H_20$,0.1% yeast extract	pH 6.5, 55 °C, 40 h, 150 rpm	1.2% colloidal chitosan	data not shown

(Continued)

Table 4.3 Cont.

Strain	Fermentation			Inducer/carbon source	Enzyme activity
	Method	Media	Condition		
Aspergillusfumigatus KH-94 [54]	Liquid fermentation	1% tryptone, 0.5% glucose, 0.1% K_2HP0_4, 0.05%$MgS0_4 \cdot 7H_20$, 0.05% KCl, and 0.001% $FeS0_4 \cdot 7H_20$, 1% lactic acid	pH 6.2, 72 h, 40 °C, 150 rpm	1.5% chitosan	–
M. chitosanotabidus 3001 [55]	Liquid fermentation	0.5% $MgSO_4$, 0.3% KH_2PO_4, 0.7% K_2HPO_4, 0.25% yeast extracts, 0.25% polypeptone	pH 7, 30 °C, 4 days, 200 rpm	0.4% (w/v) colloidal chitosan	–
Streptomyces N174 [56]	Liquid fermentation	0. 2% (w/v) olive oil, 0.035% antifoam agent	30 °C, 108 h, 250 rpm	10g chitosan/L	–
Aspergillusfumigatus S-26 [26]	Liquid fermentation	1% tryptone, 0.1% K_2HPO_4, 0.05%$MgSO_4 \cdot 7H_2O$, 0.05% KCl, 0.001% $FeSO_4 \cdot 7H_2O$, 0.025% $CaCl_2$, 0.2% Tween 80, 1% NaCl	pH 7, 37 °C, 3 days, 180 rpm	0.5%colloidal chitosan	–
Aspergillusflavus IAM2044 [39]	Liquid Fermentation	Czapek-Dox (CD) medium (K_2HPO_4 1.0 g, KCl 0.5 g, $NaNO_3$ 2.0 g, $MgSO_4 \cdot 7H_2O$ 0.5 g, $FeSO_4 \cdot 7H_2O$ 0.02 g, DW 1,000 ml) and 0.5% peptone, 0.5% yeast extract	pH 5.0, 28 °C, 3 days	Soluble chitosan 0.5%,	101.5 U/ml

4.6 Applications of Chitin, Chitosan, Chitooligosaccharides and Their Composite Materials

Chitosan has specific properties such as environmentally friendly, biodegradability, biocompatibility, bioactivity, selective permeability, chelation, ion exchange properties, antitumor and antimicrobial activity, absorption capacity, and biopolymer studies. In agricultural sector, chitosan has been used in seed, leaf, fruit and vegetable coating; as fertilizer to control release of agrochemical; to increase plant product; and to protect plants against microorganisms; and chitooligosaccharide has been used for stimulation of growth and increases yield of plants as well as induce the immune plant system and control the plant diseases [20, 11, 40]. Chitosan can be applied in agriculture without use chemical fertilizer that can increase the microbial population by large numbers, and transforms organic nutrient into inorganic nutrient, which is easily absorbed by the plant roots [11]. The application of chitosan for edible coatings on fruit and vegetable is one of the most innovative methods to extend the commercial shelf-life of fruits and vegetables by acting as a gas barrier and having a similar effect as the storage under modified atmosphere [40]. Effect in quality of preserving on coating chitosan on seed, leaf, fruit, and vegetables may vary depending on the features of coating, their species, and their maturity as well as storage conditions [40]. To control pests and improve plant defenses against microorganisms, the chitosan has been used as seed coating agents and plants by dipping or spraying. Exclusively, chitosan and their derivatives have plant protective properties that are highly desirable as they offer protection against fungi, viruses, bacterial diseases, and nematodes [23]. In addition, chitosan application is a worldwide trend to explore new methods that reduce the negative effect of abiotic stress and increase yield and quality of crops [41]. It has been reported that chitosan can reduce disease severity in orchids, possibly by increasing the activity of phenylalanine ammonia-lyase (PAL) and polyphenol oxidase (PPO), lignification resulting from increased biosynthesis of phenolic compounds or induces secondary metabolites and systemic acquired resistance (SAR) [43].

Plant tissue culture is particular suitable to study the effect of chitosan on plants that grow and multiply slowly, like orchids. Orchid roots sprayed with a much diluted chitosan solution show stimulation of growth, renewed flower production, and enhanced resistance against fungi and virus [20, 42]. Nge *et al.*, used chitooligosaccharide, fungal chitosan (10 kDa), and shrimp chitosan (100 kDa) to propagate the plantlet from orchid protocorm in liquid and on solid tissue culture medium. The highest number of plantlets was observed in the presence of 20 ppm using either 10 kDa fungal chitosan or 1 kDa chitooligosaccharide. High molecular weight chitosan (100 kDa) had no stimulating effect. The 10 kDa fungal chitosan was more effective compared with 1 kDa oligomer. Fungal chitosan and shrimp oligomer chitosan of 15 ppm and 20 ppm are most effective on protocorm propagation in liquid and on agar orchid plant tissue culture media, respectively [20]. Chitin, chitosan, and chitooligosaccharides have a promising future in development of sustainable agricultural practices. Recently, some researchers reported that chitosan and chitooligosaccharide used in the agriculture sector (Table 4.4).

In addition, the application of chitin, chitosan, and chitooligosaccharides has been found in many other fields such as for water instance, wastewater treatment, photography, cosmetics, ophthalmology, paper fishing, drug-delivery systems, and food industry [44].

In 1999, the design of chitin-based composite materials is still in its infancy. Numerous composite materials could principally be constructed by imitation of various natural biomaterials [45]. Phisalaphong *et al.*, showed that biosynthesis of cellulose-chitosan composite is a good candidate for biomedical applications such as wound dressing materials and Nagahama *et al.*, prepared chitin/gelatin membranes used for tissue engineering because the membranes are bioactive and are suitable for cell adhesion [47, 48]. Table 4.5 shows application of chitosan composite materials.

4.7 Conclusion

Production of chitin, chitosan, and chitooligosaccharides with enzymes and microorganisms has been studied for decades. Green technologies encompass various aspects of technology which help us reduce the human impact on the environment and create ways of sustainable development. So, the direct fermentation of raw biomaterials like crab and shrimp shells presents another opportunity in the production of chitin and chitooligosaccharides. Chitin, chitosan, and their derivatives are valuable compounds that meet the needs of consumers for natural products that impact positively on health. There is a great deal of interest in these polysaccharides due to their appealing properties of biodegradability, biocompatibility, and nontoxicity. Therefore, they can be used in a wide range of industrial applications in biomedicines, pharmaceuticals, agriculture, foods, cosmetics, textiles, and enzymes-immobilization. The current problem on production of green technology in which the yield of product is still lower and the cost is higher

Table 4.4 Agriculture sector application of chitosan and chitooligosaccharides.

Source of agriculture input	Form	Type of crop	Application of agriculture input
Chitosan [42]	Solution	Orchid	Plant tissue culture
Chitosan [57]	Solution	Tomato	Plant elicitor
Chitosan [58]	Solution	Okra (lady's finger)	Fruit yield improvement
Chitosan [59]	Solution	Sweet pepper plants (*Capsicum annuum*L.)	Enhancement of growth and production
Chitosan [60]	Solution	Bean (Phaseolus vulgaris super stryke)	Plant stimulator
Chitosan [61]	Solution	Cowpea (*Vignaunguiculata*(L) Walp)	Improving growth and yield
Chitosan [62]	Powder	Radish Plants (*Raphanussativus*, L. var. sativus)	Soil treatment for counteracting of cadmium levels
Chitosan [63]	Solution	Tomato (*Lycopersicumesculentum* Mill.)	Stimulation of the immunity of plants to protect plants against microorganisms and to simulation of roots, shoots, leaves and chlorophyll content and photosynthetic rate
Chitosan [64]	Solution	Anise (*Pimpinellaanisum*L.)	Produced the highest germination and seedling growth

(*Continued*)

Table 4.4 Cont.

Source of agriculture input	**Form**	**Type of crop**	**Application of agriculture input**
Chitosan oligomers [19]	Solution	Coffee (*CoffeacanephoravarRobusta*)	Growth of coffee seedlings, enhancing content of chlorophylls, carotenoid and mineral nutrient uptake; increasing area of leaves, height and stem diameter of plants, decreased intensity of transpiration of the leaves, increased mineral uptake and the growth of braches
Shrimp Chitosan (100 kDa) [20]	Solution	Orchid (*Dendrobiumphalaenopsis*)	Plant tissue culture
Chitooligosaccharides [65]	Solution	Rice	Biostimulator
Chitosan (600-900 kDa) [66]	Solution	Chilli (*Capsicum annuum* L.)	Growth of chilli seedlings
Chitosan Oligosaccharides [67]	powder	Purple Passionfruit (Passifloraedulis Sims var. edulis)	Soil conditioner
Chitosan [68]	Powder	*Eustoma grandiflorum*	Soil treatment
	Solution		Seed coating
Chitosan [41]	Solution	Isabgol (*Plantago ovate Forsk.*)	Promote germination and plant growth and reduce the harmful salinity

Chitosan [12]	Solution	Soybean (*Glycine max*)	Enhance the productivity of sprouts without adverse effects on nutritional and postharvest characteristics
Chitosan [69]	Solution	Mungbean (*Vigna radiate* L.)	Growth stimulating and highest seed yielding agent
Chitosan (45 kDa) [14]	Solution	*Dendrobium orchid* "Sureepeach"	Growth promoter
Chitosan (50 kDa and 970 kDa) [70]	Solution	Potted freesia	Plant stimulator for growth, flowering and corms yield
Chitosan [71]	Solution	Wheat (*Triticumaestivum* L.)	Controlled natural seed-borne *F. graminearum* infection and increased seed germination
Chitosan [52]	Solution	Soybean	Chemical pesticides replacement and seeds germination acceleration
Chitosan [72]	Solution	Guava (*Psidiumguajava* L.)	Fruit Coating
Chitosan [73]	Solution	Litchi (*Litchi chinensis*Sonn.)	Fruit Coating for improved preservation effect
Chitosan [74]	Solution	Apricot (*Pruusarmeniaca* L.)	Fruit Coating for maintaining fruit quality
Chitosan [40]	Solution	Plums (*Prunusdomestica* L.)	Fruit coating to extend the storage life without any considerable deterioration of the quality of the fruits
Chitosan [75]	Solution	Sunflower	Seed coating for increasing both growth and quality of sprouts

(Continued)

Table 4.4 Cont.

Source of agriculture input	Form	Type of crop	Application of agriculture input
Chitosan [76]	Solution	Peach, Japanese pear and kiwifruit	Fruit coating for storage condition
Chitosan [77]	Solution	Apples, cv. Gala (*Malusdomestica*)	Fungistatic agent
Chitosan [78]	Solution	Papaya	Fruit coation to control postharvest diseases of papaya
Chitosan [79, 80]	Solution	Strawberries	Fruit coating to extend storability
Chitosan [43]	Solution	Orchid (*Dendrobium*)	Plant growth promotor
Chitosan [81]	Solution	Peeled Litchi (Litchi chinensis Sonn.)	Fruit coating for maintain quality and extended shelf life

Table 4.5 Applications of chitosan composite materials

Composite	Application	References
Chitosan/hydroxypatite(CTS/HAp)	Tissue Engineering	[82]
Chitosan/sericin	Wound Healings	[83]
Bacterial cellulose/chitosan	Dressing material	[84]
Chitosan/cellulose	Removal of microcystin	[85]
Aloe vera blended collagen-chitosan	Tissue engineering	[86]
Nano-sized hydroxyapatite/chitosan	Bone tissue engineering	[87]

than traditional chemical methods. Reference it is expected that the production of chitin, chitosan and chitooligosaccharide using green technology will show bright light toward green environment and healthy society.

Acknowledgement

The authors are grateful for generous financial support received from PYOE PIN, Yangon, Myanmar for this work. The authors would like to thanks to Mr. Win Myo Thu, President and members of Advancing Life and Regenerating Motherland, Yangon, Myanmar for encouragement and support.

References

1. Irianto, H.E., Dewi, S.A., Giyatmi., Prospective utilization of fishery by-products in Indonesia, In: *Seafood Processing By-products: Trends and Applications*, S.K. Kim, Ed., pp. 21–34, Springer, New York, 2014.
2. Francisco, F.C., Simora, R.M.C., Nuñal, S.N., Deproteination and demineralization of shrimp waste using lactic acid bacteria for the production of crude chitin and chitosan. *AACL Bioflux.*, 8, 107–115, 2015.
3. Ghorbel-Bellaaj, O., Hajji, S., Younes, I., Chaabouni, M., Nasri, M., Jellouli, K., Optimization of chitin extraction from shrimp waste with *Bacillus pumilus* A1 using response surface methodology. *Int. J. Biol. Macromole.*, 61, 243–250, 2013.
4. Sedaghat, F., Yousefzadi, M., Toiserkani, H., Najafipour, S., Chitin from *Penaeusmerguiensis* via microbial fermentation processing and antioxidant activity. *Int. J. Biol. Macromol.*, 2015.

5. Prameela, K., Mohan, Ch. M., Smitha, P.V., Hemalatha, K.P.J., Bioremediation of shrimp biowaste by using natural probiotic for chitin and carotenoid production an alternative method to hazardous chemical method. *Int. J. Appl. Biol. Pharm. Technol.*, 1, 903–910, 2010.
6. Khorrami, M., Najafpour, G.D., Younesi, H., Hosseinpour, M.N., Production of chitin and chitosan from shrimp shell in Batch Culture of *Lactobscillusplantarum. Chem. Biochem. Eng. Q.*, 26, 217–223, 2012.
7. Paul, T., Halder, S.K., Das, A., Ghosh, K., Mandal, A., Payra, P., Barman, P., Mohapatra, P. K. D., Pati, B. R., Mondal, K. C., Production of chitin and bioactive materials from Black tiger shrimp (*Penaeusmonodon*) shell waste by the treatment of bacterial protease cocktail. *3 Biotech.*, 5, 483–493, 2015.
8. Ghaly, A., Mahmoud, N., Solid state deproteinization of shrimp shells by *Aspergillusniger* with galactose as a bioavailable carbon supplement. *J. Bioprocess. Biotech.*,5, 1–11, 2015.
9. Sini, T.K., Santhosh, S., Mathew, P.T., Study on the production of chitin and chitosan from shrimp shell by using *Bacillus subtilis*fermentation. *Carbo. Res.*, 342, 2423–2429, 2007.
10. Boonlertnirun, S., Boonraung, C., Suvanasara, R., Application of chitosan in rice production. *J. Met. Mater. Min.*, 18, 47–52, 2008.
11. Bhaskar, N., Suresh, P.V., Shkhare, P.Z., Sachindra, N.M., Shrimp biowaste fermentation with Pediococcus acidolactici CFR2182: Optimization of fermentation conditions by response surface methodology and effect of optimized conditions on deproteination/demineralization and carotenoid recovery. *Enz. Microbial. Technol.*, 40, 1427–1434, 2007.
12. Lee, Y.S., Kim, Y.H., Kim, S.B., Changes in the respiration, growth, and vitamin C content of soybean sprouts in response to chitosan of different molecular weights. *Hort. Sci.*, 40, 1333–1335, 2005.
13. Ahmed, A.B.A., Kim, S.K., Chitin, chitosan derivatives induce the production of secondary metabolites and plant development through *in vitro* and *in vivo* techniques, In: *Chitin, chitosan, oligosaccharides and their derivative*, S.K. Kim, Ed., pp. 589–604, CRC Press, New York, 2010.
14. Charoenwattana, P., Petprapai, U., Effects of chitosan and lotus extracts as growth promoter in *Dendrobium* orchid. *Intl. J. Enviro. Rural. Develo.*, 4, 133–137, 2013.
15. Toan, N.V., Hanh, T.T., Application of chitosan solutions for rice production in Vietnam. *Afr. J. Biotechnol.*, 12, 382–384, 2013.
16. Gao, X.A., Zhang, Y.F., Park R.D., Huang, X., Zhao, X.Y., Xie, J, Jin, R.D., Preparation of chitooligosaccharides from chitosan using crude enzyme of *Bacillus cereus* D-11. *J. Appl. Biol. Chem.*, 55, 13–17, 2012.
17. Choi, Y.J., Kim, E.J., Yun, Y. C., Shin, Y. C., Purification and characterization of chitosanase from *Bacillus* sp. strain KCTC 0377BP and its application for the production of chitosan oligosaccharides. *Appl. Envi. Micro.*, 70, 4522–4531, 2004.

18. Gao, X.A., Ju, W.T., Zhang, Y.F., Park R.D., Purification and characterization of chitooligosaccharides from chitosan using crude enzyme of chitosanase from *Bacillus cereus* D-11. *Carbo. Poly.*, 72, 513–520, 2008.
19. Dzung, N.A., Khanh, V.T.P., Dzung, T.T., Research on impact of chitosan oligomers on biophysical characteristics, growth, development and drought resistance of coffee. *Carbo. Poly.*, 84, 751–755, 2011.
20. Nge, K.L., Nwe, N., Chandrkrachang, S., Stevens, W.F., Chitosan as a growth stimulator in orchid tissue culture. *Plant. Sci.*, 170, 1185–1190, 2006.
21. Nwe, N., Furuike, T., Tamura, H., Production, properties and applications of fungal cell wall polysaccharides: chitosan and glucan. *Adv. Polym. Sci.*, 244, 187–208, 2011.
22. Jung, W.J., Jo, G.H., Kuk, J.H., Kim, K.Y., Park, R.D., Extraction of chitin from red crab shell waste by cofermentation with *Lactobacillus paracasei subsp. tolerans* KCTC-3074 and *Serratiamarcescens* FS-3. *Appl. Microbiol. Biotechnol.*, 71, 234–237, 2006.
23. Hamed, I., Özogul, F., Regenstein, J.M., Industrial applications of crustacean by-products (chitin, chitosan and chitooligosaccharides): A review. *Trends Food. Sci. Techno.*, 48, 40–50, 2016.
24. Mahilum-Tapy, L.M., The importance of microbial culture collection and gene banks in biotechnology. In: *Biotechnology. Encyclopedia of life support systems (EOLSS)*, H.W. Doelle, J.S.Rokem, M. Berovic, Eds., , 1, 2009.
25. Oh, K.T., Kim, Y.J., Nguyen, V.N., Jung, W.J., Park, R.D., Demineralization of crab shell waste by *Pseudomonas aeruginosa* F722. *Process. Biochem.*, 42, 1069–1074, 2007.
26. Jung, W.J., Kuk, J.H., Kim, K.Y., Jung, K.C., Park, R.D., Purification and characterization of exo-β-D-glucosaminidase from *Aspergillusfumigatus* S-26. *Prot. Expr. Purif.*, 45, 125–131, 2005.
27. Rao, M.S., Stevens, W.F., Fermentation of shrimp biowaste under different salt concentrations with amylolytic and non-amylolytic*Lactibacillus* strains for chitin production. *Food Technol. Biotechnol.*, 44, 83–87, 2006.
28. Khanafari, A., Marandi, R., Sanatei, Sh., Recovery of chitin and chitosan from shrimp waste by chemical and microbial methods. *Iran. J. Environ. Health Sci. Eng.*, 5, 19–24, 2008.
29. Chisti, Y., Fermentation (industrial): Basic considerations, In: *Encyclopedia of food microbiology*, R. Robinson, C. Batt, P. Patel, Eds., pp. 663–674, Academic Press, London, 1999.
30. Kandra, P., Challa, M.M., Jyothi, H.K.P., Efficient use of shrimp waste: present and future trends. *Appl. Microbiol. Biotechnol.*, 93, 17–29, 2012.
31. Win, N.N., Stevens, W.F., Shrimp chitin as substrate for fungal chitin deacetylase. *Appl. Microbial. Biotechnol.*, 57, 334–341, 2001.
32. Aye, K.N., Karuppuswamy, R., Ahamed, T., Stevens, W.F., Peripheral enzymatic deacetylation of chitin and reprecipitated chitin particles. *J. Bioresource. Tech.*, 97, 577–582, 2006.

33. Suresh, P.V., Sachindra, N.M., Bhaskar, N., Solid state fermentation production of chitin deacetylase by *Colletotrichumlindemuthianum* ATCC 56676 using different substrates. *J. Food. Sci. Technol.*, 48, 349–356, 2011.
34. Kim, S.K., Je, J.Y., Continuous production of chitooligosaccharides by enzymatic hydrolysis, In: *Chitin, chitosan, oligosaccharides and their derivative,* S.K. Kim, Ed., pp. 589–604, CRC Press, New York, 2010.
35. Wang, S.L., Liu, C.P., Liang, T.W., Fermented and enzymatic production of chitin/chitosan oligosaccharides by extracellular chitinases from *Bacillus cereus* TKU027. *Carbohydr. Poly.*, 90, 1305–1313, 2012.
36. Pagnoncelli, M.G.B., Araújo, N.K.D., Silva, N.M.P.D., Assis, C.F.D., Rodrigues, S., Macedo, G.R.D., Chitosanase production by *Paenibacillusehimensis* and its application for chitosan hydrolysis. *Braz. Arch. Biol. Technol.*, 53, 1461–1468, 2010.
37. Jung, W.J., Park, R.D., Bioproduction of chitooligosaccharides: present and perspectives. *Mar. Drugs.*, 12, 5328–5356, 2014.
38. Alam, J., Mathur, A., Antibacterial potential of chitin and chitin-based derivatives against pathogenic and drug-resistant bacterial strains. *Int. J. Pharm. Pharm. Sci.*, 3, 1698–1707, 2014.
39. Ji, Hoon, J., Yang, J.S., Hur, J.W., Purification and characterization of the exo-β-D-glucosaminidase from *Aspergillusflavus* IAM2044. *J. Microbiol. Biotechnol.*, 13, 269–275, 2003.
40. Bal, E., Postharvest application of chitosan and low temperature storage affect respiration rate and quality of plum fruits. *J. Agr. Sci. Tech.*, 15, 1219–1230, 2013.
41. Mahdavi, B., Seed germination and growth responses of Isabgol (*Plantago ovate Forsk*) to chitosan and salinity. *Intl. J. Agri. Crop. Sci. Vol.*, 5, 1084–1088, 2013.
42. Restanto, D.P., Santoso, B., Kriswanto, B., Supardjono, S., The application of chitosan for protocorm like bodies (PLB) induction of orchid (*Dendrobiumsp*) *In Vitro. Agri. Agri. Sci. Proc.*, 9, 462–468, 2016.
43. Uthairatanakij, A., Silva, J.A.T., Obsuwan, K., Chitosan for improving orchid production and quality. *Orchid. Sci. Biotechno.*, 1, 1–5, 2007.
44. Ravi Kumar, M.N.V., A review of chitin and chitosan applications. *React. Funct. Polym.*,46, 1–27, 2000.
45. Peter, M., Applications and environmental aspects of chitin and chitosan. *J.M.S.-Pure. Appl. Chem.*, 32, 629–640, 1995.
46. Younes, I., Riaudo, M., Chitin and chitosan preparation from marine sources. Structure, properties and applications. *Mar. Drugs.*, 13, 1133–1174, 2015.
47. Phisalaphong, M., Jatupaiboon, N., Kingkaew, J., Biosynthesis of cellulose-chitosan composite, In: *Chitin, chitosan, oligosaccharides and their derivative,* S.K. Kim, Ed., pp. 589–604, CRC Press, New York, 2010.
48. Nagahama, H., Divya Rani, V.V., Shalumon, K.T., Jayakumar, R., Nair, S.V., Koiwa, S., Furuike, T., Tamura H., Preparation, characterization, bioactive and cell attachment studies of α-chitin/gelatin composite membranes. *Int. J. Biol. Macromol.*, 44, 333–337, 2009.

49. Cira, L.A., Huerta, S., Hall, G.M., Shirai, K., Pilot scale lactic acid fermentation of shrimp wastes for chitin recovery. *Process Biochem.,* 37, 1359–1366, 2002.
50. Ploydee, E., Chaiyanan, S., Production of high viscosity chitosan from biologically purified chitin isolated by microbial fermentation and deproteinization. *Int. J. Polym. Sci.,* 1–8, 2014.
51. Shimosaka, M., Nogawa, M., Wang, X.Y., Kumehara, M., Okazaki, M., Production of two chitosanases from a chitosan-assimilating bacterium, *Acintobacter* sp. Strain CHB101. *Appl. Enviro. Micro.,* 61, 438–442, 1994.
52. Zeng, D., Luo, X., Tu, R., Application of bioactive coatings based on chitosan for soybean seed protection. *Inter. J. Carbo. Chem.*, 1–5, 2012.
53. Yoon, Geun, H., Ha, A.C., Lim, Y.H., Cho, H.Y., New thermostable chitosanase from *Bacillus*sp. :purification and characterization. *J. Microbiol. Biotechnol.,* 8, 449–454, 1998.
54. Kim, Young, S., Shon, D.H., Lee, K.H., Purification and characterization of two types of chitosanases from *Aspergillusfumigatus* KH-94. *J. Microbiol. Biotechnol.,*8, 568–574, 1998.
55. Park, J.K., Shimono, K., Ochiai, N., Shigeru, K., Kurita, M., Ohta, Y., Tanaka, K., Matsuda, H., Kawamukai, M., Purification, characterization, and gene analysis of a chitosanase (ChoA) from *Matsuebacterchitosanotabidus* 3001. *J. Bacterio.,* 181, 6642–6649, 1999.
56. Boucher, I., Dupuy, A., Vidal, P., Neugebauer, W.A., Brzezinski, R., Purification and characterization of a chitosanase from *Streptomyces* N174. *Appl. Microbiol. Biotechnol.* 38, 188–193, 1992.
57. Mondal, M.M.A., Puteh, A.B., Dafader, N.C., Foliar application of chitosan improved morpho-physiological attributes and yield in summer tomato (*Solanumlycopersicum*). *Pak. J. Agri. Sci.,* 53, 339–344, 2016.
58. Mondal, M.M.A., Malek, M.A., Puteh, A.B., Ismail, M.R., Ashrafuzzaman, M., Naher, L., Effect of foliar application of chitosan on growth and yield in okra. *Austr. J. Crop. Sci.,* 6, 918–921, 2012.
59. Ghoname, A.A, El-Nemr, M.A., Abdel-Mawgoud, A.M.R., El-Tohamy, W. A., Enhancement of sweet pepper crop growth and production by application of biological, organic and nutritional solutions. *Res. J. Agric. & Biol. Sci.,* 6, 349–355, 2010.
60. Sheikha, S. A. A. K., Physiological studies for different concentration from Biochikol 020 PC (chitosan) on bean plant. *J. Asian. Sci. Res.*, 1, 73–86, 2011.
61. Farouk, S., Amany, R.A.E.M., Improving growth and yield of cowpea plant by foliar application of chitosan under water stress. *J. Plant Prod., Mansoura Univ.,* 2, 1341–1358, 2011.
62. Farouk, S., Mosa, A.A., Taha, A.A., Ibrahim, H.M., EL-Gahmery, A.M., Protective Effect of humic acid and chitosan on radish (*Raphanussativus*, L. var. sativus) plants subjected to cadmium stress. *J. Stress. Physio. Biochem.*, 7, 99–116, 2011.

63. El-Tantawy, E.M., Behavior of tomato plants as affected by spraying with chitosan and aminofort as natural stimulator substances under application of soil organic amendments. *Pak. J. Biol. Sci.*, 12, 1164–1173, 2009.
64. Ohta, K., Taniguchi, A., Konishi, N., Hosoki, T., Chitosan treatment affects plant growth and flower quality in Eustoma grandiflorum. *Hort. Sci.*, 34, 233–234, 1999.
65. Nwe, N., Furuike, T., Tamura, H., Production of fungal chitosan by enzymatic method and applications in plant tissue culture and tissue engineering: 11 years of our progress, present situation and future prospects, In: *Biopolymers*, M. Elnashar, Ed., pp. 135–162, INTECH Open Access Publisher, 2010.
66. Chookhongkha, N., Miyagawa, S., Jirakiattikul, Y., Photchanachai, S., Chili growth and seed productivity as affected by chitosan. *Int. Conf. Agri. Tech. Food. Sci.*, 146–149, 2012.
67. Utsunomiya, N., Kinai, H., Matsui, Y., Takebayashi, T., The effect of chitosan oligosaccharides soil conditioner and nitrogen fertilizer on the flowering and fruit growth of purple passion fruit (*Passifloraedulis*Sims var. *edulis*). *J. Japan. Soc. Hort. Sci.*, 67, 567–571, 1998.
68. Mahdavi, B., Effects of Priming Treatments on Germination amd Seedling Growth of Anise (*Pimpinella anisum L.*). *Agric. sci. dev.*, 5, 28–32, 2016.
69. Mondal, M.M.A., Malek, M.A., Puteh, A.B., Ismail, M.R., Ashrafuzzaman, M., Naher, L., Foliar application of chitosan on growth and yield attributes of mungbean (*Vignaradinta*L. Wilczek). *Bangl. J. Bot.*, 42, 179–183, 2013.
70. Salachna, P., Zawadzińska, A., Effect of chitosan on plant growth, flowering and corms yield of potted freesia. *J. Ecolo. Engin.* 15, 97–102, 2014.
71. Bhaskara Reddy, M.V., Arul, J., Angers, P., Couture, Chitosan treatment of wheat seeds induce resistance to *Fusariumgraminearum* and improve seed quality. *J. Agric. Food. Chem.*, 47, 1208–1216, 1999.
72. Krishna, K.R., Rao, D.V.S., Effect of chitosan coating on the physicochemical characteristics of guava (*Psidiumguajava* L.) fruit during storage at room temperature. *Ind. J. Sci. Techno.*, 7, 554–558, 2014.
73. Sun, D., Liang, G., Xie, J., Lei, X., Mo, Y., Improved preservation effects of litchi fruit by combining chitosan coating with ascorbic acid treatment during postharvest storage. *Afr. J. Biotech.*, 9, 3272–3279, 2010.
74. Ghasemnezhad, M., Shiri, M.A., Sanavi, M., Effect of chitosan coating on some quality indices of apricot (*Prunusarmeniaca* L.) during cold storage. *Casp. J. Env. Sci.*, 8, 25–33, 2010.
75. Cho. M.H., No. H.K., Prinyawiwatkul, W., Chitosan treatments affect growth and selected quality of sunflower sprouts. *J. Food. Sci.*, 73, 70–77, 2008.
76. Du, J., Gemma, H., Iwahori, S., Effect of chitosan coating on the storage of peach, Japanese pear and kiwifruit. *J. Japan. Soc. Hort. Sci.*, 66, 15–22, 1997.
77. Assis, O.B.G., The effect of chitosan as a fungistatic agent on cut apples. *Rev. Iber. Tecnologia Postcosecha.*, 9, 148–152, 2008.
78. Bautista-Baños, S., Hernández-López, M., Bosquez-Molina, E., Wilson, C.L., Effects of chitosan and plant extracts on growth of

Colletotrichumgloeosporioides, anthracnose levels and quality of papaya fruit. *Crop. Protec.*, 22, 1087–1092, 2003.
79. Ghaouth, A.E., Arul, J., Ponnampalam, R., Boulet, M., Chitosan coating effect on storability and quality of fresh strawberries. *J. Food. Sci.*, 56, 1618–1620, 1991.
80. Devlieghere, F., Vermeulen, A., Debevere, J., Chitosan: antimicrobial activity, interactions with food components and applicability as a coating on fruit and vegetables. *Food Microbiol.*,21, 703–714, 2004.
81. Dong, H., Cheng, L., Tan, J., Zheng, K., Jiang, Y., Effects of chitosan coating on quality and shelf life of peeled litchi fruit. *J. Food. Engineer.*, 64, 355–358, 2004.
82. Venkatesan, J., Kim, S-W., Chitosan composites for bone tissue engineering-an overview. *Mar. Drugs*, 8, 2252–2266, 2010.
83. Ahmed, S., Ikram, S., Chitosan based scaffolds and their applications in wound healing. *Achiev. Life Sci.*, 10, 27–37, 2016.
84. Ciechańska, D., Multifunctional bacterial cellulose/chitosan composite materials for medical applications. *Fibres Text. East. Eur.*, 12, 69–72, 2004.
85. Tran, C.D., Duri, S., Delneri, A., Franko, M., Chitosan-cellulose composite materials: preparation, characterization and application for removal of microcystin. *J. Hazard. Mater.*, 355–366, 2013.
86. Jithendra, P., Rajam, S.M., Kalaivani, T., Mandal, A.B., Rose, C., Preparation and characterization of aloe vera blended collagen-chitosan composite scaffold for tissue engineering applications. *ACS Appl. Mater. Interf.*, 5, 7291–7298, 2013.
87. Kim, H.L., Jung, G.Y., Yoon, J.H., Han, J.S., Park, Y.J., Kim, D.G., Zhang, M., Kim, D.J., Prepartion and characterization of nono-sized hydroxyapatite/alginate/chitosan composite scaffolds for bone tissue engineering. *Mater. Sci. Eng.*, 54, 20–25, 2015.

5

Chitosan and Its Derivatives: Overview of Commercial Applications in Diverse Fields

Tanvir Arfin

Environmental Materials Division, CSIR-NEERI, Nehru Marg, Nagpur, India

Abstract
Chitosan is the multiversatile polymer next cellulose with tailor-made structural, chemical, and biological properties for myriad life and health sciences applications along with other general uses. Properties like nontoxicity, biodegradability, bioactivity, low allergenicity, antimicrobial activity, bioadhesivity, etc., not only render chitosan compatible to physiological environment, but also enable its use in a number of biomedical and pharmaceutical applications. Moreover, chemical and biological modification is an effective tool to engineer chitosan for very specific applications. The virtues of parent chitosan also traverse to its derivatives conferring them easy employment in novel and specific wastewater, seafood, textile, water treatment, electrochemical sensor, food science, biomedical, and pharmaceutical applications. This chapter presents the biomedical aspects of chitosan and its derivatives and briefly discusses their applications in drug delivery, wound healing, gene delivery, etc., with selected examples and recent developments. Noteworthy findings and results are also highlighted in this regard.

Keywords: Chitosan, chitosan derivatives, applications, biomedical, biosensor, drug delivery, seafood, science, wound healing

5.1 History

Chitosan is general name of complete or incomplete deacelyted chitosan in a group form. In 1811, Henri Braconnot discovered chitosan. He noticed

Corresponding author: tanvirarfin@ymail.com

Shakeel Ahmed and Saiqa Ikram (eds.) Chitosan, (115–150) © 2017 Scrivener Publishing LLC

the chitin in mushrooms rather hard in H_2SO_4 solution. On the basis of investigation given by Bracannot, various others research has been carried out to discover other uses of chitin which was found in variety of forms in nature.

5.2 Synthesis of Chitosan

Initially, the chitosan was prepared from shrimp shells. It was washed in the cold water then sand was removed from mud and further it was dried. Finally, Chitosan was formed in the form of flakes. The overall preparation of chitosan is given below in Figure 5.1.

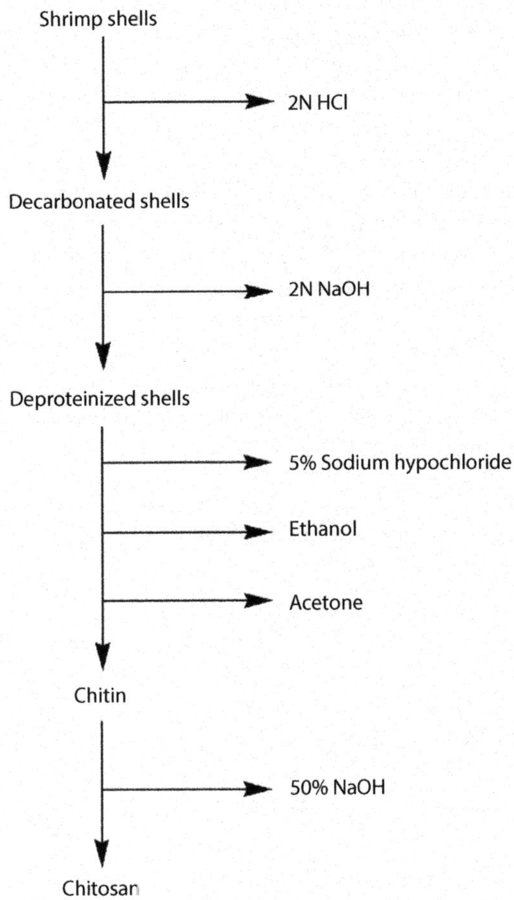

Figure 5.1 Preparation of chitosan from shrimp shells.

5.3 General Properties

Chitosan is a form of polysaccharide comprising of differentquantities of (1-4)-glycosidic bonds connecting theglucosamine and *N*-acetyl-glucosamine [1].The functional group, namely amino, acetyl amino and hydroxyl groups are found in the sequence is responsible for making a chitosan owing numerous of chemical features which is shown in Figure 5.2 [2].The amino groups enable the chitosan to be a natural polyelectrolyte dissolving easily in acidic media. The charge density is dependent on the various factors such as degree of acetylation (DA) and pH. It is crystalline in nature because of inter and intramolecular hydrogen bonds. Its chain is hydrophilic in nature, but the availability of *N*-acetyl groups it resembles hydrophobic [3]. Due to hydrogen bonds and hydrophobic interaction between them, it forms aggregate and making it hard for dissolving in neutral media. But it is capable of dissolving in dilute acid due to ionization occurred through amino groups.

5.4 Biological Properties

The system enabling to resemble as vaccine adjuvant during the intranasally organization is not well developed but yet chitosan molecule possesses variety of *in vivo* influences adding to adjuvant performance. The overall biological properties are shown in Figure 5.3.

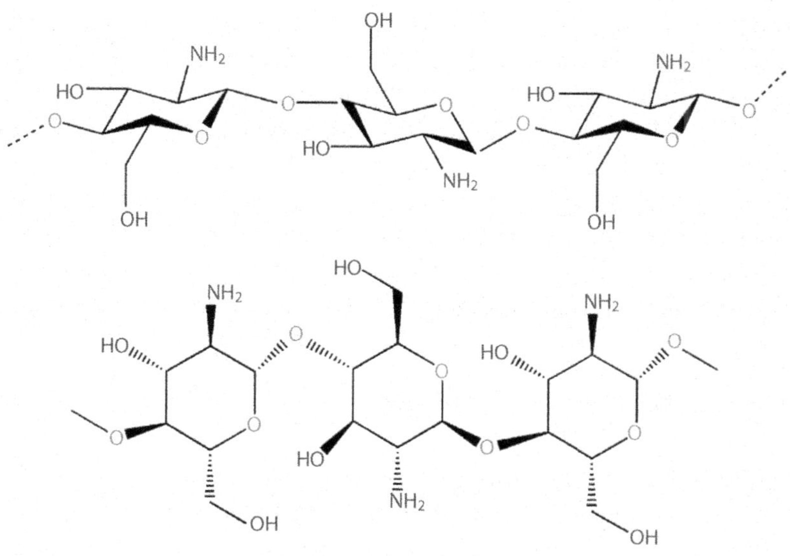

Figure 5.2 Chemical structure of chitosan.

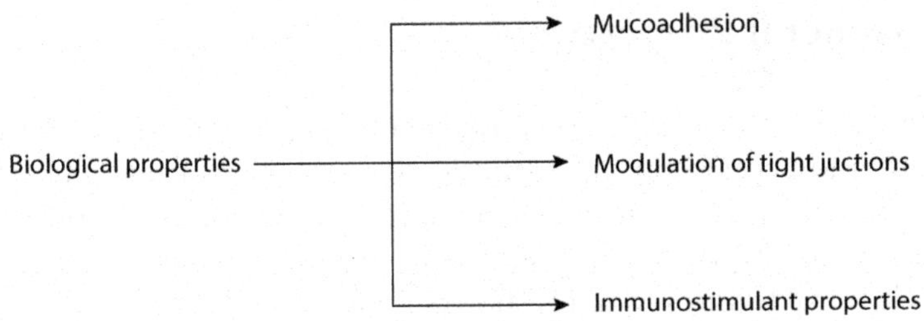

Figure 5.3 Biological properties.

5.5 Physicochemical Aspects

Chitosan is easily accessible in different classes of purity, degrees of deacetylation, chain lengths, salt-forms, charge densities, viscosities, and water retention values from various suppliers. The above stated properties mostly influence its physicochemical characteristics, by which all of its applications are regulated. Therefore, the choice of the appropriate class for its utility is associated to its utilizations.

5.6 Molecular Weight

Some of the significance of chitosan and its derivatives by the chemical and physical processes are still not known in detail, information gathered from evidences shows that most of their functional properties and physiological activities depend on their MW [4]. It is necessary to sufficiently describe the specific polymer manufactured due to the influence of MW range and polymer composition on the different physicochemical features. Some methods like gel permeation chromatography, light scattering spectrophotometry, and viscometry are required for determining the MW of chitosan [5]. Unlike monodisperse substances to delineate the sample numerically, no accurate MW is showed, but number of different means is defined.

5.7 Stability

As Ahmed *et al.*, [6] had given an evidence stating that their reaction is carried out without radical, but their outcomes was very much coincided with the research carried out by other scientist which is represented in Figure 5.4.

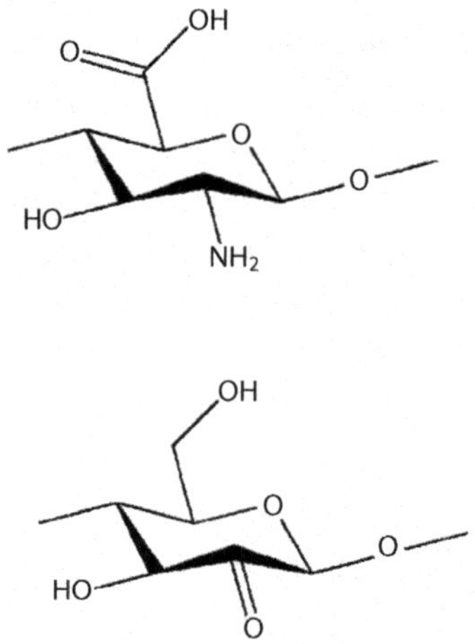

Figure 5.4 Chitosan under oxidative condition.

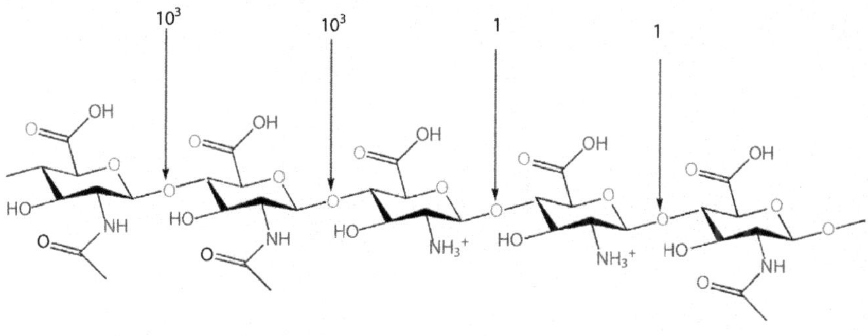

Figure 5.5 Order of monomer arrangement in acidic condition.

It is important to note that cleavage at amide monomeric unit forms three order of degree which is faster in comparison to that of amine monomeric unit. The result reveals that chance of random scission is less than that of the end group acetylated which is shown in Figure 5.5. The main cause for such outcome was the varied amount of water content and shifting SN_2 to SN_1.

5.8 Fabrication

5.8.1 Mild Acidic Conditions

It was capable to dissolve and converted into other forms in mild acidic media. The process is shown in Figure 5.6.

5.8.1 Other Acidic Conditions

The pH values below its pKa makes the chitosan positively charged one. Electrostatic forces undergo a significant task to interact among the protonated amines and the negatively charged DNA. Hence, chitosan polymers along with the negatively charged genes occurred complex nanostructures through the self-assembly process as shown in Figure 5.7.

5.9 Self-Assembly

5.9.1 Molecular Design

It possesses two sites such as hydrophobic and cationic which are already discussed making it fundamental amphiphile and polyelectrolyte. Due to the availability of two different groups –OH and $-NH_2$ in the backbone,

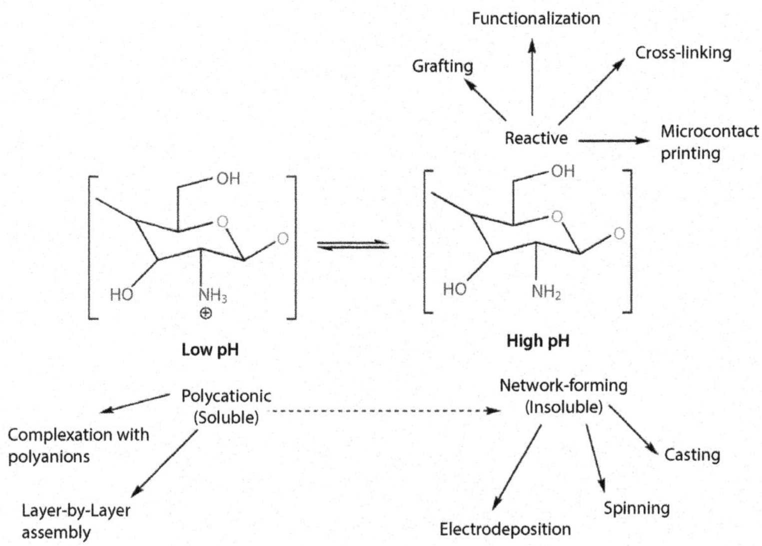

Figure 5.6 Schematic diagram of chitosan to form fabrication.

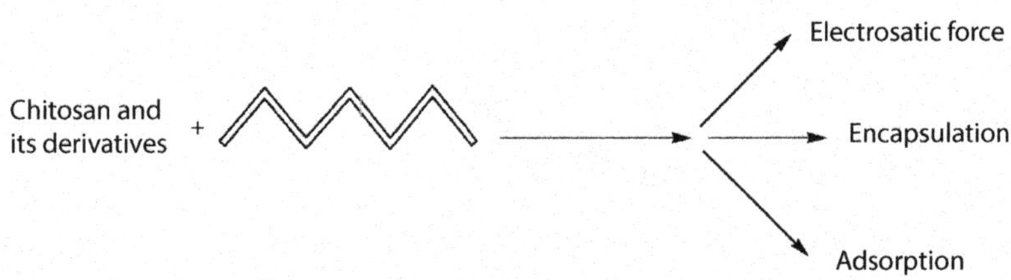

Figure 5.7 Various mechanism by chitosan based derivatives.

the hydrogen bonds such as intra and intermolecule lead for the self-assembling. The molecular weight and DA serve as a significant feature regulating in the self-assembly [7]. The low charge density and high hydrophobicity are the outcomes of high DA which is not suitable for hydrogen bond mode and electrostatic force mode of self-assembly [8]. Higher molecular weight shows steady and huge self-assembly [9].

5.9.2 Self-Assembly Mechanism

The self-assembly is categorized into two forms such as monocomponent and multicomponent. Monocomponent scheme is the form where self-assembly is obtained from either chitosan or its derivatives but in the multicomponent scheme one molecule at a time actively get involved.

5.10 Strategies Self-Assembly

5.10.1 Solution Self-Assembly

In the solution state, amphiphilic chitosan shows self-assemble as nanoaggregates in the form of polymeric micelles. In the case of higher solubility, self-assembly methods are attained by straight dissolution followed by continuous stirring [10]. It can easily dissolve in organic media. There are two factors which affect it, namely the pH and its ionic strength.

5.10.2 Template Self-Assembly

Most of self-assembles take place on the liquid–solid interface where the templates are induced. The template includes the various types of substances such as quartz, glass, etc. When the concentration of micelle is more than critical micelle concentration, the derivatives get self-assembled which is shown in Figure 5.8.

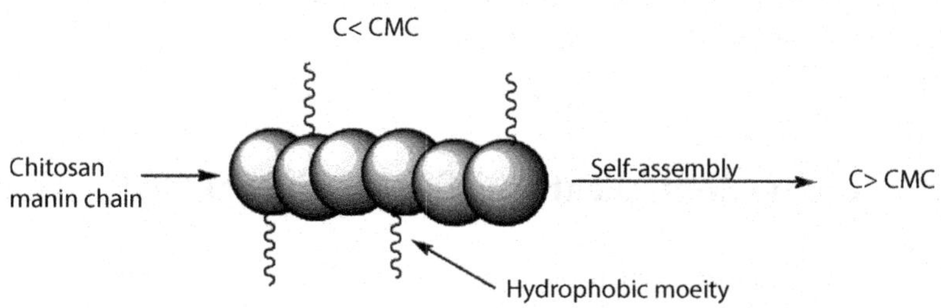

Figure 5.8 Self-assembly of amiphilic chitosan derivatives.

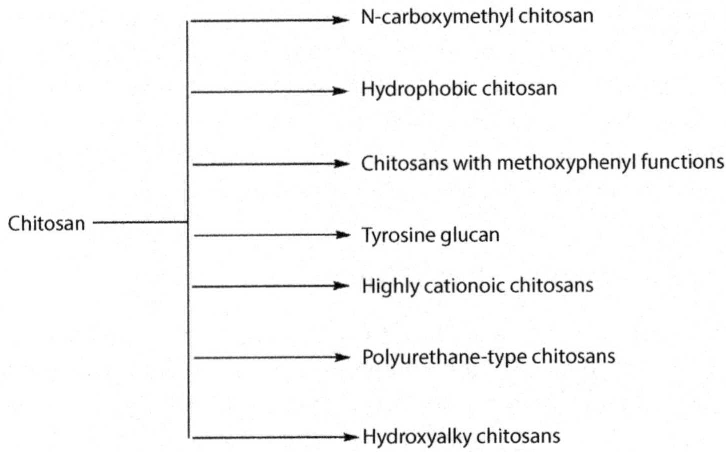

Figure 5.9 Chief significant of chitosan based material.

5.11 Chief Significance

The types of chitosan derivatives possessing significant role in biomedicine field are shown in Figure 5.9.

5.12 Various Forms

Chitosan shows versatile polymer which are extensively employed in various form. Different forms of chitosan along with the sub division and its uses are shown in Figure 5.10.

5.13 Chemical Modification

Till date various publications have been done on chemical modification of chitosan. But yet this polymer is continuously being modified yielding different derivatives with advanced features. The present work is mostly based on the studies related to chemical modification of chitosan as mentioned in Figure 5.11.

5.14 Technologic Features for Medicinal Utilization

The drugs broadcasted within the matrix are discharged through destruction or change of state and when they are surrounded by polymer, they are

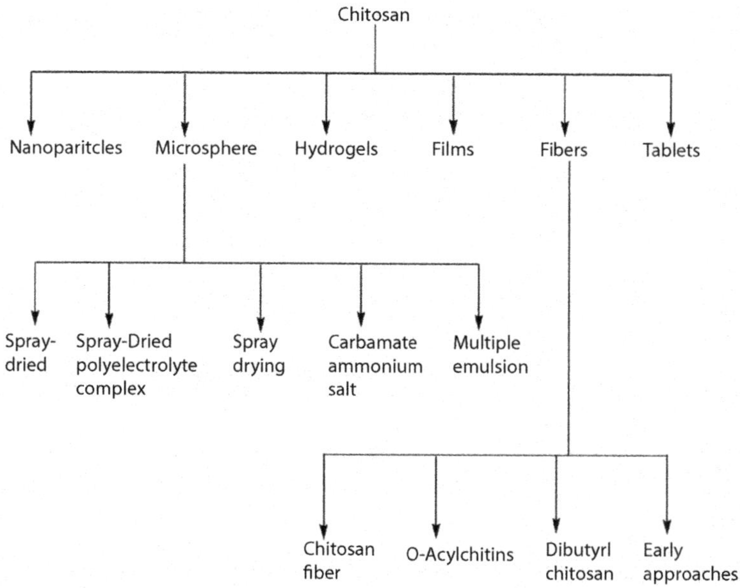

Figure 5.10 Various forms of chitosan.

discharged with the help of discharge through the process of diffusion. The general processes are listed in Figure 5.12.

5.15 Synthetic Procedure of Chitosan Nanoparticles

The chitosan nanoparticles are developed through the process of ionic gelation and coacervation by making the use of sodium tripolyphosphate. Figure 5.13 shows the different other straightforward process can be formulated that is from DNA and complexation process.

5.16 Modified Chitosan

To overcome the challenges and obtain appropriate various approaches were formulated for hydrophilic and hydrophobic alterations. The overall scheme of modified chitosan is shown in Figure 5.14.

5.17 Carboxymethyl Chitosan (CMC)

Carboxymethyl chitosan (CMC) is the modified form of chitosan manufactures by the process of esterification. The product so formed is utilized in the healthcare purposes and cosmetics which is shown in Figure 5.15.

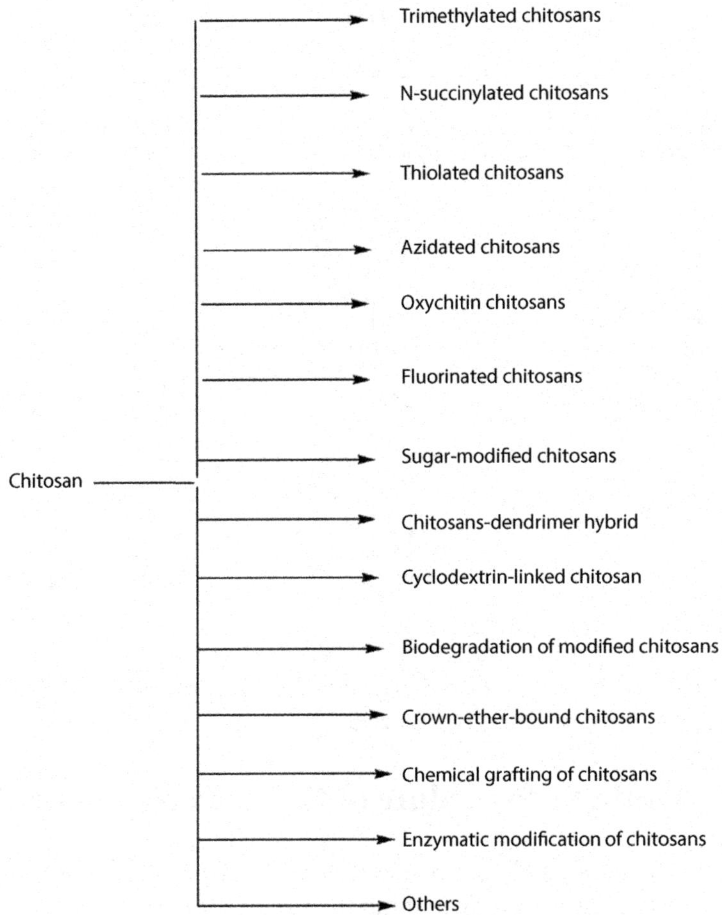

Figure 5.11 Chemically modification of chitosan.

5.18 Michael Reaction

Sashiwa *et al.*, [11] successfully composed the Michael reaction of chitosan with variety of reagents through the H_2O. The schematic reaction is given below in Figure 5.16. The scheme can be helpful for modification in preparing water-soluble derivatives.

5.19 Antioxidant

Chitosan possesses an ability to show antioxidant feature. The fungal chitosan is of two types, namely *B* and *C* which is composed from alkaline

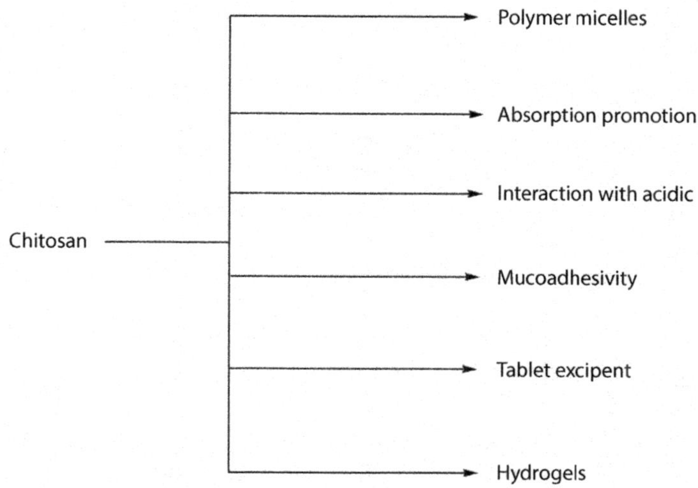

Figure 5.12 Medical utilization of chitosan.

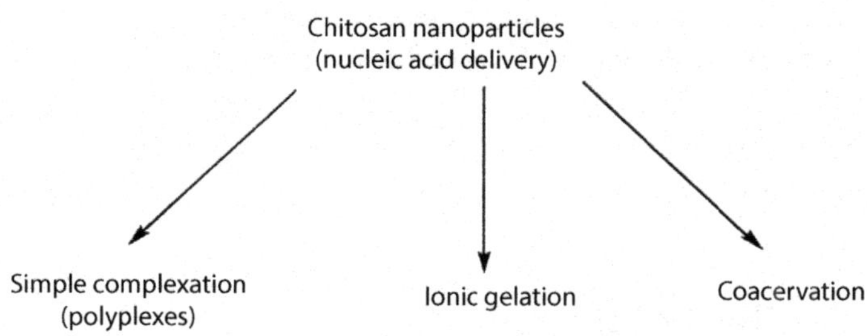

Figure 5.13 Development of chitosan nanoparticles.

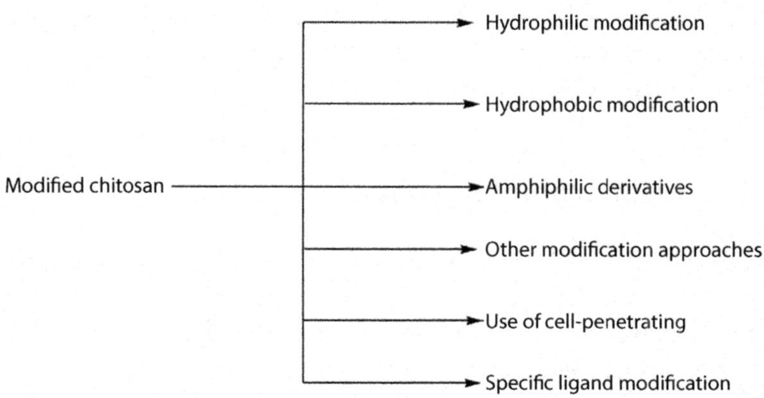

Figure 5.14 Overall modification of chitosan.

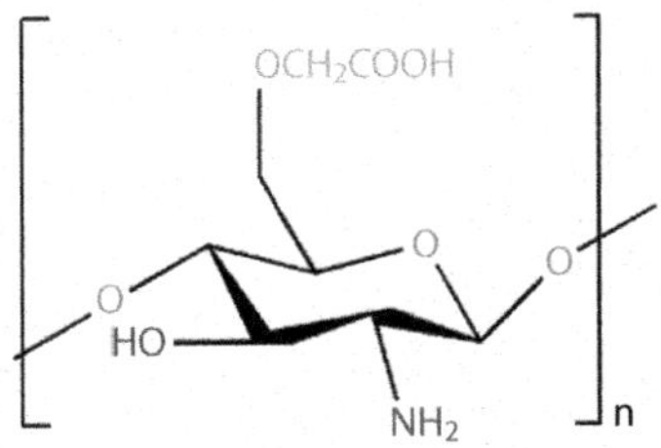

Figure 5.15 Chemical structure of carboxymethyl chitosan.

N-deacetylation as *B* and *C* at the time period of 1 h, 1 and ½ h, and 2 h respectively [12].

Wan *et al.*, [13] suggested about different mechanism of antioxidant and scheme is shown in Figure 5.17. It has been widely used for examine of antioxidant capabilities of chitosan molecule with HMW. It has been proved to be very suitable method in the field of medicine and food.

Human beings are continuously in contact with free radical from internal body metabolism and also from nearby area through ROS liberated in the active organism by metabolic activities. The oxidative stress may form involuntary enzyme activity as well as oxidative damage to macromolecules which in turns may cause cardiovascular, cancer, etc. The state is as shown in Figure 5.18.

5.20 Antibacterial Properties

The behavior of variety of food constituent, namely whey protein, oil, and starch explored on antibacterial outcome in present context. The coating on fruits and vegetables with chitosan was also done. The process indicated that gram-negative bacteria was sensitive toward chitosan which was variable to large extend [14].

5.21 Antimicrobial Activity

Chitosan is very well known as an antimicrobial agent in opposition to various bacteria and fungi, which has been quantified by a numerous of authors. The polycationic nature provides this elite property, to make possible its application fields, including food science.

It shows good response toward antimicrobial activity in comparison to chitin due to the positive charge on C-2 of glucosamine monomer to be less

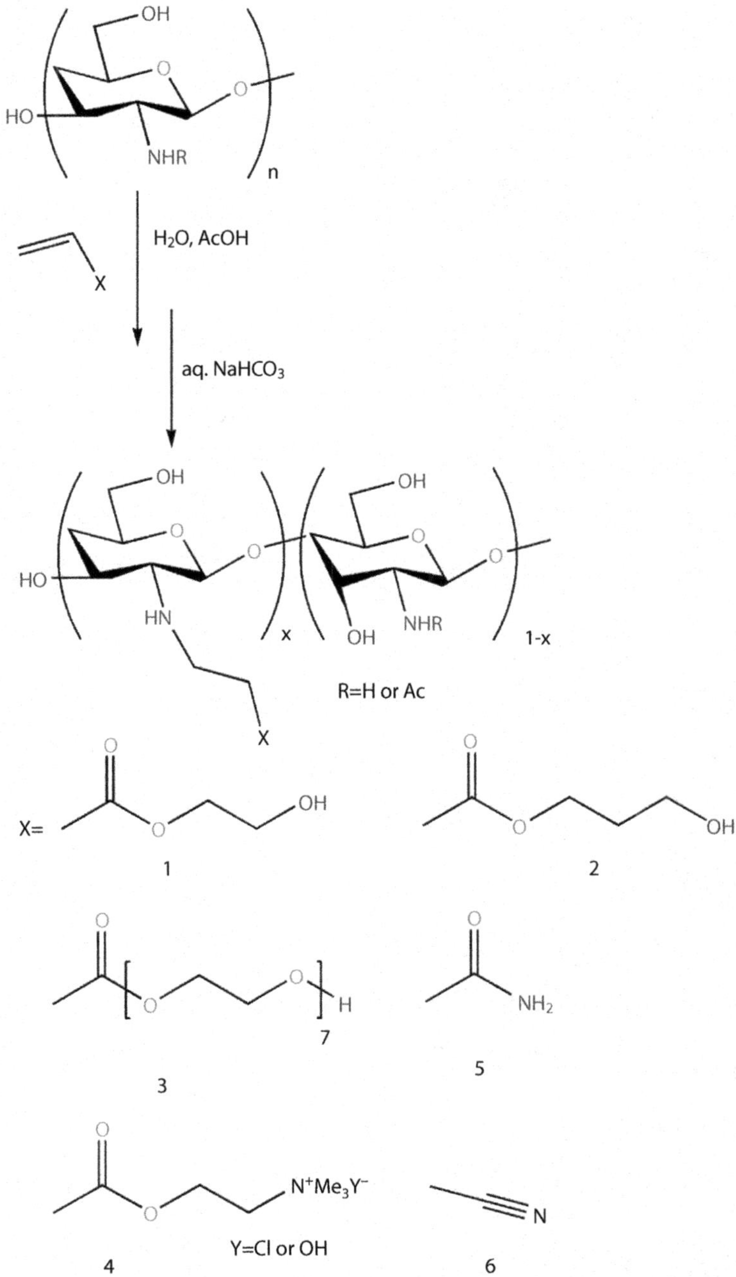

Figure 5.16 Michael reaction of chitosan.

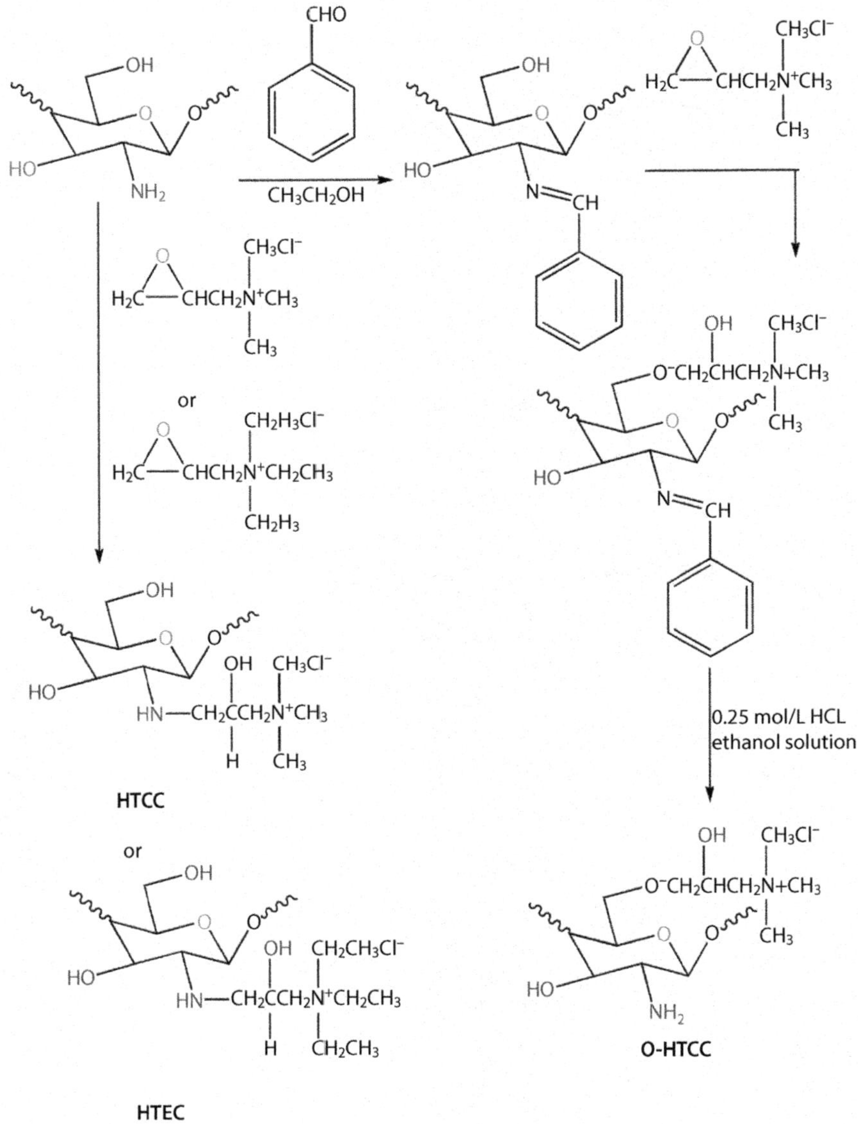

Figure 5.17 Synthetic path of *N,O*-quaternized.

than pH 6 [15]. But the precise antimicrobial mechanism of the biopolymer is yet to known completely. As the positive charge is interconnected to negativee charge of cell membrane, they form outflow of protein and various other fundamentals. Chitosan can have represented the development of *Aspergillus niger* accounted by Fang *et al.*, [16]. With the concentration

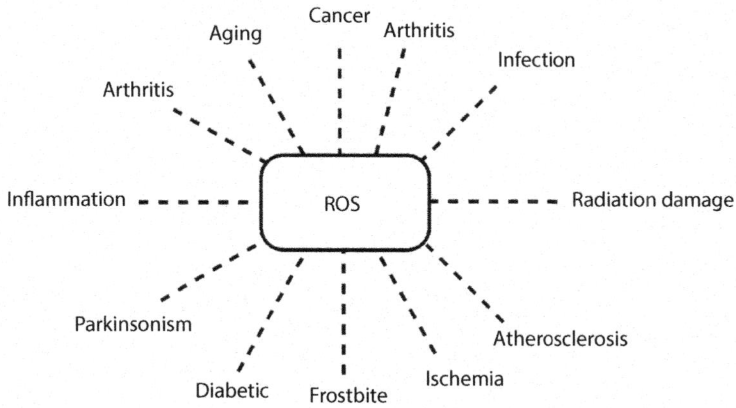

Figure 5.18 Reactive oxygen species.

of 5 mg/ml, the liberation of UV-absorption from *A. niger* and discharge of material containing protein was facilitated at pH 4.8.

5.22 Antiviral Activity

The antiviral activity could be done through different mechanism such by impulsion of resistance for viral disease in plants, infection in animals and it could be used for preventing accumulation of bacteriophages in the diseased cultures.

For extinguishing the viral infection in plant, the capability is not dependent on the kind of virus infected. Indeed, the chitosan possesses effect on plant by initiating resistance toward viral infection.

5.23 Biological Adhesive

On the initiation of azide and lactose moieties within the chitosan, the water solubility responded well at neutral pH which is extensively utilized in soft tissue in the form of biological adhesive. It acts as a photo-cross-linked material on the action of UV light liberating insoluble hydrogel on time limit of 60s.

5.23.1 Photo-Cross-Linkable Chitosan

The structure of photo-cross-linkable Az-CH-LA is shown in Figure 5.19. By the two-step condensation process, lactose and azide component was

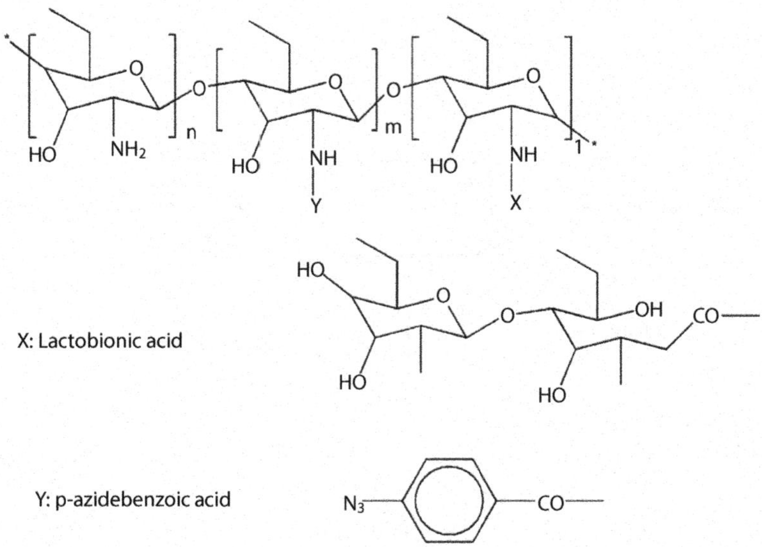

Figure 5.19 General structure of photo-cross-linkable-Az-CH-LA.

imported in chitosan where the molecular weight was 800–1,000 kDa, and degree of deactylation was found to be 80%. Ono *et al.,* [17] investigated about the photo-cross-linkable chitosan which was capable enough to be possessed as a novel tissue adhesive employed for pharmaceutical purpose.

5.24 Bonding Purposes

As the acidic pH, generally where the charge density is higher, the interaction between chitosan and adherend is strong [18]. The present scenario clarifies that various formulations have been carried out for the study. The cross-linking of different polysaccharide to the chitosan molecule is calculated to study the binding properties. The nature of binding purpose is shown in Figure 5.20.

5.25 Biodegradation

Biodegradability is an indispensable technique required for biological purposes. The chitosan molecule is resolved as enzymatic perversion and chemical in the human being. This technique of chitosan molecule is not clarified yet, but some of its process are undertaken to yield degradation fabrication which is shown in Figure 5.21.

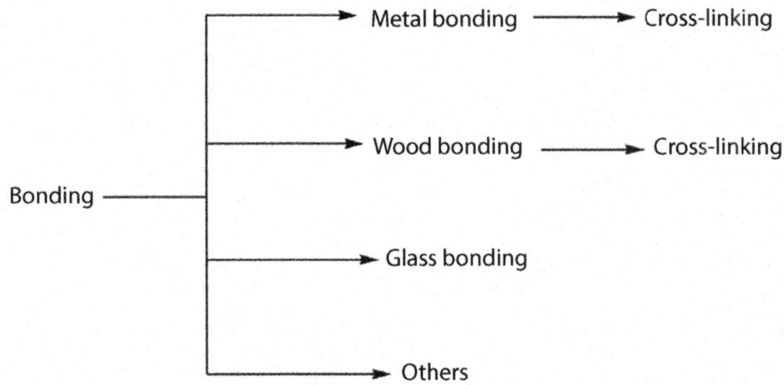

Figure 5.20 Binding purpose.

5.26 Parameter Moving Transfection Competence

Transfection is the mechanism to acquaint nucleic acid within the cell. There are various parameters affecting the transfection in the chitosan-based structure such as availability of serum. They are communicated with maternal management. The transfection features are dependent on various factors as shown in Figure 5.22.

5.27 Conjugation

The important aspects of tyrosinase-initiated are shown in Figure 5.23. It reveals that the range of tyrosinase's substrate is not obstructed toward the phenol of low molecular weight. The enzymes oxidize protein residue in tyrosine. Hence the above study makes it clear that tyrosinase stimulates the activity of protein for assembling into the chitosan through conjugation.

5.28 Functionalization of Chitosan

It is well known facts that chitosan is good absorbent and act chelating agent, it is a primary importance to propose strong ligand for binding the metal for catalytic execution.

5.29 Schiff's Base Formation

The reaction of aldehyde to produce imine is the moderate and wise route for functionalizing the primary amine. The process is applied for the

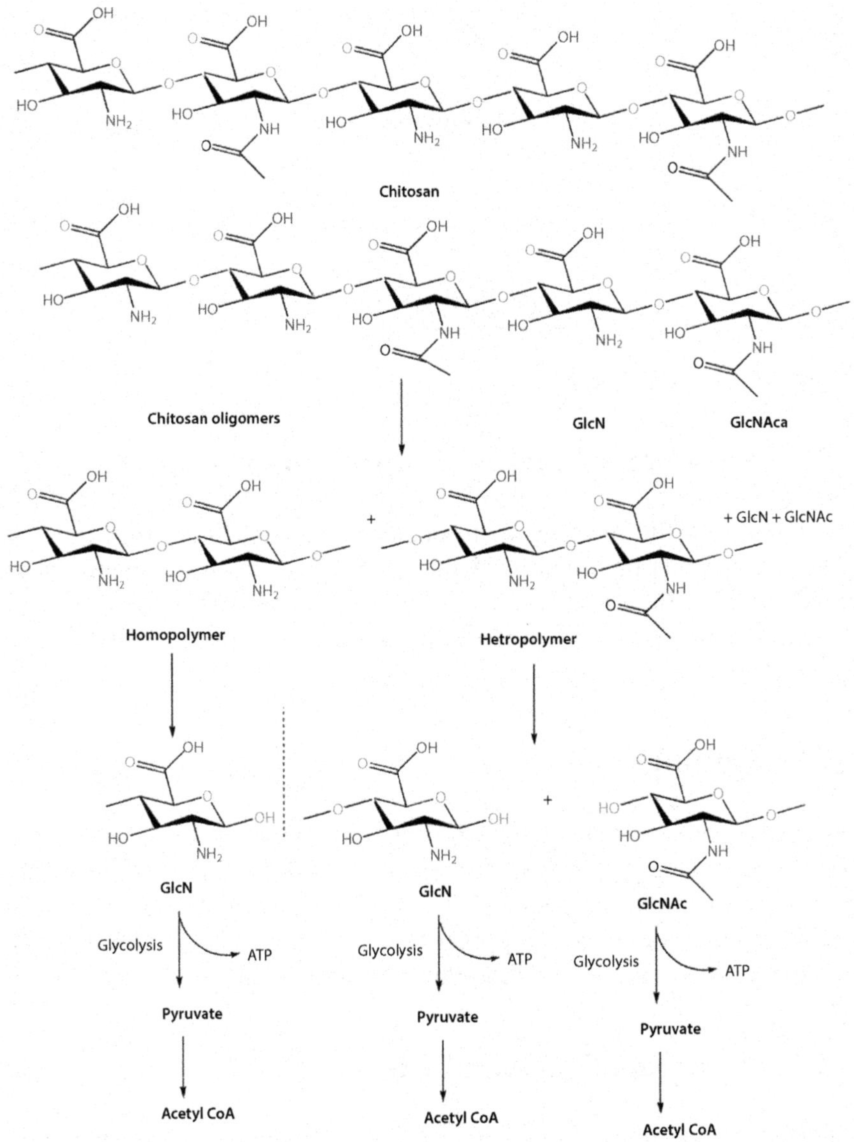

Figure 5.21 Chitosan biodegradation.

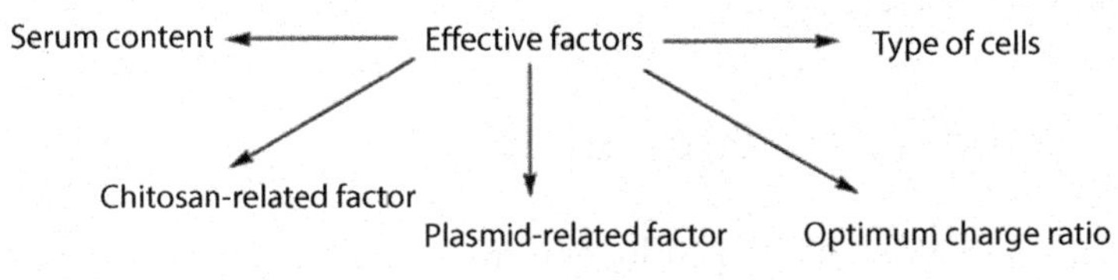

Figure 5.22 Factor affecting of chitosan based.

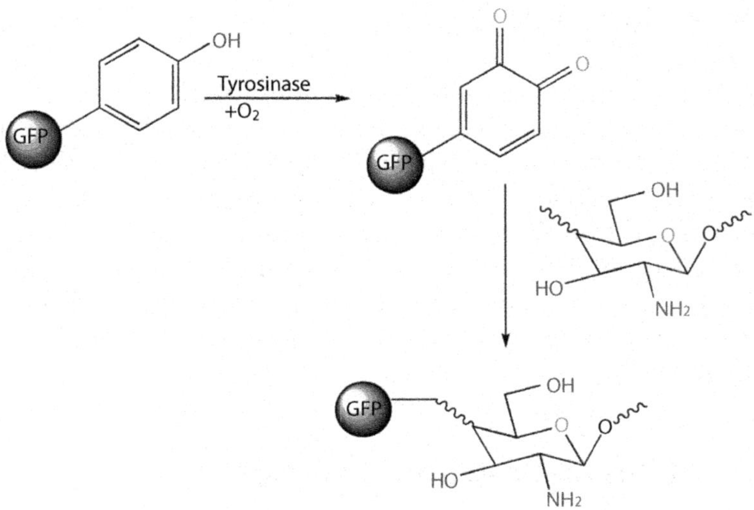

Figure 5.23 Enzymatic assembly reaction.

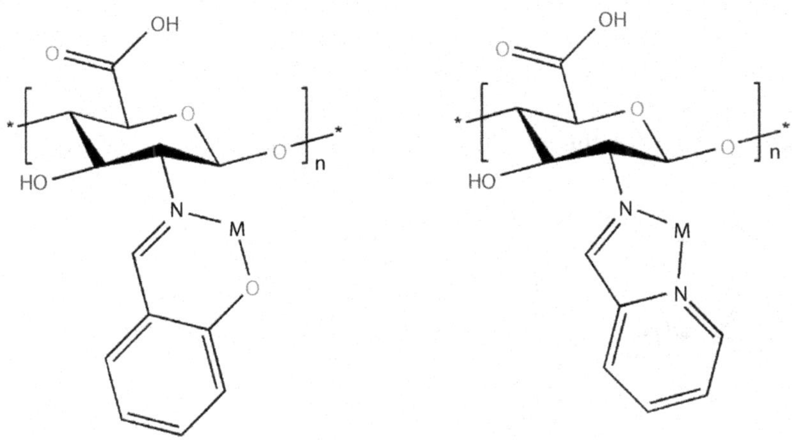

Figure 5.24 Chitosan via the imine linkage.

affinity of groups such as aromatic aldehyde and is also used for preparation of catalyst and adsorbent, adjusting the solution phase and solubility feature of the polymer.

The mechanism of bidentate complex was much better than the straightforward form of C-M complex which is shown in Figure 5.24. The cross-linking of glutaraldehyde is the significant method in the study of chitosan which forms constant derivative since amine groups in chain binds together through the di-imine.

5.30 Reductive Amination

The reductive amination process liberates hydrolytically constant amine in spite of the sensitive imine linkage. Efforts have been undertaken to elaborate the area of study where addition of water molecule in the reduced amine linkage provide a good behavior. The degree of substitution was attained to be 0.27 in all the groups which is as shown in Figure 5.25.

Strasdeit *et al.,* [19] developed the sulfide derivatives consisting of N-CH_2-S by employing the same technique. The outcome of the study binded Cadmium in the availability of nickel and zinc salts at pH 3 which is given in Figure 5.26.

5.31 Chitosan–Proteins Interaction

The study of the binding energetics among the chitosan and protein molecule was performed at pH 6.0 and was noticed that preparation temperature responded positively toward colloidal constancy of molecules. Kasimova *et al.,* [20] were successful in relating two particular temperatures. The rapid re-organization for equilibrium attained after injecting revealed that binding process does not show any effect due to slow aggregation. At the room temperature, the injecting activity was exothermic in nature and the

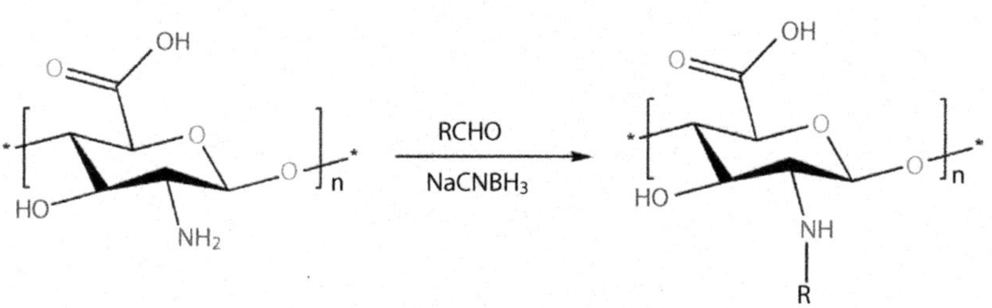

Figure 5.25 Amination of chitosan.

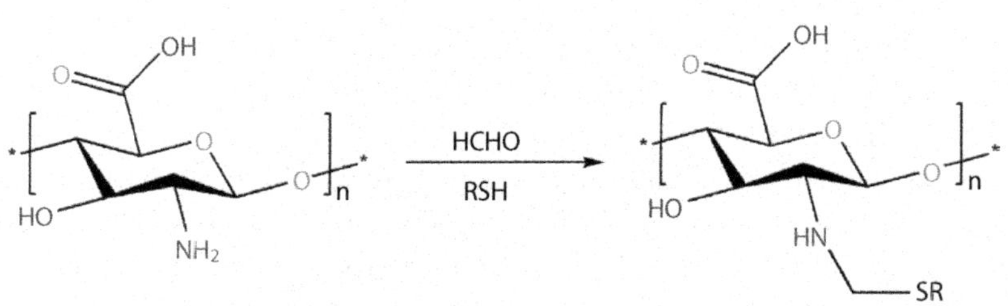

Figure 5.26 Chitosan via the Mannich.

value obtained was very low but at 50 °C, the stating rate of exothermic reaction was noticed along with the endothermic nature as well. The outcome made it clear that the above effect so not responded toward positive heat of dilution rather it was mainly due to thermally induced changes which finally lead to the formation of other protein binding sites.

5.32 Absorption Enhancer

Chitosan is regarded good absorption enhance for the hydrophilic drugs in two part of the body such as intestinal and nasal mucosa. Schipper *et al.*, (1996) suggested about the effect of molecular weight and composition on the poisonousness and intestinal penetrability by making the use of Caco-2 cells and 14C mannitol. The chitosan possessing high molecular weight and low DA acted as energetic absorption enhancer. On the other hand, the dose dependent toxicity was observed [21].

5.33 Chitosan–Starch Blends

As chitosan and starch show exceptional features possessing high benefits. It indicates that combination of starch–chitosan material form eco-friendly film and it can be high enhanced the features such as low water permeability and antibacterial activity in comparison to film of starch in absence of chitosan molecule.

Since the chitosan molecule is present in large amount in coating solution when compared with film, it states that antibacterial mechanism is dependent on technical purpose. The solubility and permeability of starch polymer are lessened on addition chitosan molecule. The starch-based film shows physical properties such as tasteless, transparent, etc., which is similar to the synthetic polymer [22].

5.34 Application in Various Fields

5.34.1 Administration Routes

The factors for designing ideal junction method for delivery are described in the context. Firstly, for the fruitful treatment of diseases, the therapeutic dose and delivery rate should be acknowledged. Later the system for delivering the particular dose in accurate time period needs to be constructed.

To attain the high material ratio of drug, the system shows high loading ability for the desired drug.

Secondly, the mode of enforcement should be of primary important including implantation as film, injection as microsphere. Lastly, the circumstances for site organization should be assayed along with the volume and pH. The gel and film accomplished for delivery of drug need to produce distortion. Hence, the calculation of mechanical power of the implant at the condition of pH that is 7.4 and room temperature is to be done for indicating the system to sustain *in vivo*. To avoid damage of nearby tissue the softness of the tools used should be regularly measured.

5.34.2 Drug Delivery

Chitosan has been vastly utilized. For the different quality to this polymer (comparison with an acidic biopolymer) the amino groups are accountable. The depiction of polymer of both chitin and chitosan is very essential following to the structure-properties connection, to define an attainable industrial application.

5.34.3 Oral Route

The most significant system in the field of drug delivery is the oral transformation where chitosan is considered for oral mucosa and colon to attain better drug delivery.

Oral mucosa is possible for decline in gastrointestinal location. The use of oral mucosa is restricted due to low penetrability and short time. It is reflected as multiversatile polymer related for gastric retention in drug delivery and is appropriate for stomach based infections [23]. The oral regulation of chitosan is quite secure for human being. It has an ability to join bile acids in the form of chitosan orotate salt [24].

5.34.4 Nasal Route

Nasal delivery has been served to gain consideration for complete drug route system owing to the features of fast arrival of performance other than injection, wide surface area, and low enzymatic performance.

Chitosan is employed for intranasal absorption ability. The absorption is impact by the structure as well as molecular weight. The nasal tissue renders good absorption ability as it is extremely vascularized. Protein absorption and peptide drugs in the nasal tissue are strengthen through chitosan molecule [25].

5.34.5 Cancer

Cancer is the disease including abnormal growth of cell invading to various body parts. The most common site of cancer in humans is lungs, stomach, breast, etc. In the presence situation of cancer prominent world, the nanomedicine plays a significant role by assimilating nanotechnology, medicine and biology aspect for detecting and eliminating the cancer at their early stage to kill or restrict the tumor cells.

Mansur *et al.*, [26] analyzed a colloidal nanohybrid structure which was created by combining CdS and anti-CD20 showing bioaffinity ligands to detect NHL cancer. The system is widely used for meeting the various tasks in area of oncology and nanotechnological aspect.

5.34.6 Neurosurgeries

Neurotmesis, the critical class of nerve damage, is responsible to diverting proximal and distal nerve stems. The nerve remedial and regaining could not be possible without the improvement of anatomical endurance [27]. Hence, it is the first priority to reconnect the serious peripheral nerve for promoting the patient health for treatment of transected nerve.

The Az-chitosan is cytocompatible and nonpoisonous in nature. The diagram of the process is shown in Figure 5.27. The workout was carried to elevate the gel mechanical features and estimates the competence of Az-chitosan adhesive [28].

5.34.7 Skin Repair

Skin is made up of 3 inter-related layer epidermis, dermis, and subcutaneous fat. Epidermis the outer most layers is stratified enclosed as barrier and it consists the layers of keratinocytes at stages of differentiation as well as melanocytes. The dermis is the next layer separated through basement membrane. The subcutaneous fat layer is attached the dermis to muscles and bones.

Tchemtchoua *et al.*, [29] carried an experiment and provided the evidence which revealed the significance of nanofibrillar structure. It stated that the device showed biocompatibility and was capable enough to be effective in wound healing. The Figure 5.28 showing skin repairing process is given below:

5.34.8 Gene Delivery

To go through the concept of molecular foundation of infection, circumstances are put forward for modulating and restraining the dysfunctional

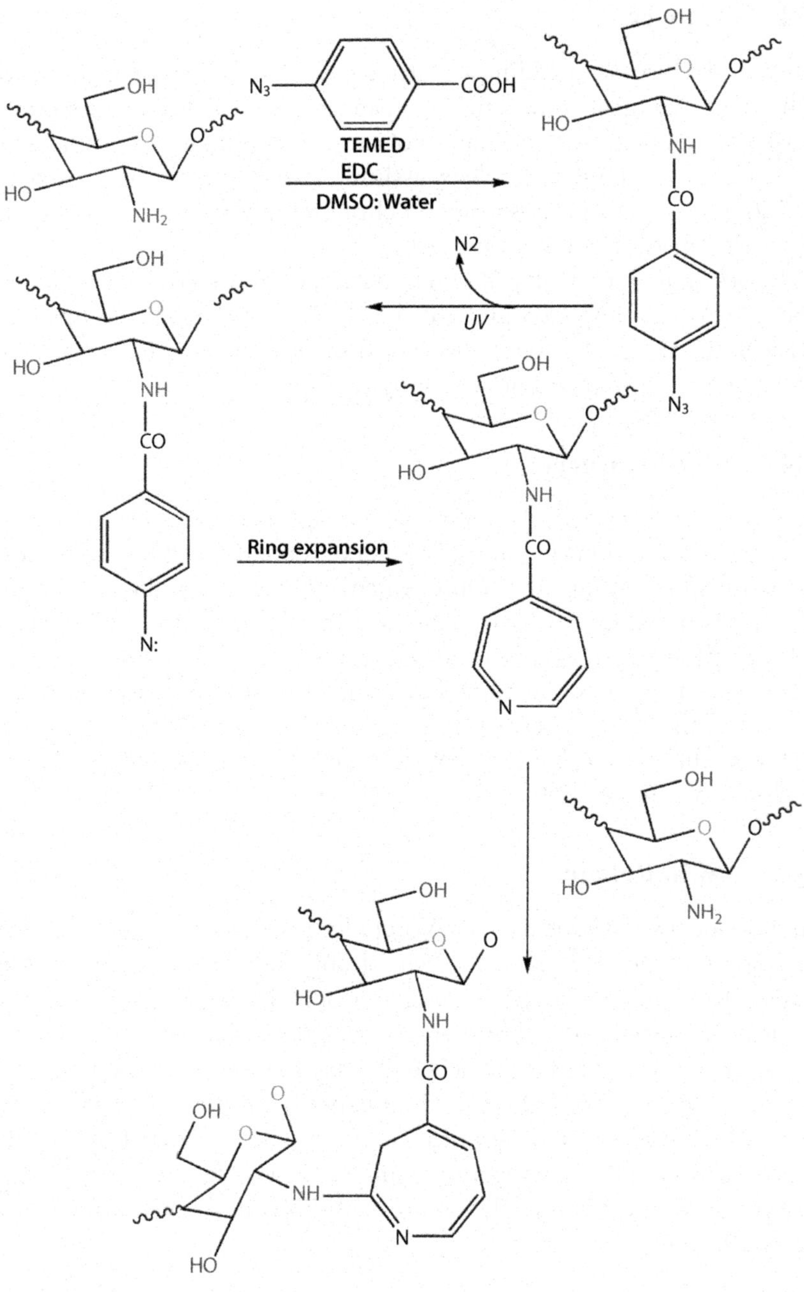

Figure 5.27 Diagram of Az-chitosan chains.

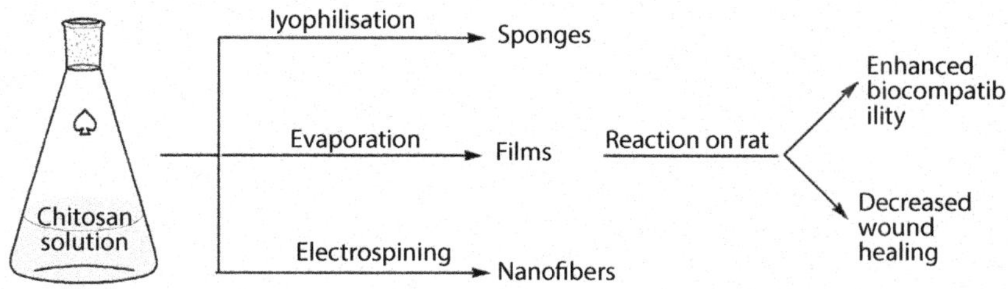

Figure 5.28 Schematic diagram of skin repair.

gene. Hence, particular vector is framed for transfection of the target cell *in vivo*. In the last decade, it is extensively noticed that cationic polymer is capable to form complex DNA which is an ample proof that it is used a nonviral vector. Felbner *et al.*, [30] suggested involvement with DNA *in vitro* contribute to the electrostatic self-assembling of polymer based DNA complexes.

5.34.9 Gastric Ulcers

Meshali *et al.*, were successful in examined the influence of chitosan molecule on the gastric ulcer formed by nonsteroidal anti-inflammatory drugs [31]. It showed positive response as it could decrease the ulcerogenic mechanism of oxyphenbutazone and glafenine noticed in albino rats. The solubility in acidic state, demulcent and strengthening activity of stomach mucosa was responsible for showing rebate in ulcerogenic mechanism of acidic and basic drug. Acikgoz *et al.*, [32] suggested about the capacity of chitosan molecule responsible for weakening of gastric mucosal disease linked to the working of diclofenac sodium.

5.34.10 Wound Dressing

The Alginate fibers are widely applied for would dressing owing to the ion-exchange and gel forming competency. The insoluble Ca^{2+} salt, that is, alginate fiber in contact with wound effluence leads to replacement of Na^{+} ion in wound effluence along with the Ca^{2+} ion [33].

The most common thing we noticed is that hydrolyzed chitosan fragment in the alginate fiber to enlargement which other hand build up the tensile feature of fiber to be used in wound dressing. The ionic interaction is shown in Figure 5.29 [34].

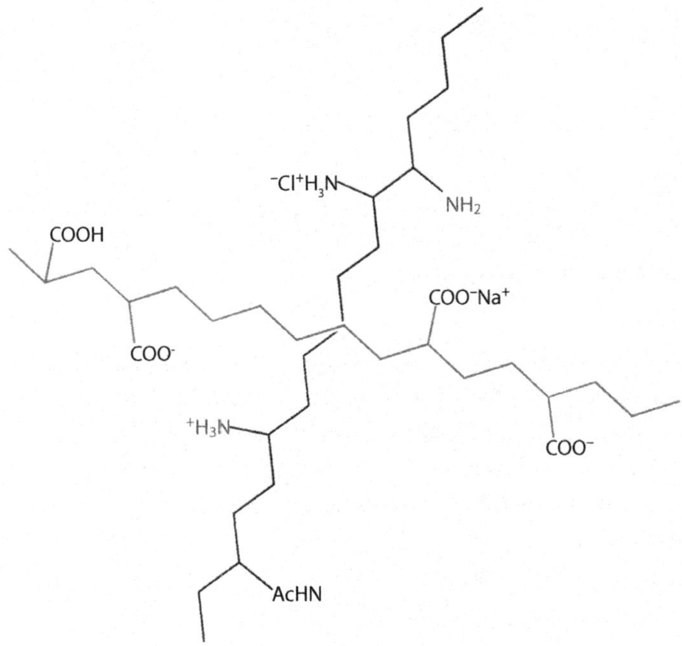

Figure 5.29 Ionic interaction.

5.34.11 Dye Adsorption

With the enhancement of textile industries, the use of dyes had aided to be the major source of water pollution where the discharge of effluents in the form of colorant has proved to be threat for human health and marine species as well. The dyes are composed of compounds such as chromophores and auxochromes where they possess an ability to diagnose the dye color and the color intensity respectively [35].

Zhu *et al.*, [36] developed an advanced bead of chitosan merged with kalolin and g-Fe_2O_3 which showed adequate absorption ability by adsorbing around 70% of methyl orange in acidic condition.

5.34.12 Biosensors

The study and research for biosensor have become significant by various researcher as it includes variety of good features such as miniaturization, movability, and less expensive when compared with other methods. Actually, biosensor is an analytical tool for detecting and converting biological responses, heat and light into electrical signals as analog and digital demonstrative.

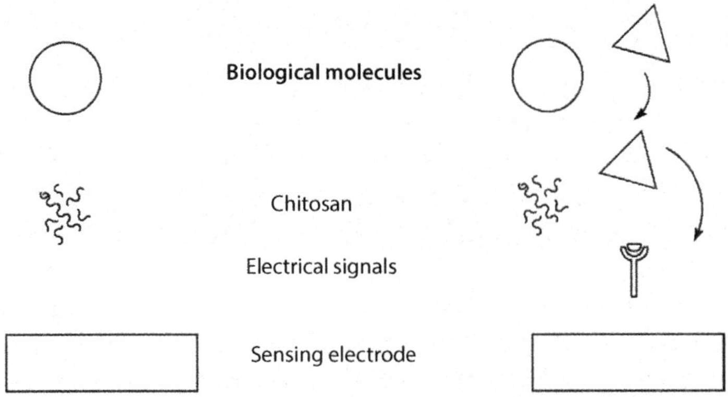

Figure 5.30 Schematic diagram of chitosan-based biosensor.

GOD enzyme along with glucose magnitude is responsible for clarifying to detect electrochemical signal transduction.

$$\beta\text{–D-glucose} + \text{GOD/FAD} \longrightarrow \beta\text{–D-glucose-}\delta\text{-lactone} + \text{GOD/FADH}_2 \quad (1)$$

$$\text{GOD/FADH}_2 + O_2 \longrightarrow \text{GOD/FAD} + H_2O_2 \quad (2)$$

$$\text{D-glucose-}\delta\text{-lactone} \xrightarrow{H_2O} \text{D-gluconic acid} \quad (3)$$

There are various factors due to which chitosan is used for biosensor activities such as surface modification, inert ability, gel formation, linking capability toward protein, less expensive etc. The schematic diagram of chitosan-based biosensor is shown in Figure 5.30.

5.34.13 Electrochemical Sensor

Zhang *et. al.* [37] developed carbon nantubes-chitosan (CNT-CHIT) biocomposite system which stands for the growth of electrochemical sensors. They utilized novel nanomaterial CNT which shows interesting intrinsic features with enhanced electrochemical action. The CNT-CHIT system investigated by scientists displays a basic and utilitarian way to deal with the mix of dehydrogenases and terminals, which can give scientific access to a vast gathering of proteins for extensive variety of bioelectrochemical purposes.

5.34.14 Enzymatic Assembly

The use to enzymes is limited as most of the material does not show response toward the enzyme which makes the synthetic polymer to be nonbiodegradable. The enzymes assigned for biofabrication is in the form of substrate.

The motor proteins are assimilated in the microfabricated equipment but the protein is not used for fabrication technique. However, the protein entitled is used as interesting attempt for executing biological system to infuse the energy as ATP into mechanical activity [38]. It shows the capability of molecular biology to limits the enzymatic assembly as mentioned in Figure 5.31.

The large extent of phenols is converted into o-quinonseses by the action of oxidative enzymes known as Tyrosinases. It is very easy to use tyrosinases enzymes because molecular oxygen acts as an oxidant and complex cofactors are not mandatory for the use. The o-quinone liberated from tyrosinases reacts with amine to form chitosan derivatives as grafted and cross-linked one [39]. The schematic reaction is shown in Figure 5.32.

5.34.15 Electrochemical Enzyme Immobilization

The Immobilization technique accomplishes the modification in the coordination sphere of Ru atom in the matrix of $Ru(bpy)_2X_2)$ as bpy is bipyridine and X is chlorine and water, respectively. It takes place on the cathodic reduction leading to restoration of the weak ligand that is chlorine or water

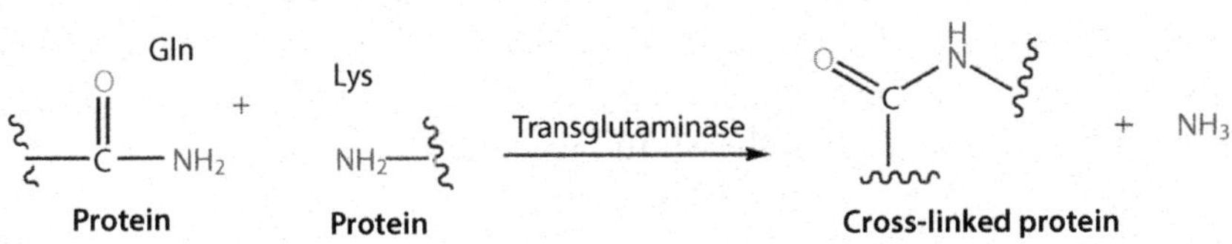

Figure 5.31 Role of enzyme activity.

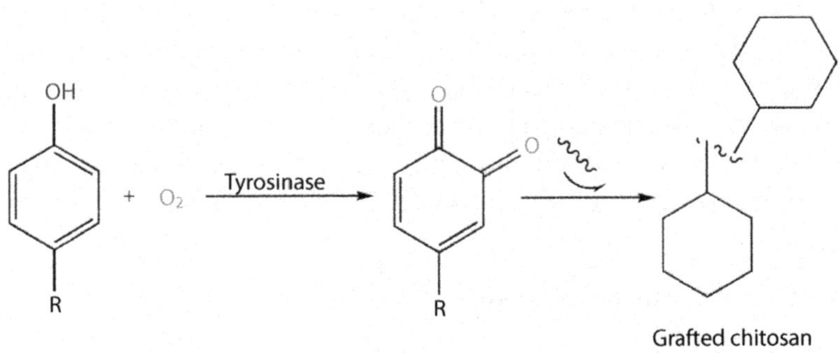

Figure 5.32 Reaction of tyrosinase.

through strong ligand as amino group present in chlorine and protein. Figure 5.33 exhibits the interconnection among the free diffused Ru centers and amino group and enzyme. It also reflects the application of the interaction for forming 3D network with enzyme.

5.34.16 Plasma Treatment in Textile

Plasma treatment is an innovative and environmentally balanced method carried out for the reinforcement of textile. It modification in this field have achieved credit in the textile industry owing to its features such as less

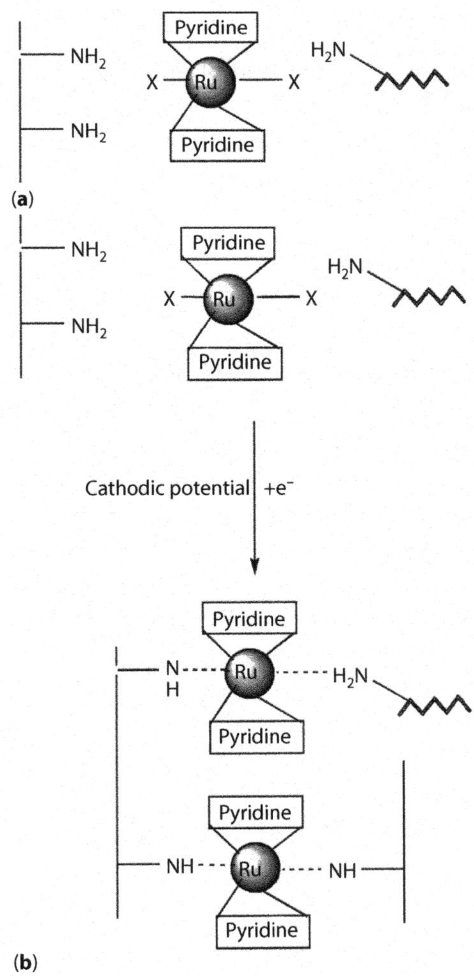

Figure 5.33 Enzyme immobilization.

energy utilization negligible use of chemicals, environmentally balanced techniques in comparison to other converting methods.

Uygun *et al.*, [40] utilized RF hydrazine modifying the nanopowder of chitosan molecule. In their experiment they investigated that RF hydrazine showed tremendous behavior toward the antibacterial activity in contrast with gram positive strains.

In some of the dyeing textile industry, the low temperature action was performed where the influence in availability acting as a caustic exhibited outstanding outcome which is employed as supplement for metallic caustic.

5.34.17 Fiber Technology

In the era of development in the field of fiber technology, the most interesting area of research in bioactive textile is applying biopolymer initiate alginate and hyaluronan. The structure and configuration of hyaluronan are shown in Figure 5.34.

5.34.18 Viral Infections

5.34.18.1 In Plants

The studies carried out in the year 1989 revealed that variety of necroses occurring through AMV would be deceases by treating bean leaf with the solution containing chitosan [41]. It was clear from the investigation that chitosan shows viral infection [42]. It was quite noticeable point that when chitosan was treated on plant chenopodiaceae, solanaceae, and amaranthaceaeetc, it was able to present infection with virus showing their structure and genomic expression. Therefore, it can be concluded that chitosan abolishes the infection independent on the form of virus.

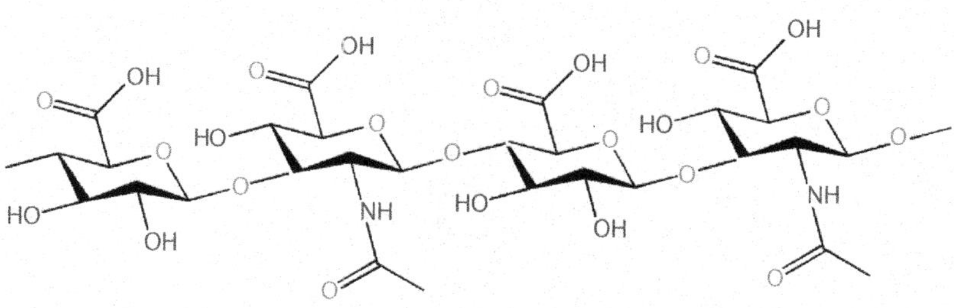

Figure 5.34 The structure of hyaluronan.

5.34.18.2 In Animals

The information related to chitosan stated that it regenerates the functional mechanism of auxiliary cell in the immune response, namely macrophages. Since the subcutaneous implantation takes place chitosan is responsible for chemotaxis in dog macrophages and it produces nitric oxide. It is even engaged for motivating leukouytosis at peripheral blood.

The consequences activity on the different blood cells well as constituent of the immune method was investigated where it was able to produce neutrophil chemotaxis in dogs [43]. Molecular weight was related through straightforward and complement mediated chemotactic functions and when it was compared with chitin, it showed higher responsiveness [44].

5.34.19 Seafood Industry

With an increase in the demand and needs of the people, the nutrition and fitness issue has become significant. Therefore, human beings are mostly focusing on marine food as they believe that sea food is healthy as well as complete diet for them. Seafood is regarded as exceptional food with high protein content and lipids along with the more quantity of PUFAs. The nutrient leads to the development in securing the human health as it overcomes the cardiovascular infection as well as hypertension. The marine foods are good sources which can be digested easily as it contains all the vital nutrients. Currently, the sea food is recommended as nutraceutical as well as the functional food. As stated by Japan in year 1980, functional food is one which has numerous benefits showing variety of function at a time in human being [45].

Chitosan molecule possesses variety of features, namely gel improvement, wound dressing, etc., acknowledging such properties, it can be used in seafood products to enumerate the quality and improve the nutrition.

5.34.20 Wastewater Treatment

Due to high amount of functional group such as hydroxyl and amino, chitosan molecule acts as a high adsorbent which is regarded as good application in drinking and waste water and also in industry. This polysaccharide, chitosan a part of chitin indicates high capable application in terms of dye adsorption and metal ions. Thus, it is a potential material for removing organic and inorganic effluent from drinking and waste water.

5.35 Conclusion

Chitin forms the significant constituent in most of living organism occurring everywhere in the biosphere where chitosan is the commercially formed derivatives established for various purposes in the field of pharmacology, agriculture, food, and industry. Many outstanding literatures have been notified till date supporting the progress of the polymer for preparing deport system for health related issue. Hence, chitosan is regarded as an assuring and appealing polymer for framing a progressive depot-based drug delivery route. For the current text, it is clear that immobilization of enzyme is not known completely where as it possesses tremendous of application in the future development. Till date work based on chitosan and its capability is not flourished to last film, membranes, fiber, etc. But in this field we need to enhance the research work, as catalytic reactors are facing the challenges on worldwide.

References

1. Agrawal, P., Strijkers, G.J., Nicolay, K., Chitosan-based systems for molecular imaging. *Adv. Drug Deliv. Rev.*, 62, 42, 2010.
2. Yi, H., Wu, L-Q., Bentley, W.E., Ghodssi, R., Rubloff, G.W., Culver, J.N., Payne, G.F., Biofabrication with chitosan. *Biomacromolecules*, 6, 2881, 2005.
3. Arfin, T., Mohammad, F., Chemistry and structural aspects of chitosan towards biomedical applications, In: *Natural polymers: derivatives, blends and composites*, S. Ikram., S. Ahmed, Eds., pp. 265–280, Nova Publishers, New York, 2016.
4. Rabea, E.I., Badawy, M.E.-T., Stevens, C.V., Smagghe, G., Steurbaut, W., Chitosan as antimicrobial agent: applications and mode of action. *Biomacromolecules*, 4, 1457, 2004.
5. Kumar, M.N.V.R., A review of chitin and chitosan applications, *React. Funct. Polym.*, 46, 1, 2000.
6. Ahmed, G. A.-W., Khairou, K. S., Hassan, R. M., Kinetics and mechanism of oxidation of chitosan polysaccharide by permanganate ion in aqueous perchlorate solutions. *J. Chem. Res., Synop.*, 4, 182, 2003.
7. Payne, G.F., Biopolymer-based materials: the nanoscale components and their hierarchical assembly. *Curr. Opin. Chem. Biol.*, 11, 214, 2007.
8. Yu, S., Hu, J., Pan, X., Yao, P., Jiang, M., Stable and pH-sensitive nanogels prepared by self-assembly of chitosan and ovalbumin. *Langmuir*, 22, 2754, 2006.
9. Wang, Y., Ke, C.-Y., Weijie, B.C., Liu, S.-Q., Goh, S.-H., Yang, Y.-Y., The self-assembly of biodegradable cationic polymer micelles as vectors for gene transfection. *Biomaterials*, 28, 5358, 2007.

10. Hammer, M.U., Anderson, T.H., Chaimvich, A., Shell, M.S., Israelachvili, J., The search for the hydrophobic force law. *Faraday Discuss*, 146, 299, 2010.
11. Sashiwa, H., Yamamori, N., Ichinose, Y., Sunamoto, J., Aiba, S.-I., Michael reaction of chitosan with various acryl reagents in water. *Biomacromolecules*, 4, 1250, 2003.
12. Yen, M.T., Tseng, Y.H., Li, R.C., Mau, J.L., Antioxidant properties of fungal chitosan from shiitake stipes. *LWT Food Sci. Technol.*, 40, 255, 2007.
13. Wan, A., Xu, Q., Sun, Y., Li, H., Antioxidant activity of high molecular weight chitosan and N,O-quaternizedchitosans. *J. Agric. Food Chem.*, 61, 6921, 2013.
14. Devlieghere, F., Vermeulen, A., Debevere, J., Chitosan: antimicrobial activity, interactions with food components and applicability as a coating on fruit and vegetables. *Food Microbiol.*, 21, 703, 2004.
15. Chen, C.S., Liau, W.Y., Tsai, G.J., Antibacterial effects of N-sulfonated and N-sulfobenzoyl chitosan and application to oyster preservation. *J. Food Prot.*, 61, 1124, 1998.
16. Fang, S.W., Li, C.F., Shih, D.Y., Antifungal activity of chitosan and its preservative effect on low-sugar candied kumquat. *J. Food Prot.*, 57, 136, 1994.
17. Ono, K., Saito, Y., Yura, H., Ishikawa, K., Kurita, A., Akaike, T., Ishihara, M., Photo-cross-linkable chitosan as a biological adhesive. *J. Biomed., Mater. Res.*, 49, 289, 2000.
18. Peter, M.G., Applications and environmental aspects of chitin and chitosan. *J. Macromol. Sci.*, 32, 629, 1995.
19. Becker, T., Schlaak, M.,Strasdeit, H., Adsorption of nickel- (II), zinc(II) and cadmium(II) by new chitosan derivatives. *React. Funct. Polym.*, 44, 289, 2000.
20. Kasimova, M.R., Velazquez-Campoy, A., Nielsen, H.M., On the temperature dependence of complex formation between chitosan and proteins. *Biomacmolecules,* 12, 2534, 2011.
21. Schipper, N.G., Varum, K.M., Artursson, P., Chitosan as absorption enhancers for poorly absorbable drugs. 1: Influence of molecular weight and degree of acetylation on drug transport across human epithelial (Caco-2) cells. *Pharm. Res.,* 13, 1686, 1996.
22. Nısperos-Carriedo, M.O., Edible coatings and films based on polysaccharides, In: *Edible coatings and films to improve food quality*, J.M. Krochta, E.A., Baldwin, M.O.Nisperos-Carriedo, Eds., pp. 305–335, Technomic, Lancaster, 1994.
23. Tavakoli, N., Varshosaz, J., Dorkoosh, F., Motaghi, S., Tamaddon, L., Development and evaluation of a monolithic floating drug delivery system for acyclovir. *Chem. Pharm. Bull.*, 60, 172, 2012.
24. Murata, Y., Kojima, N., Kawashima, S., Functions of a chitosan-orotic acid salt in the gastrointestinal tract. *Biol. Pharm. Bull.,* 26, 687, 2003.
25. Gogev, S., de Fays, K., Versali, M.F., Gautier, S., Thiry, E., Glycol chitosan improves the efficacy of intranasally administrated replicated defective human adenovirus type 5 expressing glycoprotein D of bovine herpesvirus 1. *Vaccine*, 22, 1946, 2004.

26. Mansur, A.A.P., Mansur, H.S., S-Araujo, A., Lobato, Z.I.P., Fluorescent nano-hybrids based on quantum dot-chitosan-antibody as potential cancer biomarkers, *ACS Appl. Mater. Interfaces*, 6, 11403, 2014.
27. Jiao, H., Yao, J., Yang, Y., Chen, X., Lin, W., Li, Y., Gu, X., Wang, X., Chitosan/polyglycolic acid nerve grafts for axon regeneration from prolonged axotomized neurons to chronically denervated segments. *Biomaterials*, 30, 5004, 2009.
28. Rickett, T.A., Amoozgar, Z., Tuchek, C.A., Park, J., Yeo, Y., Shi, R., Rapidly photo-cross-linkable chitosan hydrogel for peripheral neurosurgeries, *Biomacromolecules*, 12, 57, 2011.
29. Tchemtchoua, V.T., Atanasova, G., Aqil, A., Filee, P., Garbacki, N., Vanhooteghem, O., Deroanne, C., Noel, A., Jerome, C., Nusgens, B., Poumay, Y., Colige, A., Development of a chitosan nanofibrillar scaffold for skin repair and regeneration. *Biomacromolecules*, 12, 3194, 2011.
30. Felgner, P.L., Barenholz, Y., Behr, J.P., Cheng, S.H., Cullis, P., Huang, L., Jessee, J.A., Seymour, L., Szoka, F., Thierry, A.R., Wagner, E., Wu, G., Nomenclature for synthetic gene delivery systems. *Hum. Gene Ther.*, 8, 511, 1997.
31. Meshali, M.M., el-Dien, E.Z., Omar, S.A., Luzzi, L.A., A new approach to encapsulating nonsteroidal anti-inflammatory drugs. IV. Effect of cellulose derivatives with different functional groups on the bioavailability and gastric ulcerogenic activity of acidic as well as basic nonsteroidal anti-inflammatory drugs. *J. Microencap.*, 6, 355, 1989.
32. Acikgoz, M., Kas, H.S., Hascelik, Z., Milli, U., Hincal, A.A., Chitosan microspheres of diclofenac sodium. II. *In vitro* and *in vivo* evaluation. *Pharmazie*, 50, 275, 1995.
33. Qin, Y., Absorption characteristics of alginate wound dressings. *J. Appl. Polym. Sci.*, 91, 953, 2004.
34. Qin, Y., Hu, H., Luo, A., The conversion of calcium alginate fibers into alginic acid fibers and sodium alginate fibers. *J. Appl. Polym. Sci.*, 101, 4216, 2006.
35. Moussavi, G., Mahmoudi, M., Removal of azo and anthraquinone reactive dyes by using MgO nanoparticles. *J. Hazard. Mater.*, 168, 806, 2009.
36. Zhu, H.Y., Jiang, R., Xiao, L., 2010. Adsorption of an anionic dye by chitosan/kaolin/γ-Fe_2O_3 composites. *Appl. Clay Sci.*, 48, 522, 2010.
37. Zhang, M., Smith, A., Gorski, W., Carbon nanotube-chitosan system for electrochemical sensing based on dehydrogenase enzymes. *Anal. Chem.*, 76, 5045, 2004.
38. Mavroidis, C., Dubey, A., Yarmush, M.L., Molecular machines. *Annu. Rev. Biomed. Eng.*, 2004, 6, 363, 2004.
39. Aberg, C.M., Chen, T.H., Olumide, A., Raghavan, S.R., Payne, G.F., Enzymatic grafting of peptides from casein hydrolysate to chitosan. potential for value-added byproducts from food-processing wastes. *J. Agric. Food Chem.*, 52, 788, 2004.

40. Uygun, A., Kiristi, M., Oksuz, L., Manolache, S., Ulusoy, S., RF hydrazine plasma modification of chitosan for antibacterial activity and nanofiber applications. *Carbohydr. Res.*, 346, 259, 2011.
41. Pospieszny, H., Atabekov, J.G., Effect of chitosan on the hypersensitive reaction of bean to alfalfa mosaic virus. *Plant Sci.*, 62, 29, 1989.
42. Chirkov, S.N., The antiviral activity of chitosan. *Appl. Biochem. Microbiol.*, 38, 1, 2002.
43. Usami, Y., Okamoto, Y., Minami, S., Matsuhashi, A., Kumazawa, N.H., Tanioka, S., Shigemasa, Y., Migration of canine neutrophils to chitin and chitosan. *J. Vet. Med. Sci.*, 56, 1215, 1994.
44. Usami, Y., Minami, S., Okamoto, Y., Matsuhashi, Shigemasa, Y., Influence of chain length of N-acetyl-D-glucosamine and D- gluosamine residues on direct and complement-mediated chemotactic activities for canine polymorphonuclear cells. *Carbohyd. Polym.*, 32, 115, 1997.
45. Ross, S. (2000). Functional foods: the food and drug administration perspective. *Am. J. Clin. Nutr.*, 71, 1735s, 2000.

6

Chitin and Chitosan-Complexes and Their Applications

Suresh Kumar Arya[1*], Murli Manohar[2], Gulshan Singh[3] and Waseem Ahmad Siddiqui[2]

[1]*Department of Chemistry, Jawaharlal Nehru Rajkeeya Mahavidyalaya (Pondicherry University), Port Blair, India*
[2]*Department of Biochemistry, Jamia Hamdard (Hamdard University), New Delhi, India*
[3]*Institute for Water and Wastewater Technology (IWWT), Durban University of Technology, Durban, South Africa*

Abstract

Chitin is a natural biopolymer and it is made up of 1–4 linked 2-acetamido-2-deoxy-β-D-glucopyranose. Chitin has been isolated from various sources such as crustaceans, squids, centric diatoms, algae, and fungi. Chitin undergoes the process of deacetylation and produces a novel and highly potential molecule termed as "Chitosan." Chitosan is more promising than chitin because it is biocompatible, nonantigenic, nontoxic, and multi-biofunctional molecule. Due to the presence of functional amino ($-NH_2$) and carboxyl ($-COOH$) groups on chitosan, it is able to interact with other polyelectrolytes and generates suitable complexes with improved physiological properties. These complexes can be exploited for a variety of beneficial purposes in human life in terms of medical, food, agriculture, and wastewater treatment. This chapter briefly summarizes the source and extraction of chitin and synthesis of chitosan. Further, chemical and biological properties of chitosan have been discussed. Additionally, the uses of several naturally occurring polymers such as alginate, pectin, xanthan gum, and hyaluronic acid and their complexes with chitosan have been described in brief. Finally, several applications of chitosan have been discussed. This chapter provides better understanding in-depth knowledge of various complexes of chitosan and their applications.

Keywords: Chitin, chitosan, polyelectrolyte complexes, biocompatible, biopolymer

**Corresponding author*: skalcc@gmail.com

Shakeel Ahmed and Saiqa Ikram (eds.) Chitosan, (151–166) © 2017 Scrivener Publishing LLC

6.1 Introduction

Chitin is a natural biopolymer and after the cellulose it is one of the most important natural polymer in the world [1, 2]. Chitin and Chitosan are mainly acquired from insects (Ovipositors, Cuticle, and Beetle cocoon), Crustaceans (Crab shell and Shrimp shell), Squid (Ommastrephes pen and Loligo stomach wall), Centric Diatoms (*Thalassiosira fluviatilis*) algae and fungi (*Mucor rouxi* and *Aspergillis nidulans*). Chitin is made up of 1–4 linked 2-acetamido-2-deoxy-β-d-glucopyranose and after the process of deacetylation it can be converted into "chitosan" (Figure 6.1A and B) [3, 4]. Chitosan is biocompatible, nonantigenic, nontoxic, and multi-biofunctional molecule [5]. Both chitin/chitosan and their modified derivatives are mainly useful in agriculture, medicine, food, and nonfood industries as well. They are known as a new class of physiological materials of highly sophisticated functions [3, 6]. Chitosan is more promising than chitin in term of its diverse useful applications. Hence, Chitosan is novel subject of interest due to its use as a polymeric drug carrier material in dosage form design because it has dynamic as well as promising properties, for example, biodegradability, biocompatibility, low toxicity, and comparatively lower production cost from profuse natural sources. The main drawback of chitosan in modified release dosage forms for oral intake is higher rate of dissolution in the stomach. At lower pH (below its pKa value), Chitosan is positively charged therefore, impulsively reacts with polyanions in solution and generates polyelectrolyte complexes [5]. Interestingly, these chitosan-based polyelectrolyte complexes have favorable physicochemical properties with conservation of chitosan's biocompatible characteristics. The purpose of this chapter is to briefly explain the several complexes of chitosan along with various natural polyanions and their use in drug delivery systems like nano- and micro-particles, beads, and fibers.

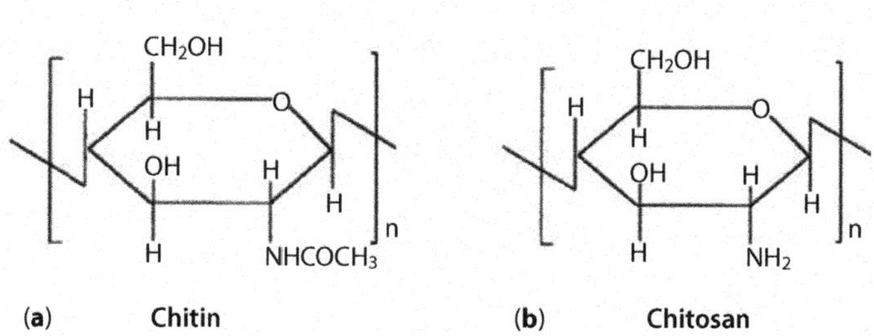

Figure 6.1 Structure of chitin (a) and chitosan (b).

6.2 Synthesis of Chitosan from Chitin

The commercial production of Chitosan can be performed by deacetylation of chitin. In brief, these steps have been described in Figure 6.2. The average molecular weight of commercially produced Chitosan may vary from 3.8 kDa to 20 kDa. In general, Chitosan is synthesized by deacetylation of chitin using sodium hydroxide (NaOH) in excess as a reagent and water as solvent and yield is up to 98% of chitosan.

Chitin has been isolated from crustacean shells by the chemical process that involves several steps as shown in Figure 6.3. These steps of isolation

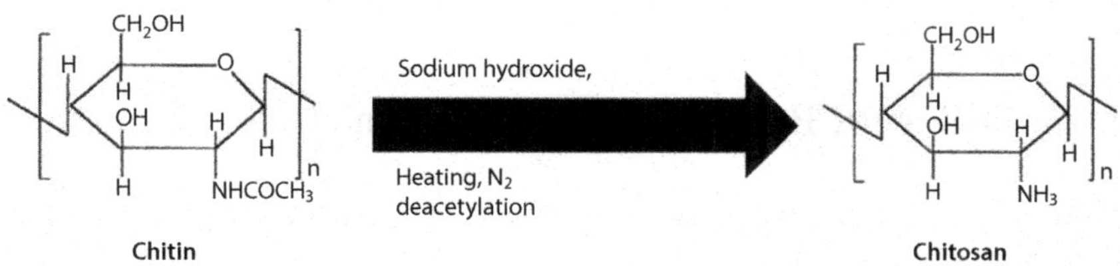

Figure 6.2 Deacetylation of chitin produces chitosan.

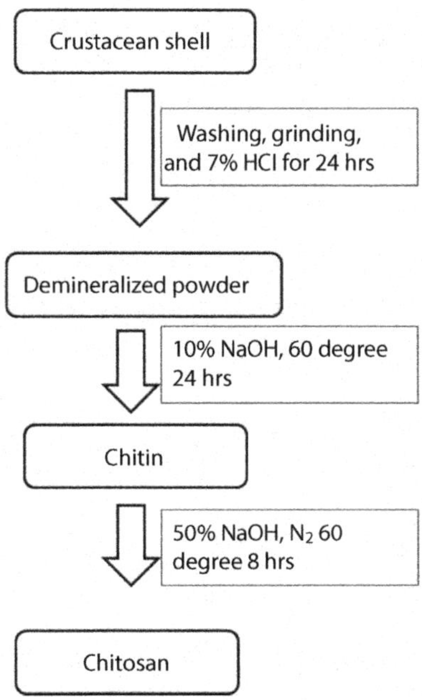

Figure 6.3 Different steps involved in preparation of chitosan from the shell of crustaceans. This method is adopted from RiZvezdova *et al.*, [7].

of chitin have been adopted from the studies by RiZvezdova *et al.*, [7], and Pokhrel *et al.*, [8]. There are three major steps are involved in the isolation of chitin from the crustacean shells:

Step 1: Demineralization: This step involves treatment with acid (HCl) that separates inorganic matters mainly calcium carbonate ($CaCO_3$).

Step 2: Deproteinization: This step involves the extraction of protein content in alkaline medium (with NaOH).

Step 3: Decolorization: This step involves bleaching of the product by chemical reagents in order to obtain the bleached product (colorless).

6.3 Different Properties of Chitosan

Chitosan is not soluble in water and other common organic solvents. However, it can be dissolved in some specific solvents such as hexafluoro-2-propanol/N,N-dimethylacetamide or hexafluoroacetone. Chitosan preparations may differ in size (average molecular weight, MW), on the basis of the source of natural chitin and the conditions of its production. It has important consideration not only as an underutilized resource but also as a new useful material of great potential in several fields, and the latest work in chitin chemistry is quite remarkable.

6.3.1 Chemical Properties of Chitosan

Chitosan revealed many chemical properties (Figure 6.4) such as it is a linear polyamine. Interestingly, chitosan possesses two major reactive groups:

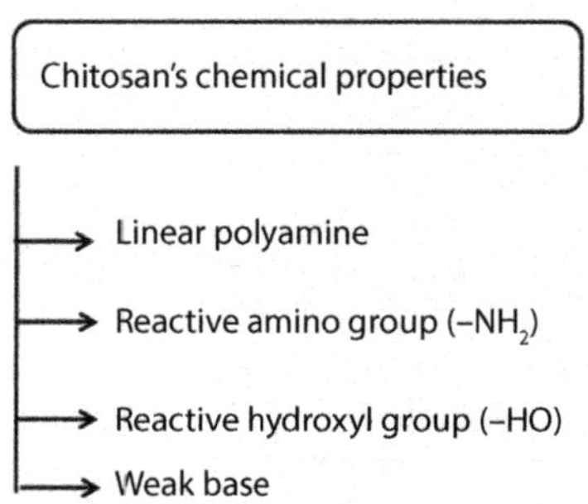

Figure 6.4 Various chemical properties of chitosan.

an amino group (–NH_2) and a hydroxyl group (–OH) that provide capability to chitosan to chelate with a variety of transition metals [9]. Chitosan has very low basic property and considered as weak base; hence it is not soluble in water. Very interestingly, chitosan can be dissolved in dilute aqueous acidic solutions with pKa value below ~6.3 that further leads the conversion of its –NH_2 group of chitosan into the soluble protonated form (–NH_3^+).

However, the solubility of chitosan can be altered by a variety of factors like: molecular weight, degree of acetylation, and biological origin [10]. Since, the chitosan is easily soluble in dilute acidic solutions, several films can be voluntarily made by casting or dipping that result in dense and porous structure [11]. The most important characteristic feature of chitosan is the degree of acetylation (DA) as well as its molecular weight (MW). Commercially available chitosan exists with >85% deacetylated units (DA<15%), with molecular weights range between 100 kDa and 1000 kDa. Currently, no specific standard is available to define the molecular weight of chitosan, but it is widely accepted that there are three categories of chitosan categorized on the basis of its molecular weight. These categories are (1) Low MW<50 kDa, (2) Medium MW 50–150 kDa, and (3) High MW>150 kDa [12].

6.3.2 Biological Properties of Chitosan

Apart from the several chemical properties, chitosan has also shown diverse promising biological properties that are summarized in Figure 6.5 [9].

1. Chitosan is having the property of biocompatibility because it is a natural polymer with no toxic effects and easily biodegradable.
2. It can easily bind to mammalian and microbial cells aggressively because of its positive charge that helps it to bind to negative charged membrane of the cells.
3. It has a regenerative effect on connective gum tissues.
4. Chitosan accelerates bone formation with the formation of osteoblast cells.
5. It also causes bleeding to stop due to its hemostatic property.
6. It has ability to inhibit the growth of several fungi and widely used as fungicide.
7. It also plays a role in birth control due to its spermicidal property.
8. It shows antitumor or anticancer property (inhibits the growth of tumor cells).

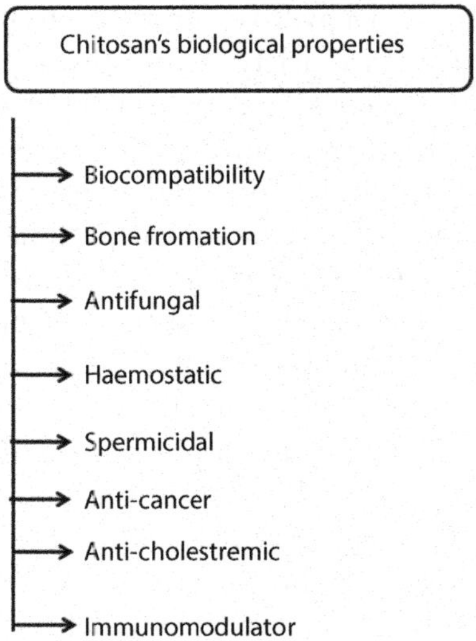

Figure 6.5 Biological properties of chitosan.

9. It has anticholestremic property and acts as a cholesterol lowering agent.
10. Chitosan slows down the brain activity, therefore can be used as central nervous system depressant.
11. It also possesses immunoadjuvant property and is involved in the regulation and improvement of immune response. It has been reported that the degree of acetylation also affects the biological properties of chitosan [13].

6.4 Polyelectrolyte Complexes of Chitosan

Chitosan has been used as a polymeric drug carrier because of its attractive and promising properties such as biocompatibility, biodegradability, low toxicity, and relatively low cost of production from natural sources. The chitosan is a positively charged molecule at low pH values (below its pKa value), and has spontaneous great ability of association with polyions in the solution to make various polyelectrolyte complexes. The chitosan-based polyelectrolyte complexes show several favorable physicochemical properties with preservation of chitosan's biocompatible characteristics. Therefore, these chitosan complexes emerge as novel materials for the design of different types of dosage forms [14]. Variety of polyanions from a biological origin, such as alginate, pectin, carrageenan, xanthan gum, and

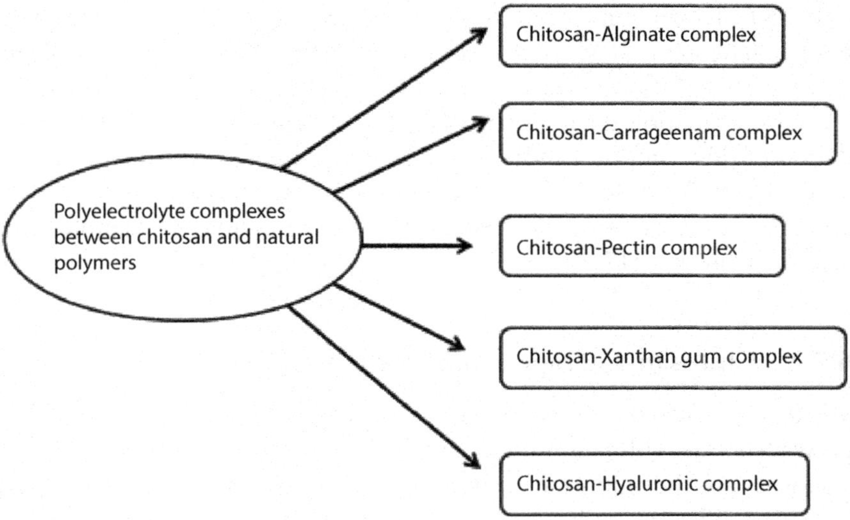

Figure 6.6 Different types of chitosan-based polyelectrolyte complexes.

hyaluronic acid, have been widely used to generate polyelectrolyte complexes, in combination with chitosan, with improved physicochemical properties to design targeted drug delivery systems [14, 15].

6.5 Complexes of Polyelectrolyte between Chitosan and Naturally Occurring Polymers

A number of polyelectrolyte complexes with chitosan and anionic natural polymers are known in literature. Some of the polyelectrolyte complexes with chitosan are alginate, carrageenan, pectin, xanthan gum, and hyaluronic acid [14]. In brief, the details about these polyelectrolyte complexes of chitosan have been discussed (Figure 6.6).

6.5.1 Chitosan-Alginate Polyelectrolyte Complex

Alginate is a linear, natural, biodegradable polysaccharide consists of 1, 4-linked β-D-mannuronic acid and α-L-guluronic acid monomers in unstable proportions. The carboxylic acid groups (negatively charged) of manuronic and guluronic acid units of alginate mainly react electrostatically with the positively charged amino groups ($-NH_2$) of the chitosan to make a polyelectrolyte complex. Alginate is valuable anionic polyelectrolytes complex of chitosan because of its high biodegradable nature, biocompatibility and it is mechanically stable at low pH values where chitosan dissolves easily [16]. The polyelectrolyte complexes exhibit partial degradation during hydrolysis. Interestingly, the rate of biodegradation of this complex can be regulated by changing the ratio of alginate and chitosan. Therefore, alginate-chitosan polyelectrolyte complexes have a great impact value in the field of tissue engineering for scaffolds and support materials [17].

6.5.2 Chitosan-Carrageenan Polyelectrolyte Complex

Another polysaccharide is termed as carrageenan and this is sulphated polysaccharide with high molecular weight. The varieties of species of red seaweeds are the important source of carrageenan. Carrageenan mainly has three different kinds of members such as kappa carrageenan (κ), iota carrageenan (ι), and lambda carrageenan (λ) [18]. In term of mechanical strength these carrageenan have been graded in the order λ->ι->κ-carrageenan. Interestingly, ι- and κ-carrageenan have ability to form more strong gels due to the formation of more cross-links with double helix secondary structures, and the gels were also more brittle and temperature sensitive [19].

6.5.3 Chitosan-Pectin Polyelectrolyte Complex

The plant cell wall contains several polysaccharides such as cellulose, hemicelluloses, and pectin. Pectin, also high molecular weight linear polysaccharide, consists of α-1, 4-linked D-galacturonic acid units. Mixing of acidic chitosan and pectin solution forms a homogenous solution with no ionic interactions between them, however, the polyelectrolyte complexes can be synthesized by adjusting the pH 5.5 of this mixture that leads favorable electrostatic interactions between negatively charged carboxylic acid groups (–COOH) of pectin and positively charged amino groups ($-NH_2$) of chitosan. Therefore, the degree of interaction can be regulated by the pH of the medium [20]. The best example of pH regulation of these interaction is the polyelectrolyte complexes formed between chitosan and pectin, converted into tablets with vancomycin (an antibiotic drug used for the treatment of intestinal infections) and these tablets revealed a pH sensitive swelling capability with drug release performance based on enzyme-dependent degradation in the presence of β-glucosidase, which shows potential for colon-specific delivery of vancomycin [21].

6.5.4 Chitosan-Xanthan Gum Polyelectrolyte Complex

Xanthan gum is the exopolysaccharide secreted from *Xanthomonas campestris*. It contains mainly a cellulosic backbone, that is, β-(1,4)-D-glucopyranose glucan, with a trisaccharide side chain: (3,1)-α-D-mannopyranose-(2,1)-β-D-glucuronic acid-(4,1)-β-D-mannopyranose, on each second glucose residue. Chitosan-xanthan gum polyelectrolyte complexes are being used as a matrix for enzyme immobilization [22, 23], but an exclusive importance of this matrix in the pharmaceutical field is to improve the dissolution of water insoluble drugs such as fenofibrate, ursodeoxycholic acid, nifedipine, and indomethacin [24].

6.5.5 Chitosan-Hyaluronic Acid Polyelectrolyte Complex

Hyaluronic acid is also termed as hyaluronan or hyaluronate. Hyaluronic acid is single nonsulfated glycosaminoglycan found in extracellular matrix throughout the connective tissues, epithelial cells, and neural tissues. It is synthesized by bacterial fermentation of a variety of species of *Streptococcus* or it can also be extracted from rooster combs, umbilical cords, synovial fluids, or vitreous humor for the ample of uses for human life as well as have great medicinal uses. These complexes of Chitosan-hyaluronic acids find their application in eyes surgery, arthritis treatment, tissue bioengineering, wound healing, and making the implants devices [4, 25]. Interestingly, it has been found that these complexes are protected against enzymatic hydrolysis, but only at pH different from the optimal pH of the enzyme [26]. During an *in vitro* drug release study, when hyaluronic acid was included into hydrogel prepared from chitosan and poly(N-isopropylacrylamide) it disallowed disintegration of the system and thereby facilitated the controlled release of nalbuphine (an analgesic drug used for the treatment of severe pain) [27]. In another study, the chitosan-hyaluronic acid polyelectrolyte complex nanoparticles loaded with heparin were effectively internalized by the mast cells (i.e., a cell of immune system) and showed the almost similar activity in terms of inhibiting histamine release compared to free heparin [28].

6.6 Various Useful and Important Applications of Chitosan

Chitosan shows variety of applications in diverse areas including cosmetics, medical, agricultural, food processing, nutritional enhancement, and wastewater treatment (Figure 6.7) [29, 30].

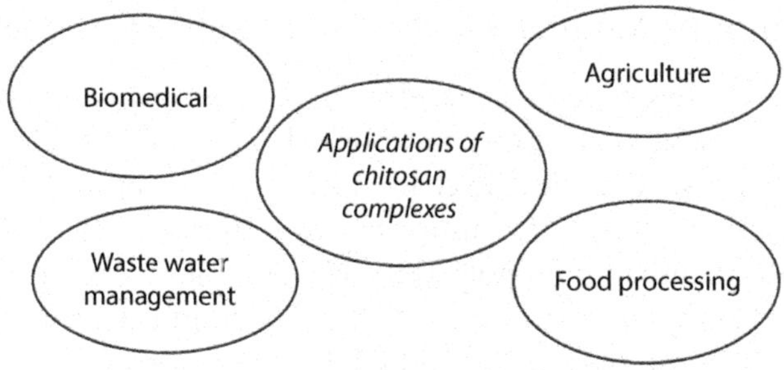

Figure 6.7 Different potential bio-applications of chitosan complexes.

6.6.1 Chitosan Application in Agriculture

The natural abundance of Chitosan, its biodegradability and nontoxic nature, makes it suitable for agricultural applications since it can be used without any concerns of disposal, harm to the consumers if ingested, and it is safe in terms of no pollution effects. The important applications in agriculture are seed coating, leaf coating, fertilizers, and time released drug or fertilizer response. The use of chitosan in these areas has increased the production of crops by means of improving the germination, rooting, leaf growth, seed yield, and also soil moisture retention, with reducing the fungal infections and diseases [31]. Chitosan has been reported to prevent the systemic spread of viruses and viroids throughout the plant and also provides hypersensitive response against viral infection in plants. The degree of inhibition of viral infections may changes as per molecular weight of chitosan. Interestingly, same results have been cited in literature indicating tobacco mosaic and necrosis viruses, and cucumber mosaic virus propagation can be stopped by chitosan treatment. Additionally, Chitosan has ample ability to prevent viability of variety of bacteria and several species of fungi and oomycetes that are harmful for various crops [32].

6.6.2 Application of Chitosan in Wastewater and Contaminated Water Treatment

Chitosan is a suitable natural component that can be effectively used as the coagulating agent specially for organic compounds, as a chelating polymer to bind toxic heavy metals like copper (Cu), lead (Pb), mercury (Hg), and uranium (U) [29]. Chitosan also has adsorption ability for dyes and small concentrations of phenols and polychlorinated biphenyls found in the industrial wastewaters. For above specific applications, the chitosan is more effective as compared to other polymers like synthetic resins, activated charcoal, and even chitin itself. In addition, the amino group ($-NH_2$) in chitosan is an major effective functional group that might be changed for production of other chitosan derivatives for the purpose of beneficial characteristics as effective absorptive agents [31, 33]. A study has shown that chitosan can also be utilized to breakdown the protein of food diet and other negatively charged chemicals present in the wastewater streams and processing outlets [31]. A very recent report by Al-Manhel *et al.,* [34] indicates that increasing doses of chitosan in drinking water cause reduced turbidity, electrical conductivity, and pH. Chitosan was found to improve the water quality by sedimenting most of the salts. Additionally, chitosan was found more effective against gram-negative bacteria when compared to gram-positive bacteria. Therefore, these unique and novel properties of chitosan had made it, a promising agent for the purification of water to make it suitable for drinking use for human [34].

6.6.3 Chitosan Use in Food Industries

The chelating property and higher functionality make chitosan a very important and novel molecule for the food industry, for example, binding with and removing certain elements, particles and materials such as dyes and fats from foods [35]. Chitosan has ability to easily make a coating on the fruits and vegetables, therefore, it lowers down the respiration rate of fruits and vegetables by regulating the permeability of CO_2 and O_2. The chitosan coating has a lot of advantages to the preservation of postharvest fruits and vegetables, but for specific fruit or vegetable, single chitosan coating sometimes shows certain defect which includes limited inhibition to certain microorganism that leads fruit to decay and poor coating structure to adjust the permeability of CO_2 and O_2 [36]. For effective chitosan coating, it was combined with other substances. In addition, the single chitosan coating was often combined with physical methods such as short heating, short gas fumigation, modified atmosphere packaging, and so on [37, 38]. The antiviral, antibacterial, and antifungal properties of chitosan have been utilized during the storage and preservation of several food items [29, 31, 39].

6.6.4 Chitosan and Biomedical Applications

Chitosan shows many areas of interest within the medical industry including orthopedic and periodontal applications, tissue engineering, wound healing, and drug delivery [30, 31, 40–44] due to its ability to function in many forms. Its important biomedical applications are in artificial skin, surgical sutures, artificial blood vessels, controlled drug release, contact lens, eye humor fluid, bandages, sponges, burn dressings, blood cholesterol control, anti-inflammatory, tumor inhibition, anti-viral, dental plaque inhibition, bone healing treatment, wound healing accelerator, hemostatic agent, antibacterial agent, antifungal agent, and weight loss effect.

6.7 Conclusion

This chapter briefly summarized the introduction of chitin and chitosan and discussed the synthesis of chitosan from chitin. Different chemical and biological properties of chitosan were discussed. Further, various polyelectrolyte complexes form with positively charged chitosan and negatively charged natural polymers such as alginate, pectin, xanthan gum, and hyaluronic acid are discussed. Finally, this chapter describes the various applications of chitosan such as medical, agricultural, food processing, nutritional enhancement, cosmetics, and wastewater treatment. This chapter provides the better understanding of chitosan to the reader and in-depth knowledge of various complexes of chitosan and their potential applications.

Acknowledgement

The authors would like to thank to the Principal JNRM, Port Blair, for his valuable suggestions during the preparation of this manuscript.

References

1. Muzzarelli, R., Tarsi, R., Filippini, O., Giovanetti, E., Biagini G., Varaldo, P.E., Antimicrobial properties of N-carboxybutyl chitosan. *Antimicrob. Agents. Chemother.*, 34, 2019, 1990.
2. Muzzarelli, R.A., Mehtedi, M.El., and M. Mattioli-Belmonte, Emerging biomedical applications of nano-chitins and nano-chitosans obtained via advanced eco-friendly technologies from marine resources. *Mar. Drugs,* 12, 5468, 2014.
3. Younes, I., Rinaudo, M., Chitin and chitosan preparation from marine sources. *Mar. Drugs,* 13, 1133, 2015.
4. Rinaudo, M., Properties and degradation of selected polysaccharides: hyaluronan and chitosan. *Corros, Eng, Sci, Technol.*, 42, 324, 2007.
5. Tharanathan, R.N., Kittur, F.S., Chitin–the undisputed biomolecule of great potential. *Crit. Rev. Food. Sci. Nutr.*,43, 61, 2003.

6. Azuma, K., Ifuku, S., Osaki, T., Okamoto, Y., Minami, S., Preparation and biomedical applications of chitin and chitosan nanofibers. *J. Biomed. Nanotechnol.*, 10, 2891, 2014.
7. RiZvezdova, D., Synthesis and characterization of chitosan from marine sources in black sea. Scientific Works of the Rousse University by Bulgaria, 49, 65, 2010.
8. Pokhrel, S., Yadav, P.N., Adhikari, R., Applications of chitin and chitosan in industry and medical science: a review. *Nep. J. Sci. Technol.*, 16, 99, 2015.
9. Dutta, P., Rinki, K., Dutta, J., Chitosan: a promising biomaterial for tissue engineering scaffolds. Chitosan for biomaterials II. In: *Advances in polymer science*. Heidelberg: Springer Berlin, R. Jayakumar, M. Prabaharan, R.A.A. Muzzarelli, Eds.45, 2011.
10. Shepherd, R., Reader, S., Falshaw, A., Chitosan functional properties. *Glycoconjugate J.*, 14, 535, 1997.
11. Assis, O.B.G., Silva, V.L., Caracterização Estrutural e da Capacidade de Absorção de Água em Filmes Finos de Quitosana Processados em Diversas Concentrações. *Polímeros. Ciência. E. Tecnologia.*, 13, 223, 2003.
12. Goy R.C., de Britto, D., Assis, O.B.G., A review of the antimicrobial activity of chitosan. *Polímeros. Ciência. E. Tecnologia.*, 19, 241, 2009.
13. Chatelet, C., Damour, O., Domard, A., Influence of the degree of acetylation on some biological properties of chitosan films. *Biomaterials*, 22, 261, 2001.
14. Hamman, J.H., Chitosan based polyelectrolyte complexes as potential carrier materials in drug delivery systems. *Mar. Drugs*, 8, 1305, 2010.
15. Berger, J., Reist, M., Mayer, J.M., Felt, O., Gurny, R., Structure and interactions in chitosan hydrogels formed by complexation or aggregation for biomedical applications. *Eur. J. Pharm. Biopharm.*, 57, 35, 2004.
16. Hein, S., Wang, K., Stevens, W.F., Kjems, J., Chitosan composites for biomedical applications: status, challenges and perspectives. *Mater. Sci. Technol.*, 24, 1053, 2008.
17. Li, X., Xie, H., Lin, J., Xie, W., Ma, X., Characterization and biodegradation of chitosan-alginate polyelectrolyte complexes. *Polym. Degrad. Stab.*, 94, 1, 2009.
18. Beneke, C.E., Viljoen, A.M., Hamman, J.H., Polymeric plant-derived excipients in drug delivery. *Molecules*, 16, 2602, 2009.
19. Shumilina, E.V., Shchipunov, Y.A., Chitosan-carrageenan gels. *Colloid. J.*, 64, 372, 2002.
20. Bernabe, P., Peniche, C., Argüelles-Monal, W., Swelling behavior of chitosan/pectin polyelectrolyte complex membranes. Effect of thermal cross-linking. *Polym. Bull.*, 55, 367, 2005.
21. Bigucci, F., Luppi, B., Cerchiara, T., Sorrenti, M., Bettinetti, G., Rodriguez, L., Zecchi, V., Chitosan/pectin polyelectrolyte complexes: selection of suitable preparative conditions for colon-specific delivery of vancomycin. *Eur. J. Pharm. Sci.*, 18, 435, 2008.
22. Dumitriu, S., Magny, P., Montane, D., Vidal, P.F., Chornet, E., Polyionic hydrogels obtained by complexation: their properties as support for enzyme immobilization. *J. Bioact. Compat. Polym.*, 9, 184, 1994.

23. Dumitriu, S., Chornet, E., Vidal, P.F., Moresoli, C., Polyionic hydrogels as support for immobilization of lipase. *Biotechnol. Tech.*, 9, 833, 1995.
24. Magnin, D., Lefebvre, J., Chornet, E., Dumitriu, S., Physicochemical and structural characterisation of a polyionic matrix of interest in biotechnology, in the pharmaceutical and biomedical fields. *Carbohydr. Polym.*, 55, 437, 2004.
25. Kim, S. J., Shin, S.R., Lee, S.M., Kim, I.Y., Kim, S.I., Thermal characteristics of polyelectrolyte complexes composed of chitosan and hyaluronic acid. *J. Macromol. Sci.*, A40, 807, 2003.
26. Denuziere, A., Ferrier, D., Damour, O., Domard, A., Chitosan-chondroitin sulphate and chitosan-hyaluronate polyelectrolyte complexes: biological properties. *Biomaterials*, 19, 1275, 1998.
27. Fang, J.Y., Chen, J.P., Leu, Y.L., Hu, J.W., Temperature-sensitive hydrogels composed of chitosan and hyaluronic acid as injectable carriers for drug delivery. *Eur. J. Pharm. Biopharm.*, 68, 626, 2008.
28. Oyarzun-Ampuero, F.A., Brea, J., Loza, M.I., Torres, D., Alonso, M,J., Chitosan-hyaluronic acid nanoparticles loaded with heparin for the treatment of asthma. *Int. J. Pharm.*, 381, 122, 2009.
29. Hudson, S.M., Jenkins, D.W., Chitin and chitosan, encyclopedia of polymer science and technology, NJ: Wiley Interscience, 2001.
30. Şenel, S., McClure, S.J., Potential applications of chitosan in veterinary medicine. *Adv. Drug Deliv. Rev.*, 56, 1467, 2004.
31. Li, Q., Dunn, E.T., Grandmaison, E.W., Applications and properties of chitosan. *J. Bioact. Compat. Polym.*, 7, 370, 1992.
32. Hadrami El, A., Adam L.R., El Hadrami, I., Daayf, F., Chitosan in plant protection. *Mar. Drugs*, 8, 968, 2010.
33. No, H.K., Meyers, S.P., Application of chitosan for treatment of wastewaters. *Rev. Environ. Contam. Toxicol.*, 163, 1, 2000.
34. Al-Manhel, A.J., Saeed Al-Hilphy, A.R., Niamah, A.K., Extraction of chitosan, characterization and its use for water purification. *J. Sau. Soc. Agr Sci.*, 2016, http://dx.doi.org/10.1016/j.jssas.2016.04.001.
35. Vargas, M., González-Martínez, C., Recent patents on food applications of chitosan. *Recent. Pat. Food. Nutr. Agric.*, 2, 121, 2010.
36. Kumar M.N.V.R., A review of chitin and chitosan applications. *React. Funct. Polym.*, 46, 1, 2000.
37. Shao, X.F., Tu, K., Tu, S., Tu, J., A combination of heat treatment and chitosan coating delays ripening and reduces decay in "Gala" apple fruit. *J. Food Quality*, 35, 83, 2012.
38. Jianglian, D., Shaoying, Z., Application of chitosan based coating in fruit and vegetable preservation: a review. *J. Food Process. Technol.*, 4, 227, 2013.
39. Khor, E., Lim, L.Y., Implantable applications of chitin and chitosan. *Biomaterials*, 24, 2339, 2003.
40. Subramanian, A., Vu, D., Larsen, G.F., Lin, H.Y., Preparation and evaluation of the electrospun chitosan/PEO fibers for potential applications in cartilage tissue engineering. *J. Biomater. Sci. Polym. Ed.*, 16, 861, 2005.

41. Di Martino, A., Sittinger, M., Risbud, M.V., Chitosan: a versatile biopolymer for orthopaedic tissue-engineering. *Biomaterials*, 26, 5983, 2005.
42. Jayakumar, R., Prabaharan, M., Reis, R.L., Mano, J.F., Graft copolymerized chitosan—present status and applications. *Carbohydr. Polym.*, 62, 142, 2005.
43. Prabaharan, M., Mano, J., Chitosan-based particles as controlled drug delivery systems. *Drug. Deliv.*, 12, 41, 2004.
44. Ueno, H., Mori, T., Fujinaga, T., Topical formulations and wound healing applications of chitosan. *Adv. Drug. Deliv. Rev.*, 52, 105, 2001.

7

Enzymes from Genetically Modified Microorganisms for Production of Chitin, Chitosan, and Chitooligosaccharide

Htun Inzali[1,2*], Sai Pyae Sone Aung[1,2], Khin Chaw Win[1,2], Wint Le Nu[1], Hla Ngwe[2] and Nitar Nwe[1*]

[1]*Ecological Laboratory, Advancing Life and Regenerating Motherland, Yangon, Union of Myanmar*
[2]*Department of Chemistry, University of Yangon, Yangon, Myanmar*

Abstract

To obtain large amount of enzymes for production of chitin, chitosan, and chitooligosaccharide, target DNA from the microorganisms which produce enzymes for production of chitin, chitosan, and chitooligosaccharide is used to clone into host microorganisms. In this process, there is need to select best microorganism for enzyme production and to extract DNA from this microorganism. In the cloning process of extracted DNA, method for purification and identification of target DNA, selection of plasmid, method for ligation of target DNA to selector plasmid, method for transformation of plasmid DNA to host cell, selection of recombinant clone, method for expression of transformant clone, method for production of enzyme using transformant clone, and method for determination of enzyme activity are important to obtain high yield of enzyme from host microorganisms. According to literature, enzymes related to production of chitin, chitosan, and chitooligosaccharide such as protease, chitin deacetylase, chitinase, and chitosanase have been produced using various genetic engineering technology and the resultant amount of enzymes increases more than wild type strains. This review chapter described microorganisms for enzymes production to produce chitin, chitosan, and chitooligosaccharide, cloning process of target DNA

**Corresponding authors*: htuninzali14@gmail.com; nitarnwe@gmail.com

Shakeel Ahmed and Saiqa Ikram (eds.) Chitosan, (167–182) © 2017 Scrivener Publishing LLC

to host microorganisms, production of chitin, chitosan, and chitooligosaccharide using enzymatic method and characterization of enzyme activities using various chitinous materials.

***Keywords*:** Chitin, chitosan, enzymes, chitooligosachharides

7.1 Introduction

Chitin is a copolymer of ß-1,4-linked *N*-acetyl-D-glucosamine and D-glucosamine. Naturally, chitin is a major constituent of the exoskeletons of insects and crustaceans [1–3]. Chitosan is usually derived from chitin by partial or complete deacetylation in the presence of alkali and can be found only in fungi, yeasts, molds, and insect cuticle of limited groups in nature. Chemical structures of chitin and chitosan are shown in Figure 7.1. Nowadays, commercially, chitin and chitosan are produced from biowastes obtained from aquatic organisms. The production of chitin and chitosan from biowastes of aquatic organisms in industrial scale appears in inconsistent physiochemical characteristics of products because of seasonal and variable supply of raw materials as well as variability and difficulties of process conditions [4].

Chitinases and chitosanases are used to produce chitooligosaccharide (CTO) by degradation of chitin and chitosan, respectively [2, 3]. Different organisms produce a wide variety of hydrolytic enzymes that exhibit different substrate specificities and other properties useful for various functions. In bacteria, chitinases play roles in nutrition

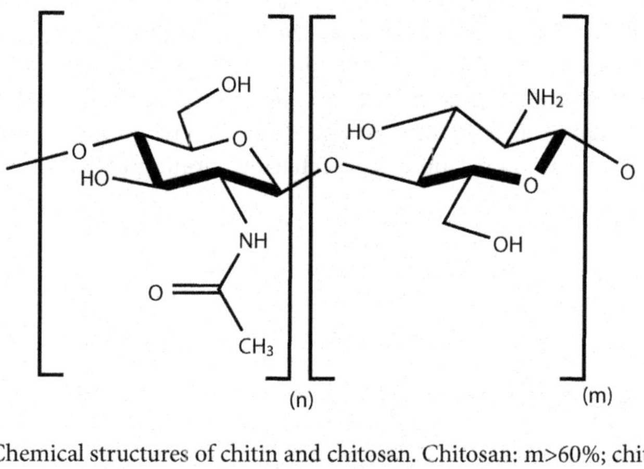

Figure 7.1 Chemical structures of chitin and chitosan. Chitosan: m>60%; chitin: m<60%.

and parasitism whereas in fungi, protozoa, and invertebrates they are also involved in morphogenesis. Chitinases are involved in the defense mechanism of plants and vertebrates. Baculoviruses, which are used for biological control of insect pests, also produce chitinases for pathogenesis [5]. In which, yields of products produced by biotechnological process are very slow and difficult to obtain products to get large amount of enzyme for biotechnology process. To overcome those problems, genetic engineering technology has been used. In genetic engineering techniques, gene cloning is the process of copying DNA fragments to use for many purposes such as to produce genetically modified organisms. Therefore, this chapter reviews the genes to be cloned, their sources and their enzymatic activities. The sources of microorganisms which can produce chitin and CTO and production steps are shown in Table 8.1. In Table 8.2, the genes to be cloned from the microorganisms and hosts are described. And enzyme activities and their treatment conditions are shown in Table 8.3.

7.2 Enzymes for Production of Chitin/Chitosan, and Chitooligosaccharide

In the production of chitin, protease is used to digest protein from shells of shrimp and crabs. In the production process, crustacean shells are firstly demineralized and deproteinization by organic acids and proteases produced by microorganisms like *Chaetomium thermophilum, Aspergillus niger* [6]. After that chitosan was produced by partial or complete deacetylation in the presence of alkali or chitin deacetylases produced by microorganisms like *Colletotrichum lindemuthianum* ATCC 56676, *Colletotrichum lindemuthianum* UPS9, *Rhizopus circinans*, and *Aspergillus nidulans* [7, 8]. Chitooligomers have been produced by acid hydrolysis, enzymatic hydrolysis, oxidative degradation, ultrasonic degradation, chemo-enzymatic and recombinant approaches [9]. Chitinases and chitosanases produced by the microorganisms from various sources are used.

7.3 Enzyme and DNA Technology for Production of Chitin, Chitosan, and CTO

After isolating the microorganisms from sources, microorganisms were grown in the appropriate medium and harvested the microorganisms

Table 7.1 List of enzymes for production of chitin, chitosan, and chitooligosaccharide (CTO).

Sources of miccroorganisms	Microorganisms	Enzymes	Substrate	Products
Compost, dung, and soil in Beijing	*Chaetomium thermophilum* [6]	Protease	–	–
–	*Colletotrichum lindemuthianum* ATCC 56676 [2]	Chitin deacetylase	Chitin	–
–	*C. lindemuthianum* UPS9 [10]	Chitin deacetylase	Chitin	–
–	*Rhizopus circinans* [11]	Chitin deacetylase	Chitin	–
–	*Aspergillus nidulans* [5]	Heat-stable chitin deacetylase	Chitin	–
Fresh water in Lake Suwa (Nagano Prefecture, Japan)	*Aeromonas hydrophila* strain SUWA-9 [12]	Chitinase	Colloidal chitin	Monomer and dimer of *N*-acetylchitooligomers
Sediment sample collected at Sagami Bay	*Alteromonas* sp. strain 0–7 [13]	Chitinase	–	–
Mushroom (*Agaricus bisporus*)	*Ewingella americana* [14]	Chitinase	–	–
Shrimp shell-enriched soil	*A. hydrophila* [15]	Chitinase	chitin or colloidal chitin	–

Soil sample obtained from a black-pepper plantation on Bangka island, Indonesia	*A. caviae* WS7b [16]	Chitinase	–	–
A naturally dead *Helicoverpa armigera*larva	*Serratia marcescens*[17]	Chitinase	Colloidal chitin	–
Tea	[18]	Chitinase	Colloidal chitin	–
Soil samples in Yunnan Province	*Lecanicillium psalliotae* [19]	Chitinase Lpchi1	Colloidal chitin	–
Barley	[20]	Chitinase	Colloidal chitin	–
Soybean paste in Korea	*Bacillus* sp. DAU101 [7]	Chitosanase	Soluble chitosan	Biofunctional oligomer
Soil	*Aspergillus fumigatus* JXSD-97 [21]	Chitosanase	chitosan	–
Soil in Korea	*Bacillus* sp. MET 1299 [22]	Chitosanase	90% deacetylated colloidal chitosan, ß-glucan	Biofunctional oligomer

Figure 7.2 Target DNA in DNA of microorganisms.

at the end of the fermentation. The target DNA was extracted from the freeze-dried microorganisms powder (Figure 7.1). Target gene was isolated from the microorganisms and amplified the target DNA by PCR (Figure 7.2). The resulted DNA was inserted to the plasmid DNA to get the recombinant plasmid with target gene. The resultant recombinant plasmid was transformed to the host microorganisms. Plasmid DNA containing target gene was extracted from the transforming cell and used in these series of experiments to study the cloning and expression of target gene into host microorganisms. Table 7.2 and Figure 7.3 show the gene to be cloned, their microorganism sources and host microorganisms for these genes.

7.4 Determinations of Enzyme Activities Using Chitinous Materials

The transformant clones were grown in the specific medium to extract the protein. Purified enzyme was extracted from the crude enzyme solution using precipitation method, etc. The activities of extracted enzymes were determined by treatment with specific substrate. Table 7.3 describes the treatment for the enzymes at various conditions and their activities. To get large scale, the treatment for the enzymes is important and also their

Table 7.2 Microorganisms for target genes and host for gene expression.

Microorganisms for target genes	Target genes	Host for gene expression
Chaetomium thermophilum [6]	Protease gene	*Escherichia coli* BL21(DE3) *Pichiapastoris*
Aeromonas hydrophila strain SUWA-9 [12]	Chitinase gene	*E. coli*
Beauveria bassiana BCC3653 [23]	Chitinase gene	*E. coli* TOP10
Alteromonas sp. strain 0–7 [13]	Chitinase gene	*E. coli* JM109
Streptomyces plicatus [24]	Chitinase-63 gene	*E. coli*
Ewingella americana [14]	Chitinase gene	*E. coli*
Bacillus thuringiensis [25]	Endochitinase chiA74 gene	*E. coli* DH5&F́
Bacillus circulans WL-12 [26]	Chitinase gene	*E. coli*
Bacillus thuringiensis (*Bt*) serovar*morrisoni*strain 3023 [27]	Endochitinase (chi3023) gene	*E. coli*
Aeromonashydrophila [15]	Chitinase gene	*E. coli*
Aeromonascaviae WS7b [16]	Chitinase gene	*E. coli*
Serratia marcescens KCTC 2172 [32]	Chitinase gene (pCHi58)	*E. coli*
Streptomyces lividans 66 [3]	Chitinase gene	Streptomyces *coelicolor* M130 *E. coli*

(*Continued*)

Table 7.2 Cont.

Microorganisms for target genes	**Target genes**	**Host for gene expression**
Streptomyces sp.(strain J-13–3) [1]	Chitinase gene	*E. coli* BL21(DE3)
Tea [18]	Chitinase gene	*Pichiapastoris*
Lecanicillium psalliotae [19]	Chitinase (Lpchi1) gene	*Pichiapastoris*
Chitinolyticbacter meiyuanensis SYBC-H1 [28]	*CHI2* chitinase gene	*E. coli* BL21
Barley [20]	Chitinase gene	*E. coli*
Aeromonas hydrophila, Pseudomonas maltophila [29]	Chitinase gene	*Bacillus thuringiensis var. istraelensis*
Bacillus licheniformis [29]	Chitinase (TP-1) gene	*B. thuringiensis Sub sp. Aizawai*
Bacillus circulans [29]	Chitinas (ChiA1) gene	*E. coli*
Enterobacter agglomerans [29]	Endochitinase (ChiA) gene	*E. coli*
Xanthomonas spp. [29]	Chitinase (ChiA) gene	*E. coli*
Kurthiazopfii [29]	Chitinase (chiSH1) gene	*E. coli*
Aphanocladium album [29]	Endochitinase gene	*Trichodermareesei*
Coccidioides immitis [29]	2 Chitinases (CTS1, CTS2) genes	*E. coli*
Brugiamalayi [29]	Chitinase (Chi) gene	*E. coli*

Rice [29]	Chitinase (Rcht 2) gene	*E. coli*
Bacillus sp. DAU101 [7]	Chitosanase gene	*E. coli* JM109
*Aspergillus fumigatus*JXSD-97 [21]	Chitosanase gene	*Pichiapastoris* strain GS115
Penicillium sp. D-1 [8]	Chitosanase gene	*E. coli*
Bacillus sp. strain K17 [30]	Chitosanase gene	*E. coli*
Bacillus cereus H-1 [31]	Chitosanase gene	*E. coli* DH5&
Bacillus sp. MET 1299 [22]	Chitosanase gene	*E. coli*
Aspergillus nidulans [5]	Chitin deacetylase gene	*E. coli* BL21
*Colletotrichum lindemuthianum*ATCC56676 [2]	Chitin deacetylase gene	*E. coli*
*Colletotrichum lindemuthianum*UPS9 [10]	Chitin deacetylase gene	*Pichiapastoris*
Rhizopus circinans [11]	Chitin deacetylase gene	*Pichiapastoris*
Vibrio furnissii [29]	*N*-Acetylglucosaminidase (*exo II*) gene, chitodextrinase (*endo 1*) gene	*E. coli*

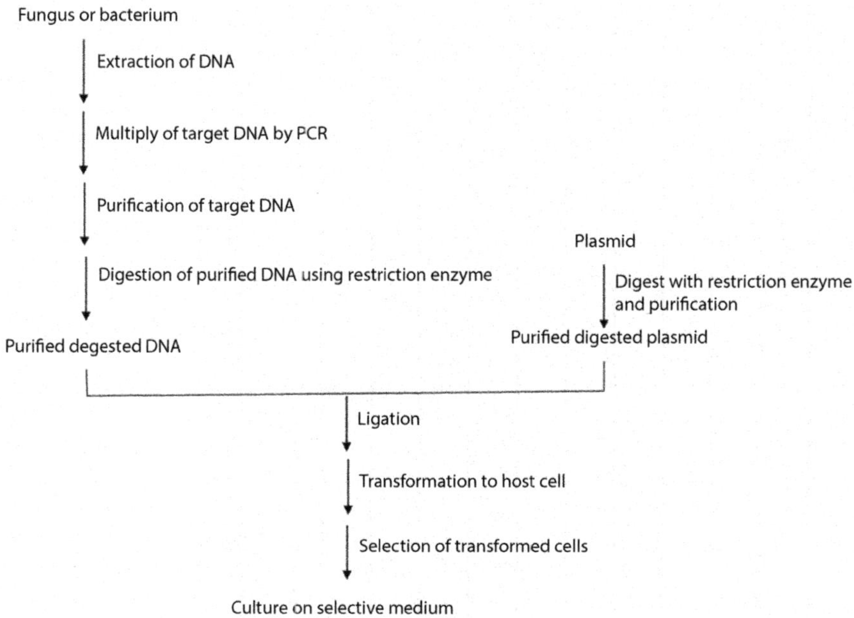

Figure 7.3 Steps for cloning.

activities. For example, chitin deacetylase enzyme was assayed at pH 8 and 50 °C by using glycol chitin. The resultant activity is 0.77 U/ml.

7.5 Conclusion

Enzymes have been produced from the microorganisms like bacteria, fungi, etc. The production rate of enzyme produced from these microorganisms is very low due to growth rate was slow and specific medium was required. Therefore, target enzyme-encoded DNA from these microorganisms was extracted and cloned to host microorganisms. In which *E. coli* was most used as host microorganisms due to its growth rate faster, easy to culture, and simple fermentation condition were used. The activities of enzyme extracted from transformant clones were nearly same as that from native strain. Therefore, these enzymes are expected to use in commercial scale production of chitin, chitosan, and CTO in near future.

Table 7.3 Determinations of enzyme activities using chitinous materials.

Enzyme		Substrate	Enzyme treatment conditions		Enzyme activities
			pH	Temp	
Chitinase [12]		Colloidal chitin	5–7	42 °C	–
Chitinase [13]	Chi-A(Native)	Chitin	8–9	50 °C	10 U/mg of protein
	Chi-78 (Clone)				9.8 U/mg of protein
	Chi-85 (Clone)				9.9 U/mg of protein
Chitinase [15]		Medium containing chitin or colloidal chitin			31 U/ml
Chitinase [32]		swollen chitin, chitobiose, and chitotriose	5	50 °C	
Chitinase [17]	Chi A	Colloidal chitin	7–11	37 °C	–
	Chi B			35 °C	
	Chi C			33 °C	
Chitinase [1]		Colloidal chitin	5.5–6.0	35–40 °C	30 mU/mg of protein
Chitinase Lpchi1 [19]		Colloidal chitin	7	37.6 °C	12.3 mU/ml
CHI2 Chitinase [28]		Colloidal chitin	6.5	40 °C	0.38 U/ml

(*Continued*)

Table 7.3 Cont.

Enzyme	**Substrate**	**Enzyme treatment conditions**		**Enzyme activities**
		pH	**Temp**	
Endochitinase chiA74 [25]	–	6	57.2 °C	5.98 U/mg
Chitosanase [7]	Soluble chitosan	7.5	50 °C	112.3 µmol/min/mg
Chitosanase [21]	1% chitosan containing medium	6–7		100 U/ml
Chitosanase [8]	Colloidal Chitosan	4	48 °C	26.4 U/mg
Chitosanase [31]	Soluble chitosan or glycol chitosan	4–11	50 °C	
Chitosanase [22]	90% deacetylated colloidal chitosan, ß-glucan	5.5	60 °C	
Chitin deacetylase [5]	Glycol chitin	8	50 °C	0.77 U/ml

References

1. Okazaki, K., Yamashita, Y., Noda, M., Sueyoshi, N., Kameshita, I., Hayakawa, S., Molecular cloning and expression of the gene encoding family 19 chitinase from *Streptomyces* sp. J-13-3. *Biosci. Biotechnol. Biochem.*, 68 (2), 341–351, 2004.
2. Tokuyasu, K., Ohnishi-Kameyama, M., Hayashi, K., Mori, Y., Cloning and expression of chitin deacetylase gene from a Deuteromycete, *Colletotrichum lindemuthianum. J BiosciBioeng.*, 87, 418–423, 1999.
3. Yashitat, K., Fujii, T., Awada, Y., Molecular cloning and characterization of chitinase genes from *Streptomyces lividans 66. J. Gen. Microbio.*, 137, 2065–2072, 1991.
4. Nwe, N., Furuike, T., *et al.*, Chitin and chitosan from terrestrial Organisms, In: Chitin, chitosan, oligosaccharides and their derivatives, S. Kim, Ed., pp. 11–23, CRC, 2010.
5. Wang, Y., Song, J., Yang, Q., Liu, Z., Huang, X., Chen, Y., Cloning of a heat-stable chitin deacetylase gene from *Aspergillus nidulans* and its functional expression in *Escherichia coli. Appl. Biochem. Biotechnol.*, 162, 843–854, 2010.
6. Li, A., Li, D., Cloning, expression and characterization of the serine protease gene from *Chaetomium thermophilum. J. Appl. Microbio.*, 106, 369–380, 2009.
7. Lee, Y.S., Yoo, J.S., Chung, S.Y., Lee, Y.C., Cho, Y.S., Choi, Y.L., Cloning, purification, and characterization of chitosanase from *Bacillus* sp. DAU101. *Appl. Microbiol. Biotechnol.*, 73, 113–121, 2006.
8. Zhu, X.F.,Tan, H.Q., Zhu, C., Liao, L., Zhang, X.Q., Wu, M., Cloning and overexpression of a new chitosanase gene from *Penicillium* sp. D-1. *AMB Express*, 2, 2012.
9. Kim, S., Je, J., Continous production of chitololigosaccharides by enzymatic hydrolysis, In: *Chitin, chitosan, oligosaccharides and their derivatives*, S. Kim, Ed., pp. 47–51, CRC, 2010
10. Shrestha, B., Blondeau, K., Stevens, W.F., Hegarat, F.L., Expression of chitin deacetylase from *Colletotrichum lindemuthianum* in *Pichiapastoris. Protein Expr. Purif.*, 38, 196–204, 2004.
11. Gauthier, C., Clerisse, F., Dommes, J., Jaspar-Versali, M.F., Characterization and cloning of chitin deacetylases from *Rhizopus circinans. Protein ExprPurif.*, 59, 127–137, 2008.
12. Lan, X., Zhang, X., Hu, J., Shimosaka, M., Cloning, expression and characterization of a chitinase from the chitinolytic bacterium *Aeromonas hydrophila* strain SUWA-9. *Biosci., Biotechnol., Biochem.*, 70(10), 2437–2442, 2006.
13. Tsujibo, H., Orikoshi, H., Tanno, H., Fujimoto, K., Miyamoto, K., Imada, C., Okami, Y., Inamori, Y., Cloning, sequence, and expression of a chitinase gene from a marine bacterium, *Alteromonas* sp. Strain 0-7. *J. Bacteriol.*, 175, 176–181, 1993.

14. Inglis, P.W., Peberdy, J.F., Socket, R.E., Cloning of a chitinase gene from *Ewingella Americana*, a pathogen of the cultivated mushroom, *Agaricusbisporus. Genet. Mol. Biol.*, 23, 3, 685–688, 2000.
15. Chen, J.P., Nagayama, F., Chang, M.C., Cloning and expression of a chitinase gene from *Aeromonas hydrophila* in *Escherichia coli. Appl. Environ. Microbiol.*, 57, 2426–2428, 1991.
16. Malik, A., Wenuganen, S., Suwanto, A., Tjahjono, B., Cloning, DNA sequence, and expression of *Aeromonas carviae* WS7b chitinase gene. *Mol. Biotechnol.*, 23, 2003.
17. Danismazoglu, M., Demir, I., Sezen, K., Muratoglu, H., Nalcacioglu, R., Cloning and expression of chitinase A, B, and C (chiA, chiB, chiC) genes from *Serratia marcescens* originating from *Helicoverpaarmigera* and determining their activities. *Turk. J. Biol.*, 39, 78–87, 2015.
18. Roy, S.C., Molecular cloning and expression of tea chitinase gene in *Pichiapastoris. IJBR*,5, 612–618, 2014.
19. Gan, Z., Yang, J., Tao, N., Liang, L., Mi, Q., Li, J., Zhang, K., Cloning of the gene *Lecanicillium psalliotae chitinase Lpchi1* and identification of its potential role in the biocontrol of root-knot nematode *Meloidogyne incognita. Appl. Microbiol. Biotechnol.*,5, 612–618, 2014.
20. Kirubakaran, S.I., Sakthivel, N., Cloning and overexpression of antifungal barley chitinase gene in *Escherichia coli. Protein Expr. Purif.*, 52, 159–166, 2007.
21. Li, Y., Ping, Y., Manying, W., Expression of chitosanase gene from *Aspergillus fumigatus* JXSD-97 in *Pichiapastoris. Afr. J. Biochem. Res.*, 4(8), 204–209, 2010.
22. Kim, P.I., Kang, T.H., Chung, K.J., Kim, I.S., Chung, K., Purification of a constitutive chitosanase produced by *Bacillus* sp. MET 1299 with cloning and expression of the gene. *FEMS Microbiol. Lett.*, 240, 31–39, 2004.
23. Songjang, K., Donchai, T., Chaiyawat, P., Meyer, C.R., Cloning and expression of chitinase gene isolated from insect pathogenic fungi, *Beauveria bassiana* in *Escherichia coli. Chiang Mai J. Sci.*, 33(3), 347–355, 2006.
24. Robbins, P.W., Albright, C., Benfield, B., Cloning and expression of a *Streptomyces plicatus* chitinase (Chitinase-63) in *Escherichia coli. J. Biol. Chem.*, 263, 443–447, 1988.
25. Barboza-Corona, J.E., Nieto-Mazzocco, E., Velazquez-Robledo, R., Salcedo-Hernandez, R., Bautista, M., Jimenez, B.E. Ibarra, J.E, Cloning, sequencing, and expression of the chitinase gene *chiA74* from *Bacillus thuringiensis. Appl. Environ. Microbiol.*, 69, 1023–1029, 2003.
26. Watanabe, T., Suzuki, K., Oyanagi, W., Ohnishis, K., Tanaka, H., Gene cloning of chitinase al from *bacillus circulans* wl-12 revealed its evolutionary relationship to *Serratia* chitinase and to the type III homology units of fibronectin. J. *Biol. Chem.*, 265, 15659–15665, 1990.
27. Okay, S., Ozcengiz, G., Molecular cloning, characterization, and homologous expression of an endochitinase gene from B*acillus thuringiensis* serovar *morrisoni. Turk. J. Biol.*, 35, 1–7, 2011.

28. Hao, Z., Wu, H., Yang, M., Chen, J., Xi, L., Zhao, W., Yu, J., Liu, J., Liao, X., Huang, Q., Cloning, Expression and 3D Structure Prediction of Chitinase from *Chitinolyticbacter meiyuanensis* SYBC-H1. *Int. J. Mol. Sci.*, 17, 825, 2016.
29. Patil, R.S., Ghormade, V., Deshpande, M.V., Chitinolytic enzymes: an exploration. *Enzyme. Microb. Technol.*, 26, 473–483, 2000.
30. Yatsunami, R., Sakihama, Y., Suzuki, M., Fukazawa, T., Shimizu, S., Sunami, T., Endo, K., Takenaka, A., Nakamura, S., A novel chitosanase from *Bacillus sp.* Strain K17: gene cloning and expression in *Escherichia coli*. *Nucleic Acids Res. Suppl.*, 2, 227–228, 2002.
31. Jang, Hong-Ki, Yi, J., Kim, J., Lee, K., Choi, S., Purification, characterization, and gene cloning of chitosanase from *Bacillus cereus* H-1. *Kor. J. Microbiol. Biotechnol.*, 31, 216–223, 2003.
32. Gal, S.W., Lee, S.W., Choi, Y.J., Molecular cloning and characterization of 58 kDa chitinase gene from *Serratia marcescens* KCTC 2172. *Biotechnol. Bioprocess Eng.*, 7, 38–42, 2002.

Section II

CHITOSAN IN TEXTILE AND FOOD INDUSTRY

8

Chitosan Applications for the Food Industry

Tomy J. Gutiérrez

Institute of Research in Materials Science and Technology (INTEMA), National University of Mar del Plata (UNMdP) and National Council of Scientific and Technical Research (CONICET), Buenos Aires, Argentina

Abstract

The development of active biomolecules such as chitosan and its derivatives for their use in food applications in recent years has progressed significantly and rapidly. Chitosan is produced by the deacetylation of chitin, the second most abundant polymer in nature next to cellulose. Chitosan is a unique cationic polysaccharide with many functional and bioactive properties such as antioxidant, lipid-lowering and antimicrobial activities, film-forming and gelling properties, encapsulation potential, etc., which are described in this chapter. The chapter also focuses on advances in the development of chitosan and its derivatives in the last five years as regards their use in novel applications related to food science and technology. Nutritional aspects are also discussed.

Keywords: Biopolymers, chitosan, functional foods

8.1 Introduction

Chitosan is a biopolymer derived from chitin, the second most abundant natural polymer next to cellulose. Chitin can be found in many places in nature, including the exoskeletons of crustaceans, insects, mollusks, and fungi. The exoskeleton of crustaceans is currently the major industrial source of biomass for chitin production, with an estimated 1,560 million tons available worldwide from marine environments [1].

**Corresponding author*: tomy.gutierrez@fi.mdp.edu.ar; tomy_gutierrez@yahoo.es

Shakeel Ahmed and Saiqa Ikram (eds.) Chitosan, (185–232) © 2017 Scrivener Publishing LLC

Crustacean shell waste consists of protein (20–40%), calcium and magnesium salts (mainly carbonate and phosphate, 30–60%), chitin (20–30%), and lipids (0–14%). These proportions vary with the species and season [2]. Crustaceans contain α-chitin, which is by far the most common and widely available form of chitin.

In contrast with shell mollusks, a less pure source of chitin, squid skeleton comprises 40% chitin, which is nearly free of calcium salts. Squid pen is also the only important source of the polymorph, β-chitin [3].

Cellulose, chitin, and chitosan all share very similar backbone structures, as shown in Figure 8.1. The only difference among these three molecules is the functional group at the C-2 position. The molecular chains of chitin are formed by replacing the hydroxyl group at the C-2 position on the cellulose molecular chains with an acetamido group, thus generating linear structures of 2-acetamido-2-deoxy-β-D-glucose through β(1→4) linkages. Chitosan is a derivative of chitin, and is produced by the *N*-deacetylation process which results in the placement of the amino group at the C-2 position on the chitosan backbone. Like cellulose, both chitin and chitosan are considered to be naturally occurring polysaccharides, and are of particular commercial interest because of their high nitrogen content (6.89%) [4]. On the surface, both cellulose and chitin are water-insoluble polymers with very low chemical reactivity. When the degree of deacetylation reaches about 50%, however, chitosan becomes water-soluble under acidic pH conditions (lower than its pKa ~ 6.2) due to the protonation of $-NH_2$ groups, which results in its solubilization. After protonation chitosan carries positive surface charges on its d-glucosamine repeat unit, thereby becoming the only pseudo natural cationic biopolymer. Chitosan is also well-known for its biocompatibility, biodegradability, and low toxicity. Because of these unique properties, chitosan is considered as a versatile biopolymer that can be developed into various forms, such as gels, films, nano/micro-particles, beads, etc. It thus has numerous potential applications in different fields, including the food, pharmaceutical, and cosmetic sciences.

Due to the hydroxyl and amino groups on its backbone, chitosan is an amenable molecule that can be easily modified by various methods. Modifications of chitosan have been aimed at improving its physicochemical

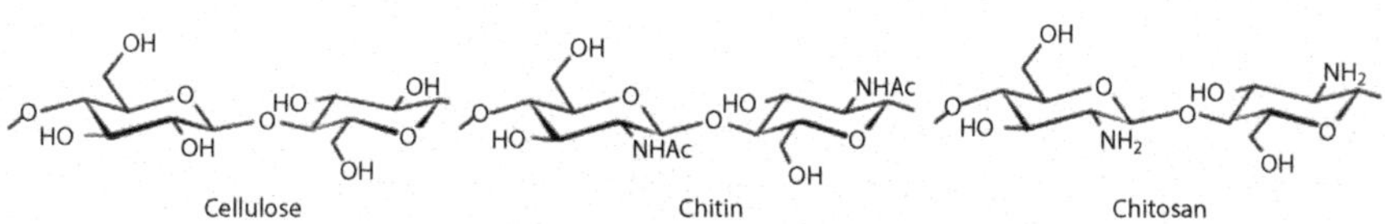

Figure 8.1 Molecular structures of cellulose, chitin, and chitosan.

properties and in order to extend its applications in different situations. Several modification methods have been well documented for chitosan, including chemical, physical, and enzymatic approaches. For instance, because its large molecular weight and high viscosity may limit its applications under certain biological conditions, chitosan is often depolymerized in order to obtain oligosaccharides and/or monomers [5]. Depolymerization is usually achieved *via* enzymatic modifications using chitonase. Many chemical means for chitosan modification have also been reported, such as quaternization, *N*-alkylation, hydroxylalkylation, carboxyalkylation, thiolation, glycation, etc. [5]. Physical modifications include electromagnetic radiation and sonication. Modified chitosans with enriched properties, such as excellent solubility in aqueous solutions at different pHs, modulated surface charges, higher absorption efficiency, new cross-linking sites, amongst others, have found novel applications and can be tailored to a particular purpose.

In the last decade, chitosan has received increasing attention and its applications have been extensively explored in many scientific fields. In the food sciences, chitosan has an advantage over synthetic polymers, as it is considered as GRAS (Generally Recognized as Safe) by the Food and Drug Administration (FDA). This review focuses on the most recent advances in the study and applications of chitosan and its derivatives in food science and technology, including their nutritional and antimicrobial properties, uses as edible coatings, food packaging, and emulsion and encapsulation materials, and potential as agents for enzyme immobilization and water purification.

8.2 Biological Activities of Chitosan and Its Derivatives

8.2.1 Antioxidant Properties

The antioxidant activities of chitosan have been extensively studied both *in vitro* and *in vivo* using different methodologies, and are reported to be correlated to its structural characteristics, including Molecular Weight (MW) and Degree of Deacetylation (DD), as well as the source material. The MW and DD may also have some synergistic effects on the biological activities of chitosan, and different results have been obtained with different free radical generating systems. The most recent advances regarding the antioxidant properties of various chitosan-based products are summarized in Table 8.1.

Table 8.1 Antioxidant properties of chitosan and its derivatives (2008–2013).

Chitosan and derivatives	**MW/DD**	***In vitro* studies - free radicals scavenging and metal chelation assays**	**References**
Chitosan from crabs	Unknown/83.3-93.3%	DPPH•, reducing power, hydroxyl radicals, Fe^{2+} chelation	[6]
Chitosan from the larvae of the housefly, *Musa domestica*	426 kDa/90%	DPPH•, hydroxyl and superoxide anion radicals, reducing power, Fe^{2+} chelation	[7,8]
Phosphorylated chitosan from squid gladius	Unknown	DPPH•, ABTS, Fe^{2+} chelation, hydroxyl radicals	[9]
Chitosan supplement (Chitosamin®)	100 kDa/90%	DPPH•, ABTS	[10]
Water soluble chitosan (Chitosan-up®)	20 kDa/95%	Human serum albumin, DPPH•, ABTS	[11]
N-acyl chitosan oligosaccharide	7310 kDa/–	Hydroxyl and superoxide anion radicals, reducing power	[12]
N-(2-hydroxyl-3-trimethylammonium) propyl chitosan chloride	9641 (weight average molar mass)/80%	β-carotenen linoleic acid activity, DPPH•, reducing power, hydrogen peroxide radical	[13]
Quatermized chitosan conjugated with gallic acid or caffeic acid	70 kDa/97%	Hydroxyl radical, superoxide radical, DPPH•	[14]
2-[phenylhydrazine (or hydrazine)-thiosemicarbazone]-chitosan	200, 8 kDa/96%	Hydroxyl and superoxide anion radicals, reducing power, Fe^{2+} chelation	[15]

Callic acid grated chitosan	307 kDa/80%; 950 kDa/92%	DPPH•, reducing power, horseradish peroxidase, carbon-centered alkyl radicals, hydroxyl radicals	[16,17]
Ferulic acid-grafted chitosan	200 kDa/90%	DPPH•	[18]
Caffeic acid-grafted chitosan	47, 198, 544 kDa/ unknown	DPPH•, reducing power	[19]
Chitosan-glucose conjugate	Unknown/78–82%; 105 kDa/90%	DPPH•, reducing power, superoxide radical, β-carotenen linoleic acid activity	[20–22]
Chitosan-xylan conjugate	810 kDa/90% 118 kDa/95%	Lipid peroxidation *In vivo* studies – oxidative stress	[23]
Chitosan	750 kDa/85–87%	Age-associated dyslipidemic rats - attenuated oxidative stress in the heart tissue of aged rats	[24]
Chitosan	Unknown	Hyperlipidemic rats – elevated serum levels of antioxidative enzymes	[25,26]
Chitosan supplement (Chitosamin®)	100 kDa/90%	Healthy human subjects and nephrectomized rats - reduced albumin oxidation and increased plasma antioxidant activity	[11,27]
Water soluble chitosan (Chitosan-up®)	20 kDa/95%	Healthy human subjects – reduced albumin oxidation and increased plasma antioxidant activity	[11]

Abbreviations: MW – molecular weight; DD – degree of deacetylation; DPPH• – 1, 1-diphenyl-2-picrylhydrazyl; ABTS – 2, 2′-azino-bis (3-ethylbenzothiazoline-6-sulphonic acid)

8.2.2 Antioxidant Properties of Native Chitosan and Its Oligomers

Yen *et al.*, investigated the antioxidant properties of chitosan prepared from crab shells [6] and shiitake stipes [28]. Both studies revealed that the antioxidant activities of chitosan increased with an increase in DD during its preparation. Longer *N*-deacetylation times resulted in more amino groups at the C-2 positions which contributed significantly to the antioxidant activities. Hydroxyl radicals scavenging, conjugated diene formation, and reducing power assays produced the greatest antioxidant activity values for the chitosans from both sources with EC50s of lower than 1.5 mg/ml. Less satisfactory activities were found for (1,1-diphenyl-2-picrylhydrazyl) DPPH• radicals and ferrous ion chelating ability tests, which gave EC50s as high as 16.3 mg/ml. EC50 is defined as the effective concentration that produces 50% antioxidant activity. The MW of chitosan has also been reported to have a major effect on its antioxidant activities, with low MW chitosans having more pronounced scavenging effects on superoxide and hydroxyl radicals than those with high MW [29]. Another study investigated the antioxidant properties of four chitosan oligomers obtained by H_2O_2 degradation, and confirmed that the MW is negatively correlated with their antioxidant activities [30]: the chitosan oligomer with the lower MW, 2300 Da, showed greater antioxidant activities against superoxide anion and hydroxyl radicals than the higher Mw oligomer. This is because high MW chitosans are more compact in structure due to stronger intramolecular hydrogen bonding, such as between N_2–O_6 and O_3–O_5, than low MW ones. The hydroxyl and amino groups in the lower MW chitosan were thus more able to react with the free radicals, and hence exhibited higher free radical scavenging properties.

8.2.3 Antioxidant Properties of Modified Chitosan

In order to further improve the antioxidant properties of chitosan by overcoming its solubility limitation, chitosan molecules have been modified in several ways. Sulfation has generated an extremely important family of chitosan derivatives with enhanced biological activities, especially as regards their antioxidant properties. Sulfated and sulfanilamide chitosans have been shown to possess significantly improved free radical scavenging activities compared to native chitosan [29, 31] mainly due to their enhanced water solubility and hence increased availability to free radicals. Many other modifications that have improved the free radical scavenging activities of chitosan have also been reported, for example, the Schiff bases reaction [32], quaternization [33, 34] carboxymethylation [32, 35] and acylation [36]. Recently, the conjugation of small antioxidant molecules with chitosan has been developed as a new approach to improve its antioxidant properties. This type of modification is aimed at upgrading the functionality of chitosan with the incorporation of natural plant antioxidants to its polysaccharide backbone. Chemical cross-linking methods are generally adopted to graft antioxidants to chitosan molecules, including essential oils [37, 38] and phenolics [17, 39]. As a result the antioxidant properties of grafted chitosan have been significantly improved, especially as regards DPPH• scavenging and metal ions chelating activities, where native chitosan shows very little activity. Unfortunately, the toxic and irritating reagents are required for the chemical modification process have made this method less popular. However, another recent, novel and attractive method, enzymatic modification, has received increasing interest as an alternative to chemical cross-linking. Using this procedure, several bioactives have been successfully grafted to chitosan molecules, including flavonoids [40, 41] and phenolics [42, 43]. Finally, physical modifications, such as ionizing radiation [44] and irradiation [45], have also been reported to improve the antioxidant activities of chitosan.

8.2.4 *In Vivo* Studies of the Antioxidant Properties of Chitosan

The antioxidant properties of chitosan have not only been demonstrated from *in vitro* antioxidant assays but also by many *in vivo* studies using various animal models as well as clinical trials. Anraku *et al.*, investigated the effects of high MW chitosan on antioxidant stress and chronic renal failure using nephrectomized rats [27]. The study showed that the ingestion of chitosan over a 4-week period not only resulted in a significant decrease in the ratio of oxidized to reduced albumin and an increase in the biological antioxidant potential, but also alleviated renal failure. Another recent study also revealed the antiaging effect of high MW chitosan in a glutathione-dependent antioxidant system in rats [24]. The oral administration of chitosan was reported to significantly attenuate oxidative stress in the heart tissue of aged rats by maintaining antioxidant enzymes. In addition, the antioxidant activities of high MW chitosan have been recently demonstrated in a clinical trial [10]. The antioxidant properties of low MW chitosan, especially oligosaccharides, have also been widely studied in many *in vivo* experiments. For instance, chitosan oligosaccharides have been shown to protect mice from liposaccharide-induced sepsis by attenuating organ dysfunction, increasing the levels of antioxidative enzymes, and preventing redox imbalance [46]. Similar effects of chitosan oligosaccharides have also been observed in streptozotocin-induced diabetic rats [47]. As discussed above, *in vitro* assays show that low MW chitosan possesses greater antioxidant activities than high MW chitosan. However, the comparison of MW-dependent antioxidant activities has not yet been completely elucidated. Furthermore, the *in vivo* evaluation of chitosan derivatives is still not well-established, especially as regards their toxicity. This is important as one or more toxic chemicals are usually involved in the modification process which may pose a potential risk of poisoning to animals or humans.

8.2.5 Lipid-Lowering Effects

The lipid-lowering activities of chitosan have been observed since the 1980s [48, 49] in animal models, and were reported for the first time in humans without any side effects as early as 1993 [50]. Since then, the increasing evidence of the potent lipid-lowering effects of chitosan has been consistently reported in the literature. The pronounced capacity of chitosan to lower plasma and hepatic triacylglycerol, as well as total cholesterol levels, is well documented and has been observed in a large number of animal studies. Currently, however, more and more attention has been given to exploring the mechanisms underlying the hypolipidemic activity of chitosan, and several have been recently proposed. Firstly, chitosan binds strongly to fat, cholesterol, and bile salt. This binding effect results partly from hydrophobic interactions and hydrogen bonding between the lipids and chitosan [51], and partly from the electrostatic attraction between the positively charged amino groups of chitosan and negatively charged carboxyl groups of fatty acid and bile salts. This last leads to the entrapment of fat droplets in the stomach and their later precipitation in the small intestine, thus delaying the digestion of fat [52]. Secondly, chitosan can help the body maintain levels of antioxidant enzymes such as superoxide dismutase (SOD), catalase (CAT), and glutathione peroxidase (GSH-PX), all of which play important roles in lipid peroxidation [25, 26]. Thirdly, chitosan has a profound impact on plasma adipocytokines, significantly reducing the adiposity index. This means that it can regulate the level of circulating triglycerides and counteract some inflammatory disorders and metabolic alterations [53].

Different chitosans have different physicochemical properties which significantly affect the efficacy of their lipid-lowering activities *in vivo*. The DD of a particular chitosan determines the number of free amino groups on its backbones. The more free amino groups available, the higher the positive surface charge of the chitosan, thus strengthening its capacity to electrostatically bind to fatty acids and bile salts. The MW also plays an important role in lipid-lowering efficacy. High MW chitosans are more viscous than low Mw chitosans, which can help the binding and entrapment of fat droplets in the gastrointestinal tract, thus preventing their absorption. This phenomenon has been verified in a recent study which showed that high MW chitosan exhibited a more pronounced effect on the increase of fecal fat and cholesterol in mice, while low MW chitosan seemed to be more effective in elevating lipoprotein lipase activities in mouse plasma and liver [54]. Additionally, the particle size of chitosan powders greatly influences their lipid-lowering effects. Nanopowdered chitosan prepared by ultrafine milling has been reported to be more effective than ordinary chitosan, due to the larger surface area available for interaction with lipids, resulting in their increased excretion in the feces [55]. Several studies have also suggested that the modification of chitosans by different methods can alter their lipid-lowering mechanisms [56–58].

8.2.6 Other Biological Activities

Chitosan has many other well-documented biological activities and health benefits [59–61] including the prevention of renal failure, wound healing, the reduction of gastric ulcers (anti-inflammatory effect), antigenotoxic, and anticancer activities, etc. As novel modification methods have continued to be developed in recent years, more and more studies have focused on the biological activities of chitosan derivatives, especially those with low MW. For instance, chitosan oligosaccharides prepared from enzymatic degradation/depolymerization exhibit many new biological functions, such as immune-modulating and hemostasis effects [62].

8.3 Chitosan and Its Derivatives for Food Applications

8.3.1 Antimicrobial Applications

Chitosan is widely recognized for its potent and broad spectrum antimicrobial activity and high killing rate, coupled with its low toxicity toward mammalian cells. Although the mode of antimicrobial action of chitosan is not completely understood, it is well established that its molecular structure is a prerequisite for its antimicrobial properties [63, 64]. The polycationic behavior of chitosan in an acidic medium is the main factor contributing to its antimicrobial activity. Due to its positive surface charges under acidic conditions, chitosan interacts with anionic components on the bacteria surface: for example, negatively charged lipopolysaccharides in the outer membrane of gram-negative bacteria, and peptidoglycan and teichoic acid in the cell walls of gram-positive bacteria. These electrostatic interactions produce the release of most of the proteinaceous materials from the cells. A recent study suggested that chitosan could have a profound effect on the negative charges of the cryotococcal cellular membranes of fungi, consequently interfering with surface colonization and cell–cell interactions during biofilm formation [65]. As shown in Figure 8.2, chitosan significantly reduced capsule size and the number of sheared capsular polysaccharides in crytococcal cells, as well as clipping the exopolymeric matrix from biofilms, which together explained the underlying mechanisms of chitosan in preventing the formation of fungal biofilms. This potent effect was ascribed to the electrostatic interactions between the chitosan molecules and microbial cell membranes which led to the leakage of proteinaceous materials, consequently increasing chitosan penetration to the nucleus and binding to the DNA, thus inhibiting mRNA synthesis.

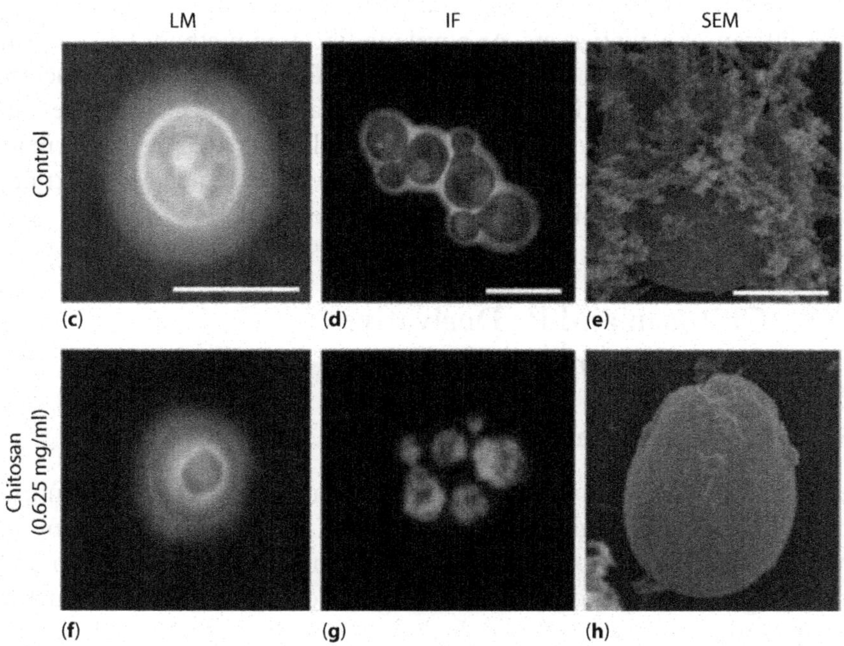

Figure 8.2 Effects of chitosan on capsule size (e and f) and the number of capsular polysaccharides (d and g) in *Cryptococcus neoformans* cells, and the exopolymeric matrix of biofilms (e and h). Scale bars in all pictures represent 2 μm [65].

The antimicrobial efficacy of chitosan can be affected in many ways. Intrinsic factors include DD, MW, and chemical modifications, whereas extrinsic factors are related to environmental conditions, especially the pH and ionic strength of the medium, and the target bacterial species. Different DDs produce different numbers of amino groups on the chitosan backbones. This determines the charge density and thus the intensity of electrostatic interactions with microbial cell membranes. The MW reflects the compactness of the chitosan molecular structure, which may affect the availability of functional groups that react with microbial cells. Modifications of chitosan that increase the number of amino groups or decrease MW are known to change its molecular structure and hence increase its antimicrobial activity. For instance, the asparagines *N*-conjugated chitosan oligosaccharide, with two positively charged sites and a low MW, interacts more strongly with carboxyl-negative charges on bacterial cell walls [66] not only due to the higher number of amino groups, but also the greater flexibility of the chitosan structure after a decrease in the MW. Other possible modes of action of antimicrobial activity are hydrophobic interactions and metal chelating effects, particularly in the case of water soluble chitosan [67] or when the pH of the medium is above its pKa [63].

Due to its biodegradability, nontoxicity and intrinsic antibacterial effects, chitosan has been widely used as an antimicrobial agent to improve food quality and extend shelf life. The desired antimicrobial effects may be achieved by using chitosan and its derivatives either alone or blended with other ingredients. For example, chitosan has been combined with biocontrol yeast and/or calcium chloride to control blue mold in pears [68]. The synergistic effects of mixing chitosan, biocontrol yeast and calcium chloride produced a more effective and stable reduction in fungal decay than treatment with either chitosan or biocontrol yeast alone. In addition to bacterial and antifungal activity, chitosan has been recently tested for its efficacy against foodborne viruses, including human norovirus and enteric virus surrogates [69]. In some cases, chitosan has been modified to improve its antimicrobial efficacy in specific food systems. For example, a chitosan–glucose complex coating was found to be the most effective treatment for protecting mushrooms from microbial degradation and improving postharvest quality, compared to treatment with a chitosan or glucose coating alone [70]. The chitosan–glucose coating maintained mushroom tissue firmness, inhibited an increase in the respiration rate, reduced microorganism counts (tested pseudomonads, yeasts, and molds), and delayed changes in ascorbic acid concentrations, whilst maintaining overall sensory quality. The antimicrobial activities of water soluble chitosan have also been well characterized [71, 72]. A recent study developed chitosan- and carboxymethyl chitosan–zinc complexes to compare their antimicrobial activities [73]. The study revealed that the carboxymethyl chitosan–zinc complex exhibited much greater antimicrobial activity against both gram-positive and gram-negative bacteria than the chitosan–zinc complex. The authors deduced that the carboxymethyl groups greatly improved the water solubility of chitosan enhancing its diffusivity, thus making higher concentrations available at the site of action.

8.3.1.1 Antimicrobial Activity of Modified Chitosan

Although chitosan has attracted attention as a biopolymer with distinctive physicochemical properties and biological activities, it is currently only utilized to a limited extent. This delay in the application of chitosan is partly attributable to the difficulty in controlling the modifications due to its insolubility in organic solvents, and multifunctionality. Many types of modification reactions have, however, been explored with the aim of increasing its antimicrobial properties. The growing demand for a more accurate control of polysaccharide properties has prompted the development of numerous techniques for selective modification. The amino and two hydroxyl groups present in the repeating unit of chitosan are the most common targets for different chemical modifications [74] as they significantly affect the functionality of linear polysaccharides according to the presence, level and distribution of substituents along the main chain. As specified by Rabea *et al.*, [64], chitosan and chitin are commercially interesting compounds because of their high nitrogen content (6.89%) compared to synthetically substituted cellulose (1.25%). This makes chitosan a useful chelating agent. However, these naturally abundant materials are also limited in their reactivity and processability.

Several alkylated chitosans have been synthesized. Kim *et al.*, [75] prepared *N*-alkyl chitosan derivatives by introducing alkyl groups into the amine groups of chitosan *via* Schiff's base intermediates. In fact, in homogeneous mediums such as acetic acid and methanol the free amine groups of the chitosan reacted with the aldehydes to give Schiff's base (Figure 8.3) [76].

Long alkyl chains (until C12) can be introduced onto chitosan. The quaternization of *N*-alkyl chitosan derivatives have been achieved using a halogenoalkane in the presence of sodium hydroxide [77, 78]. As shown in Figure 8.4, MeI was used to produce water-soluble cationic polyelectrolytes and novel chitosan derivatives, with quaternary ammonium salts [75, 78].

The antibacterial activities of this modified chitosan against *S. aureus* increased with an increase in the chain length of the alkyl substituent when tested by the viable cell count method in acetate buffer at pH 6, and

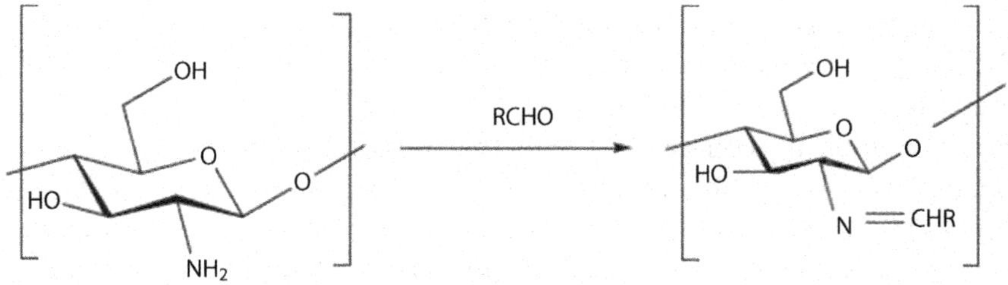

Figure 8.3 Schiff's base obtained from the reaction between free amino groups of chitosan and aldehydes.

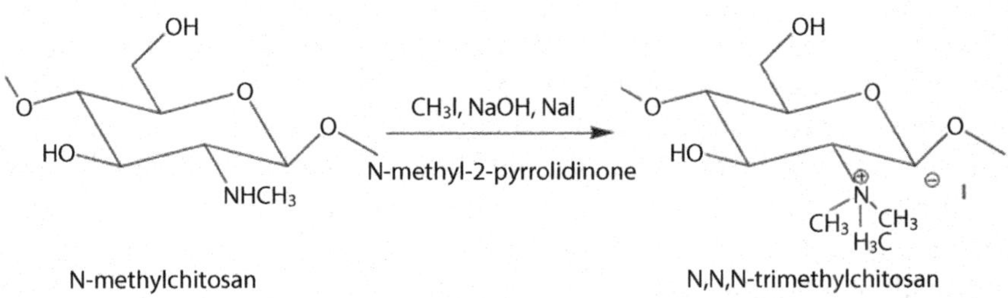

Figure 8.4 Synthesis of *N,N,N*-trimethylchitosan [78].

this increased activity was ascribed to the contribution of the heightened hydrophobic properties of the derivatives. Avadi *et al.*, [79] prepared a quaternized chitosan, *N*-diethylmethyl chitosan (DEMC) using a modified two-step process. The DEMC exhibited greater antibacterial activity than native chitosan against *Escherichia coli* at a degree of quaternization of 79%. However, the antimicrobial effects of both compounds were pH dependent such that an increase in the concentration of acetic acid produced a significant decrease in the minimum inhibitory concentration (MIC), as determined by the turbidimetric method. As a result, the antibacterial activities of the native chitosan and DEMC were higher in 1% acetic acid than in lower concentrations of acetic acid, 0.25%, for example. In addition, the MIC of the DEMC decreased from 500 mg/ml to 62.5 mg/ml when the medium was changed from water to a 1% acetic acid solution. The authors suggested that this is due to the polycation target site, i.e. the negatively charged surface of the bacteria cell. Thus, the high charge density polycated DEMC interacts with the bacteria more strongly than chitosan alone. Other chitosan derivatives such as *N,N,N*-trimethyl chitosan, *N*-propyl-*N,N*-dimethyl chitosan and *N*-furfuryl-*N,N*-dimethyl chitosan were prepared and tested for their activities against *E. coli* [77]. The antibacterial activity of quaternary ammonium chitosan against this bacterium was stronger in an acetic acid medium than in water and also stronger than the antibacterial activity of chitosan alone. More recently, Chi *et al.*, [80] prepared chitosan-*N*-2-hydroxypropyl trimethyl ammonium chloride by reacting chitosan with glycidyl trimethylammonium chloride. The chitosan derivatives, with different MWs, near 1.7 kDa, 35.7 kDa, 90.2 kDa, and 415.5 kDa, showed biocidal activity against, *Staphylococcus epidermidis, S. aureus, Bacillus subtilis,* and *Candida albicans*. However, the chitosan with a MW of 415.5 kDa exhibited a slightly lower biocidal activity against *C. albicans* than the other species tested. Nevertheless, the biocidal activities of the chitosan derivatives against the gram-positive bacteria tested were positively correlated with their MWs between 1.7 kDa and 90 kDa. In contrast, these same derivatives did not exhibit any biocidal effects against *E. coli* or *Pseudomonas aeruginosa* even at concentrations up to 10 mg/ml. In another study, Kim *et al.*, [81] synthesized diethylaminoethyl-chitosan (DEAE-chitosan) through the deacetylation of a diethylaminoethyl-chitin (DEAE18 chitin) by introducing DEAE groups onto the chitin C(6)-OH. The deacetylation was conducted by heating in aqueous 10% sodium hydroxide containing sodium borohydride. The DEAE-chitin was then quaternized to produce triethylaminoethyl-chitin (TEAE-chitin). The antibacterial activities of these compounds against *S. aureus* and *E. coli* were evaluated by colony count using the shake flask method, and were found to increase in the following order: DEAE-chitin, DEAE-chitosan and TEAE-chitin. In order to obtain copolymers with zwitterionic properties, Jung *et al.*, [82] prepared water-soluble anionic chitosan moieties and investigated their antimicrobial activities. Mono(2-methacryloyl oxyethyl) acid phosphate (chitosan-MAP) and vinylsulfonic acid sodium salt (chitosan-VSS) were grafted onto chitosan (Figure 8.5). Chitosan-MAP and chitosan-VSS exhibited inhibition values of about 95% and 75%, respectively, on the growth of *C. albicans*. However, both chitosan derivatives showed the same high pH-dependence, such that if the pH rose from 5.7 to 6.2, their antimicrobial activities dropped by 10–15%: lower than those of the parent chitosan. In addition, MAP grafting had a low negative impact on antimicrobial activity against *Trichophyton rubrum*, whereas the anionic chitosan showed much enhanced bioactive action against *T. violaceum*, compared to unmodified chitosan. The authors suggested that this selectivity could result from the structural affinity between the cell walls of microbial strains and chitosan or its derivatives. A possible reason for this might be that the cell walls of microorganisms may consist of either chitin, chitosan, or β-glycan.

Other chitosan derivatives are sulfonated and sulfobenzoyl chitosans. The antibacterial activities of these chemically modified chitosans were

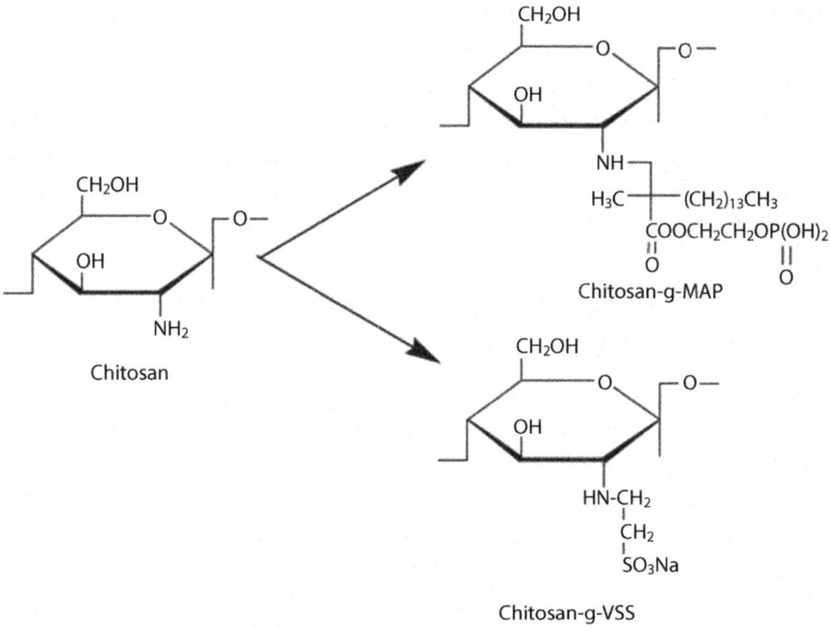

Figure 8.5 Mono (2-methacryloyl oxyethyl) acid phosphate (chitosan-MAP) and vinylsulfonic acid sodium salt (chitosan-VSS) grafted onto synthesized chitosan by Jung *et al.* [82].

evaluated and compared with those of 69% deacetylated chitosan by Chen *et al.*, [83]. The MIC values of sulfonated chitosan (0.63% sulfur content) against *Shigella dysenteriae*, *Aeromonas hydrophila*, *S. typhimurium*, and *Bacillus cereus* were found to be lower than those of the deacetylated chitosan. Thus the high sulfur content of the sulfonated chitosan adversely influenced its antibacterial effects. Sulfobenzoyl chitosan exhibited excellent water solubility and an antibacterial effect comparable to that of sulfonated chitosan. As regards their antifungal properties, Cuero *et al.*, [84, 85] showed that aqueous solutions of *N*-carboxymethylchitosan suppressed both the growth of, and aflatoxin production by *Aspergillus flavus* and *A. parasiticus* in submerged cultures. In another study, five chemically modified chitosans were tested for their antifungal activities against *Saprolegnia parasitica* [86] using the radial growth assay in chitosan-bearing agar and the fungal growth assay in chitosan-bearing broth (members of the *Saprolegnia* genus are responsible for infections of fish and eggs in aquaculture facilities). The results indicated that methylpyrrolidinone chitosan, *N*-carboxymethyl chitosan and *N*-phosphonomethyl chitosan exerted effective fungistatic action against the target strain. Electron microscopy observations provided evidence of ultrastructural alterations, damaged fungal structures, the uptake of modified chitosans, and hyphal distortion and retraction. In conclusion, chemical modifications of chitosan amino sites are numerous and relatively easy due to the reactivity of the primary amine. However, in order to preserve the amino groups and maintain their bioactive properties, modification procedures that yield C-6 modified chitosan derivatives are the ones most commonly employed.

8.3.2 Chitosan-Based Emulsions for Food Quality Preservation

Chitosan is considered to be an excellent emulsifier among food grade hydrocolloids and can be used to stabilize oil in water (O/W) without the addition of any other surfactant. This is because, under acidic conditions, the amino groups are protonated giving chitosan a positive surface charge. This converts it into an amphiphilic substance that can adsorb bioactive substances at oil/water interfaces, thus facilitating the formation of an emulsion by lowering interfacial tension. The emulsifying capacity of chitosan, however, is highly dependent on its MW and DD [87, 88]. Low MW chitosan was reported to exhibit better emulsifying properties than high MW chitosan, while chitosan with either a low (60%) or a high DD (86%) showed better emulsifying properties and produced a unimodal oil droplet size, compared to chitosan with intermediate DD (65–77%). The emulsion-stabilizing effects of chitosan are also concentration-dependent, especially as regards chitosan with intermediate DD [89]. The hydrophile-lipophile balance for chitosan was reported as 36.7, demonstrating its high hydrophilic properties [90]. Although the surface activity of chitosan is limited, using higher concentrations results in a higher viscosity emulsion which stabilizes the oil droplets by forming a denser and thicker polyelectrolitic brush at the water side of the interface [91].

Chitosan-based emulsions in the form of edible coatings or films have been widely reported as a novel approach to improve food quality. A mineral oil-chitosan emulsion has been recently studied for its use as a coating material on chicken eggs to maintain quality and prolong shelf-life during storage. The coating minimized weight loss and significantly preserved the albumen and yolk quality of eggs stored at room temperature for at least 3 weeks longer than those without the coating, and this effect was more pronounced under refrigeration [92]. Sensory discrimination tests indicated that the mineral oil-chitosan emulsion coating on the eggs did not affect their overall appearance and that purchase intent was the same as for uncoated eggs [93]. Similar beneficial effects have also been recently reported for a soybean oil-chitosan emulsion coating on eggs [94]. The ability of a lemon essential oil–chitosan emulsion coating to preserve strawberry quality was investigated by Perdones *et al.*, [95]. The coating maintained the high quality of strawberries stored at 5 °C by slowing down the respiration rate and enhancing antifungal activity. Various other essential oil–chitosan emulsions have also been characterized and developed with varying results, suggesting that the particular essential oil used greatly affects the emulsion film forming properties and antimicrobial activities [96, 97]. As well as essential oils, other lipid–chitosan emulsions have also been studied for their applications in the food industry. For instance, a novel application of an oleic acid-chitosan emulsion in osmotic dehydration during food processing was investigated by García *et al.*, [98]. These authors evaluated the effects of chitosan emulsion coatings on the osmotic dehydration of scalded-cut papaya at two different ripening stages, green and ripe. The immersion of papaya cubes in the oleic acid-chitosan emulsion significantly improved the efficiency of the osmotic dehydration process at both ripening stages, by increasing water loss and decreasing solids gain, compared with uncoated samples. Finally, the quality maintaining capacity of chitosan has been further improved by preparing nanoemulsions by ultrasonication. The nanosized chitosan emulsions exhibited greater efficiency as a biofungicide for controlling the anthracnose of tropical fresh fruits, compared with a conventional chitosan emulsion [99].

8.3.3 Edible Film/Coating and Food Packaging Applications

Over the last decade, there has been an increasing interest by the food industry in the development of novel active films/coatings as food packaging materials, not only due to their natural origin, but also because they can enhance food safety and quality during storage. Among these biomaterial candidates, chitosan is one of the most promising thanks to the combination of its excellent film-forming properties and antimicrobial activities.

Biopolymer-based edible films for food packaging applications are generally characterized by several parameters, including mechanical properties (elongation, tensile strength and breaking force), thickness, water vapor and oxygen permeability, as well as moisture content and color. Chitosan films are usually prepared by casting a chitosan solution onto a particular plate. Since chitosan is only soluble under acidic conditions, the type of organic acid used is crucial to the mechanical properties of chitosan films. Although acetic acid has been considered as a solvent for chitosan (due to its easy availability) it has been reported to have adverse effects on chitosan films, compared with many other organic acids, such as malic, lactic and citric acid [100]. Acetic acid produced the toughest films and lowest elongation rate of all the acids tested, making it too brittle for coating applications when used alone. This means that a plasticizer is necessary to prepare adequate chitosan films. A recent study showed that both the plasticizer concentration and the drying methods used significantly affected the mechanical properties of chitosan films [101] due to changes in its glass transition temperature under different conditions.

Due to its hydrophilic nature, films prepared from chitosan alone do not exhibit adequate water vapor/oxygen barrier properties or antimicrobial activities during long-term storage. This has led to a new trend in the development of chitosan films or coatings whereby they are prepared as active films/coatings by combining them with natural antimicrobials or incorporating one or more bioactive compounds. Active chitosan-based films have been used to coat various food products, especially fruits, fresh produce, and meat. For instance, the incorporation of green tea extract to produce active chitosan films has been reported to improve both their mechanical and water vapor barrier properties, mainly due to the enhanced intermolecular interactions between the phenolic compounds and the chitosan molecules [102]. These green tea extract-chitosan films have been further applied as active packaging materials for extending the shelf life of pork sausages by reducing the oxidation rate, minimizing color changes, and limiting microbial growth during storage at 4 °C [103]. Xiao *et al.*, [104] developed edible coatings by incorporating sodium chloride into chitosan and carboxymethyl chitosan and investigating their synergistic effects on the quality of fresh-cut d'Anjou pears [104]. These results revealed that the combination of sodium chloride and chitosan adversely affected the quality of pear slices by accelerating the discoloration of the cut surfaces and increasing polyphenol oxidase activity, while combining sodium chloride with carboxymethyl chitosan showed desirable effects. In addition to the incorporation of small bioactive molecules into chitosan active films, an emerging trend is to develop composite films using chitosan and other natural polymers, including both proteins and polysaccharides [105]. The aim of using composite films as food packaging materials is to combine the beneficial effects of different polymers, including their mechanical properties, water vapor permeability, solubility, thermal properties, appearance, antimicrobial activities, etc. For example, incorporating gelatin into chitosan films resulted in more flexible films with a lower tensile strength and higher elongation rate [106]. In addition, increasing the ratio of gelatin to chitosan produced more translucent films due to the enhanced light transmission. The translucent appearance of these films will greatly expand the applications of active chitosan-based films as edible coatings.

8.3.4 Encapsulation and Nutrient Delivery Applications

Encapsulation is defined as a process that entraps one substance within a certain matrix. This technology has been widely applied in the pharmaceutical and biomedical industries in order to, firstly, protect drugs from the gastric environment and hence improve their absorption in the intestine, and secondly, to provide the targeted delivery of encapsulated drugs after they enter the blood circulation system. Biodegradable polymeric nanoparticles are of particular interest in this field [107]. The identification of more and more natural biodegradable polymers with encapsulation and drug delivery capabilities has led scientists to explore the applications of this emerging technology in the food industry. Encapsulation may be useful for a variety of reasons [108], such as (1) to protect labile bioactive foods from harsh processing conditions, e.g. heat, oxygen, high pressure, etc; (2) to protect nutraceuticals susceptible to low pH conditions from forming isomers or oligomers in the stomach; (3) to render fat-soluble, but water insoluble, bioactives, such as lipophilic vitamins and flavonoids, soluble in water by encapsulating them into a hydrophilic polymeric matrix, thus greatly expanding their applications in the food industry; (4) to mask the undesirable flavors or smells of certain active compounds and hence improve their overall acceptance as food products; (5) to provide targeted delivery systems and prolonged blood circulation time required by food derived bioactive compounds (as for many drugs) so that they can exert their functions; and (6) to provide probiotic living cells with adequate protection from adverse environmental conditions by microencapsulation. This last is currently receiving considerable interest in the food industry.

8.3.4.1 Chitosan-Based Delivery Systems for Food Applications

Among the many food biopolymers that have been investigated for their use in encapsulation and delivery systems, chitosan is one of the most promising, as it is already considered as extremely versatile in drug delivery [109]. In the food industry, chitosan has also been extensively studied for its potential as an encapsulant for delivering various nutrients. The cationic nature of chitosan under acidic conditions greatly simplifies the formulation of encapsulation and delivery systems under different guises, including nano/micro-particles, hydrogel beads, emulsions, fibers, films and membranes. Chitosan-based polymeric delivery systems for biomedical applications have been developed using a variety of approaches [109]. In this chapter, however, the focus is on the most commonly used formulation methods and their applications in the food sciences, which are briefly introduced in the following sections. In addition, recent advances of chitosan-based particulate systems for the encapsulation of nutrients are summarized in Table 8.2. In general, electrostatic interactions between chitosan and anionic molecules are the main driving forces behind the methods used to assemble chitosan delivery systems, although hydrogen bonding and hydrophobic interactions are also involved in some cases.

8.3.4.2 Nano/Micro-Particles

Chitosan nano/microparticles can be easily prepared using sodium TriPolyPhosphate (TPP) which carries five negative charges per molecule. In this method, TPP solution is added to a chitosan solution dropwise, and the spontaneously formed particles are then spray dried. As shown in Figure 8.6, pure chitosan has a rough membrane-like morphology. When TPP is added, however, spherical nanoparticles are formed. Particle sizes can range from 100 nm to several μm depending on the concentrations of the chitosan and TPP, their mass ratios, and the drying methods used. Due to the simplicity of this method, chitosan/TPP particles have been widely adopted for the encapsulation of both hydrophilic and hydrophobic nutraceuticals. For this, the bioactive compound (dissolved in water or ethanol) is added either to the chitosan or the TPP solution before they are mixed together. Several chitosan derivatives are also able to form nano/microparticles with oppositely charged molecules. For instance, glycol chitosan [126] and chitosan hydrochloride [131] are both positively charged chitosan derivatives that can form nano/micro-particles with TPP, while negatively charged derivatives, e.g. carboxymethyl chitosan [132] form nano/microparticles with calcium.

Alishahi *et al.*, investigated the encapsulation of vitamin C in chitosan/TPP nanoparticles in order to extend its shelf life and achieve controlled release [133]. Particle size and encapsulation efficiency were greatly affected by the MW of the chitosan used. Nanoparticles prepared with low MW chitosan were smaller and more uniform, whereas greater encapsulation efficiency was achieved by using high MW chitosan. In all cases, however, the shelf-life of the vitamin C was greatly improved, compared with unencapsulated samples. The controlled release profile of the vitamin C was found to be pH responsive, with a faster release rate at pH 7.4, and a slower release rate in an acidic medium. The effect of digestive enzymes on the release rate was not, however, tested. The encapsulation of sodium selenite, a nutraceutical supplement of the trace element selenium, in chitosan/TPP nanoparticles was studied by Luo *et al.*, [114]. Particle size, zeta potential, and encapsulation efficiency were all greatly influenced by the concentrations of chitosan and TPP, as well as the loading concentration of selenite.

Table 8.2 Chitosan-based particulate systems for the encapsulation of nutrients (2008–2013).

Chitosan and derivatives	Preparation methods	Particulate systems	Nutrients encapsulated	Beneficial effects	References
Chitosan	Emulsification, ionic gelation with TPP	Nanoparticles, 280–400 nm	Essential oils	Controlled release	[110]
Chitosan	Ionic gelation with peptides	Nanoparticles, 150 nm	EGCG	Enhanced bioavailability, lower toxicity	[111,112]
Chitosan	Ionic gelation with TPP	Nanoparticles, 125–300 nm	Selenium	Controlled release, enhanced bioactivity, improved bioavailability	[113,114]
Chitosan	Ionic gelation with TPP	Nanoparticles, 180–580 nm	Vitamin C	Prolonged shelf life, enhanced *in vivo* delivery	[115,116]
Chitosan	Coating on zein nanoparticles	Nanoparticles, 200–800 nm	Vitamin E	Controlled release in GI track	[117]
Carboxymethyl chitosan	Coating on zein nanoparticles and ionic gelation with calcium	Nanoparticles, 80–250 nm	Indole, Vitamin D3	Controlled release, improved stability	[118,119]
Carboxymethyl chitosan	Ionic gelation with calcium and soy protein	Nanoparticles, 160–240 nm	Vitamin D3	Controlled release	[120]
Linoleic acid-modified chitosan	Ionic gelation with β-lactoglobulin	Nanoparticles, 170–350 nm	Quercetin	–	[121]

(*Continued*)

Table 8.2 Cont.

Chitosan and derivatives	Preparation methods	Particulate systems	Nutrients encapsulated	Beneficial effects	References
Chitosan	Spray drying with tween	Nanoparticles, 320–360 nm	Curcumin	Solubilization, controlled release	[122]
Chitosan	Emulsification, ultrasonication, and freeze drying	Microparticles, 3 μm	Fish oil	–	[123]
Chitosan	Emulsification, lay-by-lay deposition with caseinate and pectin	Microparticles, 0.2–2 μm	Corn oil	Controlled release, delayed digestion in GI track	[124]
Chitosan	Spray drying	Microparticles, 3 μm	Betalains	Improved stability, retained bioactivity	[125]
Chitosan hydrochloride	Ionic gelation with TPP and spray drying	Microparticles, 5 μm	Antioxidant extract	Controlled release	[126]
Chitosan	Ion gelation with alginate and calcium	Hydrogel beads, 10–>100 μm	Lipids	Controlled digestion in GI track	[127]
Chitosan	Emulsification and cross-linking with vanillin	Microparticles, 53–311 μm	Resverarol	Controlled release, improved stability	[128]
Chitosan	Ion gelation with alginate and calcium	Hydrogel beads, 780 μm–1 mm	Plant extracts	Controlled release	[129]
Carboxymethyl chitosan	Ion gelation with calcium in alcohol solution	Hydrogel beads 1 mm	Vitamin D3	High loading capacity, controlled release	[130]

*Abbreviations: TPP: Sodium TriPolyPhosphate; EGCG: (–)-EpiGalloCatechin-3-Gallate.

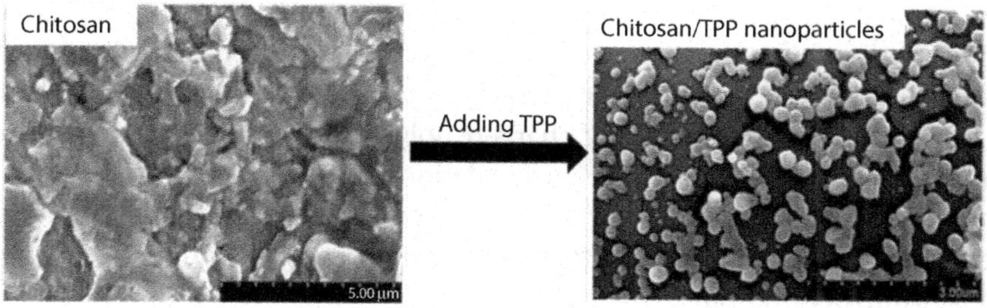

Figure 8.6 Scanning Electron Microscope (SEM) photographs of chitosan and chitosan/TPP nanoparticles [114].

The encapsulation of selenite in the chitosan nanoparticles enhanced its antioxidant properties, which was attributed to the antioxidant activities of chitosan. Although a fast release of selenite was observed in both PBS and simulated gastrointestinal fluids containing digestive enzymes, the release profile was significantly improved after the chitosan/TPP nanoparticles were coated with zein, a water insoluble protein. Cellular evaluation of the encapsulated selenite demonstrated that encapsulation not only improved selenium cellular uptake but also protected cells from selenium-induced DNA damage response [113]. The encapsulation of nutraceuticals in zein nanoparticles coated with carboxymethyl chitosan/calcium has also been recently studied by Luo *et al.,* [118, 119]. The carboxymethyl chitosan/calcium coating increased encapsulation efficiency, retarded controlled release, and improved the thermal and photo-stabilities of the encapsulated compounds: vitamin D3, indole-3-carbinal, and diindolylmethane.

Chitosan nanoparticles can also be developed by cross-linking with oppositely charged biopolymers. Nanoparticles prepared by this method are usually considered as polyelectrolyte complexes. This trend has received increasing attention in recent years, since more and more research provides evidence that nanoparticles prepared with two or more polymers are able to better protect encapsulated compounds against environmental conditions [134]. Many biopolymers have been reported to form polyelectrolyte complex nanoparticles with chitosan, including alginate, hyaluronic acid, carrageenan, gum Arabic, carboxymethyl cellulose, etc. Complex nanoparticles can also be prepared by cross-linking negatively charged chitosan derivatives and biopolymers together with cationic ions, such as calcium. Teng *et al.*, [120] developed carboxymethyl chitosan-soy protein complex nanoparticles for nutrient delivery. In this method, carboxymethyl chitosan and soy protein complex were cross-linked by calcium. The complex nanoparticles exhibited remarkable encapsulation efficiencies and improved the release profile of vitamin D3 under simulated gastrointestinal conditions, compared with nanoparticles prepared with single ingredients. Another promising development is to prepare nanoparticles using chitosan and negatively charged polypeptides, which are nontoxic and edible. For example, the encapsulation of tea catechins in chitosan-poly(r-glutamic acid) nanoparticles [135]. By using this method, negatively or positively charged nanoparticles can be prepared depending on the chitosan/polypeptide ratio. The antioxidant activity of the tea catechins was mostly retained under simulated gastric conditions, due to their encapsulation in nanoparticles. It was shown that the positively charged nanoparticles could transiently open the tight junction between the Caco-2 cells and thus increase the paracellular transport of the tea catechins. Chitosan-caseinophosphopeptide nanoparticles that improve the bioavailability of epigallocatechin gallate have also been developed [111, 112].

Chitosan microparticles are developed using similar procedures to those used for the preparation of nanoparticles, but are dried by the spray drying method. Typical particle sizes range from 2 μm to 20 μm, depending on the cross-linker type and chitosan concentration. Unlike chitosan nanoparticles, microparticles have been widely studied in the pharmaceutical industries to encapsulate drugs, however, the encapsulation of nutrients for food applications has not yet been investigated [136]. Vitamin C encapsulation in spray dried chitosan/TPP microparticles has been systematically studied by the Park group [137–139] including the effects of chitosan concentration, the chitosan/TPP mass ratio, chitosan MW, and spray drying parameters, all of which influenced vitamin C encapsulation efficiency and the release profile. A recent study investigated the encapsulation of orange oil in chitosan emulsion microparticles by spray drying [140]. The prepared chitosan microparticles were aimed at aiding the retention of orange oil in fabrics after washing in a detergent solution. By choosing the appropriate formulation, small microparticles with uniform particle size that effectively deposited orange oil in cotton fabrics were successfully prepared, and the encapsulated orange oil remained stable under storage due to its slow release. These findings may lead to new applications of chitosan microparticles in textiles.

8.3.4.3 Hydrogel Beads

Hydrogels are defined as hydrophilic polymer networks that can absorb a significant amount of water, from 10% to thousands of times their dry weight [141], as they swell but do not dissolve in water. Hydrogels are formed when a three-dimensional polymeric network is loosely cross-linked, either chemically or physically. Among the various hydrogels, hydrogel bead is one of the most investigated types, due to the ease of preparation. Hydrogel beads are usually formed spontaneously by adding a highly concentrated polymer solution dropwise into a cross-linking solution containing oppositely charged ions or polyelectrolyte polymers. No further cutting or shaping procedure is needed.

Chitosan and its derivations are some of the most studied polyelectrolytes, due to their abundance, nontoxicity, and biodegradability. To prepare hydrogel beads, a concentrated chitosan solution (as high as 3%) is added to a TPP solution through a syringe needle to ensure a small particle size. Chitosan/TPP hydrogel beads have been proven to possess pH-sensitive swelling behavior, and the encapsulation efficiency and release profile are dependent on the preparation parameters [142]. Chitosan hydrogel beads can also be prepared by cross-linking chitosan with negatively charged polymers, such as alginate [143] and pectin [144]. Chitosan-based hydrogel beads not only have a variety of applications in the biomedical and pharmaceutical fields [145], but also are potentially very promising for the food sciences [146, 147].

8.3.4.4 *The Encapsulation and Delivery of Nutrients*

Chitosan-based hydrogel beads are believed to be a useful tool to encapsulate and deliver nutrients. Due to their large size, hydrogel beads are considered to have several advantages over nano/microparticle delivery systems, such as a higher loading capacity and encapsulation efficiency, a more controlled release mechanism, and simpler and easier preparation and collection procedures. Chitosan–alginate beads prepared by ionic gelation were used to encapsulate several polyphenolic antioxidants from different plant extracts, including raspberry leaf, hawthorn, ground ivy, yarrow, nettle, and olive leaf [129]. Encapsulation efficiency and the kinetic release profile were greatly affected by the constituents of the plant extracts tested. However, the hydrogel beads did not provide adequate protection against the rapid degradation and loss of the antioxidative stability of the encapsulated nutrients. This was possibly because the hydrogel beads used in the study were not dried during storage and the excess water caused the deterioration of their bioactivities. Chitosan/alginate hydrogel beads were also studied for controlling lipid digestion [14]. Hydrogel beads with diameters over 100 μm were more effective than smaller hydrogel beads in delaying lipid digestion when tested by an *in vitro* digestion model with lipase.

Carboxymethyl chitosan, a water soluble chitosan derivative with a negative charge, is reported as being unable to form hydrogel beads on its own

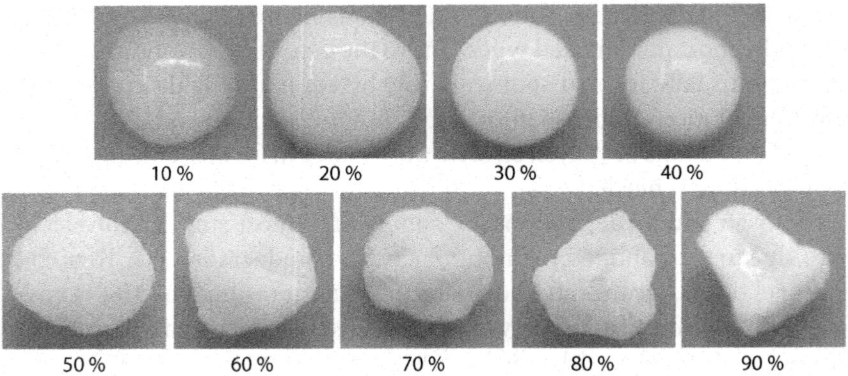

Figure 8.7 Digital photographs of carboxymethyl chitosan hydrogel beads prepared in different binary aqueous-alcohol solvents. The percentage amount below each photograph represents the concentration of alcohol used for the preparation of that bead [130].

due to its chain rigidity, and so alginate is normally used as an adjuvant polymer for the formation of hydrogel beads with this compound [148]. Recently, a novel method for the preparation of carboxymethyl chitosan hydrogel beads has been successfully developed by Manojlović *et al.*, [149]. Briefly, calcium dissolved in aqueous-alcohol was used as a cross-linking solution to prepare carboxymethyl chitosan hydrogel beads. As shown in Figure 8.7, the formation of the beads was dependent on the concentrations of the alcohol-aqueous solvent, into which calcium chloride was dissolved to physically cross-link carboxymethyl chitosan. It was found that the solvent prepared with 30% alcohol gave hydrogel beads with the most spherical shape, and that the drying method used had significant impact on the surface morphology and swelling behavior. Since the beads were prepared in an aqueous-alcohol solvent they provided an ideal encapsulation and delivery system for hydrophobic nutrients, such as vitamin D3: the encapsulation efficiency reached 97%, much higher than other delivery systems for hydrophobic nutrients.

8.3.4.5 *The Encapsulation and Delivery of Probiotics*

Probiotics are defined as a group of bacteria that can confer health benefits to the host when they are administered in adequate amounts. Probiotics can benefit human hosts in several ways, for example, by: producing nutrients and cofactors, competing with pathogens, and hence inhibiting their growth, maintaining beneficial gut microflora, stimulating the immune response, and ameliorating bowel-associated diseases [150]. Probiotics are currently used in the production of functional foods and pharmaceutical products.

In order to remain viable and preserve their health benefits, however, they need to be protected from harsh conditions. Probiotics are extremely susceptible to the highly acidic conditions of the gastric environment, which greatly reduces their viability. The immobilization of probiotics in a polymer matrix can protect them from the gastric environment before releasing them in the small intestine where they can be absorbed. The encapsulation of probiotics has received increasing attention in recent years as an emerging technology that can effectively enhance their viability [149].

Chitosan-based hydrogel beads have been widely applied for the encapsulation of these microorganisms. Specifically, chitosan is often used as a coating material on alginate beads to enhance the protection of probiotics in hostile environments. For instance, chitosan coated alginate hydrogel beads were reported to be much more effective than alginate/calcium beads for protecting the viability of probiotics in simulated gastric solutions, as well as prolonging the controlled release of viable probiotics in simulated intestinal solutions [151]. A similar system has been tested for pomegranate juice in order to explore the protective effects of chitosan/alginate beads. Chitosan coated beads significantly increased cell viability by 5.5 log CFU/ml, compared with uncoated alginate beads [152]. Carboxymethyl chitosan has also proved to be an effective coating for alginate beads to help increase the survival rate of probiotics in gastric and bile environments [153]. It is worth mentioning that before coating alginate beads with chitosan it is crucial to adjust the pH of the chitosan solution to 6, since its original pH is lower than 5 which is harmful to probiotics. Another benefit of using chitosan to coat alginate beads is that chitosan enhances their mucoadhesive properties [154]. Although alginate itself is a mucoadhesive polymer, cross-linking with calcium ions to produce hydrogel beads significantly reduced its mucoadhesivity. By coating the alginate beads with chitosan or thiolated chitosan, however, the mucoadhesivity was successfully restored. The coated beads were thus able to deliver markedly higher amounts of probiotics to the *in vitro* colonic mucosa model than uncoated beads.

8.3.5 The Immobilization of Enzymes by Chitin and Chitosan

Enzyme immobilization is a technique used to confine enzyme molecules to a distinct phase separated from the bulk phase whilst allowing exchange to take place between these two phases [155]. Several methods for enzyme immobilization, such as covalent bonding, electrostatic bonding, copolymerization, polymer entrapment, hydrophobic interaction, liposomal entrapment and encapsulation, are frequently used. The most common of these is covalent bonding onto an insoluble polymer such as cellulose or chitin. Immobilized enzymes are reusable, stable and suitable as specific industrial catalysts [155–157].

Synowiecki *et al.*, [158] documented possible mechanisms for the immobilization of α-amylase, β-amylase, glucose isomerase, and amyloglucosidase on krill chitin activated by formaldehyde. These authors proposed that the immobilization reaction was initiated by the generation of the hydrated form of formaldehyde which condenses with the free NH_2 groups of chitin forming Schiffs bases and dihydroxymethyl derivatives of aldehydes. These Schiff's bases might then react with various functional groups of the enzymes forming methylene bridges, thus causing enzyme immobilization. A similar study by Han and Shahidi [159] reported 20–29% activity retention of crude seal gastric proteases after immobilization on glutaraldehyde-treated chitin. The characteristics of the immobilized crude native seal gastric proteases were similar to those of chymosin. The immobilization of penicillin G acylase on different physical forms of chitosan, namely beads, particles, and powder was studied by Braun *et al.*, [156] who observed activity retentions of 40%, 93%, and 100%, respectively. A further study by Siso *et al.*, [157] demonstrated that microencapsulation in chitosan beads was an effective enzyme immobilization method for invertase and α-amylase.

8.3.6 Other Applications of Chitosan in the Food Sciences

The immobilization of enzymes on chitosan-based biomaterials is an important application of chitosan in the food sciences. On immobilizing onto chitosan supports, enzymes become more robust and more resistant to environmental changes. The heterogeneity of immobilized enzyme systems also enables easier recovery, multiple reuse, the faster termination of reactions, and many other benefits [160]. Chitosan-based biomaterials in different geometrical configurations have been studied for enzyme immobilization applications, including powders, flakes, beads, films, and membranes. Enzyme-immobilized chitosan biomaterials have also been developed into various biosensors for their novel applications in the food industry, for example, glucose [161], choline [162] and food-borne pathogen biosensors [163], as well as a polyphenol detection kit [164].

Another application of chitosan is the purification of waste water in the food industry. This has recently attracted great scientific and industrial interest as water pollution is one of the greatest environmental issues adversely affecting the quality of life. The purification process involves many steps, such as the removal of heavy metal ions, dyes and phenol compounds, sludge conditioning, and so on. The water industry generally uses polyelectrolyte polymers for the treatment of waste water and production of drinking water [165]. Due to its many amino groups, chitosan is a very effective adsorbent for removing water impurities and is mostly applied in the form of hydrogel beads and microspheres. It has also been widely investigated for its role in the bioconversion of phenol compounds [166], dye removal [167], and Cu^{2+} and Zn^{2+} removal [168]. In order to achieve improve the efficiency of chitosan as a water purifier, various modifications have been applied, such as functionalized chitosan membrane with carbon nanotubes [169], chitosan beads impregnated by ion imprinting for metal removal [170], water soluble chitosan for clay flocculation [171], and conjugated chitosan for the removal of phenols [172]. In addition to waste water applications, chitosan is currently being investigated for its potential to improve the quality of drinking water, by the removal of trace elements [173], defluoridation [174], and microtoxin removal [175].

8.4 Nutritional Aspects of Chitin and Chitosan

The multiple roles of chitin and chitosan in food systems are associated with their use as sources of dietary fiber and as functional ingredients. The United States Food and Drug Administration (USFDA) approved chitosan as a feed additive in 1983 [176]. In the food industry, chitosan is used as a food quality enhancer in certain countries. For example, Japan produces dietary cookies, potato chips and noodles enriched with chitosan due to its hypocholesterolemic effect [177]. Furthermore, vinegar products containing chitosan are manufactured and sold in Japan, again because of their cholesterol lowering properties [177].

Recently, Hirano *et al.*, [178] demonstrated the nutritional benefits of chitinous polymers to animals and suggested that chitin and chitosan could be employed as high performance feed additives. Normal growth patterns were observed for hens and broilers fed <1.4 g of chitosan/kg of body weight per day for up to 239 days, and rabbits fed <0.8 g of chitosan/kg of body weight per day over the same period. Furthermore, the serum cholesterol and triacylglycerol values of rabbits, hens and broilers were kept low by mixing their feed with 2% chitosan, although no effects were observed when they were fed 1% chitosan or 2% chitin. In a similar study, Razdan and Pettersson [179] observed an increase in high density lipoprotein (HDL) concentrations after feeding broiler chickens with a diet containing chitosan, possibly attributable to enhanced reverse cholesterol transport in response to intestinal losses of dietary fats. The effects of chitin, chitosan, and cellulose as dietary supplements on the growth of cultured red sea bream, Japanese eel, and yellow tail was investigated by Kono *et al.*, [180]. The maximum growth rate was obtained for fish fed with a 10% chitin supplement, indicating its use as a feed enhancer. Feed efficiency in red sea bream and Japanese eel fed with a 10% chitin supplemented diet was also the highest of all the treatments administered. Austin *et al.*, [181] reported that the growth of bifidobacteria in the guts of chickens increased with the addition of chitin to their feed. This could be extremely useful as bifidobacteria inhibit the growth of other types of microorganisms, as well as producing lactase required for the digestion of milk lactose. This last may also be of significance for humans and animals with lactose intolerance [176, 181]. For example, a digestible and highly nutritious animal feed could be formulated in conjunction with high lactose cheese whey [181]. The use of chitin as a functional ingredient in dough fermentation for bread was reported by Knorr and Betschart [182, 183]. These authors found that the loaf volume of wheat bread increased when up to 2% microcrystalline chitin was included in the mixture.

The nondigestibility, high viscosity, polymeric nature and high water binding properties of dietary fibers in the upper gastrointestinal tract, together with their low water binding properties in the lower gastrointestinal tract, are all responsible for the effective hypocholesterolemic potential of these nutrients [184, 185]. Chitosan has most of these characteristics and in addition has a highly desirable property not easily found in other dietary plant fibers: due to its ability to form ionic bonds at low pH, it can bind *in vitro* to different types of anions such as bile acids or free fatty acids [184]. Large proportions of these bound lipids are thus excreted. In addition, bound triacylglycerols escape hydrolysis by lipase, thus promoting the excretion of fatty materials including cholesterol, sterols and triacylglycerols [185, 186]. Within the digestive tract, chitosan forms micelles with cholesterol (both endogenous and from dietary sources) in the alkaline fluids in the upper part of the intestine, resulting in a decrease in the absorption of dietary cholesterol and an increase in the circulation of cholic acid to the liver. The formation of cholic acid from blood cholesterol in the liver tends to decrease blood cholesterol concentrations. Large intestinal microbials secrete chitinases that can digest the micelles formed by the chitosan, resulting in the excretion of bile acids and sterols as free forms into faeces without absorption [187]. Furthermore, Deuchi *et al.*, [188] proposed that chitosan is solubilized in the stomach to form an emulsion of intragastric oil droplets and begins to precipitate in the small intestine at pH 6–6.5. With the aggregation of polysaccharide chains, the oil droplets are entrapped in their matrices thereby passing through the lumen and emptying into the faeces.

8.5 Preparation of Chitin and Chitosan Oligomers and Their Applications in the Food Industry as Health Supplements

Research into the preparation and physiological activities of chitin and chitosan oligomers has continuously attracted much attention in the food and pharmaceutical industries due to their versatile antitumor activities [189, 190], immuno-enhancing effects [191, 192], protective effects against some infectious pathogens in mice [193, 194], and antifungal [195, 196] and antimicrobial activities [195, 197].

Two hydrolytic methods are used to prepare chitin and chitosan oligomers: acid hydrolysis and enzymatic hydrolysis. Acid hydrolysis with inorganic acids leads to the formation of oligomers with a low degree of polymerization (DP), with only monomers to trimers showing quantifiable yields. However, the yields of desirable, biologically active oligomers, i.e. those with relatively higher DPs (tetramers to heptamers), are low. In contrast, the enzymatic hydrolysis of chitin and chitosan by chitinase, chitosanase, lysozyme, or cellulase readily results in the production of high DP oligomers for different applications. Detailed descriptions of the methods used for the preparation of chitin chitosan oligomers are available in the literature [198–208].

Research into the functional properties of chitin and chitosan oligomers has clearly revealed their high dependency on the DP [195, 189]: oligomers with high DPs (from pentamer to heptamer) have better functional characteristics than relatively low DP oligomers [209].

Several studies have demonstrated that chitin and chitosan oligomers possess antitumorigenic properties that inhibit the growth of tumor cells *via* an immuno-enhancing effect [189]. Suzuki *et al.*, [210] also revealed that *N*-acet-ylchitooligosaccharides, from $(GlcNAc)_4$ to $(GlcNAc)_7$, displayed strong attracting responses to peritoneal exudate cells in BALB/c mice, whereas chitooligosaccharides, from $(GlcN)_2$ to $(GlcN)_6$, did not. Suzuki *et al.*, [211] also found that the chitin and chitosan oligomers, $(GlcNAc)_6$ and $(GlcN)_6$, respectively, showed a tumor growth-inhibitory effect in allogenic and syngeneic mouse systems, including a sarcoma 180 solid tumor and a MM46 solid tumor. The authors concluded that this effect was host-mediated, i.e. not caused by direct cytocidal action on the tumor cells. Tokoro *et al.*, [192] also showed that the two oligosaccharides, $(GlcNAc)_6$ and $(GlcN)_6$, had a growth-inhibitory effect against Meth-A solid tumors transplanted into BALB/c mice. The antitumor mechanism was assumed to involve the increased production of lymphokines, including interleukins

1 and 2, sequentially, leading to the manifestation of the antitumor effect through the proliferation of cytolytic T-lymphocytes. Tsukada *et al.*, [190] reported a significant antimetastatic effect for $(GlcNAc)_6$ in mice bearing Lewis lung carcinoma. Suzuki *et al.*, [191] analyzed changes in the spleen cells of tumor-bearing mice administered with chitooligosacchairdes, such as $(GlcNAc)_6$, in order to unravel the tumor inhibitory mechanism and cell growth produced by the immuno-enhancing effects of the oligomers. The increase in the activity of cytotoxic T-lymphocytes caused by the acceleration in the differentiation of helper T-cells was remarkable and paralleled a decrease in the activity of suppressor T-cells.

Chitin and chitosan oligomers are also responsible for protecting against some pathogenic infections in mice. Tokoro *et al.*, [194] demonstrated the protective effect of a chitin oligomer in mice infected with *Listeria monocytogenes*, based on the fact that interferon-λ and interleukin 2 enhanced the growth-inhibitory effect of $(GlcNAc)_6$-treated macrophages on this pathogen. Yamada *et al.*, [193] showed that $(GlcNAc)_6$ induced phytoalexin formation in suspension-cultured rice cells, whereas GlcNAc oligomers smaller than trimers and a series of deacetylated oligomers showed almost no activity.

Chitosan and chitosan oligomers have also been shown to inhibit the growth of several fungi and bacteria especially pathogens [197, 208]. Hirano and Nagao [195] studied the relationship between the DP of chitosan and the inhibition effect. They showed that chitosan oligomers (DP 2-8), as well as low MW chitosan, had a stronger growth inhibitory effect than high MW chitosan against several phytopathogens including *Alternaria alternata*, *Phomopsis fukushi*, and *Fusarium oxyporum*, among others. Kendra *et al.*, [196] explained that some of the chitosan oligomers with biological activity present in the pea/*Fusarium* interface appear to inhibit fungal growth. Uchida *et al.*, [197] found that oligomers with higher MWs, slightly hydrolyzed with chitosanase, had greater antifungal and antibacterial activities than native chitosan and low MW oligomers. Jeon and Kim [208] reported that among the three fractions of oligomers produced and separated using an ultrafiltration membrane enzymatic reactor system, the highest MW oligomers (MW 5,000–10,000 Da) showed the strongest bactericidal and fungicidal activities against most of the pathogens they tested. Muraki and Aiba [212] found that partially derivatized *N*-lauroyl (PNL) chitooligosaccharides (DP 7–8), PNL-$(GlcN)_7$ and PNL-$(GlcN)_8$, with a degree of *N*-lauroylation of about 50%, showed fairly strong antibacterial activity against *E. coli*, compared to all $(GlcN)_n$ and PNL-$(GlcN)_n$ with a chain length shorter than seven residues.

8.6 Future Trends: Chitosan-Based Nanotechnology for Food Applications

There is no doubt that chitosan-based nanotechnologies are becoming ever more popular and important. Current research efforts, however, are mainly confined to the biomedical fields, with studies on their applications in the food industry still being very limited. Although chitosan has been granted GRAS status for its use as a food additive by the FDA, the use of chitosan at a nanoscale in food/medical products has not yet been approved. Nevertheless, chitosan composite nanomaterials have been developed for their potential applications in the food industry, for example, chitosan-TiO_2 nanotube films [213] for food packaging and chitosan-silver nanoparticle composites for drinking water filtration applications [214]. The safety of these materials when they come into direct contact with food products is, however, still unknown. In addition to chitosan, many chitosan derivatives have also been reported to show numerous benefits, and may be more appropriate than native chitosan for some food applications. Compared with chitosan, however, little information on the toxicity and safety of chitosan derivatives is available, and so far none of those developed have been GRAS approved for their use in food products. This is mainly due to the difficulty of removing toxic solvent residues produced during the modification process. Thus, in addition to developing chitosan-based nanomaterials, there is an urgent need to evaluate their toxicity and safety when used with real food products, in order to facilitate the approval required for their application in the food industry.

8.7 Conclusion

A deacetylated product of chitin, chitosan is a versatile food biopolymer that has found a variety of applications in all areas of the food sciences. As a functional biopolymer, chitosan has many intrinsic nutritional values, such as antioxidant properties and health-promoting bioactivities against many chronic diseases, including hypercholesterolemia, hypertension, inflammation, immune diseases, etc. Chitosan also possesses promising broad spectrum antimicrobial activities, and hence has been widely studied as a food preservative to improve food quality and extend the shelf life of perishable food products. In addition, chitosan has excellent emulsifying properties and is used to stabilize various oil-in-water emulsions in the food industry in order to avoid the use of synthetic surfactants. All of these intrinsic properties vary with the MW and DD, which are the most important characteristics of chitosan. Due to the high number of amino groups, chitosan carries a high positive charge in acidic media, and as a result has becomes a popular biopolymer for the development of encapsulation and delivery systems, such as nano/micro-particles, hydrogel beads and nanocomposites, for their use in the food industry. With the appropriate modifications, the functional properties and biological activities of chitosan could be further enhanced leading to the development of even more applications.

Acknowledgements

The authors would like to thank the Consejo Nacional de Investigaciones Científicas y Técnicas (CONICET) (Postdoctoral fellowship internal PDTS-Resolution 2417), Universidad Nacional de Mar del Plata (UNMdP), for the financial support and to Dr. Mirian Carmona-Rodríguez.

References

1. Cauchie, H.M., An attempt to estimate crustacean chitin production in the hydrosphere. *Adv. Chit. Sci.*, 2, 32, 1997.
2. No, H.K., Meyers, S.P., Chitin Composition. In: *Chitin handbook*, R.A.A. Muzzarelli, M.G. Peter, Eds, 475, 1997.
3. Kurita, K., Tomita, K., Tada, T., Ishii, S., Nishimura, S.I., Shimoda, K., Squid chitin as a potential alternative chitin source: deacetylation behavior and characteristic properties. *J. Polym. Sci. Pol. Chem.*, 31, 485, 1993.
4. Kumar, M.N.R., A review of chitin and chitosan applications. *React. Funct. Polym.*, 46, 1, 2000.
5. Mourya, V.K., Inamdar, N.N., Chitosan-modifications and applications: opportunities galore. *React. Funct. Polym.*, 68, 1013, 2008.
6. Yen, M.T., Yang, J.H., Mau, J.L., Antioxidant properties of chitosan from crab shells. *Carbohydr. Polym.*, 74, 840, 2008.
7. Ai, H., Wang, F., Xia, Y., Chen, X., Lei, C., Antioxidant, antifungal and antiviral activities of chitosan from the larvae of housefly, *Musca domestica* L. *Food Chem.*, 132, 493, 2012.
8. Ai, H., Wang, F., Yang, Q., Zhu, F., Lei, C., Preparation and biological activities of chitosan from the larvae of housefly, *Musca domestica*. *Carbohydr. Polym.*, 72, 419, 2008.
9. Subhapradha, N., Ramasamy, P., Sudharsan, S., Seedevi, P., Moovendhan, M., Srinivasan, A., Shanmugam, V., Shanmugam, A., Preparation of phosphorylated chitosan from gladius of the squid *Sepioteuthis lessoniana* (Lesson,

1830) and its *in vitro* antioxidant activity. *Bioact. Carbohydr. Diet. Fibre*, 1, 148, 2013.

10. Anraku, M., Fujii, T., Kondo, Y., Kojima, E., Hata, T., Tabuchi, N., Tsuchiya, D., Goromaru, T., Tsutsumi, H., Kadowaki, D., Maruyama, T., Otagiri, M., Tomida, H., Antioxidant properties of high molecular weight dietary chitosan *in vitro* and *in vivo*. *Carbohydr. Polym.*, 83, 501, 2011.
11. Anraku, M., Fujii, T., Furutani, N., Kadowaki, D., Maruyama, T., Otagiri, M., Gebicki, J.M., Tomida, H., Antioxidant effects of a dietary supplement: Reduction of indices of oxidative stress in normal subjects by water-soluble chitosan. *Food Chem. Toxicol.*, 47, 104, 2009.
12. Sun, T., Zhu, Y., Xie, J., Yin, X., Antioxidant activity of N-acyl chitosan oligosaccharide with same substituting degree. *Bioorg. Med. Chem. Lett.*, 21, 798, 2011.
13. Zhang, X., Geng, X., Jiang, H., Li, J., Huang, J., Synthesis and characteristics of chitin and chitosan with the (2-hydroxy-3-trimethylammonium)propyl functionality, and evaluation of their antioxidant activity *in vitro*. *Carbohydr. Polym.*, 89, 486, 2012.
14. Ren. J., Li, Q., Dong, F., Feng, Y., Guo, Z., Phenolic antioxidants-functionalized quaternized chitosan: Synthesis and antioxidant properties. *Int. J. Biol. Macromol.*, 53, 77, 2013.
15. Zhong, Z., Zhong, Z., Xing, R., Li, P., Mo, G., The preparation and antioxidant activity of 2-[phenylhydrazine (or hydrazine)-thiosemicarbazone]-chitosan. *Int. J. Biol. Macromol.*, 47, 93, 2010.
16. Schreiber, S.B., Bozell, J.J., Hayes, D.G., Zivanovic, S., Introduction of primary antioxidant activity to chitosan for application as a multifunctional food packaging material. *Food Hydrocolloid.*, 33, 207, 2013.
17. Pasanphan, W., Buettner, G.R., Chirachanchai, S., Chitosan gallate as a novel potential polysaccharide antioxidant: an EPR study. *Carbohydr. Res.*, 345, 132, 2010.
18. Woranuch, S., Yoksan, R., Preparation, characterization and antioxidant property of water-soluble ferulic acid grafted chitosan. *Carbohydr. Polym.*, 96, 495, 2013.
19. Aytekin, A.O., Morimura, S., Kida, K., Synthesis of chitosan-caffeic acid derivatives and evaluation of their antioxidant activities. *J. Biosci. Bioeng.*, 111, 212, 2011.
20. Rao, M.S., Chawla, S.P., Chander, R., Sharma, A., Antioxidant potential of Maillard reaction products formed by irradiation of chitosan-glucose solution. *Carbohydr. Polym.*, 83, 714, 2011.
21. Ying, G.Q., Xiong, W.Y., Wang, H., Sun, Y., Liu, H.Z., Preparation, water solubility and antioxidant activity of branched-chain chitosan derivatives. *Carbohydr. Polym.*, 83, 1787, 2011.
22. Kosaraju, S.L., Weerakkody, R., Augustin, M.A., Chitosan-glucose conjugates: Influence of extent of Maillard reaction on antioxidant properties. *J. Agr. Food Chem.*, 58, 12449, 2010.

23. Li, X., Shi, X., Jin, Y., Ding, F., Du, Y., Controllable antioxidative xylan-chitosan Maillard reaction products used for lipid food storage. *Carbohydr. Polym.*, 91, 428, 2013.
24. Anandan, R., Ganesan, B., Obulesu, T., Mathew, S., Asha, K.K., Lakshmanan, P.T., Zynudheen, A.A., Antiaging effect of dietary chitosan supplementation on glutathione-dependent antioxidant system in young and aged rats. *Cell Stress and Chaperones*, 18, 121, 2013.
25. Xia, W., Liu, P., Zhang, J., Chen, J., Biological activities of chitosan and chitooligosaccharides. *Food Hydrocolloid.*, 25, 170, 2011.
26. Osman, M., Fayed, S.A., Ghada, I.M., Romeilah, R.M., Protective effects of chitosan, ascorbic acid and *Gymnema sylvestre* against hypercholesterolemia in male rats. *Austr. J. Basic Appl. Sci.*, 4, 89, 2010.
27. Anraku, M., Tomida, H., Michihara, A., Tsuchiya, D., Iohara, D., Maezaki, Y., Uekama, K., Maruyama, T., Otagiri, M., Hirayama, F., Antioxidant and renoprotective activity of chitosan in nephrectomized rats. *Carbohydr. Polym.*, 89, 302, 2012.
28. Yen, M.T., Tseng, Y.H., Li, R.C., Mau, J.L., Antioxidant properties of fungal chitosan from shiitake stipes. *LWT-Food Sci. Technol.*, 40, 255, 2007.
29. Xing, R., Liu, S., Guo, Z., Yu, H., Wang, P., Li, C., Li, Z., Li, P., Relevance of molecular weight of chitosan and its derivatives and their antioxidant activities *in vitro*. *Bioorgan. Med. Chem.*, 13, 1573, 2005.
30. Sun, T., Zhou, D., Xie, J., Mao, F., Preparation of chitosan oligomers and their antioxidant activity. *Eur. Food Res. Technol.*, 225, 451, 2007.
31. Zhong, Z., Ji, X., Xing, R., Liu, S., Guo, Z., Chen, X., Li, P., The preparation and antioxidant activity of the sulfanilamide derivatives of chitosan and chitosan sulfates. *Bioorgan. Med. Chem.*, 15, 3775, 2007.
32. Guo, Z., Xing, R., Liu, S., Yu, H., Wang, P., Li, C., Li, P., The synthesis and antioxidant activity of the Schiff bases of chitosan and carboxymethyl chitosan. *Bioorg. Med. Chem. Lett.*, 15, 4600, 2005.
33. Guo, Z., Liu, H., Chen, X., Ji, X., Li, P., Hydroxyl radicals scavenging activity of *N*-substituted chitosan and quaternized chitosan. *Bioorg. Med. Chem. Lett.*, 16, 6348, 2006.
34. Xing, R., Liu, S., Guo, Z., Yu, H., Zhong, Z., Ji, X., Li, P., Relevance of molecular weight of chitosan-*N*-2-hydroxypropyl trimethyl ammonium chloride and their antioxidant activities. *Eur. J. Med. Chem.*, 43, 336, 2008.
35. Sun, T., Yao, Q., Zhou, D., Mao, F., Antioxidant activity of *N*-carboxymethyl chitosan oligosaccharides. *Bioorg. Med. Chem. Lett.*, 18, 5774, 2008.
36. Sun, T., Zhu, Y., Xie, J., Yin, X., Antioxidant activity of *N*-acyl chitosan oligosaccharide with same substituting degree. *Bioorg. Med. Chem. Lett.*, 21, 798, 2011.
37. Chen, F., Shi, Z., Neoh, K.G., Kang, E.T., Antioxidant and antibacterial activities of eugenol and carvacrol-grafted chitosan nanoparticles. *Biotechnol. Bioeng.*, 104, 30, 2009.
38. Jung, B.O., Chung, S.J., Lee, S.B., Preparation and characterization of eugenol-grafted chitosan hydrogels and their antioxidant activities. *J. Appl. Polym. Sci.*, 99, 3500, 2006.

39. Cho, Y.S., Kim, S.K., Je, J.Y., Chitosan gallate as potential antioxidant biomaterial. *Bioorg. Med. Chem. Lett.*, 21, 3070, 2011.
40. Sousa, F., Guebitz, G.M., Kokol, V., Antimicrobial and antioxidant properties of chitosan enzymatically functionalized with flavonoids. *Process Biochem.*, 44, 749, 2009.
41. Fras-Zemljič, L., Kokol, V., Čakara, D., Antimicrobial and antioxidant properties of chitosan-based viscose fibres enzymatically functionalized with flavonoids. *Text. Res. J.*, 81, 1532, 2011.
42. Aytekin, A.O., Morimura, S., Kida, K., Synthesis of chitosan-caffeic acid derivatives and evaluation of their antioxidant activities. *J. Biosci. Bioeng.*, 111, 212, 2011.
43. Torres, E., Marín, V., Aburto, J., Beltrán, H.I., Shirai, K., Villanueva, S., Sandoval, G., Enzymatic modification of chitosan with quercetin and its application as antioxidant edible films. *Appl. Biochem. Micro.*, 48, 151, 2012.
44. El-Rehim, H.A.A., El-Sawy, N.M., Hegazy, E.S.A., Soliman, E.S.A., Elbarbary, A.M., Improvement of antioxidant activity of chitosan by chemical treatment and ionizing radiation. *Int. J. Biol. Macromol.*, 50, 403, 2012.
45. Feng, T., Du, Y., Li, J., Hu, Y., Kennedy, J.F., Enhancement of antioxidant activity of chitosan by irradiation. *Carbohydr. Polym.*, 73, 126, 2008.
46. Qiao, Y., Bai, X.F., Du, Y.G., Chitosan oligosaccharides protect mice from LPS challenge by attenuation of inflammation and oxidative stress. *Int. Immunopharmacol.*, 11, 121, 2011.
47. Yuan, W.P., Liu, B., Liu, C.H., Wang, X.J., Zhang, M.S., Meng, X.M., Xia, X.K., Antioxidant activity of chito-oligosaccharides on pancreatic islet cells in streptozotocin-induced diabetes in rats. *World J. Gastroenterol.*, 15, 1339, 2009.
48. Sugano, M., Fujikawa, T., Hiratsuji, Y., Nakashima, K., Fukuda, N., Hasegawa, Y., A novel use of chitosan as a hypocholesterolemic agent in rats. *Am. J. Clin. Nutr.*, 33, 787, 1980.
49. Sugano, M., Watanabe, S., Kishi, A., Izume, M., Ohtakara, A., Hypocholesterolemic action of chitosans with different viscosity in rats. *Lipids*, 23, 187, 1988.
50. Maezaki, Y., Tsuji, K., Nakagawa, Y., Kawai, Y., Akimoto, M., Tsugita, T., Takekawa, W., Terada, A., Hara, H., Mitsuoka, T., Hypocholesterolemic effect of chitosan in adult males. *Biosci., Biotechnol. Biochem.*, 57, 1439, 1993.
51. Wydro, P., Krajewska, B., Hac-Wydro, K., Chitosan as a lipid binder: a Langmuir monolayer study of chitosan-lipid interactions. *Biomacromolecules*, 8, 2611, 2007.
52. Zhou, K., Xia, W., Zhang, C., Yu, L.L., *In vitro* binding of bile acids and triglycerides by selected chitosan preparations and their physico-chemical properties. *LWT-Food Sci. Technol.*, 39, 1087, 2006.
53. Neyrinck, A.M., Bindels, L.B., De Backer, F., Pachikian, B.D., Cani, P.D., Delzenne, N.M., Dietary supplementation with chitosan derived from mushrooms changes adipocytokine profile in diet-induced obese mice, a phenomenon linked to its lipid-lowering action. *Int. Immunopharmacol.*, 9, 767, 2009.

54. Zhang, J., Zhang, W., Mamadouba, B., Xia, W., A comparative study on hypolipidemic activities of high and low molecular weight chitosan in rats. *Int. J. Biol. Macromol.*, 51, 504, 2012.
55. Zhang, W., Zhang, J., Jiang, Q., Xia, W. The hypolipidemic activity of chitosan nanopowder prepared by ultrafine milling. *Carbohydr. Polym.*, 95, 487, 2013.
56. Liu, X., Zeng, A., Song, T., Li, L., Yang, F., Wang, Q., Wu, B., Liu, Y., Zhi, X., Hypocholesterolemic effects of *N*-[(2-Hydroxy-3-*N*, *N*-Dimethylhexadecyl ammonium) propyl] chitosan chloride in high-fat-diet-induced rats. *J. Biomat. Sci.-Polym. E.*, 23, 1107, 2012.
57. Liu, X., Zhi, X., Liu, Y., Wu, B., Sun, Z., Shen, J., Effect of chitosan, O-carboxymethyl chitosan, and *N*-[(2-hydroxy-3-*N*, *N*-dimethylhexadecyl ammonium) propyl] chitosan chloride on overweight and insulin resistance in a murine diet-induced obesity. *J. Agr. Food Chem.*, 60, 3471, 2012.
58. Liu, X., Yang, F., Song, T., Zeng, A., Wang, Q., Sun, Z., Shen, J., Effects of chitosan, O-carboxymethyl chitosan and *N*-[(2-hydroxy-3-*N*,*N*-dimethylhexadecyl ammonium) propyl] chitosan chloride on lipid metabolism enzymes and low-density-lipoprotein receptor in a murine diet-induced obesity. *Carbohydr. Polym.*, 85, 334, 2011.
59. Singla, A.K., Chawla, M., Chitosan: some pharmaceutical and biological aspects-an update. *J. Pharm. Pharmacol.*, 53, 1047, 2001.
60. Xia, W., Liu, P., Zhang, J., Chen, J., Biological activities of chitosan and chitooligosaccharides. *Food Hydrocolloid.*, 25, 170, 2011.
61. Kim, S.K., Rajapakse, N. Enzymatic production and biological activities of chitosan oligosaccharides (COS): a review. *Carbohydr. Polym.*, 62, 357, 2005.
62. Zhang, J., Xia, W., Liu, P., Cheng, Q., Tahi, T., Gu, W., Li, B., Chitosan modification and pharmaceutical/biomedical applications. *Mar. Drugs*, 8, 1962, 2010.
63. Kong, M., Chen, X.G., Xing, K., Park, H.J., Antimicrobial properties of chitosan and mode of action: A state of the art review. *Int. J. Food Microbiol.*, 144, 51, 2010.
64. Rabea, E.I., Badawy, M.E.T., Stevens, C.V., Smagghe, G., Steurbaut, W., Chitosan as antimicrobial agent: Applications and mode of action. *Biomacromolecules*, 4, 1457, 2003.
65. Martinez, L.R., Mihu, M.R., Han, G., Frases, S., Cordero, R.J., Casadevall, A., Friedman, A.J., Friedman, J.M., Nosanchuk, J.D., The use of chitosan to damage *Cryptococcus neoformans* biofilms. *Biomaterials*, 31, 669, 2010.
66. Kim, S.K., Effect of antimicrobial activity by chitosan oligosaccharide *N*-conjugated with asparagine. *J. Microbiol. Biotechn.*, 11, 281, 2001.
67. Chanthateyanonth, R., Ruchirawat, S., Srisitthiratkul, C., Preparation of new water-soluble chitosan containing hyperbranched-vinylsulfonic acid sodium salt and their antimicrobial activities and chelation with metals. *J. Appl. Polym. Sci.*, 116, 2074, 2010.
68. Yu, T., Yu, C., Chen, F., Sheng, K., Zhou, T., Zunun, M., Abudu, O., Yang, S., Zheng, X., Integrated control of blue mold in pear fruit by combined

application of chitosan, a biocontrol yeast and calcium chloride. *Postharvest Biol. Tec.*, 69, 49, 2012.
69. Davis, R., Zivanovic, S., D'Souza, D.H., Davidson, P.M., Effectiveness of chitosan on the inactivation of enteric viral surrogates. *Food Microbiol.*, 32, 57, 2012.
70. Jiang, T., Feng, L., Li, J., Changes in microbial and postharvest quality of shiitake mushroom (*Lentinus edodes*) treated with chitosan-glucose complex coating under cold storage. *Food Chem.*, 131, 780, 2012.
71. da Silva, L.P., de Britto, D., Seleghim, M.H.R., Assis, O.B., *In vitro* activity of water-soluble quaternary chitosan chloride salt against *E. coli*. *World J. Microbiol. Biotechnol.*, 26, 2089, 2010.
72. Feng, Y., Xia, W., Preparation, characterization and antibacterial activity of water-soluble O-fumaryl-chitosan. *Carbohydr. Polym.*, 83, 1169, 2011.
73. Patale, R.L., Patravale, V.B., O,*N*-carboxymethyl chitosan–zinc complex: A novel chitosan complex with enhanced antimicrobial activity. *Carbohydr. Polym.*, 85, 105, 2011.
74. Hirano, S., Sato, N., Yoshida, S., Kitagawa, S., Chemical modification of chitin and chitosan and their novel applications. In: *Industrials polysaccharides: genetic, engineering, structure/property relations and applications*, M. Yalpani, Ed., p. 163, Amsterdam, The Netherlands: Elsevier, 1987.
75. Kim, C.H., Choi, J.W., Chun, H.J., Choi, K.S., Synthesis of chitosan derivatives with quaternary ammonium salt and their antibacterial activity. *Polym. Bull.*, 38, 387, 1997.
76. Hirano, S., Chitin and chitosan: Molecular and biological functions newly generated by chemical modifications. In: *Advances in chitin science*. Vol. 7(II), A.Domard, A.F.Roberts, K.M. Varum, Eds., p. 10, Jacques Andre, Lyon (France), 1997.
77. Jia, Z., Shen, D., Xu, W., Synthesis and antibacterial activities of quaternary ammonium salt of chitosan. *Carbohydr. Res.*, 333, 1, 2001.
78. Belalia, R., Grelier, S., Benaissa, M., Coma, V., New bioactive biomaterials based on quaternized chitosan. *J. Agr. Food Chem.*, 56, 1582, 2008.
79. Avadi, M.R., Sadeghi, A.M.M., Tahzibi, A., Bayati, K.H., Pouladzadeh, M., Zohuriaan-Mehr, M.J., Rafiee-Tehrani, M., Diethylmethyl chitosan as an antimicrobial agent: Synthesis, characterization and antibacterial effects. *Eur. Polym. J.*, 40, 1355, 2004.
80. Chi, W., Qin, C., Zeng, L., Li, W., Wang, W., Microbiocidal activity of chitosan-*N*-2-hydroxypropyl trimethyl ammonium chloride. *J. Appl. Polym. Sci.*, 103, 3851, 2007.
81. Kim, C.H., Kim, S.Y., Choi, K.S., Synthesis and antibacterial activity of water-soluble chitin derivatives. *Polym. Advan. Technol.*, 8, 319, 1997.
82. Jung, B.O., Kim, C.H., Choi, K.S., Lee, Y.M., Kim, J.J., Preparation of amphiphilic chitosan and their antimicrobial activities. *J. Appl. Polym. Sci.*, 72, 1713, 1999.
83. Chen, C.S., Liau, W.Y., Tsai, G.J., Antibacterial effects of *N*-sulfonated and *N*-sulfobenzoyl chitosan and application to oyster preservation. *J. Food Protect.*, 61, 1124, 1998.

84. Cuero, R.G., Duffus, E., Osuji, G., Pettit, R., Aflatoxin control in preharvest maize: Effects of chitosan and two microbial agents. *J. Agr. Sci.*, 117, 165, 1991.
85. Cuero, R.G., Osuji, G., Washington, A., *N*-carboxymethylchitosan inhibition of aflatoxin production: Role of zinc. *Biotechnol. Lett.*, 13, 441, 1991.
86. Muzzarelli, R.A., Muzzarelli, C., Tarsi, R., Miliani, M., Gabbanelli, F., Cartolari, M., Fungistatic activity of modified chitosans against *Saprolegnia parasitica*. *Biomacromolecules*, 2, 165, 2001.
87. Del Blanco, L.F., Rodriguez, M.S., Schulz, P.C., Agullo, E., Influence of the deacetylation degree on chitosan emulsification properties. *Colloid Polym. Sci.*, 277, 1087, 1999.
88. Li, X., Xia, W., Effects of concentration, degree of deacetylation and molecular weight on emulsifying properties of chitosan. *Int. J. Biol. Macromol.*, 48, 768, 2011.
89. Rodríguez, M.S., Albertengo, L.A., Agulló, E., Emulsification capacity of chitosan. *Carbohydr. Polym.*, 48, 271, 2002.
90. Schulz, P.C., Rodríguez, M.S., Del Blanco, L.F., Pistonesi, M., Agulló, E., Emulsification properties of chitosan. *Colloid Polym. Sci.*, 276, 1159, 1998.
91. Payet, L., Terentjev, E.M., Emulsification and stabilization mechanisms of O/W emulsions in the presence of chitosan. *Langmuir*, 24, 12247, 2008.
92. Torrico, D.D., No, H.K., Prinyawiwatkul, W., Janes, M., Corredor, J.A., Osorio, L.F., Mineral oil-chitosan emulsion coatings affect quality and shelf-life of coated eggs during refrigerated and room temperature storage. *J. Food Sci.*, 76, S262, 2011.
93. Torrico, D.D., No, H.K., Sriwattana, S., Ingram, D., Prinyawiwatkul, W., Effects of initial albumen quality and mineral oil-chitosan emulsion coating on internal quality and shelf-life of eggs during room temperature storage. *Int. J. Food Sci. Tech.*, 46, 1783, 2011.
94. Wardy, W., Torrico, D.D., Jirangrat, W., No, H.K., Saalia, F.K., Prinyawiwatkul, W., Chitosan-soybean oil emulsion coating affects physico-functional and sensory quality of eggs during storage. *LWT-Food Sci. Technol.*, 44, 2349, 2011.
95. Perdones, A., Sánchez-González, L., Chiralt, A., Vargas, M., Effect of chitosan-lemon essential oil coatings on storage-keeping quality of strawberry. *Postharvest Biol. Tec.*, 70, 32, 2012.
96. Bonilla, J., Atarés, L., Vargas, M., Chiralt, A., Effect of essential oils and homogenization conditions on properties of chitosan-based films. *Food Hydrocolloid.*, 26, 9, 2012.
97. Wang, L., Liu, F., Jiang, Y., Chai, Z., Li, P., Cheng, Y., Jing, H., Leng, X., Synergistic antimicrobial activities of natural essential oils with chitosan films. *J. Agr. Food Chem.*, 59, 12411, 2011.
98. García, M., Díaz, R., Martínez, Y., Casariego, A., Effects of chitosan coating on mass transfer during osmotic dehydration of papaya. *Food Res. Int.*, 43, 1656, 2010.

99. Zahid, N., Ali, A., Manickam, S., Siddiqui, Y., Maqbool, M., Potential of chitosan-loaded nanoemulsions to control different *Colletotrichum* spp. and maintain quality of tropical fruits during cold storage. *J. Appl. Microbiol.*, 113, 925, 2012.
100. Park, S.Y., Marsh, K.S., Rhim, J.W., Characteristics of different molecular weight chitosan films affected by the type of organic solvents. *J. Food Sci.*, 67, 194, 2002.
101. Thakhiew, W., Devahastin, S., Soponronnarit, S., Effects of drying methods and plasticizer concentration on some physical and mechanical properties of edible chitosan films. *J. Food Eng.*, 99, 216, 2010.
102. Siripatrawan, U., Harte, B.R., Physical properties and antioxidant activity of an active film from chitosan incorporated with green tea extract. *Food Hydrocolloid.*, 24, 770, 2010.
103. Siripatrawan, U., Noipha, S., Active film from chitosan incorporating green tea extract for shelf life extension of pork sausages. *Food Hydrocolloid.*, 27, 102, 2012.
104. Xiao, Z., Luo, Y., Luo, Y., Wang, Q., Combined effects of sodium chlorite dip treatment and chitosan coatings on the quality of fresh-cut d'Anjou pears. *Postharvest Biol. Tec.*, 62, 319, 2011.
105. Aider, M. Chitosan application for active bio-based films production and potential in the food industry: Review. *LWT-Food Sci. Technol.*, 43, 837, 2010.
106. Hosseini, S.F., Rezaei, M., Zandi, M., Ghavi, F.F., Preparation and functional properties of fish gelatin-chitosan blend edible films. *Food Chem.*, 136, 1490, 2013.
107. Kumari, A., Yadav, S.K., Yadav, S.C., Biodegradable polymeric nanoparticles based drug delivery systems. *Colloid. Surf. B*, 75, 1, 2010.
108. Nedovic, V., Kalusevic, A., Manojlovic, V., Levic, S., Bugarski, B., An overview of encapsulation technologies for food applications. *Procedia Food Sci.*, 1, 1806, 2011.
109. Dash, M., Chiellini, F., Ottenbrite, R.M., Chiellini, E., Chitosan-A versatile semi-synthetic polymer in biomedical applications. *Prog. Polym. Sci.*, 36, 981, 2011.
110. Hosseini, S.F., Zandi, M., Rezaei, M., Farahmandghavi, F., Two-step method for encapsulation of oregano essential oil in chitosan nanoparticles: Preparation, characterization and *in vitro* release study. *Carbohydr. Polym.*, 95, 50, 2013.
111. Hu, B., Ting, Y., Yang, X., Tang, W., Zeng, X., Huang, Q., Nanochemoprevention by encapsulation of (−)-epigallocatechin-3-gallate with bioactive peptides/chitosan nanoparticles for enhancement of its bioavailability. *Chem. Commun.*, 48, 2421, 2012.
112. Hu, B., Ting, Y., Zeng, X., Huang, Q., Cellular uptake and cytotoxicity of chitosan-caseinophosphopeptides nanocomplexes loaded with epigallocatechin gallate. *Carbohydr. Polym.*, 89, 362, 2012.
113. Zhang, S., Luo, Y., Zeng, H., Wang, Q., Tian, F., Song, J., Cheng, W.H., Encapsulation of selenium in chitosan nanoparticles improves selenium

availability and protects cells from selenium-induced DNA damage response. *J. Nutr. Biochem.*, 22, 1137, 2011.

114. Luo, Y., Zhang, B., Cheng, W.H., Wang, Q., Preparation, characterization and evaluation of selenite-loaded chitosan/TPP nanoparticles with or without zein coating. *Carbohydr. Polym.*, 82, 942, 2010.
115. Alishahi, A., Mirvaghefi, A., Tehrani, M.R., Farahmand, H., Shojaosadati, S.A., Dorkoosh, F.A., Elsabee, M.Z., Shelf life and delivery enhancement of vitamin C using chitosan nanoparticles. *Food Chem.*, 126, 935, 2011.
116. Alishahi, A., Mirvaghefi, A., Tehrani, M.R., Farahmand, H., Koshio, S., Dorkoosh, F.A., Elsabee, M.Z., Chitosan nanoparticle to carry vitamin C through the gastrointestinal tract and induce the non-specific immunity system of rainbow trout (*Oncorhynchus mykiss*). *Carbohydr. Polym.*, 86, 142, 2011.
117. Luo, Y., Zhang, B., Whent, M., Yu, L.L., Wang, Q., Preparation and characterization of zein/chitosan complex for encapsulation of α-tocopherol, and its *in vitro* controlled release study. *Colloid. Surface. B*, 85, 145, 2011.
118. Luo, Y., Teng, Z., Wang, Q., Development of zein nanoparticles coated with carboxymethyl chitosan for encapsulation and controlled release of vitamin D3. *J. Agr. Food Chem.*, 60, 836, 2012.
119. Luo, Y., Wang, T.T., Teng, Z., Chen, P., Sun, J., Wang, Q., Encapsulation of indole-3-carbinol and 3,3′-diindolylmethane in zein/carboxymethyl chitosan nanoparticles with controlled release property and improved stability. *Food Chem.*, 139, 224, 2013.
120. Teng, Z., Luo, Y., Wang, Q., Carboxymethyl chitosan-soy protein complex nanoparticles for the encapsulation and controlled release of vitamin D 3. *Food Chem.*, 141, 524, 2013.
121. Ha, H.K., Kim, J.W., Lee, M.R., Lee, W.J., Formation and characterization of quercetin-loaded chitosan oligosaccharide/β-lactoglobulin nanoparticle. *Food Res. Int.*, 52, 82, 2013.
122. O'Toole, M., Henderson, R.M., Soucy, P.A., Fasciotto, B.H., Hoblitzell, P.J., Keynton, R.S., Ehringer, W.D., Gobin, A.S., Curcumin encapsulation in submicrometer spray-dried chitosan/tween 20 particles. *Biomacromolecules*, 13, 2309, 2012.
123. Klaypradit, W., Huang, Y.W., Fish oil encapsulation with chitosan using ultrasonic atomizer. *LWT-Food Sci. Technol.*, 41, 1133, 2008.
124. Hu, M., Li, Y., Decker, E.A., Xiao, H., McClements, D.J., Impact of layer structure on physical stability and lipase digestibility of lipid droplets coated by biopolymer nanolaminated coatings. *Food Biophys.*, 6, 37, 2011.
125. Gandía-Herrero, F., Cabanes, J., Escribano, J., García-Carmona, F., Jiménez-Atiénzar, M., Encapsulation of the most potent antioxidant betalains in edible matrixes as powders of different colors. *J. Agr. Food Chem.*, 61, 4294, 2013.
126. Harris, R., Lecumberri, E., Mateos-Aparicio, I., Mengíbar, M., Heras, A., Chitosan nanoparticles and microspheres for the encapsulation of natural antioxidants extracted from *Ilex paraguariensis*. *Carbohydr. Polym.*, 84, 803, 2011.

127. Li, Y., McClements, D.J., Controlling lipid digestion by encapsulation of protein-stabilized lipid droplets within alginate–chitosan complex coacervates. *Food Hydrocolloid.*, 25, 1025, 2011.
128. Peng, H., Xiong, H., Li, J., Xie, M., Liu, Y., Bai, C., Chen, L., Vanillin cross-linked chitosan microspheres for controlled release of resveratrol. *Food Chem.*, 121, 23, 2010.
129. Belščak-Cvitanović, A., Stojanović, R., Manojlović, V., Komes, D., Cindrić, I.J., Nedović, V., Bugarski, B., Encapsulation of polyphenolic antioxidants from medicinal plant extracts in alginate-chitosan system enhanced with ascorbic acid by electrostatic extrusion. *Food Res. Int.*, 44, 1094, 2011.
130. Luo, Y., Teng, Z., Wang, X., Wang, Q., Development of carboxymethyl chitosan hydrogel beads in alcohol-aqueous binary solvent for nutrient delivery applications. *Food Hydrocolloid.*, 31, 332, 2013.
131. Trapani, A., Sitterberg, J., Bakowsky, U., Kissel, T., The potential of glycol chitosan nanoparticles as carrier for low water soluble drugs. *Int. J. Pharm.*, 375, 97, 2009.
132. Jayakumar, R., Prabaharan, M., Nair, S.V., Tokura, S., Tamura, H., Selvamurugan, N. Novel carboxymethyl derivatives of chitin and chitosan materials and their biomedical applications. *Prog. Mater. Sci.*, 55, 675, 2010.
133. Alishahi, A., Mirvaghefi, A., Tehrani, M.R., Farahmand, H., Shojaosadati, S.A., Dorkoosh, F.A., Elsabee, M.Z., Shelf life and delivery enhancement of vitamin C using chitosan nanoparticles. *Food Chem.*, 126, 935, 2011.
134. Hamman, J.H., Chitosan based polyelectrolyte complexes as potential carrier materials in drug delivery systems. *Mar. Drugs*, 8, 1305, 2010.
135. Tang, D.W., Yu, S.H., Ho, Y.C., Huang, B.Q., Tsai, G.J., Hsieh, H. Y., Sung, H.W., Mi, F.L., Characterization of tea catechins-loaded nanoparticles prepared from chitosan and an edible polypeptide. *Food Hydrocolloid.*, 30, 33, 2013.
136. Estevinho, B.N., Rocha, F., Santos, L., Alves, A., Microencapsulation with chitosan by spray drying for industry applications - a review. *Trends Food Sci. Tech.*, 31, 138, 2013.
137. Desai, K.G.H., Park, H.J., Encapsulation of vitamin C in tripolyphosphate cross-linked chitosan microspheres by spray drying. *J. Microencapsul.*, 22, 179, 2005.
138. Desai, K.G., Park, H.J., Effect of manufacturing parameters on the characteristics of vitamin C encapsulated tripolyphosphate-chitosan microspheres prepared by spray-drying. *J. Microencapsul.*, 23, 91, 2006.
139. Desai, K.G., Liu, C., Park, H.J., Characteristics of vitamin C encapsulated tripolyphosphate-chitosan microspheres as affected by chitosan molecular weight. *J. Microencapsul.*, 23, 79, 2006.
140. Li, Y., Ai, L., Yokoyama, W., Shoemaker, C.F., Wei, D., Ma, J., Zhong, F., Properties of chitosan-microencapsulated orange oil prepared by spray-drying and its stability to detergents. *J. Agr. Food Chem.*, 61, 3311, 2013.
141. Hoffman, A.S., Hydrogels for biomedical applications. *Adv. Drug Deliver. Rev.*, 64, 18, 2012.

142. Sun, P., Li, P., Li, Y.M., Wei, Q., Tian, L.H., A pH-sensitive chitosan-tripolyphosphate hydrogel beads for controlled glipizide delivery. *J. Biomed. Mater. Res. B*, 97, 175, 2011.
143. Takka, S., Gürel, A., Evaluation of chitosan/alginate beads using experimental design: formulation and *in vitro* characterization. *AAPS Pharm. Sci. Tech.*, 11, 460, 2010.
144. Maestrelli, F., Cirri, M., Mennini, N., Bragagni, M., Zerrouk, N., Mura, P., Influence of cross-linking agent type and chitosan content on the performance of pectinate-chitosan beads aimed for colon-specific drug delivery. *Drug Dev. Ind. Pharm.*, 38, 1142, 2012.
145. Bhattarai, N., Gunn, J., Zhang, M., Chitosan-based hydrogels for controlled, localized drug delivery. *Adv. Drug Deliver. Rev.*, 62, 83, 2010.
146. Muzzarelli, R.A.A., Muzzarelli, C., Chitin and chitosan hydrogels. *Handbook of hydrocolloids*, 849, 2009.
147. Liu, L.S., Kost, J., Yan, F., Spiro, R.C., Hydrogels from biopolymer hybrid for biomedical, food, and functional food applications. *Polymers*, 4, 997, 2012.
148. Lin, Y.H., Liang, H.F., Chung, C.K., Chen, M.C., Sung, H.W., Physically cross-linked alginate/N, O-carboxymethyl chitosan hydrogels with calcium for oral delivery of protein drugs. *Biomaterials*, 26, 2105, 2005.
149. Manojlović, V., Nedović, V.A., Kailasapathy, K., Zuidam, N.J., Encapsulation of probiotics for use in food products. In *Encapsulation technologies for active food ingredients and food processing*, pp. 269–302, Springer, New York, 2010.
150. Cook, M.T., Tzortzis, G., Charalampopoulos, D., Khutoryanskiy, V.V., Microencapsulation of probiotics for gastrointestinal delivery. *J. Contr. Rel.*, 162, 56, 2012.
151. Cook, M.T., Tzortzis, G., Charalampopoulos, D., Khutoryanskiy, V.V., Production and evaluation of dry alginate-chitosan microcapsules as an enteric delivery vehicle for probiotic bacteria. *Biomacromolecules*, 12, 2834, 2011.
152. Nualkaekul, S., Lenton, D., Cook, M.T., Khutoryanskiy, V.V., Charalampopoulos, D., Chitosan coated alginate beads for the survival of microencapsulated *Lactobacillus plantarum* in pomegranate juice. *Carbohydr. Polym.*, 90, 1281, 2012.
153. Li, X.Y., Chen, X.G., Sun, Z.W., Park, H.J., Cha, D.S., Preparation of alginate/chitosan/carboxymethyl chitosan complex microcapsules and application in *Lactobacillus casei* ATCC 393. *Carbohydr. Polym.*, 83, 1479, 2011.
154. Chen, S., Cao, Y., Ferguson, L.R., Shu, Q., Garg, S., Evaluation of mucoadhesive coatings of chitosan and thiolated chitosan for the colonic delivery of microencapsulated probiotic bacteria. *J. Microencapsul.*, 30, 103, 2013.
155. Trevan, M.D., *Immobilized enzymes: an introduction and applications in biotechnology*. John Wiley & Sons Inc, 1980.
156. Braun, J., Le Chanu, P., Le Goffic, F., The immobilization of penicillin G acylase on chitosan. *Biotechnol. Bioeng.*, 33, 242, 1989.
157. Siso, M.G., Lang, E., Carreno-Gomez, B., Becerra, M., Espinar, F.O., Méndez, J.B., Enzyme encapsulation on chitosan microbeads. *Process Biochem.*, 32, 211, 1997.

158. Synowiecki, J., Sikorski, Z.E., Naczk, M., Piotrzkowska, H., Immobilization of enzymes on krill chitin activated by formaldehyde. *Biotechnol. Bioeng.*, 24, 1871, 1982.
159. Han, X.Q., Shahidi, F., Extraction of harp seal gastric proteases and their immobilization on chitin. *Food Chem.*, 52, 71, 1995.
160. Krajewska, B., Application of chitin- and chitosan-based materials for enzyme immobilizations: a review. *Enz. Microb. Tech.*, 35, 126, 2004.
161. Shi, W., Ma, Z., Amperometric glucose biosensor based on a triangular silver nanoprisms/chitosan composite film as immobilization matrix. *Biosens. Bioelectron.*, 26, 1098, 2010.
162. Dai, H., Chi, Y., Wu, X., Wang, Y., Wei, M., Chen, G., Biocompatible electrochemiluminescent biosensor for choline based on enzyme/titanate nanotubes/chitosan composite modified electrode. *Biosens. Bioelectron.*, 25, 1414, 2010.
163. Lin, H., Lu, Q., Ge, S., Cai, Q., Grimes, C.A., Detection of pathogen *Escherichia coli* O157:H7 with a wireless magnetoelastic-sensing device amplified by using chitosan-modified magnetic Fe_3O_4 nanoparticles. *Sensor. Actuat. B-Chem.*, 147, 343, 2010.
164. Diaconu, M., Litescu, S.C., Radu, G.L., Laccase-MWCNT-chitosan biosensor-A new tool for total polyphenolic content evaluation from *in vitro* cultivated plants. *Sensor. Actuat. B-Chem.*, 145, 800, 2010.
165. Bolto, B., Gregory, J., Organic polyelectrolytes in water treatment. *Water Res.*, 41, 2301, 2007.
166. Yamada, K., Akiba, Y., Shibuya, T., Kashiwada, A., Matsuda, K., Hirata, M., Water purification through bioconversion of phenol compounds by tyrosinase and chemical adsorption by chitosan beads. *Biotechnol. Progr.*, 21, 823, 2005.
167. Shen, C., Song, S., Zang, L., Kang, X., Wen, Y., Liu, W., Fu, L., Efficient removal of dyes in water using chitosan microsphere supported cobalt (II) tetrasulfophthalocyanine with H_2O_2. *J. Hazard. Mater.*, 177, 560, 2010.
168. Zhang, C., Jia, Z., Preparation of porous chitosan microsphere absorbent and research on its absorption ability for Cu^{2+} and Zn^{2+}. *Int. J. Chem.*, 2, 113, 2010.
169. Salehi, E., Madaeni, S.S., Rajabi, L., Vatanpour, V., Derakhshan, A.A., Zinadini, S., Sh. Ghorabi, S., Monfared, H.A., Novel chitosan/poly (vinyl) alcohol thin adsorptive membranes modified with amino functionalized multi-walled carbon nanotubes for Cu (II) removal from water: Preparation, characterization, adsorption kinetics and thermodynamics. *Sep. Purif. Technol.*, 89, 309, 2012.
170. Liu, Y., Kang, Y., Huang, D., Wang, A., Cu^{2+} removal from aqueous solution by modified chitosan hydrogels. *J. Chem. Technol. Biot.*, 87, 1010, 2012.
171. Rojas-Reyna, R., Schwarz, S., Heinrich, G., Petzold, G., Schütze, S., Bohrisch, J., Flocculation efficiency of modified water soluble chitosan *versus* commonly used commercial polyelectrolytes. *Carbohydr. Polym.*, 81, 317, 2010.
172. Saitoh, T., Asano, K., Hiraide, M., Removal of phenols in water using chitosan-conjugated thermo-responsive polymers. *J. Hazard. Mater.*, 185, 1369, 2011.

173. Bleiman, N., Mishael, Y.G., Selenium removal from drinking water by adsorption to chitosan-clay composites and oxides: Batch and columns tests. *J. Hazard. Mater.*, 183, 590, 2010.
174. Jagtap, S., Yenkie, M.K.N., Labhsetwar, N., Rayalu, S., Defluoridation of drinking water using chitosan based mesoporous alumina. *Micropor. Mesopor. Mat.*, 142, 454, 2011.
175. Zhang, H., Zhu, G., Jia, X., Ding, Y., Zhang, M., Gao, Q., Hu, C., Xu, S., Removal of microcystin-LR from drinking water using a bamboo-based charcoal adsorbent modified with chitosan. *J. Environ. Sci.*, 23, 1983, 2011.
176. Knorr, D., Nutritional quality, food processing, and biotechnology aspects of chitin and chitosan: a review. *Process Biochem.*, 21, 90, 1986.
177. Hirano, S., Production and application of chitin and chitosan in Japan. In: *Chitin and chitosan*, G. Skjak-Braek, T. Anthonsesn, P. Sandford, Eds., pp. 51–69, Elsevier Applied Science, London, UK. 1989.
178. Hirano, S., Itakura, C., Seino, H., Akiyama, Y., Nonaka, I., Kanbara, N., Kawakami, T., Chitosan as an ingredient for domestic animal feeds. *J. Agr. Food Chem.*, 38, 1214, 1990.
179. Razdan, A., Pettersson, D., Effect of chitin and chitosan on nutrient digestibility and plasma lipid concentrations in broiler chickens. *Brit. J. Nutr.*, 72, 277, 1994.
180. Kono, M., Matsui, T., Shimizu, C., Effect of chitin, chitosan, and cellulose as diet supplements on the growth of cultured fish. *Nippon Suisan Gakk.*, 53, 125, 1987.
181. Austin, P.R., Brine, C.J., Castle, J.E., Zikakis, J.P., Chitin: new facets of research. *Science*, 212, 749, 1981.
182. Knorr, D., Betschart, A.A., The relative effect of an inert substance and protein concentrates upon loaf volume of breads. *LWT-Food Sci. Technol.*, 11, 198, 1978.
183. Knorr, D., Betschart, A.A., Water absorption and loaf volume of protein-fortified breads. *LWT-Food Sci. Technol.*, 14, 306, 1981.
184. Muzzarelli, R.A., Chitosan-based dietary foods. *Carbohydr. Polym.*, 29, 309, 1996.
185. Muzzarelli, R.A.A., De Vincenzi, M., Chitosans as dietary food additives. In: *Applications of chitin and chitosan*. Lancaster, Pa: Technomic Publishing, 115, 1997.
186. Zacour, A.C, Silva, M.E, Cecon, P.R., Bambirra, E.A., Vieira, E.C., Effect of dietary chitin on cholesterol absorption and metabolism in rats. *J. Nutr. Sci. Vitaminol.*, 38, 609, 1992.
187. Hirano, S., Akiyama, Y., Absence of a hypocholesterolaemic action of chitosan in high-serum-cholesterol rabbits. *J. Sci. Food Agr.*, 69, 91, 1995.
188. Deuchi, K., Kanauchi, O., Imasato, Y., Kobayashi, E., Decreasing effect of chitosan on the apparent fat digestibility by rats fed on a high-fat diet. *Biosci., Biotechnol. Biochem.*, 58, 1613, 1994.

189. Suzuki, S., Studies on biological effects of water-soluble lower homologous oligosaccharides of chitin and chitosan. *Korean Chitin Chitosan J.*, 50, 1996.
190. Tsukada, K., Matsumoto, T., Aizawa, K., Tokoro, A., Naruse, R.S., Suzuki, S., Suzuki, M., Antimetastatic and growth-inhibitory effects of *N*-acetylchitohexaose in mice bearing lewis lung carcinoma. *Jpn. J. Cancer Res.*, 81, 259, 1990.
191. Suzuki, S., Watanabe, T., Mikami, T., Matsumoto, T., Suzuki, M., Immuno-enhancing effects of *N*-acetylchitohexaose. In: *Advances in chitin and chitosan*, 277, 1992.
192. Tokoro, A., Tatewaki, N., Suzuki, K.O., Mikami, T., Suzuki, S., Suzuki, M., Growth-inhibitory effect of hexa-*N*-acetylchitohexanse and chitohexaose against Meth-A solid tumor. *Chem. Pharm. Bull.*, 36, 784, 1988.
193. Yamada, A., Shibuya, N., Kodama, O., Akatsuka, T., Induction of phytoalexin formation in suspension-cultured rice cells by *N*-acetyl-chitooligosaccharides. *Biosci. Biotech. Bioch.*, 57, 405, 1993.
194. Tokoro, A., Kobayashi, M., Tatewaki, N., Suzuki, K., Okawa, Y., Mikami, T., Suzuki, S., Suzuki, M., Protective effect of *N*-acetyl chitohexaose on *Listeria monocytogenes* infection in mice. *Microbiol. Immunol.*, 33, 357, 1989.
195. Hirano, S., Nagao, N., Effects of chitosan, pectic acid, lysozyme, and chitinase on the growth of several phytopathogens. *Agr. Biol. Chem. Tokyo*, 53, 3065, 1989.
196. Kendra, D.F., Christian, D., Hadwiger, L.A., Chitosan oligomers from *Fusarium solani*/pea interactions, chitinase/β-glucanase digestion of sporelings and from fungal wall chitin actively inhibit fungal growth and enhance disease resistance. *Physiol. Mol. Plant P.*, 35, 215, 1989.
197. Uchida, Y., Izume, M., Ohtakara, A., Preparation of chitosan oligomers with purified chitosanase and its application, In: Proceeding from the 4th International Conference on Chitin and Chitosan, Trondheim, Norway, 373, 1988.
198. Takiguchi, Y., Shimahara, K., Isolation and identification of a thermophilic bacterium producing *N,N′*-diacetylchitobiose from chitin. *Agr. Biol. Chem. Tokyo*, 53, 1537, 1989.
199. Usui, T., Hayashi, Y., Nanjo, F., Sakai, K., Ishido, Y., Transglycosylation reaction of a chitinase purified from Nocardia orientalis. *BBA-Gen. Sub.*, 923, 302, 1987.
200. Usui, T., Matsui, H., Isobe, K., Enzymic synthesis of useful chito-oligosaccharides utilizing transglycosylation by chitinolytic enzymes in a buffer containing ammonium sulfate. *Carbohydr. Res.*, 203, 65, 1990.
201. Aiba, S.I., Preparation of *N*-acetylchitooligosaccharides by hydrolysis of chitosan with chitinase followed by *N*-acetylation. *Carbohydr. Res.*, 265, 323, 1994.
202. Aiba, S.I., Preparation of *N*-acetylchitooligosaccharides from lysozymic hydrolysates of partially *N*-acetylated chitosans. *Carbohydr. Res.*, 261, 297, 1994.

203. Somashekar, D., Joseph, R., Chitosanases-properties and applications: a review. *Biores. Technol.*, 55, 35, 1996.
204. Sakai, K., Nanjo, F., Taichi, U.S.U.I., Production and utilization of oligosaccharides from chitin and chitosan. *J. Jap. Soc. Starch Sci.*, 37, 79, 1990.
205. Izume, M., Ohtakara, A., Preparation of D-glucosamine oligosaccharides by the enzymatic hydrolysis of chitosan. *Agr. Biol. Chem. Tokyo*, 51, 1189, 1987.
206. Jeon, Y.J., Park, P.J., Byun, H.G., Kim, S.K., Song, B.K., Production of chitosan oligosaccharides using chitin-immobilized enzyme. *Korean J. Biotechnol. Bioeng. (Korea Republic)*, 13, 147, 1998.
207. Kim, S.K., Jeon, Y.J., Production of chitosan oligosaccharide using enzyme in an ultrafiltration membrane reactor, In: Proceedings of the International Symposium and 1997 Spring Meeting of the Japanese Society for Chitin and Chitosan. *Shizuoka, Japan*, 144, 1997.
208. Jeon, Y.J., Kim, S.K., Bioactivities of chitosan oligosaccharides and their derivatives in abstracts of 3rd Asia Pacific Chitin and Chitosan Symposium, Kilung, Taiwan, 64, 1998.
209. Kendra, D.F., Hadwiger, L.A., Characterization of the smallest chitosan oligomer that is maximally antifungal to *Fusarium solani* and elicits pisatin formation in *Pisum sativum*. *Exp. Mycol.*, 8, 276, 1984.
210. Suzuki, K., Tokoro, A., Okawa, Y., Suzuki, S., Suzuki, M., Enhancing effects of *N*-acetyl-chito-oligosaccarides on the active oxygen-generating and microbicidal activities of peritoneal exudate cells in mice. *Chem. Pharm. Bull.*, 33, 886, 1985.
211. Suzuki, K., Mikami, T., Okawa, Y., Tokoro, A., Suzuki, S., Suzuki, M., Antitumor effect of hexa-*N*-acetylchitohexaose and chitohexaose. *Carbohydr. Res.*, 151, 403, 1986.
212. Muraki, E., Aiba, S., Antibacterial activity of partially *N*-lauroyl chitooligosaccharides in 2nd Asia Pacific Chitin Symposium. *Bangkok, Thailand*, 283, 1996.
213. Díaz-Visurraga, J., Meléndrez, M.F., García, A., Paulraj, M., Cárdenas, G., Semitransparent chitosan-TiO_2 nanotubes composite film for food package applications. *J. Appl. Polym. Sci.*, 116, 3503, 2010.
214. Saifuddin, N., Nian, C.Y., Zhan, L.W., Ning, K.X., Chitosan-silver nanoparticles composite as point-of-use drinking water filtration system for household to remove pesticides in water. *Asian J. Biochem.*, 6, 142, 2011.

9

Chitosan: Sustainable and Environmental-Friendly Resource for Textile Industry

Mohd Shabbir, Luqman Jameel Rather and Faqeer Mohammad*

Department of Chemistry, Jamia Millia Islamia, New Delhi, India

Abstract
With the industrial developments in recent times, textile industry also needs sustainable and environmental-friendly resources. Today's world has been overburdened with the use of synthetic or hazardous materials in day-to-day life. Natural products on the other side, having never ending sources and advantageous over synthetic ones, meet the needs of present demand for industrial applications. Chitosan polymers obtained from sea animals and fungi, having a lot of properties beneficial to mankind without being hazardous to environment and human, are currently gaining popularity for research and development all over the globe. Functional applications on textiles with the help of chitosan are new trends in textile industrial sectors. Chitosan having more adsorption properties also finds its application in wastewater or textile effluent treatments. This chapter discusses applications of chitosan on textiles and highlights its application for wastewater treatment.

Keywords: Textile industry, natural products, chitosan, sustainable, water treatment

9.1 Introduction

Population explosion and high awareness of modern human being about beneficial part of natural products accelerated the research in the field of utilization of natural products for welfare of mankind. Natural products obtained from plants, animals, and other kingdoms such as fungi and

**Corresponding author*: faqeermohammad@rediffmail.com

Shakeel Ahmed and Saiqa Ikram (eds.) Chitosan, (233–252) © 2017 Scrivener Publishing LLC

microorganisms are very much popular in today's era for applications in textile industry as well as other various industries. Natural dyes obtained from natural resources such as plants and animals are very much known for functionalization of textiles with coloration simultaneously [1, 2]. Better compatibility to human being and environment, never ending availability and biodegradability of natural products make them dominant over synthetic products [3]. A lot of synthetic chemical compounds are utilized till now for functionalization of textiles, which are toxic to human beings, aquatic ecosystem and cannot easily be degraded. These toxic chemicals may be inorganic salts, organometallics, iodo-phors, phenols and thiophenols, onium salts, nitro compounds, ureas and related compounds, amines [4].

Textile industry is one of the industries which interferes into day-to-day life of human being. A large area of economy of a particular state depends on textile industry. From early developments of textiles to cover the body and in recent times smart textiles are covering the textile industry developments [5]. Textile industry is evolving speedily on the effect of increased choices of people in case of clothing on the factors such as colorful, comfortable, functional, and protective clothing. To meet the sophisticated needs of modern human being, a lot of research and development is gaining pace in search of new resources and technologies. Natural products and resources are best fulfilling the needs with advantageous properties such as biocompatibility and biodegradability [6, 7].

Chitosan is among the natural products can be utilized for textile industry development, as it is biocompatible and environment friendly, and with these properties having other functional properties for textiles such as antimicrobial, UV protection, etc. (Figure 9.1). Chitin, chitosan, and other natural products have been much studied resources in recent decades worldwide for human welfare applications [8].

9.2 Chitosan and Chitosan Resources

French Professor, Henri Braconnot in 1811, first discovered chitin in mushrooms which was also isolated from insects in 1820s. Chitin is the most abundant natural amino polysaccharide produced after cellulose and very much similar to cellulose, which is of great interest due to its functional applications in various fields. Its structure contains 2-acetamido-2-deoxy-β-D-glucose through a β-(1–4) linkage. Chitin can be found in shells of crabs, lobsters and shrimps, squids and cell walls of fungi in three crystalline forms α-chitin, β-chitin, and γ-chitin based on their molecular

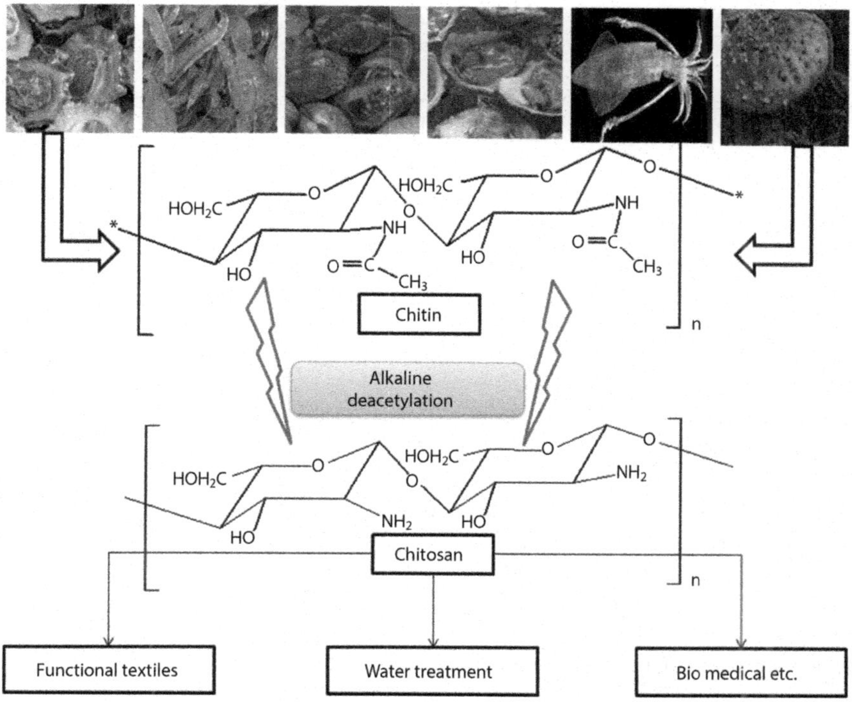

Figure 9.1 Chitosan chain from resources to applications.

arrangement [9]. Chitin is the most important inexpensive polysaccharide occurring in nature which has gained much attention for its industrial applications. Chitin is not directly utilized for industrial applications, but it produces chitosan on chemical modification which has more perspectives in industries due to biodegradability, biocompatibility, antimicrobial activity, nontoxicity, and versatile chemical and physical properties.

Chitosan is a unique basic linear polysaccharide obtained from the deacetylation of chitin which comprises an unbranched chain consisting of β-(1,4)-2-amino-2-deoxy-d-glucopyranose. Chitosan finds its applications in a variety of areas, such as pharmaceutical and medical applications, textiles, wastewater treatment, biotechnology, cosmetics, food processing, and agriculture [8, 10]. A unique property of chitosan is due to its chemical structure possesses amino group at C-2 position allows easy chemical modifications or derivatization.

Chitin occurs as a structural component of the exoskeleton of insects and crustaceans as well as in the cell wall of yeasts and fungi. Chitosan also occurs naturally in several fungi, especially *Mucor* species. Lobster, krill, shrimp, crawfish, cuttlefish, crab, and prawn are some of the examples

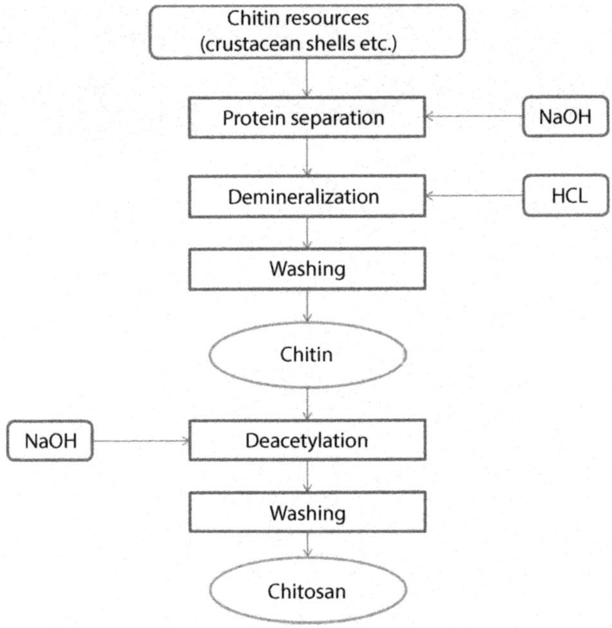

Figure 9.2 Chitosan production steps.

of chitin resources [11, 12]. Different processing conditions are required according to the resources for production of chitosan from chitin.

Chitosan is deacetylated product of chitin polymer obtained from crustaceans or other sea animals. Production of chitosan from resources requires processing steps shown in Figure 9.2. Crustacean shells processed first for removal of proteins by sodium hydroxide and dissolution of calcium carbonate by acid (HCl), which are present in crab shells in high concentrations. Resulting chitin is deacetylated in sodium hydroxide at 120 °C for 1–3 h. Alkali helps in removal of protein and deacetylation of chitin simultaneously [8]. After washing in the last step chitosan can be utilized for further applications in textiles.

9.3 Chitosan Treatment of Textiles

Various textiles or textile substrates can be treated with chitosan polymers for wide range of functionalities such as antimicrobial, superior dyeability, and shrink resistance. A lot of research work has been published in recent times on chitosan application for textile functionalization on wool, silk, cotton, and other synthetic fiber/fabric materials (Figure 9.3).

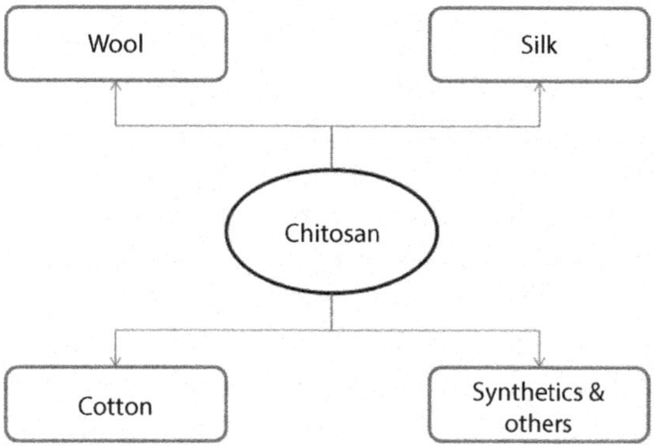

Figure 9.3 Textile materials functionalized with chitosan.

9.3.1 Wool

Wool has been used as textile material since long time for its inherent properties of thermoregulation, high dyeability, flame retardancy, and hygroscopicity. Proteinous nature of wool with large number of functional groups on surface makes them susceptible for most of the functional agents. Improvements in these properties with other functionalities simultaneously have been worked upon with the help of natural and synthetic agents. Chitosan among natural agents is satisfactorily justifying its importance for functionalization of textiles. In one of the studies chitosan was used on wool for making it highly durable shrink resistance. Before applying chitosan, wool was exposed to plasma treatment for creating more anionic groups over the surface of wool resulting high chitosan adsorption [13].

The effect of chitosan application on the dyeing properties of wool fabrics was studied by Dev *et al.*, with henna dye. The antimicrobial properties of chitosan and natural dye independently and collectively on fabrics were assessed and proved that the chitosan treated wool fabrics showed increased dye uptake of fabrics with higher antimicrobial characteristics [14]. The effect of chitosan on the resistance of wool fabric to felting on washing has been studied using nine structurally different samples of chitosan. Effects of structural differences such as molecular weight, level of *N*-acetylation, and the nature and concentration of homologous *N*-acyl groups on antifelting behavior of wool studied and found no strong dependency of shrinkage on molecular weight or level of *N*-acetylation. But increased hydrophobic character of chitosan through incorporation of a number of long-chain *N*-acyl groups improved antifelting behavior [15]. For various functional properties such as antimicrobial, particularly for antifelting functionalization and high dyeability, chitosan is utilized on wool [16–18].

9.3.2 Silk

Silk is another natural textile fiber of proteinaceous nature like wool, very much used fiber in textiles because of its unique properties. Silk is considered as luxurious, comfortable, absorbent, lustrous fabric with thermoregulation properties also. Proteinaceous and moisture adsorbent nature of silk makes it susceptible for microbe's growth responsible for bad odor and deterioration of fabric. Antibacterial properties of chitosan make its introduction into the treatment of silk for antibacterial functionalization and increased dyeability. Antibacterial property of chitosan is imparted on silk textiles by grafting of chitosan on to them. Silk fiber was degummed and acylated with anhydrides for chitosan grafting, and resulted in to increased dyeability of the chitosan grafted-acylated silk fiber in comparison to the acylated and degummed silk samples with higher antibacterial property [19]. The antimicrobial activity of chitosan (a solution of 2% w chitosan in aqueous acetic acid (2% v/v) added of 2-hydroxy-2-methylphenylpropane-1-one as photoinitiator) finished fabrics were tested against *Escherichia coli*. Obtained results showed a strong antimicrobial activity conferred by the treatment, homogeneous on fabric surface, without affecting the hand properties of fabrics due to the low chitosan weight on (about 2%) with durability to domestic laundering with both anionic and nonionic detergents [20]. Dyeability with mono and bi-functional reactive dyes, and antibacterial activity against *Staphylococcus aureus* and *E. coli* on chitosan treated silk were investigated for effective results by Hosseini *et al.*, [21]. O-Carboxymethyl chitosan with a low degree of substitution and a high degree of deacetylation was prepared directly from chitin and utilized to increase the color yield of Acid Red 44 and Acid Green 25 on silk fabrics without lowering the corresponding washing fastness property [22].

9.3.3 Cotton

Cotton is a natural fiber made up of cellulose polymer. Its properties such as durability due to cellulose composition, softness, and compatibility to human skin (nonallergic) make it more universal for use in clothing. Certain limitations are there for the use of cotton textiles. Low dyeability due to less functional group on surface is one of them. A lot of research has been carried out for cotton dyeing and functionalization. Chitosan is used with citric acid by means of the conventional pad-dry-cure process as durable press and antimicrobial finishing agents for cotton. Citric acid is expected to react with hydroxyl groups in cellulose and chitosan (with amino groups in chitosan), as cross-linking agent. These properties were retained through twenty washing cycles in durability tests [23]. A chitosan derivative, *O*-acrylamidomethyl-*N*-[(2-hydroxy-3-trimethylammonium)propyl] chitosan chloride (NMA-HTCC), which is fiber reactive in nature, was applied to cotton fabrics to evaluate its use as a durable antimicrobial textile finish. The antimicrobial activities of treated cotton fabrics were evaluated quantitatively against *S. aureus* and appreciable results were found. Treated cotton at a concentration of 1% on weight of fabric showed 100% of bacterial reduction and the activity was retained over 99% upto 50 consecutive home laundering conditions [4]. Cross-linking agents are supposed to enhance the properties achieved on fabrics with the effect of chitosan treatments. Cotton fabrics were treated with cross-linking agents [butane tetracarboxylic acid (BTCA) and Arcofix NEC (low formaldehyde content)] in the presence of chitosan to provide the cotton fabrics a durable press finishing and antimicrobial properties. Cross-linking agents were found very much helpful for chitosan bonding to cellulosic material cotton by chemical linking of chitosan to the cellulose structure. The treated cotton fabrics showed broad-spectrum antimicrobial activity against gram-positive and gram-negative bacteria and fungi accordingly with the type and concentration of finishing agent in the presence of chitosan as well as the treatment conditions [24]. A water-soluble chitosan quaternary ammonium derivative, *N*-(2-hydroxy) propyl-3-trimethylammonium chitosan chloride (HTCC), was successfully used as an antimicrobial agent for cotton fabrics as it was found having lower minimum inhibition concentration (MIC) against *S. aureus*, *Klebsiella pneumoniae,* and *E. coli* compared to that of chitosan, but imparted antimicrobial activity was less durable against laundering. So, cross-linking agents were used to get durable antimicrobial treatment by immobilizing *N*-(2-hydroxy) propyl-3-trimethylammonium chitosan chloride. Cross-linkers such as dimethyloldihydroxyethylene urea (DMDHEU), butanetetracarboxylic acid (BTCA), and citric acid (CA) were used to improve the laundering durability by covalent bond formation between the cross-linker, chitosan derivative and cellulose. The treated fabrics showed durable antimicrobial activity up to several laundering cycles [25]. A water-soluble carboxymethyl derivative of chitosan was prepared with a view to develop a multifunctional finish on cotton. Treated cotton was found having better dyeability with direct and reactive dyes, and durable antimicrobial activity against *E. coli* and *S. aureus* as well as improved wrinkle recovery [26].

9.3.4 Others (Synthetic or Semisynthetic Fibers)

Textile materials such as regenerated cellulosic fibers (viscose rayon) and other synthetic fibers (polyester) are also treated by chitosan to get functionalities on them. Polyester fabrics are electrostatic and scientific attention needed for their antistatic functionalization. Hydroxyl functional groups of polyester fabrics can be cross-linked with chitosan amine groups for antistatic finishing and improvement of mechanical properties [27]. Cotton/polyester and polyester fabrics were successfully functionalized for antistatic properties via permanent fixation of chitosan onto them. This finishing with chitosan generally improves the water uptake capacity of the finished fabrics without disturbing their physico-mechanical properties. Improvement in water uptake of the finished fabrics is generally attributed to the hydroxyl groups introduced to the fabric structure through finishing, which increase the fabric's ability to absorb water and moisture from air. Presence of more water in such fabrics increases their electrical conductivity and improves their antistatic properties [28]. Dyeability of acrylic and polyester fabrics was successfully improved with the application of chitosan or chitosan derivative toward acid dyes. The color strength of the prints was found higher for treated samples and this enhancement in color strength was in correlation with *N*% of chitosan [29]. High durable antibacterial activity against both gram-positive and gram-negative bacteria was achieved on bamboo rayon also, via chitosan treatment extracted from shrimp shells [30].

9.4 Textile Functionalities Achieved

Chitosan is a polysaccharide composed of copolymers of glucosamine and *N*-acetyl-glucosamine, mainly obtained through alkaline deacetylation of chitin derived from the exoskeleton of crustaceans and arthropods [31]. Chitin, annually produced as much as cellulose, is the most abundant natural amino polysaccharide. Both chitosan and chitin have great practical utility not only as an underutilized resource but also as new and environment friendly bioresource functional material of high potential in various fields especially textile industry. Chitosan has attracted significant interests due to its various functional properties such as antimicrobial [10], antitumor, and immune enhancing effects [32–38].

Chitosan is reasonably priced, nontoxic, and biodegradable, possesses reactive amino groups, and retains its identity as a wonderful polysaccharide material in many areas of applications such as wastewater treatment, food and textile industry, drug industry and as a hydrating agent in cosmetics [23, 39–42]. The main applications of chitosan in textile industrial sectors can be categorized into two main topics: the production of manmade fibers and textile wet processing, which include dyeing, finishing,

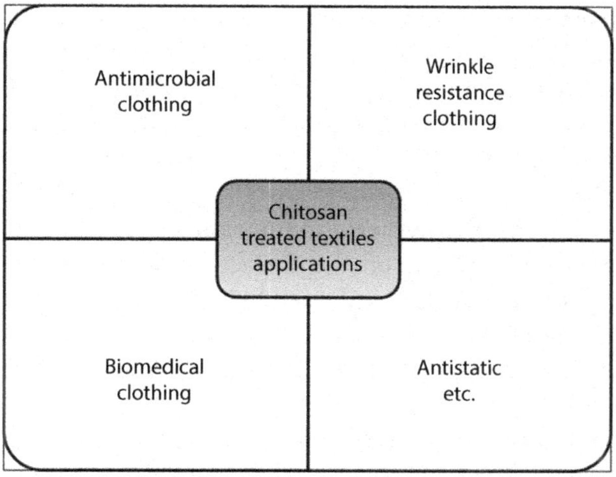

Figure 9.4 Application array for chitosan treated textiles.

and printing (Figure 9.4). Hence, in this part of the chapter, we review the functional properties of textile materials imparted through functional treatment of chitosan, improving dye-ability, print paste thickener, and recent progress in chitosan chemistry.

9.4.1 Antimicrobial Finishing and Enhanced Dyeing Effects

Functionalization of textile materials with different natural and synthetic antibacterial finishing agents is gaining popularity all over the world as they provide consumers with greater comfort, remain fresh and odor free in everyday use [43, 44]. Although, recent research studies have explored the toxic effects and nonbiodegradable nature of synthetic antibacterial agents in comparison to natural ones such as triclosan, metals and their salts, organometallics, phenols, etc. [45]. Textile industrial sectors are looking for alternative eco-friendly and biodegradable substitutes of synthetic antibacterial agents such as chitosan, chitosan derivatives, antimicrobial natural dyes, nanoparticles, and nanoparticle-biopolymer composite finishing that can eradicate ecological and environmental risks posed by synthetic ones [46]. During last few decades, textile applications of chitosan as antimicrobial finishing agent have attracted worldwide interest from various researchers.

Antimicrobial finishing of cotton fabric was presented in an economic way to prevent cross-infection and harm to human body [23, 40, 47]. Researchers have presented several proposed mechanism of action for the antimicrobial activity by chitosan:

- **Polycationic nature (structure) of chitosan:** Excessive positive charge of chitosan is expected to interact with anionic components (lipopolysaccharides and proteins) resulting in changes in membrane permeability, causing death of microorganisms through leakage of intracellular components [48–50].
- **Formation of polymeric membrane:** Chitosan can form polymer membrane on the surface of the cell, preventing nutrients from entering the cell [35, 48, 51].
- **DNA Binding:** Low molecular weight chitosan enters the cell and binds to DNA and inhibits RNA and protein synthesis [52].
- **Coagulation of intercellular components:** Higher adsorption capacity of chitosan can lead to the adsorption of electronegative substance from intercellular fluids and flocculate them, disturbing the physiological activities of the microorganism which eventually leads to the death of the cells [38].

Chitosan can be considered as a multifunctional textile finishing agent because its antimicrobial activity can be combined with other functions such a dyeing improvement, antistatic, and deodorant activity. Chitosan pretreatment enhanced the uptake of lac dye and increased lac dye sorption on cotton. Langmuir adsorption isotherm described best the increased dye adsorption of chitosan pretreated cotton fabric [53]. Adsorption characteristics of lac dye on chitosan pretreated silk fabric were studied with the aim of determination of enhanced dye adsorption ability in terms of increased color strength values with increase wash and rub fastness results [21].

However, there are some limitations for its use in textile applications, as it is only effective at relatively higher concentrations, where it forms a film on the surface of fabric, reduces air permeability, and induces stiffness in the fabric. However, lack of strong chemical bonding with textile substrate became major drawback for chitosan antibacterial finishes because of the durability problems against repeated laundering (washing) cycles [20]. Biotechnological and surface modification tools appear an emerging technology to create a new range of high performing and environmental friendly materials for traditional and innovative applications of natural polymers. Accordingly, improved technologies such as use of grafting (cross-linking) agents and nanoparticle composites are highly desirable for chitosan textile applications wherein the consumers get benefit of excellent antimicrobial attributes of chitosan without losing the textile properties. In this regard, researchers all over the world have used many cross-linking agents such as butane tetracarboxylic acid (BTCA), citric acid, cyanuric chloride, propane 1,2,3-tricarbonic acid, Arcofix NBC, hydroxydichlorotriazine, dodecylamine, and Dimethylol dihydroxy ethylene urea (DMDHEU) to enhance the antimicrobial function and enhanced dyeability of finished textile materials. Another recent environmental friendly cross-linking technology (UV-curing technology) in the field of coatings, printing, adhesives, and electrical insulation are well known. Some of the noteworthy advantages of UV-curing over conventional and hazardous cross-linking agents are low-temperature process, simple, cheap, small equipment, and high treatment speed. Antimicrobial properties were conferred against *E. coli* to cotton and silk fabrics through UV-curing technology of chitosan with satisfactory washing resistance (durability) [20].

Surface modification of silk fiber using anhydrides (succinic anhydride and phthalic anhydride) in different solvents (DMSO and DMF) grafted with the polysaccharide chitosan was studied for evaluating the dye uptake efficiency and subsequent antibacterial potential of grafted silk fibers [19]. Nanochitosan manufactured from low molecular weight chitosan (LMWC) is a natural material with excellent physicochemical properties, environmentally safe, and possesses many biological activities. In addition, nanochitosan has been used to improve the strength and washability of textiles and to confer antibacterial effects. It has been frequently used for controlled-release of drugs for gene transfer in artificial organs and for immune prophylaxis. Yang *et al.*, [54] prepared nanochitosan from LMWC for application onto wool fabric for imparting antibacterial and shrink-proofing resistance [54]. Synthesis of chitosan nanoparticles by ionic gelation with sodium tripolyphosphate and subsequently loaded silver nanoparticles (Ag-CSN composite) was used to improve bactericidal efficacy of finished polyester fabric. Mechanism of bactericidal effects can be explained through the release of Ag+ from Ag-CSN finished polyester fabric by antibacterial testing showing clear zone of inhibition [55]. Chitosan nanoparticles synthesized through similar reaction scheme as reported by Ali *et al.*, [55], were applied to *Antheraea pernyi* silk to enhance the antimicrobial, breaking strength, and wrinkle-resistance properties of treated silk filaments and fabric [56]. Nanochitosan cross-linked with methacrylic acid onto cotton textiles showed improved dye uptake capacity and thermal stability along with enhanced antimicrobial action and UV protection properties [57].

9.4.2 Wrinkle Proof Resistance

The wrinkle-proofing resistance of cellulosic textile materials confers an additional value, especially in terms of ease and comfort. Low water absorbing capacity, perspiration, etc., are major inconvenient factors which impose restrictions in the utilization of the cellulosic wrinkle-proofed materials though excluding the clothing area. However, numerous chemical substances are being used to confer resistance to wrinkle formation or good wrinkle-recovering effects of finished textile materials. Formaldehyde and its derivatives were among the first substances used as wrinkle-proofing agents [58]. The environmental concerns related to the use of formaldehyde and its derivatives along with the associated toxicity (either due to its emission in the working medium or as residues on the textile material) restrict their use in textile finishing, even though the wrinkle-proofing effects were very good. However, nonformaldehyde wrinkle proofing agents have been searched and are able to remove formaldehyde based finishes, offering wrinkle-proofing and wettability effects. Among these chemical compounds, the most used are polycarboxylic acids such as butane tetracarboxylic acid (BTCA) which result in wrinkle-proofing effects comparable to those of the urea-formaldehyde compounds [59]. β-cyclodextrin was grafted onto cellulose via BTCA cross-linking without increasing hydrophilicity because a strong network formed between cellulose and β-cyclodextrin [60]. Another nonformaldehyde wrinkle-proof finishing agent is chitosan which is used alone or in combination with other agents (bonding bridges), involved both in covalent cross-linking and ionic cross-linking resulting in excellent wrinkle-recovering angles and tensile strength [61–63].

Simultaneous hydrophilizing and wrinkle-proofing effects on cotton fabric were achieved through pad-dry-cure treatment with chitosan, tetronic, and mono chlorotriazine cyclodextrin. Quantitative influences were produced by changing concentrations of these three multifunctional agents on adsorption capacity, wrinkle recovery angles, breaking strength, durability test, immersion time, contact angle, and capillarity [64]. In an another study, aqueous low molecular weight chitosan-polyurethane emulsions were applied onto different quality plain weave poly-cotton dyed and printed fabric through pad-dry-cure method. Experimental results have revealed that the incorporation of chitosan has produced pronounced effects on air permeability, stiffness, crease recovery angle, pilling resistance, tear, and tensile strength of pure cotton and woolen fabrics [65].

9.4.3 Biomedical Applications

Exploiting bone grafting materials with microstructure to form composites of comparable mechanical properties to native bone poses a significant challenge in the area of tissue engineering. Natural bone's unique hybrid composition and hierarchical structure have presented valuable clues to achieve intelligent scaffold materials [66]. Combining the advantages of inorganic and organic components to form inorganic-organic hybrids [67] shows good biocompatibility and favorable bonding ability with the host tissues [68].

A novel hydroxyapatite/chitosan–silk fibroin (HA/CTS-SF) composite was prepared for bone repair and replacement through coprecipitation method. The resulting composite exhibited higher compressive strength as compared to the precipitated hydroxyapatite. The improvement in the interfacial bonding and enhanced mechanical property are due to the increased chemical interactions between the inorganic phase and the organic matrix. A number of studies have been reported for the use of chitosan scaffolds and membranes to treat patients with deep burns and wounds. Novel α-chitin/nanosilver composite scaffolds for wound healing were found to possess excellent antibacterial activity against *S. aureus* and *E. coli*, combined with good blood clotting ability [69].

9.5 Effluent Treatment Applications

Textile industry consumes a large amount of water for textile processing such as wet processing, dyeing, and textile finishing. A variety of chemicals of varying composition are used and streamed into water bodies from textile industries. Increased pollution in water ecosystem acquired much attention of researchers to cure the problem of wastewater effluent from textile industries. Various methods of different categories such as physical, chemical, and biological are in trend for dye effluent treatment [70]. Dyeing effluent is one of the major part of textile effluent and has hazardous effects on the environment. Adsorption via natural polysaccharides can be an eco-friendly and effective treatment for decolorization of textile dyeing effluent. Valency forces and van der Waals forces are supposed to play important role in adsorption of dyes on to chitosan polymers [71]. Adsorption mechanisms such as ion-exchange, complexation, coordination/chelation, electrostatic interactions, acid-base interactions, hydrogen bonding, hydrophobic interactions, physical adsorption, and precipitation are some reported ones [72]. Adsorption of acid dyes on chitosan was studied by many research groups for wastewater treatment via adsorption models and intraparticle diffusion mechanism. Adsorption capacities of chitosan were found affected by molecular size and the number of sulfonate groups of each dye [73–75].

Chitosan had been studied for successful results as a biosorbent to remove the anionic dyes from contaminated watercourses in both single and binary systems of pollutants [76]. Addition of amido and imino groups on chitosan backbone through grafting reactions was carried and direct correlation of adsorption capacity of adsorbents was observed, and the cross-linking method also made them capable of reuse for many cycles without loss of their capacity [77]. Many of coagulants have been found highly effective and suggested for de-colorization of textile wastewater. Natural coagulants have been in trend because of their eco-friendly nature. Coagulation and flocculation properties of chitosan polymer can be useful to remove hazardous chemical substances streamed from textile industry [78, 79].

9.6 Future Perspectives and Conclusion

Chitosan is the polymeric macromolecules obtained from nature having various physiological and biological properties. Biocompatibility, biodegradability, and bioactivity of chitosan make its importance toward human welfare. From textile finishing to textile effluent treatment, every aspects of textile industry seems having requirement of chitosan like polymers. A lot of activities (both physical and biological) are studied in past with the applications in various industries and area of applications seems increased. This area extended to bone regeneration within chitosan scaffolds, and drug delivery systems. Mucoadhesive and bioadhesive properties of chitosan directed its use in orthopaedics, dentistry, and surgical procedures. Such properties very much introduced use of chitosan in advanced applications in future [80]. Fabrication of biocompatible textile chitosan scaffolds was investigated for successful results of proliferation and differentiation of osteoblasts [81, 82]. Antimicrobial activity of chitosan gives its a large scope, and can be further extended by understanding inhibition mechanism of chitosan and factors such as intrinsic, environmental, microorganism, and physical state. A review has been published with this thought of understanding mechanism of action and depending factors [83].

Chitosan nanocomposites are new trend for enhanced functionalities. Simultaneous fabrication of nanomaterials and chitosan on textile surfaces leads to achievement of multi-functionalities imparted by both nanomaterials and chitosan. Hybrid ZnO/Chitosan are coated via sonochemical process on textiles for enhanced antimicrobial potential [84]. ZnO and Ag:ZnO nanoparticles were prepared and incorporated within chitosan matrix resulted into composite coatings, which are applied on textile for antimicrobial functionality toward medical applications [85]. Ultimately, chitosan pertaining good adsorption and antimicrobial properties with good compatibility, biodegradability, and environment friendly nature can have a long term future with lot of advancements and evolutions in wide range of sectors of textiles (biomedical and environmental).

References

1. Shahid, M., Mohammad, F., Recent advancements in natural dye applications: a review. *J. Clean. Prod.*, 53, 310–331, 2013.
2. Yusuf, M., Shabbir, M., Mohammad, F., Natural colorants: historical, processing and sustainable prospects. *Nat. Prod. Bioprospect.*, 1–23, 2017.
3. Shabbir, M., Mohammad, F., Natural polymers: scope in textile functionalization. In: *Natural polymers: derivatives, blends and composites,* Volume I, S. Ikram, S. Ahmed, Eds., 157–170, 2016.
4. Lim, S.H., Hudson, S.M., Application of a fiber-reactive chitosan derivative to cotton fabric as an antimicrobial textile finish. *Carbohydr. Polym.*, 56(2), 227–234, 2004.
5. Islam, S., Shabbir, M., Mohammad, F., Insights into the functional finishing of textile materials using nanotechnology. In: *Textiles and clothing sustainability* , 97–115, Springer Singapore, 2017.
6. Yusuf, M., Mohammad, F., Shabbir, M., Khan, M.A., Eco-dyeing of wool with *Rubia cordifolia* root extract: Assessment of the effect of *Acacia catechu* as biomordant on color and fastness properties. *Textiles Clothing Sustain.*, 2(1), 1–9, 2017.
7. Yusuf, M., Khan, S.A., Shabbir, M., Mohammad, F., Developing a shade range on wool by Madder (Rubia cordifolia) root extract with Gallnut (Quercus infectoria) as biomordant. *J. Nat. Fibers*, 1–11, 2016.
8. Kumar, M.N.R., A review of chitin and chitosan applications. *React. funct. polym.*, 46(1), 1–27, 2000.
9. Sudha, P.N., Aisverya, S., Nithya, R., Vijayalakshmi, K., Industrial applications of marine carbohydrates. *Adv. Food Nutr. Res.*, 73, 145–181, 2014.
10. Lim, S., Hudson, S.M., Review of chitosan and its derivatives as antimicrobial agents and their uses as textile chemicals. *J. Macromol. Sci.: Rev.*, C43, 223–269, (2003).
11. Peter, M.G., Applications and environmental aspects of chitin and chitosan. *J. Macromol. Sci., Part A: Pure Appl. Chem.*, 32(4), 629–640, 1995.
12. No, H.K., Meyers, S.P., Preparation and characterization of chitin and chitosan-a review. *J. Aquat. Food Prod. Technol.*, 4(2), 27–52, 1995.
13. Erra, P., Molina, R., Jocic, D., Julia, M.R., Cuesta, A., Tascon, J.M.D., Shrinkage properties of wool treated with low temperature plasma and chitosan biopolymer. *Text. Res. J.*, 69(11), 811–815, 1999.
14. Dev, V.G., Venugopal, J., Sudha, S., Deepika, G., Ramakrishna, S., Dyeing and antimicrobial characteristics of chitosan treated wool fabrics with henna dye. *Carbohydr. Polym.*, 75(4), 646–650, 2009.

15. Roberts, G.A., Wood, F.A., A study of the influence of structure on the effectiveness of chitosan as an anti-felting treatment for wool. *J. Biotechnol.*, 89(2), 297–304, 2001.
16. Davidson, R.S., Xue, Y., Improving the dyeability of wool by treatment with chitosan. *Color. Technol.*, 110(1), 24–29, 1994.
17. Pascual, E., Julia, M.R., The role of chitosan in wool finishing. *J. Biotechnol.*, 89(2), 289–296, 2001.
18. Ammayappan, L., Moses, J.J., Study of antimicrobial activity of aloevera, chitosan, and curcumin on cotton, wool, and rabbit hair. *Fibers Polym.*, 10(2), 161–166, 2009.
19. Davarpanah, S., Mahmoodi, N.M., Arami, M., Bahrami, H., Mazaheri, F., Environmentally friendly surface modification of silk fiber: chitosan grafting and dyeing. *Appl. Surf. Sci.*, 255(7), 4171–4176, 2009.
20. Periolatto, M., Ferrero, F., Vineis, C., Antimicrobial chitosan finish of cotton and silk fabrics by UV-curing with 2-hydroxy-2-methylphenylpropane-1-one. *Carbohydr. polym.*, 88(1), 201–205, 2012.
21. Hosseini, M., Montazer, M., Damerchely, R., Enhancing dye-ability and antibacterial features of silk through pre-treatment with chitosan. *Biotechnol.*, 21, p. 26, 2013.
22. Wu, Y.G., Chan, W.L., Szeto, Y.S., Preparation of O-carboxymethyl chitosans and their effect on color yield of acid dyes on silk. *J. Appl. Polym. Sci.*, 90(9), 2500–2502, 2003.
23. Chung, Y.S., Lee, K.K., Kim, J.W., Durable press and antimicrobial finishing of cotton fabrics with a citric acid and chitosan treatment. *Text. Res. J.*, 68, 772–775, 1998.
24. El-Tahlawy, K.F., El-Bendary, M.A., Elhendawy, A.G., Hudson, S.M., The antimicrobial activity of cotton fabrics treated with different cross-linking agents and chitosan. *Carbohydr. Polym.*, 60(4), 421–430, 2005.
25. Kim, Y.H., Nan, C.W., Choi, J.W., Jang, J., Durable antimicrobial treatment of cotton fabrics using N-2-hydroxy propyl-3-trimethylammonium chitosan chloride and polycarboxylic acids. *J. Appl. Polym. Sci.*, 88, 1567–1572, 2003.
26. Gupta, D., Haile, A., Multifunctional properties of cotton fabric treated with chitosan and carboxymethyl chitosan. *Carbohydr. Polym.*, 69(1), 164–171, 2007.
27. Eom, S.I., Using chitosan as an antistatic finish for polyester fabric. *AATCC Rev.*, 1(3), 2001.
28. Abdel-Halim, E.S., Abdel-Mohdy, F.A., Al-Deyab, S.S., El-Newehy, M.H., Chitosan and monochlorotriazinyl-β-cyclodextrin finishes improve antistatic properties of cotton/polyester blend and polyester fabrics. *Carbohydr. Polym.*, 82(1), 202–208, 2010.
29. Hakeim, O.A., El-Gabry, L., Abou-Okeil, A., Rendering synthetic fabrics acid printable using chitosan and binder. *J. Appl. Polym. Sci.*, 108(4), 2122–2127, 2008.

30. Teli, M.D., Sheikh, J., 2012. Extraction of chitosan from shrimp shells waste and application in antibacterial finishing of bamboo rayon. *Int. J. Biol. Macromolec.*, 50(5), 1195–1200.
31. Li, Q., Dunn, E.T., Grandmaison, E.W., Goosen, M.F., Application and properties of chitosan. In: M.F.A. Goosen, Ed.*Application of chitin and chitosan*, 3–29, Lancaster, PA: Technomic Publishing, 3–29, 1997.
32. Choi, B.K., Kim, K.Y., Yoo, Y.J., Oh, S.J., Choi, J.H., Kim, C.Y., *In vitro* antimicrobial activity of a chitooligosaccharide mixture against Actinobacillus actinomycetemcomitans and Streptococcus mutans. *Int. J. Antimicrob. Agents*, 18, 553–557, 2001.
33. Hirano, S., Nagao, N.,Effects of chitosan, pectic acid, lysozyme and chitosan on the growth of several phytopathogens. *Agric. Biol. Chem.*, 53, 3065–3066, 1989.
34. Jumaa, M., Furkert, F.H., Muller, B.W., A new lipid emulsion formulation with high antimicrobial efficacy using chitosan. *Eur. J. Pharm. Biopharm.*, 53, 115–123, 2002.
35. Liu, H., Du, Y., Yang, J., Zhu, H., Structural characterization and antimicrobial activity of chitosan/betain derivative complex. *Carbohydr. Polym.*, 55, 291–297, 2004.
36. Shon, D.H., Chitosan oligosaccharides for functional foods and microbial enrichment of chitosan oligosaccharides in soy-paste. Proceedings of the International Workshop on Bioactive Natural Products, 56–66. *The Committee on Science and Technology in Developing Countries and the Science Council*, Japan, (2001).
37. Tsukada, K., Matsumoto, T., Aizawa, K., Tokora, A., Naruse, R., Suzuki, S., Suzuki, M., Antimetastatic and growth-inhibitory effects of N-acetylchitohexose in mice bearing Lewis lung carcinoma. *Jpn. J. Cancer Res.*, 81, 259–265, (1990).
38. Zheng, L.Y., Zhu, J.F., Study on antimicrobial activity of chitosan with different molecular weights. *Carbohydr. Polym.*, 54, 527–530, 2003.
39. Davidson, R.S., Xue, Y., Improving the dyeability of wool by treatment with chitosan. *J. Soc. Dyers Colour.*, 110, 24–29, 1994.
40. Kim, Y.H., Choi, H.M., Yoon, J.H., Synthesis of a quaternary ammonium derivative of chitosan and its application to a cotton antimicrobial finish. *Text. Res. J.*, 68, 428–434, 1998.
41. Rippon, J.A., Improving the dye coverage of immature cotton fibers by treatment with chitosan. *J. Soc. Dyers Colour.*, 100, 298–303, 1984.
42. Tsuragai, K., Yoshikawa, A., Tajima, J., Ishii, Y., Preparation and antimicrobial activities of cross-linked chitosan films. *Sen-I Gakkaishi*, 47, 190–197, 1991.
43. Rather, L.J., Islam, S., Azam, M., Shabbir, M., Bukhari, M.N., Shahid, M., Khan, M.A., Haque, Q.M.R., Mohammad, F., Antimicrobial and fluorescence finishing of woolen yarn with Terminalia arjuna natural dye as an ecofriendly substitute to synthetic antibacterial agents. *RSC Adv.*, 6(45), 39080–39094, 2016.

44. Shahid, M., Ahmad, A., Yusuf, M., Khan, M.I., Khan, S.A., Manzoor, N., Mohammad, F., Dyeing, fastness and antimicrobial properties of woolen yarns dyed with gallnut (Quercus infectoria Oliv.) extract. *Dyes Pigm.*, 95(1), 53–61, 2012.
45. Shabbir, M., Islam, S.U., Bukhari, M.N., Rather, L.J., Khan, M.A., Mohammad, F., Application of Terminalia chebula natural dye on wool fiber—evaluation of color and fastness properties. *Textiles and Clothing Sustainability*, 2(1), p. 1, 2016.
46. Rather, L.J., Shabbir, M., Bukhari, M.N., Shahid, M., Khan, M.A., Mohammad, F., Ecological dyeing of Woolen yarn with Adhatoda vasica natural dye in the presence of biomordants as an alternative copartner to metal mordants. *J. Environ. Chem. Eng.*, 4(3), 3041–3049, 2016.
47. Seventekin, N., Ucarci, O., The damage caused by microorganisms to cotton fabrics. *J. Text. Inst.*, 84, 304–312, 1993.
48. Helander, I.M., Nurmiaho-Lassila, E.L., Ahvenainen, R., Rhoades, J., Roller, S., Chitosan disrupts the barrier properties of the outer membrane of gram-negative bacteria. *Inter. J. Food Microbiol.*, 71, 235–244, 2001.
49. Lim, S.H., Hudson, S.M., Synthesis and antimicrobial activity of a water-soluble chitosan derivative with a fiber-reactive group. *Carbohydr. Res.*, 339, 313–319, 2004.
50. Wang, X., Du, Y., Liu, H., Preparation, characterization and antimicrobial activity of chitosan Zn complex. *Carbohydr. Polym.* (in press).
51. Zheng, L.Y., Zhu, J.F., Sun, K.S., Antimicrobial activity of chitosan. *Mater. Sci. Eng.*, 18, 22–24, 2000.
52. Liu, X.F., Guan, Y.L., Yang, D.Z., Li, Z., Yao, K.D., Antimicrobial action of chitosan and carboxymethylated chitosan. *J. Appl. Polym. Sci.*, 29, 1324–1335, 2001.
53. Rattanaphani, S., Chairat, M., Bremner, J.B., Rattanaphani, V., An adsorption and thermodynamic study of lac dyeing on cotton pretreated with chitosan. *Dyes Pigm.*, 72, 88–96, 2007.
54. Yang, H.-C., Wang, W.-H., Huang, K.-S., Hon, M.-H., Preparation and application of nanochitosan to finishing treatment with anti-microbial and anti-shrinking properties. *Carbohydr. Polym.*, 79, 176–179, 2010.
55. Ali, S.W., Rajendran, S., Joshi, M., Synthesis and characterization of chitosan and silver loaded chitosan nanoparticles for bioactive polyester. *Carbohydr. Polym.*, 83, 438–446, 2011.
56. Lu, Y.-H., Chen, Y.-Y., Lin, H., Wang, C., Yang, Z.-D., Preparation of Chitosan Nanoparticles and Their Application to Antheraea pernyi Silk. *J. Appl. Polym. Sci.*, 117, 3362–3369, 2010.
57. Ali Hebeish, A., Sharaf, S., Farouk, A., Utilization of chitosan nanoparticles as a green finish in multifunctionalization of cotton textile. *Inter. J. Biol. Macromol.*, 60, 10–17, 2013.
58. Harifi, T., Montazer, M., Past, present and future prospects of cotton cross-linking: New insight into nano particles. *Carbohydr. Polym.*, 88, 1125–1140, 2012.

59. Fouda, M.M.G., Fahmy, H.M., Multifunctional finish and cotton cellulose fabric. *Carbohydr. Polym.*, 86, 625–629, 2011.
60. Medronho, B., Andrade, R., Vivod, V., Ostlund, A., Miguel, M.G., Lindman, B., *et al.*, Cyclodextrin-grafted cellulose: physico-chemical characterization. *Carbohydr. Polym.*, 93, 324–330, 2013.
61. Vasluianu, E., Popescu, V., Grigoriu, A., Forna, N.C., Sandu, I., Comparative study concerning the FTIR analysis and the performances of chitosan-based wrinkle-proofing agents. *Revista de Chimie* (Bucharest), 64, 1104–1115, 2013.
62. Hashem, M., Hauser, P., Smith, B., Wrinkle recovery for cellulosic fabric by means of ionic cross-linking. *Text. Res. J.*, 73, 762–766, 2003.
63. Popescu, V., Vasluianu, E., Forna, N.C., Sandu, I., Bercu, E., Comparative study of the FTIR analysis and the performances of N,N,N-trimethyl chitosan as wrinkle-proofing agent. *Revista de Chimie* (Bucharest), 64, 1284–1294, 2013.
64. Popescua, V., Vasluianua, E., Popescu, G., Quantitative analysis of the multifunctional finishing of cotton fabric with non-formaldehyde agents. *Carbohydr. Polym.*, 111, 870–882, 2014.
65. Muzaffar, S., Bhatti, I.A., Zuber, M., Bhatti, H.N., Shahid, M., Synthesis, characterization and efficiency evaluation of chitosan-polyurethane based textile finishes. *Int. J. Biol. Macromol.*, 93, 145–155, 2016.
66. Li Wang, Chunzhong Li., Preparation and physicochemical properties of a novel hydroxyapatite/chitosan–silk Wbroin composite. *Carbohydr. Polym.*, 68, 740–745, 2007.
67. Sivakumar, M., Manjubala, I., Rao, P.K., Preparation, characterization and *in-vitro* release of gentamicin from coralline hydroxyapatite – chitosan composite microspheres. *Carbohydr. Polym.*, 49, 282–288, 2002.
68. Nunes, C.R., Simske, S.J., Sachdeva, R., Wolford, L.M., Longterm ingrowth and apposition of porous hydroxylapatite implants. *J. Biomed. Mat. Res.*, 36, 560–563, 1997.
69. Madhumathi, K., Sudheesh Kumar, P.T., Abilash, S., Sreeja, V., Tamura, H., Manzoor, K., *et al.*, Development of novel chitin/nanosilver composite scaffolds for wound dressing applications. *J. Mat. Sci.: Mat. Med.*, 21, 807–813, 2010.
70. Robinson, T., McMullan, G., Marchant, R., Nigam, P., Remediation of dyes in textile effluent: a critical review on current treatment technologies with a proposed alternative. *Biores. Technol.*, 77(3), 247–255, 2001.
71. Blackburn, R.S., Natural polysaccharides and their interactions with dye molecules: applications in effluent treatment. *Environ. Sci. Technol.*, 38(18), 4905–4909, 2004.
72. Crini, G., Recent developments in polysaccharide-based materials used as adsorbents in wastewater treatment. *Prog. Polym. Sci.*, 30(1), 38–70, 2005.
73. Wong, Y.C., Szeto, Y.S., Cheung, W., McKay, G., Adsorption of acid dyes on chitosan—equilibrium isotherm analyses. *Proc. Biochem.*, 39(6), 695–704, 2004.

74. Wong, Y.C., Szeto, Y.S., Cheung, W.H., McKay, G., Equilibrium studies for acid dye adsorption onto chitosan. *Langmuir*, 19(19), 7888–7894, 2003.
75. Cheung, W.H., Szeto, Y.S., McKay, G., Intraparticle diffusion processes during acid dye adsorption onto chitosan. *Biores. Technol.*, 98(15), 2897–2904, 2007.
76. Mahmoodi, N.M., Salehi, R., Arami, M., Bahrami, H., Dye removal from colored textile wastewater using chitosan in binary systems. *Desalination*, 267(1), 64–72, 2011.
77. Kyzas, G.Z., Kostoglou, M., Vassiliou, A.A., Lazaridis, N.K., Treatment of real effluents from dyeing reactor: Experimental and modeling approach by adsorption onto chitosan. *Chem. Eng. J.*, 168(2), 577–585, 2011.
78. Renault, F., Sancey, B., Badot, P.M., Crini, G., Chitosan for coagulation/flocculation processes–an eco-friendly approach. *Eur. Polym. J.*, 45(5), 1337–1348, 2009.
79. Verma, A.K., Dash, R.R., Bhunia, P., A review on chemical coagulation/flocculation technologies for removal of colour from textile wastewaters. *J. Environ. Manag.*, 93(1), 154–168, 2012.
80. Şenel, S., McClure, S.J., Potential applications of chitosan in veterinary medicine. *Adv. Drug Deliv. Rev.*, 56(10), 1467–1480, 2004.
81. Heinemann, C., Heinemann, S., Bernhardt, A., Worch, H., Hanke, T., Novel textile chitosan scaffolds promote spreading, proliferation, and differentiation of osteoblasts. *Biomacromolecules*, 9(10), 2913–2920, 2008.
82. Heinemann, C., Heinemann, S., Lode, A., Bernhardt, A., Worch, H., Hanke, T., *In vitro* evaluation of textile chitosan scaffolds for tissue engineering using human bone marrow stromal cells. *Biomacromolecules*, 10(5), 1305–1310, 2009.
83. Kong, M., Chen, X.G., Xing, K., Park, H.J., Antimicrobial properties of chitosan and mode of action: a state of the art review. *Int. J. Food Microbiol.*, 144(1), 51–63, 2010.
84. Petkova, P., Francesko, A., Fernandes, M.M., Mendoza, E., Perelshtein, I., Gedanken, A., Tzanov, T., Sonochemical coating of textiles with hybrid ZnO/chitosan antimicrobial nanoparticles. *ACS Appl. Mat. Interf.*, 6(2), 1164–1172, 2014.
85. Buşilă, M., Muşat, V., Textor, T., Mahltig, B., Synthesis and characterization of antimicrobial textile finishing based on Ag: ZnO nanoparticles/chitosan biocomposites. *RSC Adv.*, 5(28), 21562–21571, 2015.

Section III

Chitosan in Biomedical Applications

10

Perspectives of Chitin- and Chitosan-Based Scaffolds Dressing in Regenerative Medicine

Parappurath Narayanan Sudha[1,*], Madhavan Saranya[1], Thandapani Gomathi[1], S. Gokila[1], Soundararajan Aisverya[1], Jayachandran Venkatesan[2] and Sukumaran Anil[3]

[1]*P. G and Research Department of Chemistry, D.K.M College for Women, Vellore, India*
[2]*Marine Bioprocess Research Center and Department of Marine Bio-Convergence Science, Pukyong National University, Busan, South Korea*
[3]*Department of Preventive Dental Sciences, Prince Sattam Bin Abdulaziz University, AIKharj, Saudi Arabia*

Abstract

Over the last decade, much interest has been developed in biopolymer-based materials due to their biocompatible, biodegradable, nontoxic, and nonallergenic nature. Need for scaffold dressing which aids the healing of chronically refractory wounds. To this end, we discussed scaffold supporting biomaterial chitin and chitosan. Chitosan, a natural polysaccharide, resembles with glycosaminoglycan of extracellular matrix (ECM), which makes it a promising candidate for wound healing application apart from its excellent biocompatibility and biodegradability, nontoxicity, ease of availability, nonimmunogenic beneficial biological and antimicrobial properties, and high valuable potential for wound healing are attractive for wound care. Scaffold or extra cellular matrices, made by polymers, are devices that can be utilized in wound healing as frame work which promotes cellular proliferation and growth as well as delivery system for controlled release of antibiotics to hasten the healing process by preventing the microbial infections. To perform these functions, chitin and chitosan were chosen for scaffolding material due to its potential properties. There are many techniques available to fabricate scaffolds including freeze drying, electrospinning, and rapid prototyping. Some of these techniques having much attention due to their versatility, mechanical properties,

**Corresponding author*: drparsu8@gmail.com

Shakeel Ahmed and Saiqa Ikram (eds.) Chitosan, (255–270) © 2017 Scrivener Publishing LLC

and porosity of scaffold are discussed. The general characteristics of a scaffold and applications of chitin, chitosan as wonderful dressing material, and ECM for regenerative medicine are been explained in detail in this chapter.

Keywords: Scaffold, techniques, chitosan, fabrication, regenerative medicine

10.1 Introduction

Chitosan is a linear, semicrystalline polysaccharide composed of (1-4)-2-acetamido-2-deoxyb-D-glucan (*N*-acetyl-D-glucosamine) and (1-4)-2-amino-2-deoxyb-D-glucan (D-glucosamine) units [1]. Chitosan is not extensively present in the environment – however, it can be easily derived from the partial deacetylation of a natural polymer, the chitin. The deacetylated chitin should contain at least 60% of D-glucosamine residues [2]. The deacetylation of chitin is conducted by chemical hydrolysis under severe alkaline conditions or by enzymatic hydrolysis in the presence of particular enzymes, among of chitin deacetylase. After cellulose, chitin is the second most abundant biopolymer and is not only found in invertebrates – as crustacean shells or insect cuticles – but also in some mushrooms envelopes, green algae cell walls, and yeasts [3].

The molecular weight of chitosan typically ranges from 300 kDa to 1000 kDa, depending on the source and preparation [4]. Chitin and chitosan are biocompatible polymers but there are some evidence that chitosan is more cytocompatible *in vitro* than chitin. While the number of positive charges increases, the interaction between cells and chitosan increases as well, which tends to improve biocompatibility [5]. Chitin and chitosan with beneficial biological and antimicrobial properties and high valuable potential for wound healing are attractive for wound care. Healing restores the integrity of the injured tissue and prevents organisms from deregulation of homeostasis. The treatment of the wounds has evolved from the ancient times. Initially, application of dressing material was aimed at inhibition of bleeding, protection of the wound from environmental irritants as well as water and electrolyte disturbances. Skin plays an important role in homeostasis and the prevention of invasion by microorganisms. Skin generally needs to be covered with a dressing immediately after it was damaged. At present, there are three categories of wound dressing: biologic, synthetic, and biologic-synthetic. Allo skin and pigskin are biologic dressings commonly used clinically, but they have some disadvantages, such as limited supplies, high antigenicity, poor adhesiveness, and risk of cross contamination. Synthetic dressings have long shelf life, induce minimal inflammatory reaction and carry almost no risk of pathogen transmission. In recent years, researchers have focused on biologic-synthetic dressings [6, 7] which are bilayered and consist of high polymer and biologic materials. These three categories of wound dressing are all used frequently in the clinical setting, but none is without disadvantages. An ideal dressing should maintain a moist environment at the wound interface, allow gaseous exchange, act as a barrier to microorganisms, and remove excess exudates. It should also be nontoxic, nonallergenic, nonadherent, and easily removed without trauma, and it should be made from a readily available biomaterial that requires minimal processing, possesses antimicrobial properties, and promotes wound healing. In recent years, a large number of research groups are dedicated to producing a new, improved wound dressing by synthesizing and modifying biocompatible materials [8–10].

Materials, based on chitosan, have no allergic effect of living body and nontoxic. The presence of the protonatable amino group along d-glucosamine residues allows to mucoadhesion of chitosan due to the presence of negatively charged residues (sialic acid) in the mucin – the glycoprotein that composes the mucus [11]. Mucoadhesion related to the deacetylation degree of chitosan – if it increases, the number of positive charges increases, which leads to improved mucoadhesive properties.

The polycationic nature of chitosan also allows explaining chitosan analgesic effects. Indeed, the amino groups of the d-glucosamine residues can protonate in the presence of proton ions that are released in the inflammatory area, resulting in an analgesic effect [12]. Due to its positive charges, chitosan can also interact with the negative part of cells membrane, which can lead to reorganization and an opening of the tight junction proteins, explaining the permeation enhancing the property of this polysaccharide [13].

The biodegradation of chitosan leads to the formation of nontoxic oligosaccharides of variable length. These oligosaccharides can be incorporated in metabolic pathways or be further excreted. The degradation rate of chitosan is mainly related to its degree of deacetylation, but also to the distribution of *N*-acetyl-d-glucosamine residues and the molecular mass of chitosan [14]. Chitosan has antimicrobial properties that strongly related to DD and molecular weight [15]. There are two mechanisms that explain this effect – (1) interaction of positively charged chitosan with negatively charged groups at the surface of bacterial cells and (2) binding of chitosan with the cell DNA (still via protonated amino groups), which would lead to the inhibition of the microbial RNA synthesis [16]. The protonatable amino group can be related to hemostatic activity of chitosan. Chitin shows less effective hemostatic activity than chitosan, which tends to confirm this explanation. One of the main properties of chitin and chitosan is a stimulation of regeneration. Chitosan stimulates cell proliferation and histoarchitectural tissue organization as well as affects macrophage function that helps in faster wound healing.

As a summary, chitosan is most promising material for hemostatic dressing due to direct hemostatic properties and possible improvement of wound healing. Also, chitosan has no disadvantages in biological response. Various forms of wound dressings materials based on chitin and chitosan derivatives are commercially available. The ordered regeneration of wounded tissues requires the use of chitin and chitosan in the form of nonwovens, nanofibrils, composites, films, scaffolds, and sponges. So far a number of research works have been published on chitin and chitosan as wound dressing materials. However, only a few review articles have been reported about chitin- and chitosan-based wound dressings with limited information [17–20]. In this chapter, we reviewed a recent development and applications of wound dressing materials based on chitin, chitosan, and their derivatives.

10.2 Scaffold Characteristics

The scaffolds to be used for bone tissue regeneration satisfy various macro and microstructural properties which cannot be neglected if the proper growth of the tissue is desired [21]. The following properties are important for scaffolds in bone tissue engineering.

10.2.1 Biodegradability

The material for building framework of tissues must be biodegradable in nature. Degradation rate must be such that it is in accord with the generation of the tissues. The best suited material will be the one which provides all the desired above mentioned properties and gets completely degraded at the time of complete proliferation of cells. The biodegradation rate of the system should be adequate in order to allow, for the time necessary to the cells to produce the new tissue, the mechanical stability of the vascular graft, and its complete degradation after the new vessel is formed. To develop a biomaterial with these properties, electrospinning seems to be one of the most useful technique due to the possibility to obtain nanofibrillar scaffolds starting from polymeric solution. Biodegradable polymers, on the other hand, encompass polymers (both natural and synthetic) that degrade due to microbial action. There is a lack of consensus among researchers on a universal definition of polymer biodegradation and the nature of end products that would render a polymer biodegradable, apart from a lack of correlation between timescales adopted by various researchers [22]. Biodegradability is often an essential factor since scaffolds should preferably be absorbed by the surrounding tissues without the necessity of a surgical removal [23]. Biodegradable polymers have been attractive candidates for scaffolding materials because they degrade as the new tissues are formed, eventually leaving nothing foreign to the body [24].

10.2.2 Biocompatibility

The very first criterion of any scaffold for tissue engineering is that it must be biocompatible; cells must adhere, function normally, and migrate onto the surface and eventually through the scaffold and begin to proliferate before laying down new matrix. After implantation, the scaffold or tissue engineered construct must elicit a negligible immune reaction in order to prevent it causing such a severe inflammatory response that it might reduce healing or cause rejection by the body.

10.2.3 Manufacturing Technology

The development of scalable manufacturing processes to good manufacturing practice (GMP) standard is critically important in ensuring successful translation of tissue engineering strategies to the clinic [25]. In order for a particular scaffold or tissue engineered construct to become clinically and commercially viable, it should be cost effective and possible to scale-up from making one at a time in a research laboratory to small batch production [26].

10.2.4 Mechanical Properties

Mechanical integrity of the scaffold design has to the sufficient to resist handling during implantation and *in vivo* loading. An ideal scaffold would be biomechanically similar to the type of bone being replaced in order to function quickly as a synthetic bone replacement. The primary bone tissue has relative high compressive strength that supports the body weight. So the scaffold must provide mechanical support during the reconstruction process. The development of bone *in vivo* is accompanied by regular stresses. Thus, it is mandatory for the scaffolds to bear these mechanical stresses and ensures the mechanical integrity and protection to the developing tissues. The property of the scaffold should be such that with the increase in mechanical stresses the strength of the growing cells increases [27, 28].

10.2.5 Scaffold Properties

Producing scaffolds with adequate mechanical properties is one of the great challenges in attempting to engineer bone or cartilage. For these tissues, the implanted scaffold must have sufficient mechanical integrity to function from the time of implantation to the completion of the remodeling process [29]. Biodegradability is often an essential factor since scaffolds should preferably be absorbed by the surrounding tissues without the necessity of a surgical removal [30]. Mimic the native ECM, an endogenous substance that surrounds cells, binds them into tissues, and provides signals that aid cellular development and morphogenesis. A high porosity and an adequate pore size are necessary to facilitate cell seeding and diffusion throughout the whole structure of both cells and nutrients. An optimum pore size is in the range between 100 μm and 500 μm [31]. The effects of pore size on tissue regeneration have been emphasized by experiments demonstrating optimum pore size of 5 μm for neovascularization [32]. The surface properties can be selectively modified to enhance the performance of the biomaterials. For instance, by altering the surface functionality using thin film deposition, the optimal surface, chemical, and physical properties can be attained [33, 34]. Surface properties include both chemical and topographical characteristics, which can control and affect cellular adhesion and proliferation. Should provide a 3D template for the cells to attach and to guide their growth. Should have a porous architecture with a high surface area for the maximum loading of cells, cell surface interaction, tissue in growth, and transportation of nutrients and oxygen. Should be degradable under physiological conditions and the degradation rate should match the rate of tissue regeneration to sustain tissue functionality. Finally, the production process of the scaffold with all the above unique characteristics must be accomplished in a reproducible, economical, and up-scalable manner [35]. Since each of the different cell types favors different conditions for optimal growth and differentiation, tissue engineering using multiple cell types must take these factors into account.

10.3 Fabrication Techniques

Many techniques are used in fabricating scaffold. In the body, cells and tissues are organized into three-dimensional architecture. To engineer these functional tissues and organs, scaffolds have to be fabricated by the different methodology to facilitate the cell distribution and guide their growth into three-dimensional space. The main techniques for scaffolds fabrication are mentioned here: solvent casting [36], gas foaming method [37–39], particulate-leaching techniques [40], phase separation [41], electrospinning [42–44], fiber mesh [45–48], fiber bonding [49], Self-assembly [50], rapid prototyping [51–55], melt moulding [56], membrane lamination [57], and freeze drying [58, 59].

10.4 Applications of Chitin and Chitosan as Regenerative Medicine

Chitosan is a benefit to wound healing because it stimulates hemostasis and accelerates tissue regeneration. For a material to be used for biomedical research, a natural product is preferred because these materials are more biocompatible than synthetic materials. Chitosan is metabolized by certain human enzymes, such as lysozyme, thus, it is biodegradable. Chitosan is an attractive material for a tissue engineering scaffold because it has structural similarities to glycosaminoglycans and is hydrophilic. Chitosan has the potential to be used as artificial kidney membrane, hypocholesterolemic agents, drug delivery systems, absorbable sutures and supports for immobilized enzymes. Chitosan has some advantages due to its nontoxicity and biodegradability without damaging the environment. It is a biocompatible material that breaks down slowly to harmless products that are absorbed completely in the body [60].

Recently, β-chitin/nanosilver composite scaffolds were prepared for wound tissue engineering applications using β-chitin hydrogel with silver nanoparticles [61]. The antibacterial, blood clotting, swelling, cell attachment, and cytotoxicity studies of the prepared composite scaffolds were evaluated. The prepared β-chitin/nanosilver composite scaffolds were bactericidal against *Escherichia coli* and *Staphylococcus aureus* and showed good blood clotting ability as well. Cell attachment studies using Vero (epithelial cells) showed that the cells were well attached on the scaffolds. Similarly, α-chitin/nanosilver composite scaffolds showing similar results were developed for wound tissue engineering applications using α-chitin hydrogel with silver nanoparticles [62]. These results suggested that chitin/nanosilver composite scaffolds could be potential candidates for wound tissue engineering.

Cell studies conducted on chitin/PGA and chitin/SF [63] fibrous scaffolds proved that a matrix consisting of 25% PGA or SF and 75% chitin had the best results. The chitin/PGA fibrous scaffold had a bovine serum albumin coating and was considered as a good candidate for use as a tissue engineering scaffold. The chitin/SF fibrous scaffold had the highest spreading of normal human epidermal fibroblasts (NHEF) and normal human epidermal keratinocytes (NHEK). Therefore, these scaffolds were recommended for wound tissue engineering applications.

Oliveira *et al.*, (2006) created a bilayer scaffold using the freeze-dry method and pouring a 3% chitosan solution onto a sintered HA scaffold. The scaffolds demonstrated a high connectivity, adequate water uptake and porosity, good mechanical properties, and cellular adhesion [64].

Chitosan derivative sheets and pastes were evaluated *in vitro* for possible utilization in wound dressing applications [65]. In this study, the cytotoxicity of oligo chitosan, *N,O*-carboxymethyl-chitosan (*N,O*-CMC), and *N*-carboxymethyl-chitosan (*N*-CMC) derivatives in sheet-like and paste forms was evaluated using primary normal human dermal fibroblast cultures and hypertrophic scars; a fibrotic conditions representing a model of altered wound healing with over production of extracellular matrix and fibroblast hyperproliferative activity. Cytotoxicities of these chitosan derivatives were assessed using MTT assay. The results indicated that both chitosan derivative sheets and pastes have appropriate cytocompatibility and appear promising as safe biomaterials with potential wound healing applications. *N,O*-CMC sheet exhibited highest cytocompatibility property and may be regulated by matrix metalloproteinase-13 (MMP-13) in controlling the cell growth and its expression level [66].

Biocompatible chitosan/polyethylene glycol diacrylate (PEGDA) blend films were successfully prepared by Michael addition reaction with different weight ratios as wound dressing materials. The mechanical properties and the swelling property of chitosan were found to be enhanced after the chemical modification. Indirect cytotoxicity assessment of films with mouse fibroblasts (L929) indicated that the material showed no cytotoxicity toward the growth of L929 cell and had good *in vitro* biocompatibility. SEM observation indicated that the microporous surface structure of the chitosan/PEGDA films was good to grow, proliferate, and differentiate of L929 cell. These chitosan/PEGDA films have the potential to be used as wound dressing material [67].

Denkbaş *et al.*, [67] prepared and characterized chitosan sponges encapsulated with a model antibiotic drug, norfloxacin, as wound dressing material. The cross-linked chitosan sponges were prepared by a solvent evaporation technique. The results indicated that the chitosan sponges were in the fibrillar structure. In this study, the swelling behavior, norfloxacin loading, *in vitro* release characteristics, and antibacterial activity were determined. The effects of cross-linker concentration, norfloxacin/chitosan ratio, chitosan molecular weight, and base concentration were also investigated. The most effective parameter was found to be the degree of neutralization. It was also observed that the equilibrium-swelling ratio decreased with increasing cross-linking density. The norfloxacin release was found to be swelling controlled initially and diffusion controlled at the extended release periods. It was found that the antibacterial activity was directly proportional to the release rate. These chitosan-based sponges could be used for wound dressing applications [68].

Chitosan and γ-polyglutamic acid (PGA) polyelectrolyte complexes increase the degree of the complex formation lowered the water uptake, reduced the pore size of the porous structure, decreased the *in vitro* degradation, but increased the compressive modulus of the chitosan/PGA PECs. Furthermore, the chitosan/PGA PECs provided adequate moisture, and thus, reduced the risk of dehydration in the presence of PGA. The wounds treated with the chitosan/PGA PECs healed significantly faster [69].

Dowling *et al.*, have reported that the amphiphilic biopolymer, hm-chitosan, can act as an effective hemostatic agent. It has the ability to transform whole liquid blood into a gel, and it quickly stops bleeding from both minor and severe injuries in small and large animals. The gelling mechanism is predicated on the self-assembly of hm-chitosan chains in such a way that their hydrophobes anchorin blood cell membranes and thereby bridge the cells into a three-dimensional network. The cells are thus active components (nodes or junction points) in the network rather than being physically trapped in a polymer mesh. The gelling ability of hm-chitosan is similar to that of fibrin-based sealants, but at a much lower cost and wider availability. An additional important aspect of hm-chitosan as a hemostatic agent is that its gelling effect on blood can be reversed by addition of the supramolecule α-CD. The reversal occurs because α-CD molecules sequester the hydrophobes and prevent them from attaching to blood cells. Our studies show that hm-chitosan could be an effective and safe hemostatic agent for treatment of both external and internal bleeding injuries [70].

The PVA/bentonite, PVA/Ag nanoparticles, PVA/clove extract, and PVA/cellulose nanocomposite hydrogel were reported by Gonzalez *et al.*, PVA/clove hydrogels did not exhibit a homogenous aspect, while PVA/bentonite and PVA/Ag nanocomposite hydrogels showed significant antimicrobial activity against growth of *E. coli*, good water vapor transmission rate, and adequate water absorbing capacity due to addition of clay and Ag nanoparticles as filler [71]. Heterogeneous chemical modification of the chitosan film can tune the surface properties of the film; for instance, when a stearoyl group was attached to the chitosan films, they became more hydrophobic and promoted proteins adsorption. When chitosan films were reacted with succinic anhydride or phthalic anhydride, it resulted in more hydrophilic films which promoted lysozyme adsorption [72]. Wang *et al.*, evaluated the wound healing ability of PEC membrane of CS and alginate suspensions developed through solvent casting method. The prepared membrane was nontoxic toward mouse and human fibroblast cells. *In vivo* evaluation in incisional wounds in rat proved the faster healing potential of this composite membrane [73].

The poly(vinyl alcohol)/chitosan/montmorillonite nanocomposite hydrogels were prepared by freezing–thawing method, as a biocompatible wound dressing. Improved mechanical properties of this system along with the other characteristics, such as biocompatibility, antibacterial activity, and good swelling behavior, made it a desirable candidate for wound dressing applications [67]. The sponge-like dressings based on chitosan glutamate (high molecular weight) and sericin were developed for the treatment of chronic skin ulcers. The amount of sericin in the optimized dressing is suitable to exert a protective effect on human fibroblasts against oxidative damage. Moreover, the optimized dressing is able to improve fibroblast proliferation, that is, to promote wound healing [74].

The composite sponge of curcumin/chitosan/gelatin was prepared at the various ratios of chitosan and gelatin showing improved water uptake ability, antibacterial activity, and wound closure. The higher content of gelatin in composite sponge exhibited a faster release behavior up to 240 min. These composite sponges were also found to enhance the formation of collagen and wound closure *in vivo* and therefore improved the wound healing activity [75].

Abdel-Mohsen *et al.*, have worked on novel wound dressing material fabricated from chitin/chitosan-glucan complex (ChCSGC) was used for the first time as wound dressing material. *In vivo* test with rats showed excellent wound healing ability and promoted accelerated wound closure of the rat skin. We believe that the new wound dressing sheets can be used for different biomedical applications, especially as new wound dressing/healing materials [76].

10.5 Conclusion

Chitosan is a quite unique biosourced polymer characterized by primary amines along the backbone. Such structure imparts to this polysaccharide not only highly valuable physicochemical properties but also particular interactions with proteins, cells, and living organisms. In addition, this polycation (upon protonation in acidic conditions) shows improved solubility in aqueous media which allows the processing of this material in very mild conditions in a wide variety of shapes. In this review, the biomedical applications of chitosan-based materials especially wound dressings have been covered. Chitosan has been shown to be a potential biomaterial to be used in wound healing due to its antimicrobial and anti-inflammatory nature. Effective dressings should have characteristics optimized for particular type of wound at a reasonable low cost and with minimum inconvenience to patients. Many results have been reported, but complete and standard characterization of these materials is needed. Therefore, there is need to manipulate the physical properties of identified systems to achieve such objectives. Hence, this polymer is very attractive for biomedical uses, notably in tissue engineering. The wide panel of chitin, chitosan properties, and processed materials gives to this biosourced polymer a quite promising future as biomaterial as demonstrated by emerging products on the market notably in the Regenerative Medicine field.

References

1. Rinaudo, M., Chitin and chitosan: properties and applications. *Prog. Polym. Sci.*, 31, 603, 2006.
2. Madihally, S., Matthew, H., Porous chitosan scaffolds for tissue engineering. *Biomaterials*, 20, 1133, 1999.
3. Aranaz, I., Harris, R., Heras, A., 2010. Chitosan amphiphilic derivatives. Chemistry and applications. *Curr. Org. Chem.*, 14, 308–330.
4. Liu, X., Ma, L., Mao, Z., *et al.*, Chitosan-based biomaterials for tissue repair and regeneration. Chitosan for biomaterials II. In: *Advances in polymer science*, R. Jayakumar, M. Prabaharan, R.A.A. Muzzarelli, Eds., pp. 81–127. Heidelberg: Springer Berlin, 2011.
5. Chatelet, C., Influence of the degree of acetylation on some biological properties of chitosan films. *Biomaterials*, 22, 261, 2001.
6. Bruin, P., Jonkman, M., Meijer, H., Pennings, A., A new porous polyetherurethane wound covering. *J. Biomed. Mater. Res.*, 24, 217, 1990.
7. Suzuki, S., Matsuda, K., Isshiki, N., Tamada, Y., Ikada, Y., Experimental study of a newly developed bilayer artificial skin. *Biomaterials*, 11, 356, 1990.
8. Shibata, H., Shioya, N., Kuroyanagi, Y., Development of new wound dressing composed of spongy collagen sheet containing dibutyryl cyclic AMP. *J. Biomater. Sci. Polym. Ed.*, 8, 601, 1997.
9. Draye, J., Delaey, B., Van de Voorde, A., Van Den Bulcke, A., Bogdanov, B., Schacht, E., *In vitro* release characteristics of bioactive molecules from dextran dialdehyde cross-linked gelatin hydrogel films. *Biomaterials*, 19, 99, 1998.
10. Ulubayram, K., Nur Cakar, A., Korkusuz, P., Ertan, C., Hasirci, N., EGF containing gelatin-based wound dressings. *Biomaterials*, 22, 1345, 2001.
11. He, P., Davis, S., Illum, L., *In vitro* evaluation of the mucoadhesive properties of chitosan microspheres. *Int. J. Pharm.*, 166, 75, 1998.

12. Okamoto, Y., Kawakami, K., Miyatake, K., Morimoto, M., Shigemasa, Y., Minami, S., Analgesic effects of chitin and chitosan. *Carbohydr. Polym.*, 49, 249, 2002.
13. Smith, J., Wood, E., Dornish, M., Effect of chitosan on epithelial cell tight junctions. *Pharm. Res.*, 21, 43, 2004.
14. Aiba, S., Studies on chitosan: 4. Lysozymic hydrolysis of partially N-acetylated chitosans. *Int. J. Biol. Macromol.*, 14, 225, 1992.
15. Chung, Y., Chen, C., 2008. Antibacterial characteristics and activity of acid-soluble chitosan. *Bioresour. Technol.*, 99, 2806, 2008.
16. Croisier, F., Jérôme, C., Chitosan-based biomaterials for tissue engineering. *Eur. Polym. J.*, 49, 780, 2013.
17. Ravi Kumar, M.N.V., A review of chitin and chitosan applications. *Reac. Func. Polym.*, 46, 1, 2000.
18. Kim, I., Seo, S., Moon, H., Yoo, M., Park, I., Kim, B., Cho, C., Chitosan and its derivatives for tissue engineering applications. *Biotech. Adv.*, 26, 1, 2008.
19. Muzzarelli, R., Chitins and chitosans for the repair of wounded skin, nerve, cartilage and bone. *Carbohydr. Polym.*, 76, 167, 2009.
20. Tamura, H., Furuike, T., Nair, S., Jayakumar, R., Biomedical applications of chitin hydrogel membranes and scaffolds. *Carbohydr. Polym.*, 84, 820, 2011.
21. Leong, K., Cheah, C., Chua, C., Solid freeform fabrication of three-dimensional scaffolds for engineering replacement tissues and organs. *Biomaterials,* 24, 2363, 2003.
22. Gautam, R., Bassi, A., Yanful, E., A review of biodegradation of synthetic plastic and foams. *Appl. Biochem. Biotech.*, 141, 85, 2007.
23. Gu, H., Yue, Z., Leong, W., Nugraha, B., Tan, L., Control of *in vitro* neural differentiation of mesenchymal stem cells in 3D macroporous, cellulosic hydrogels. *Regen. Med.*, 5, 245, 2010.
24. Woo, K., Chen, V., Jung, H., Kim, T., Shin, H., Baek, J., Ryoo, H., Ma, P., Comparative evaluation of nanofibrous scaffolding for bone regeneration in critical-size calvarial defects. *Tiss. Eng. Part A.*, 15, 2155, 2009.
25. Hollister, S., Scaffold engineering: a bridge to where? *Biofabrication,* 1, 012001, 2009.
26. Partap, S., *et al.*, Scaffolds & surfaces. In: *Basic engineering for medics and biologists, an ESEM primer on engineering for medicine*, T.C. Lee, P. Niederer, Eds., pp. 152–187. IOS Press, Nieuwe Hemweg, 2010.
27. Agrawal, C., Ray, R., Biodegradable polymeric scaffolds for musculoskeletal tissue engineering. *J. Biomed. Mater. Res.,* 55, 141, 2001.
28. Yang, S., Leong, K., Du, Z., Chua, C., The Design of Scaffolds for Use in Tissue Engineering. Part I. Traditional Factors. *Tiss. Eng.*, 7, 679, 2001.
29. Hutmacher, D.W., Scaffolds in tissue engineering bone and cartilage. *Biomaterials,* 21, 2529, 2000.
30. Hetal, P., Minal, B., Ganga S., Biodegradable polymer scaffold for tissue engineering. *Trends Biomater. Artif. Organs*, 25, 20, 2011.
31. Ikada, Y., Challenges in tissue engineering. *J. R. Soc. Interface.*, 3, 589, 2006.

32. Brauker, J.H., Carr-Brendel, V.E., Martinson, L.A., Crudele, J., Johnston, W.D., Johnson, R.C., Neovascularization of synthetic membranes directed by membrane micro architecture. *J. Biomed. Mater. Res.*, 29, 1517, 1995.
33. Sodhi, R.N.S., Application of surface analytical and modification techniques to biomaterial research. *J. Electron Spectrosc.*, 81, 269, 1996.
34. Brewis, D.M., Briggs, D., Adhesion to polyethylene and polypropylene. *Polymer.*, 22, 7, 1981.
35. Murugan, R., Ramakrishna, S., Design strategies of tissue engineering scaffolds with controlled fiber orientation. *Tiss. Eng.*, 13, 1845, 2007.
36. Mikos, A.G., Lu, L., Temenoff, J. S., *et al.*, Synthetic bioresorbable polymer scaffolds. In: *An introduction to material in medicine,* B.D. Ratner, A.S. Hoffman, F.J. Schoen, J.E. Lemons, Eds., p. 743, Elsevier Academic Press, US, 2004.
37. Naessens, M., Cerdobbel, A., Soetaert, W., Vandamme, E., Leuconostoc dextransucrase and dextran: production, properties and applications. *J. Chem. Technol. Biotechnol.,* 80, 845, 2005.
38. Cascone, M., Barbani, N., Giusti, P., Cristallini, C., Ciardelli, G., Lazzeri, L., Bioartificial polymeric materials based on polysaccharides. *J. Biomater. Sci. Polym. Ed.,* 12, 267, 2001.
39. Mooney, D., Baldwin, D., Suh, N., Vacanti, J., Langer, R., Novel approach to fabricate porous sponges of poly(d,l-lactic-co-glycolic acid) without the use of organic solvents. *Biomaterials,* 17, 1417, 1996.
40. Plikk, P., Målberg, S., Albertsson, A., Design of resorbable porous tubular copolyester scaffolds for use in nerve regeneration. *Biomacromolecules,* 10, 125, 2009.
41. Smith, L.A., Beck, J.A., *et al.*, Nano fibrous scaffolds and their biological effects. In: *Tissue Cell and Organ Engineering,* C. Kumar, Ed., p. 195, *Wiley-VCH,* 2006.
42. Deitzel, J., Kosik, W., McKnight, S., Beck Tan, N., Desimone, J., Electrospinning of polymer nanofibers with specific surface chemistry. *Polymer.*, 43, 1025, 2002.
43. Sill, T., von Recum, H., Electrospinning: applications in drug delivery and tissue engineering. *Biomaterials,* 29, 1989, 2008.
44. Lee, K., Kim, H., Khil, M., Ra, Y., Lee, D., Characterization of nano-structured poly(ε-caprolactone) nonwoven mats via electrospinning. *Polymer.*, 44, 1287, 2003.
45. Gassner, F., Owen, A.J, Some properties of poly(3-hydroxybutyrate)-poly(3-hydroxyvalerate)blends. *Polym. Int.,* 39, 215, 1996.
46. Sun, J.Y, Wu, J., Li, H., Chang, J., Macroporous poly(3-hydroxybutyrate-co-3-hydroxyvalerate) matrices for cartilage tissue engineering. *Eur. Polym. J.*, 41, 2443, 2005.
47. Gao, J., Crapo, P.M., Wang, Y.D., Macroporous elastomeric scaffolds with extensive micropores for soft tissue engineering. *Tiss. Eng.*, 12, 917, 2006.
48. Gomes, M., Reis, R., Mikos, A., Bone tissue engineering constructs based on starch scaffolds and bone marrow cells cultured in a flow perfusion bioreactor. *Mater. Sci. Forum.,* 514–516, 980, 2006.

49. Smith, L.A., Liu, X., Ma, P.X, Tissue engineering with nano-fibrous scaffolds. *Soft Matter,* 4, 2144, 2008.
50. Yang, S., Leong, K., Du, Z., Chua, C., The design of scaffolds for use in tissue engineering. Part II. Rapid prototyping techniques. *Tiss. Eng.,* 8, 1, 2002.
51. Baker, C.T., Bocharov, G., Paul, C., Rihan, F., Modelling and analysis of time-lags in some basic patterns of cell proliferation. *J. Math. Biol.,* 37, 341, 1998.
52. Landers, R., Pfister, A., Hübner, U.R., Fabrication of soft tissue engineering scaffolds by means of rapid prototyping techniques. *J. Mater. Sci.,* 37, 3107, 2002.
53. Naing, M., Chua, C., Leong, K., Wang, Y., Fabrication of customised scaffolds using computer-aided design and rapid prototyping techniques. *Rapid Prototyping J.,* 11, 249, 2005.
54. Yoshimoto, H., Shin, Y.M., Terai, H., Vacanti, J.P., A biodegradable nanofiber scaffold by electrospinning and its potential for bone tissue engineering. *Biomaterials,* 24, 2077, 2003.
55. Thomson, R.C., Yaszemski, M.J, Powers, J.M., Mikos, A., Fabrication of biodegradable polymer scaffolds to engineer trabecular bone. *J. Biomater. Sci-Polym. Ed.,* 7, 23, 1996.
56. Maquet, V., Jerome, R., Design of macroporous biodegradable polymer scaffolds for cell transplantation. *Mater. Sci. Forum.,* 250, 15, 1997.
57. Cordts, H.P., Karloske, J.E., Aqueous liquid filled polyurethane gels and method of making the same. Google Patents, 1985.
58. Schoof, H., Apel, J., Heschel, I., Rau, G., Control of pore structure and size in freeze-dried collagen sponges. *J. Biomed. Mater. Res.,* 58, 352, 2001.
59. Ninan, N., Muthiah, M., Park, I., Wong, T., Thomas, S., Grohens, Y., Natural polymer/inorganic material based hybrid scaffolds for skin wound healing. *Polym. Rev.,* 55, 453, 2015.
60. Sudheesh Kumar, P., Abhilash, S., Manzoor, K., Nair, S., Tamura, H., Jayakumar, R., Preparation and characterization of novel β-chitin/nanosilver composite scaffolds for wound dressing applications. *Carbohydr. Polym.,* 80, 761, 2010.
61. Madhumathi, K., Sudheesh Kumar, P., Abhilash, S., Sreeja, V., Tamura, H., Manzoor, K., Nair, S., Jayakumar, R., Development of novel chitin/nanosilver composite scaffolds for wound dressing applications. *J. Mat. Sci. Mater. Med.,* 21, 807, 2009.
62. Park, K.E., Kang, H.K., Lee, S.J., Min, B.M., Park, W.H., Biomimetic nanofibrous acaffolds: preparation and characterization of PGA/Chitin blend nanofibers. *Biomacromolecules,* 7, 635, 2006.
63. Park, K.E., Jung, S.Y., Lee, S.J., Min, B.M., Park, W.H., Biomimetic nanofibrous scaffolds: Preparation and characterization of chitin/silk fibroin blend nanofibers. *Int. J. Biol. Macromol.,* 38, 165, 2006.
64. Oliveira, J., Rodrigues, M., Silva, S., Malafaya, P., Gomes, M., Viegas, C., Dias, I., Azevedo, J., Mano, J., Reis, R., Novel hydroxyapatite/chitosan bilayered scaffold for osteochondral tissue-engineering applications: scaffold

design and its performance when seeded with goat bone marrow stromal cells. *Biomaterials*, 27, 6123, 2006.

65. Abdull Rasad, M., Halim, A., Hashim, K., Rashid, A., Yusof, N., Shamsuddin, S., *In vitro* evaluation of novel chitosan derivatives sheet and paste cytocompatibility on human dermal fibroblasts. *Carbohydr. Polym.*, 79, 1094, 2010.
66. Zhang, X., Yang, D., Nie, J., Chitosan/polyethylene glycol diacrylate films as potential wound dressing material. *Int. J. Biol. Macromol.*, 43, 456, 2008.
67. Denkbaş, E., Öztürk, E., Özdemir, N., Keçeci, K., Agalar, C., 2004. Norfloxacin-loaded chitosan sponges as wound dressing material. *J. Biomat. Appl.*, 18, 291, 2004.
68. Tsao, C., Chang, C.H., Lin, Y.Y., Wu, M.F., Wang, J.W., Young, T.H., Han, J., Hsieh, K.H., Evaluation of chitosan/γ-poly(glutamic acid) polyelectrolyte complex for wound dressing materials. *Carbohydr. Polym.*, 84, 812, 2011.
69. Dowling, M., Kumar, R., Keibler, M., Hess, J., Bochicchio, G., Raghavan, S., A self-assembling hydrophobically modified chitosan capable of reversible hemostatic action. *Biomaterials,* 32, 3351, 2011.
70. Gonzalez, J.S., Maiolo, A.S., Ponce, A.G., Alvarez, V.A., Composites based on poly(vinyl alcohol) hydrogels for wound dressing. XVIII The Argentine Congress of Bioengineering and Clinical Engineering Conference VII, SABI, pp. 1–4. 2011.
71. Tangpasuthadol, V., Pongchaisirikul, N., Hoven, V.P., Surface modification of chitosan films. Effects of hydrophobicity on protein adsorption. *Carbohydr. Res.,* 338, 937, 2003.
72. Wang, L.S., Khor, E., Wee, A., Lim, L.Y., Chitosan-alginate PEC membrane as a wound dressing: assessment of incisional wound healing. *J. Biomed. Mater. Res.*, 63, 610, 2002.
73. Noori, S., Kokabi, M., Hassan, Z.M., Nanoclay enhanced the mechanical properties of poly(vinyl alcohol)/chitosan/montmorillonite nanocomposite hydrogel as wound dressing. *Proc. Mater. Sci.*, 11, 152, 2015.
74. Mori, M., Rossi, S., Ferrari, F., Bonferoni, M., Sandri, G., Chlapanidas, T., Torre, M., Caramella, C., 2016. Sponge-like dressings based on the association of chitosan and sericin for the treatment of chronic skin ulcers. I. Design of experiments–assisted development. *J. Pharmacol. Sci.,* 105, 1180–1187.
75. Van, C.N., Van, B.N., Ming-Fa, H., Curcumin-loaded chitosan/gelatin composite sponge for wound healing application. *Int. J. Polym. Sci.,* 106570, 2013.
76. Abdel-Mohsen, A., Jancar, J., Massoud, D., Fohlerova, Z., Elhadidy, H., Spotz, Z., Hebeish, A., Novel chitin/chitosan-glucan wound dressing: Isolation, characterization, antibacterial activity and wound healing properties. *Int. J. Pharm.,* 510, 86, 2016.

11

Chitin – and Chitosan-Based Scaffolds

Upendra Bulbake, Sindhu Doppalapudi and Wahid Khan*

Department of Pharmaceutics, National Institute of Pharmaceutical Education and Research (NIPER), Hyderabad, India

Abstract

Tissue and organ failure induced as a result of injury or other damage is one of the serious problems in medicine. Tissue engineering field aims to regenerate damaged tissues by combining cells from the body with highly porous scaffold biomaterials. These biomaterials act as templates for tissue regeneration to guide the growth of new tissue. Scaffolds consisting of natural biopolymers have been developed and attained more popularity recently. For this, the contribution of chitin, chitosan, and their composites as a scaffold material is quite significant owing to their biocompatibility, biodegradability, antibacterial, and stimulation of healing properties which are essential for tissue regeneration. Furthermore, these biopolymers can be easily processed into scaffold forms for a variety of biomedical applications and also their composites have been explored widely as tissue engineering scaffolds. This chapter highlights different methods for the fabrication of chitin/chitosan scaffolds, essential properties of scaffolds required for tissue regeneration, and applications of chitin, chitosan, and their composites as scaffolds for regeneration of various tissues/organs. This chapter also discusses the ability of these biopolymer-based scaffolds for the delivery of drug and growth factors and will be helpful for the researchers working in the field of tissue engineering and regenerative medicine.

***Keywords*:** Chitin, chitosan, scaffold, fabrication, tissue engineering, drug delivery, growth factor delivery

**Corresponding author*: wahid@niperhyd.ac.in

Shakeel Ahmed and Saiqa Ikram (eds.) Chitosan, (271–310) © 2017 Scrivener Publishing LLC

11.1 Introduction

Tissue engineering is the application of principles of engineering and life sciences to the development of biological substitutes to restore, maintain, or improve function of tissue or whole organ. Development of biomedical technologies in the field of tissue engineering is accelerating rapidly from utilizing synthetic implants and tissue grafts to degradable porous material [1]. It has emerged as a new field involving the combination of cells, scaffold, and bioactive agents to fabricate functional new tissue to replace damaged tissue [2, 3]. This combination is often referred to as a tissue engineering triad (Figure 11.1).

Scaffold is a temporary material, which is a well designed three-dimensional tool to guide tissue regeneration *in vitro* and *in vivo*. These scaffolds are mainly used to support implanted material and integrated with biological cells or molecules to regenerate and maintain tissues [4]. It is important to select a proper scaffold to enable the cells to behave in the desired manner for the regeneration of tissue possessing desired

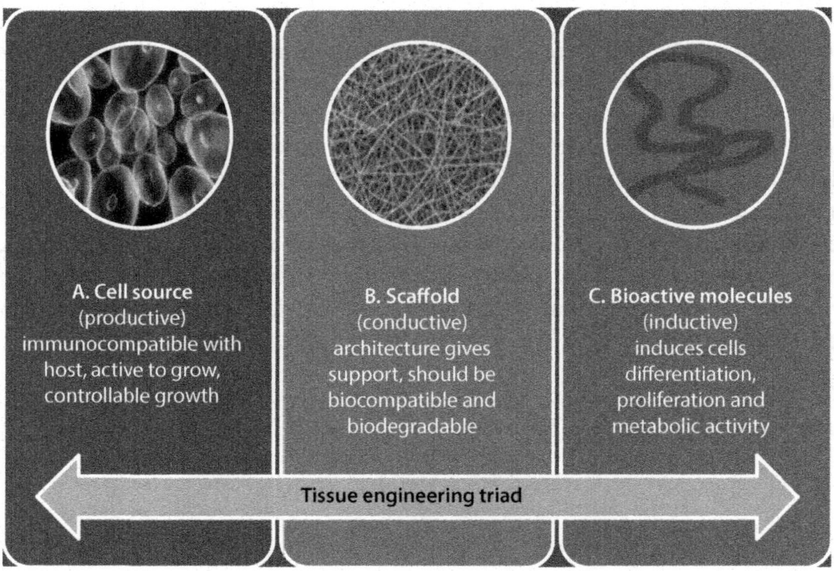

Figure 11.1 Elements of tissue engineering triad for new tissue formation by allowing cells to migrate, adhere, and produce tissue. Source of the cell act as a productive template and is able to produce large amount of cells through cell growth (a), scaffold architecture and material provide conductive template which gives proper support and strength for the growth of new tissue (b) and bioactive molecules induce proper signals for the system to mentor cells to differentiate, proliferate and perform metabolic activities (c).

shape and size. Ideal scaffold should possess following characteristics/properties:

- Possess pores of appropriate size to favor tissue integration, cell infiltration and vascularization [5, 6]
- Biocompatible to avoid unwanted host tissue responses to the scaffold [7]
- Made from material with controlled biodegradability or bioresorbability to aid the formation of new tissue [8, 9]
- Appropriate surface chemistry to favor cellular attachment, migration, differentiation, and proliferation [10–12]
- Mimic the structure and biological function of the extra cellular matrix (ECM) [13, 14]
- Adequate mechanical properties to match the intended site of implantation, handling and maintenance of the structure and function immediately after implantation [15]
- Ability to provide the bimolecular signals to the cells for the formation of ECM by promoting cellular functions [15]
- Nonantigenic, inexpensive and it should not induce any adverse response
- Easy to fabricate and easily applicable to the damaged site

Chitin (CT) and chitosan (CS) polymers are natural aminopolysaccharides and second most abundant materials available on earth. Chitin is primarily extracted from crustaceans, principally crabs and shrimps. It is a copolymer of *N*-acetyl-glucosamine and *N*-glucosamine units distributed throughout the biopolymer chain depending on the processing method used to derive the biopolymer. When the number of *N*-acetyl-glucosamine units is higher than 50%, the biopolymer is termed as chitin. Chitosan is formed through the *N*-deacetylation of CT. Chitosan is a biopolymer consisting primarily of repeating units of β-(1–4)-2-amino-2-deoxy-D-glucose.

Chitin and its derivative CS-based functional biomaterials have been given considerable attention toward the development of scaffolds, wound dressings, and novel drug delivery systems [16]. Chitin and CS polymers are biodegradable, biocompatible, nontoxic, with unique structures, multidimensional properties, highly specific functions and possess wide ranging applications in biomedical areas [17, 18]. Their biomedical applications include drug delivery, wound healing dressings, surgical sutures, artificial skin, dialysis membranes, contact lenses or eye bandages, orthopedics, haemostatic agents, and dentistry [16]. Chitin and CS can be easily fabricated into different scaffolds like hydrogels, nanofibers, nanoparticles, dressings, and also for biomedical applications such as wound healing, tissue engineering, and gene delivery.

Chitin and CS have been shown to possess mucoadhesive properties owing to the molecular attractive forces formed by electrostatic interaction between positively charged CT or CS and negatively charged mucosal surfaces. These biopolymers have free amino groups. Due to the easy availability of free amino groups in CT and CS, they carry a positive charge and thus, in turn, react with negatively charged surfaces/polymers. This cationic nature of CT and CS is primarily responsible for electrostatic interactions with negatively charged molecules like anionic glycosaminoglycans (GAGs), proteoglycans, and others for increasing the contact with mucosa. Because of this property, scaffolds incorporating a CS–GAG complex may retain and concentrate growth factors secreted by colonizing cells as large number of cytokines/growth factors are linked to GAGs (mostly with heparin and heparan sulphate) [19]. Furthermore, the presence of the *N*-acetylglucosamine moiety on CS also suggests related bioactivities. Both CT and CS are chemo attractants for neutrophils both *in vitro* and *in vivo*. As compared to CS, CT applications are limited because of its poor solubility and inert nature. On the other hand, CS can be produced in various forms and it is highly reactive. Chitosan is soluble in acidic pH and insoluble in neutral or basic pH conditions. This may be due to protonation of free amino groups in dilute acids [19].

11.2 Scaffold Components

The most important features of scaffold are its components, as they help in regeneration of new tissue in a proper way. The components of scaffold generally include (Figure 11.2):

a. **Polymers** – It provides support to the tissue for regeneration. Biodegradable polymers are particularly likely to be used in tissue engineering, for example, CT and CS.
b. **Cells/growth factors** – They are required for the stimulation of growth of the cells, for example, transforming growth factor beta (TGFβ), platelet-derived growth factor (PDGF), vascular endothelial growth factor (VEGF), basic fibroblast growth factor (bFGF), endothelial growth factor (EGF), and insulin growth factor (IGF).

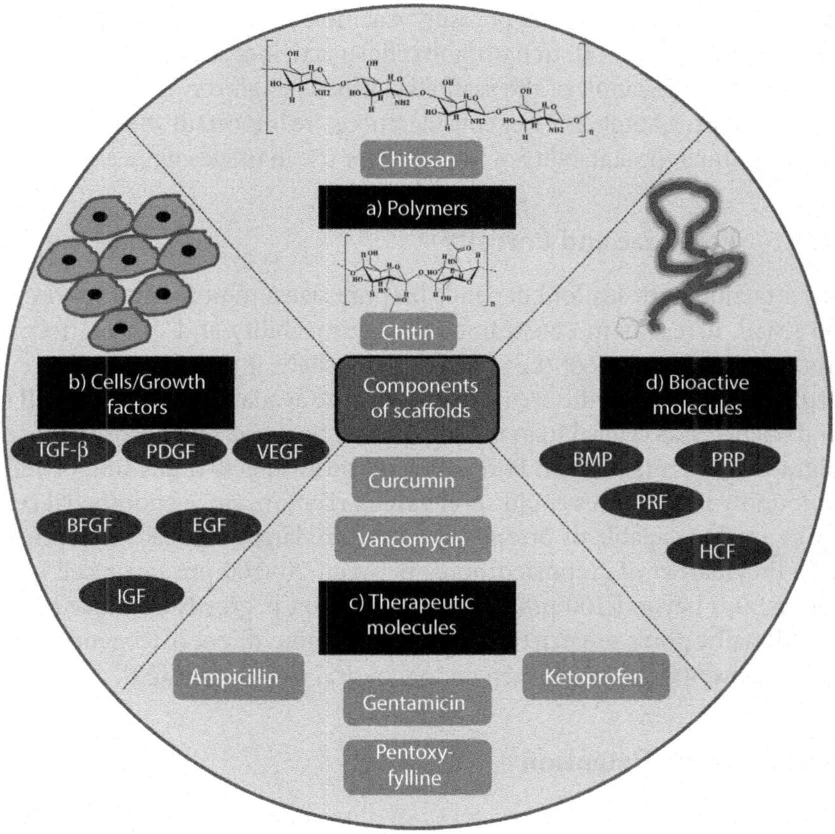

Figure 11.2 Different components of scaffold: (a) polymers, (b) cells/growth factors, (c) therapeutic molecules and (d) bioactive molecules.

c. **Therapeutic molecules** – They are essential for achieving relevant affect and maintaining the implanted site uninfected, for example, pentoxyfylline and curcumin.
d. **Bioactive molecules** – These molecules have an effect on cells or tissue which influence their growth, for example, bone morphogenetic protein (BMP), platelet-rich plasma (PRP), platelet-rich fibrin (PRF), and heparin-conjugated fibrin (HCF).

11.3 Scaffold Requirements

The success of tissue engineering scaffolds depends on its physico-chemical and biological properties of the materials. The biomechanical properties of scaffolds such as pore size and porosity, water retention, biocompatibility, biodegradation, mechanical strength, and cellular activities are important parameters for the application of the scaffolds in tissue engineering. Regardless of the tissue type, a number of key considerations are important when designing or determining the suitability of a scaffold for use in tissue engineering.

11.3.1 Pore Size and Porosity

Cell attachment to scaffold depends on pore size and porosity. If pores are too small, it results in constrained cell permeability and capsule formation around the edges of the scaffold. Oppositely, if pores are too large, it results in limited specific area and reduces the available area for the cell to bind. Hence, it is crucial to maintain optimal pore size and porosity for cell migration and attachment. Pore size vigorously affects trafficking of cells; profoundly larger pores could spoil vascularization, since endothelial cells (ECs) are not capable of bridging pores more larger than a cell diameter [20]. Freyman *et al.*, reported pores between 20–100 μm favor cell infiltration, and beyond 100 μm neovascularization is greatly enhanced [21]. Additionally, pores greater than 300 μm promote direct osteogenesis and pores below 300 μm strengthen endochondral ossification [22].

11.3.2 Water Retention

Water retention ability also known as swelling ability, refers to the ability of tissue engineering scaffolds to retain water. *In vivo* implantation of scaffolds results in the absorption of water from circumventing tissues and results in incremented pore size which avails cellular infiltration deep into the internal structures of the scaffolds [23]. The swelling of polymers depends on the ionizable groups in the structures and the circumventing medium. Chitin and CS chains swell through protonation of amine/imine groups and lead to the mechanical relaxation of coiled chains [24]. Amine groups of CT and CS chains are the key players in determining the swelling ability. Due to cationic nature of CT and CS, it would have electrostatic interaction with anionic polymers, and most of the amine groups in CT and CS would involve in polymeric complexation, leading to decremented swelling. Chitosan–gelatin scaffolds were fabricated and its swelling ratio was significantly reduced through the integration of nanobioactive glass ceramics particles [25].

11.3.3 Biocompatibility

The first and foremost criterion of any scaffold for tissue engineering is that it must be biocompatible. After implantation, the scaffold or tissue engineered construct must elicit a negligible immune reaction in order to prevent it causing such a severe immune response that it might cause rejection by the body [26]. Also cells must adhere, function normally and migrate onto the surface and eventually through the scaffold and commence to proliferate before laying down into new matrix.

11.3.4 Biodegradability

Tissue engineering permits the body's own cells, over time, to eventually supersede the implanted scaffold or tissue engineered construct. Scaffolds and constructs are not intended as permanent implants. The scaffold must, therefore, be biodegradable so as to allow cells to produce their own ECM [27]. Chitin and CS are mainly degraded by the action of lysosyme [28]. Higher degree of deacetylation of CS leads to more expeditious degradation by lysosyme [29]. The *N*-acetyl-D-glucosamines of CS are hydrolyzed by the action of lysosyme and the degraded products are amino sugars which are incorporated into GAGs and glycoprotein metabolic pathways.

11.3.5 Mechanical Strength

The scaffold should have mechanical properties consistent with the anatomical site into which it is to be implanted and, from a practical point of view, it must be strong enough to permit surgical handling during implantation. Chitin and CS have low to moderate mechanical properties which restricts its use in load bearing applications [30]. However, incorporation of various polymers increased the mechanical properties of these polymers [31].

To engineer biodegradable and bioresorbable materials into three-dimensional polymeric scaffolds of high porosity and surface area, scaffolds have to be fabricated by different methodology to facilitate the cell distribution, migration, proliferation and guide their growth into three-dimensional space.

11.4 Chitin – and Chitosan-Based Scaffolds Fabrication Techniques

There are many techniques used for making scaffolds which are divided into two categories, that is, designed manufacturing techniques and non-designed manufacturing techniques. Non-designed manufacturing techniques are mainly based on manual work so extra procedure was needed to obtain suitable shape and the microstructure. It includes freeze drying or emulsion freezing, gas foaming or high pressure processing, melt molding, solvent casting or particulate leaching, phase separation, electrospinning, and combination of these techniques. A designed manufacturing technique exploits computer-aided design software for designing or customizing the scaffold model. The scaffold layers were physically built using selective materials as specified by a computer program. It includes rapid prototyping or solid free-form technology.

11.4.1 Non-Designed Manufacturing Techniques

11.4.1.1 Phase Separation

Phase separation is the conversion of a single-phase system into a multiphase system; especially the separation of a solution into two immiscible liquids. The phase separation process can be induced thermally or by a nonsolvent to fabricate porous membranes or foams for filtration and separation purposes [32]. In this method, scaffolds designing requires temperature change that separates the polymeric solution in two phases, one with polymer-lean phase (low polymer concentration) and other with the polymer-rich phase (high polymer concentration) [33]. Phenol or naphthalene is used for dissolving the polymer and the biologically active molecule is dispersed in the solution. Subsequent to temperature change, the phase is separated and quenched to form a two phase solid and the remaining solvent has been removed by extraction, evaporation, and sublimation to give porous scaffolds [34]. Advantage of the phase separation technique is that, it can be easily combine with other technologies like rapid prototyping to create nanofibrous scaffolds for tissue engineering applications [35]. It can also be combined with particulate leaching to design three-dimensional structures with a control on pore morphology [35]. Lim *et al.*, fabricated three-dimensional porous CS-polyvinyl pyrrolidone (CS-PVP) scaffolds for tissue engineering applications via liquid–liquid or liquid–solid phase separation from a quaternary system. The CS-PVP scaffolds were obtained with a porosity of over 70%. The pore morphology on the surface and within the porous scaffolds showed the presence of homogenous open pores with excellent interconnectivity. The porous scaffold cultured with NIH-3T3 fibroblasts cells showed an improved efficiency for cell adhesion after 4 h. The cultured cells presented an increase in the cell proliferation on the porous scaffolds after 72 h [36].

11.4.1.2 Electrospinning

Electrospinning technique involves electrostatic spraying of polymer coatings known as electrostatic fiber spinning or electrospinning. A highly porous scaffold of ultra-fine fibers can be fabricated using electrospinning. Electrospinning methods fabricate scaffolds from nanoscale to microscale fibers that are similar to the size of collagens in the ECM of cartilage [37]. Electrospinning process is controlled by high intensity electric field between two electrodes with electric charges of opposite polarity. In this process, an appropriate solvent has been selected to dissolve particular polymer. A syringe is used to load the polymer solution at a constant rate by syringe pump and it gets expelled through a metal capillary. Afterwards high voltage around 10–15 kV is applied to the capillary for charging and ejecting the polymer toward a grounded collecting surface which produces the fibers. Simultaneous evaporation of the solvent at once yields thin fibers on the plate and results in a nonwoven porous scaffold. Using electrospinning technique, Deepthi *et al.*, fabricated a tendon construct of electrospun aligned poly (L-lactic acid) (PLLA) nanofibers bundles and layering PLLA fibers with CS-collagen hydrogel to mimic the GAGs of sheath ECM for tendon regeneration. An outer coating of alginate gel was given to the construct to prevent peritendinous adhesion. Both uncoated and alginate gel coated CS-collagen/PLLA scaffolds showed good cell proliferation. These constructs were proven to be nontoxic to tenocytes and also tenocytes showed good attachment and spreading over the scaffolds [38].

11.4.1.3 Solvent Casting and Particulate Leaching

Solvent casting and particulate leaching technique involve the casting of a polymer/salt/organic solvent mixture solution followed by solvent evaporation and dissolution of the salt particulates in an aqueous solution. It allows preparation of scaffold with specific porosity, surface to volume ratio, pore size and crystallinity suitable for different applications. The pore size of the scaffold will depend on the size of the porogen particles, while the polymer to porogen ratio is directly correlated to the amount of porosity of the final structure. After casting polymer solution the solvent is allowed to fully evaporate. Then, the composite structure in the mold is immersed in a solvent suitable for dissolving the porogen (water in case of sodium chloride) to obtain porous structure. Although this method does not require much polymer for scaffold fabrication, but it has problem in controlling the pore shape and openings of interpores [39] and also this technique uses toxic solvents. Lim *et al.*, prepared a three-dimensional porous CS scaffold from a mixture of acidic CS solution and sodium acetate particles as the porogen by salt-leaching method. Prepared scaffold showed good interconnectivity and improved mechanical properties. Furthermore, the CS scaffolds showed a high initial cell adhesion after 4 h cell culture and increased cell proliferation than the control [40].

11.4.1.4 *Freeze Drying or Emulsion Freezing*

Freeze drying or emulsion freezing technique is based upon the principle of sublimation. First, a synthetic polymer is dissolved into a suitable organic solvent followed by addition of aqueous phase. The two phases are homogenized in order to obtain an emulsion. Later, prepared emulsion is frozen and solvent is removed by lyophilization under the high vacuum which fabricates the scaffold with high porosity and inter connectivity [11, 12]. Unlike SCPL, this technique does not require the use of a solid porogen for the fabrication of porous scaffolds [41]. The pore size can be controlled by the rate of freezing and pH where a fast rate of freezing produces smaller pores. The advantage of freeze drying technique is that, it neither requires high temperature nor separate leaching step. The main drawback of this technique is its long processing time and it gives scaffold with smaller pore size only [42].

Pattnaik *et al.*, fabricated a biocomposite scaffold by combining CS, nano-scaled silicon dioxide and zirconia using freeze drying technique. Fabricated scaffold possesses porous nature with pore dimensions suitable for cell infiltration and colonization. The addition of zirconia nano particles substantially improved swelling, protein adsorption, and biomineralization in the CS-silicon scaffold. The CS-silicon-zirconia scaffold was also found to be nontoxic to rat osteoprogenitor cells [43].

11.4.1.5 *Gas Foaming or High Pressure Processing*

Gas foaming or high pressure processing technique has been developed, which utilizes gas as a porogen, to overcome the need to use organic solvents and solid porogens. Firstly, desired polymer has been prepared by means of compression molding using heated mold disc-shaped structures. These discs will be placed in the chamber and exposed to carbon dioxide at high pressure (800 psi) for several days to saturate the polymer with gas [34]. The pressure inside the chamber is gradually restored to atmospheric levels. Dissolved carbon dioxide becomes unstable under this condition, and it will get separated into another phase from the polymer. The carbon dioxide molecules form cluster to minimize the free energy which in turns create pore nucleation. After completion of foaming processes a three-dimensional porous scaffold is formed. The main drawback of this technique is the use of excessive heat during compression molding which prohibits the incorporation of any thermolabile material into the polymer matrix. Irmak *et al.*, developed a 17β-estradiol releasing scaffold-nanoparticle system in order to promote osteogenic differentiation of adipose tissue-derived mesenchymal stem cells (AdMSCs) in rats for bone tissue regeneration. Chitosan–hydroxyapatite (HA) scaffolds in macroporous structures with interconnected pores were prepared by combining microwave irradiation and gas-foaming techniques, and 17β-estradiol poly(lactic-co-glycolic acid) (PLGA) nanoparticles were loaded onto these scaffold. Cell culture studies indicated that the controlled delivery of 17β-estradiol from PLGA nanoparticles loaded on CS-HA scaffolds had a significant effect on the osteogenic differentiation of AdMSCs [44].

11.4.1.6 Melt Molding

Melt molding is an alternative method of constructing three-dimensional scaffolds. Highly interconnective and geometrical scaffolds can be fabricated by using melt molding method. In melt molding method, a fine polymer powder was mixed with previously sieved gelatin microspheres and poured into a teflon mold. Next, the mold is heated above the glass transition temperature of polymer and while applying pressure to the mixture [45]. The polymer/gelatin microsphere composite was then removed from the mold. Once the mold is removed, gelatin microspheres will be dissolved by immersing the mixture into water and scaffolds are then dried to achieve the geometry identical to the shape of the mold [45]. The porosity could be controlled by varying the amount of gelatin. By using different sizes of microspheres, the pore size of the scaffold could also be altered independent of the porosity. Other leachable components such as salt or other polymer microspheres could also be used as an alternative to gelatin microspheres. Advantage of melt molding method is that it does not utilize organic solvents and is carried out at relatively low temperatures; hence it has the potential for the incorporation and controlled delivery of bioactive molecules. Pinto *et al.*, processed novel melt-based CS-poly (butylenes succinate) scaffolds by compression molding, followed by salt leaching. The morphology, mechanical properties, cytocompatibility, cell proliferation, and osteogenic differentiation of a mesenchymal stem cell line (BMC9) on CS-poly (butylenes succinate) scaffolds were studied. Results have showed that the developed scaffolds had the adequate mechanical properties. Further, they disclosed a noncytotoxic behavior and simultaneously supported the growth and osteogenic differentiation of mouse mesenchymal stem cells within its structure [46].

11.4.2 Designed Fabrication Technique

11.4.2.1 Rapid Prototyping (RP)

Rapid prototyping is also called as solid free-form technique that fabricates three-dimensional scaffolds with fully interconnected porous network [47]. This technique is more advanced technique and it is computer-controlled fabrication technique [5] that enables reproducible fabrication of scaffolds with designed internal and external architecture [48]. The porosity can be tailored using algorithms within the software. The advantages of computer-controlled fabrication techniques are their capability to fabricate scaffolds with complex architectures in micron scale [49]. Kufelt *et al.*, developed chemically modified CS three-dimensional scaffold by rapid prototyping technique enabling photopolymerization with two-photon polymerization (2PP). They combined and cross-linked CS with synthetic poly(ethylene glycol) diacrylate. Also CS was covalently cross-linked with a modified VEGF for a biological adaption to the *in vivo* situation. *In vitro* studies revealed that modified CS was biocompatible and VEGF enhances CS bioactivity. Furthermore, a three-dimensional CS scaffold can be successfully seeded with cells [50].

11.5 Applications of Chitin and Chitosan for Regeneration of Various Tissues

Tissue engineering is evolved from the field of biomaterial development, whereby tissue and organ failure is addressed by implanting natural, synthetic, or semisynthetic tissue to restore the functionality of damaged tissue. Chitin and CS have shown remarkable promise in the biomedical field among the various natural and synthetic polymers currently in use [51]. Chitosan derivatives have been used to fabricate various tissue analogs including skin, bone, cartilage, liver, etc., over the last decades.

11.5.1 Bone Tissue

Bone tissue regeneration is enhanced by fabricating three-dimensional porous polymeric matrixes which facilitates parent cell migration, proliferation, and differentiation [52]. For successful development of a scaffold matrix they also fabricated with/without cells and biological matter. Chitin and CS matrices have been combined with osteogenic materials, like hydroxyapatite [53–55], calcium phosphate, and sulfate [56, 57], and others [58–60] for bone tissue engineering. These combinations simulate the bone structure to obtain organic and inorganic composites [61]. The applications of CT and CS in the field of bone tissue engineering are outlined in Table 11.1.

Chitosan has been shown to promote growth and mineral-rich matrix deposition by osteoblasts *in vitro* culture [62]. Various studies have emphasized on the use of CS–calcium phosphate (CP) composites for bone tissue

Table 11.1 Different applications of CT and CS in bone tissue engineering.

Components	Study	Cell lines used	Observation	Conclusion	Ref.
Chitosan–hydroxyapatite	The effect of HA on composite structure and cellular behavior for bone formation	MC3T3-E1 preosteoblastic cells	Chitosan–HA has enhanced the number and osteogenic differentiation ability of MC3T3 cells	Study demonstrated that CS–HA is promising candidate for bone tissue-engineering	[75]
Chitosan–gelatin	Chitosan–gelatin scaffolds were characterized for physico-chemical and biological properties	Buffalo embryonic stem (ES) cells	Chitosan–gelatin scaffolds were found to be cytocompatible to buffalo ES cells	Chitosan–gelatin scaffolds are promising candidates for ES cell-based tissue engineering	[76]
Chitosan-poly(butylene succinate) (CS–PBS)	Chitosan-based scaffolds were analyzed for cell adhesion, proliferation and differentiation	Human bone marrow mesenchymal stem cells (hBMSCs)	Chitosan-based scaffolds supported proliferation and osteogenic differentiation of hBMSCs *in vitro* also bone formation *in vivo*	Possibility of CS–PBS–hBMSC constructs to be used as implants for non-load-bearing bone defects	[77]
α – and β-chitin	Bioactivity studies were performed for chitin membranes with the simulated body fluid (SBF) solution	MG-63 osteoblast like cells	The cells got adhered and spread over the membrane after 24 h of incubation	Study indicated that chitin membranes are useful for biomedical applications	[78]
Hydroxtapatite–chitin	Hydroxyapatite–CT scaffold was loaded with osteoblasts and implanted into the intramusculature of a rat model	MSC-induced osteoblasts	Osteoblasts loaded onto the porous HA–CT matrixes showed bone regeneration with biodegradation of the HA–CT matrix	Hydroxyapatite–CT matrix is a good substrate candidate for tissue engineered bone substitute	[73]

(*Continued*)

Table 11.1 Cont.

Chitosan–alginate	Chitosan –alginate (CA) scaffolds were studied for critical size calvarial defects (diameter, 5.0 mm)	Undifferentiated MSCs	Chitosan–alginate scaffolds provided greater defect closure and bone deposition	CA scaffolds showed promise in reconstruction of critical size bone defects	[79]
Chitosan–Polylactic acid (PLLA)–μHA	Scaffold composed of PLLA, CS and HA was characterized for osteoconduction and osteoinduction properties	MC3T3-E1 preosteoblastic cells	Pre-osteoblastic cells proliferated in the macropore and micropore channels of the scaffold. They formed an ECM network and also got differentiated into mature osteoblasts	These three-dimensional porous scaffolds could be employed as a potential biomaterial for bone reconstruction as non-load-bearing implants	[80]
Chitosan–gelatin–*n*SiO_2	Nano-SiO_2 incorporated CS/gelatin scaffolds were compared to pure CS and CS/gelatin scaffolds for bone tissue engineering	MG-63 cells	Chitosan–gelatin–*n*SiO_2 scaffold was cytocompatible and allows better cell attachment and proliferation over pure CS and CS/gelatin scaffolds	Prepared nanocomposite scaffold could be a prospective candidate for bone tissue engineering	[81]

engineering [63, 64]. Calcium phosphate is composed of essentially highly soluble tricalcium phosphate and insoluble hydroxyapatite. When CS scaffolds containing CP microparticles were introduced to cell culture media, cells were quickly enveloped into mineral layer on the surface of scaffold [65]. Gentamycin-conjugated macroporous CS scaffolds supported with beta-tricalcium phosphate (βTCP) and CP have been developed by Zang *et al.*, for bone regeneration [64]. They also synthesized a macroporous CS scaffolds composed of hydroxyapatite (HA) or CP glass with an interconnected porosity of approximately 100 mm [30]. The same group developed a three-dimensional macroporous CP bioceramic scaffold incorporated with porous CS sponges [66]. They have showed enhanced mechanical strength of the ceramic phase via matrix reinforcement and preserved the osteoblast phenotype [30]. Altogether composites of CS–CP appear to have a promising clinical application.

Naturally occurring mineral form of calcium apatite powder has been combined with CS to fabricate scaffolds via phase separation [67]. Hydroxyapatite has the ability to make the scaffolds osteoinductive and allowed differentiation of local MSCs. The incorporation of HA in the polymer matrix appears to improve CS scaffolds in three ways: (1) osteogenic cells response is increased in terms of cell attachment, proliferation, and differentiation [68–70], (2) the compressive modulus of scaffold is increased, and (3) scaffold swelling is decreased [67, 71]. Changes in scaffold surface topography [68], enhanced protein binding to HA [69] and/or biological apatite formation [72] might be the reasons for the improvement in osteogenic cell response. Hydroxyapatite–CT material that was found to be osteoinductive exhibited rapid degradation and neovascularization *in vivo* for the duration of 3 months [73]. Chitosan–HA paste was fabricated and studied *in vivo* by applying on the surface of the tibia after periosteum removal. The results obtained showed the formation of new bone after 1 week and growth continued during a 20 week follow-up [74].

11.5.2 Cartilage Tissue

Cartilage tissue engineering focuses on achieving a completely healed, functional, and scar less tissue for cartilage degeneration. The causes of cartilage degeneration are trauma or disease or genetic abnormalities, which finally might require surgical intervention, mostly replacement surgeries.

For the successful tissue engineering approaches in cartilage repair, the selection of biomaterial is very important [82]. The cell-carrier substance, that is, biomaterial should be able to mimic the natural environment in the articular cartilage matrix. Hydrogels comprising of CS composites have been prepared for the inclusion of cells and molecules required for cartilage regeneration. Composites of CS with polycaprolactone [83–85], silk fibrils [86], genipin [87], chondroitin sulfate [88], and polyester [89] have been designed that resemble the properties of cartilage. They also provide a proper environment for extracellular matrix deposition, cell viability, proliferation, and differentiation.

Collagen type II and GAGs play a critical role in regulating expression of the chondrocytic phenotype and in supporting chondrogenesis *in vitro* as well as *in vivo* [90, 91]. Collagen type II is the main protein in cartilage tissue; it is an inducer agent of cartilage regeneration. It enhances the adhesion and formation of clusters of chondrocytes *in vitro* [84, 87, 92], a requirement for cartilage regeneration. Chitosan having structural resemblance with GAGs and hyaluronic acid present in articular cartilages thus stimulates chondrogenesis. The applications of CT and CS in the field of cartilage tissue engineering are outlined in Table 11.2.

The rate of biodegradation of the scaffold plays a crucial role in case of cartilage tissue engineering applications. The remnants of nonbiodegradable scaffolds often cause acute foreign body reactions evoked by the body's immune system which can result in severe inflammation and allergic reactions. Chitin and CS degrade *in vivo* mainly by enzymatic hydrolysis mediated by lysozyme, which is found abundantly in the human body. However, this enzymatic hydrolysis is dependent on type of CT or CS, pH of the environment and their preparation method.

11.5.3 Dermal Tissue

The skin damage can be caused by heat, chemicals, electricity, ultraviolet, accident, or disease. Healing a skin tissue is complicated which includes a wide range of cellular, molecular, physiological, and biological processes. Skin substitutes such as xenografts, allografts, and autografts have been employed for wound healing, full-thickness burns or deep ulcers. But these skin substitutes have many disadvantages like limited availability of skin grafts in severely burned patients, disease transmission, and immune response [97, 98]. Tissue-engineered skin equivalents are good alternatives to skin grafts for curing skin damage. Scaffolds made of polymers like PLGA, collagen, and chitosan are currently being employed for skin tissue reconstruction [99, 100]. The aim of skin tissue engineering is to rapidly regenerate the complete and functional skin, and allow the skin to fulfill its normal functions, that is, pigmentory defence against UV

Table 11.2 Applications of CT and CS in the field of cartilage tissue engineering.

Components	Study	Cell lines used	Observation	Conclusions	Ref.
Chitosan and Chondroitin Sulphate	Nanostructured three-dimensional constructs prepared and their effect was studied on cartilage tissue engineering	Bovine chondrocytes (BCH) and hMSCs	Cells got attached, proliferated and were metabolically active over the entire multilayered constructs	Nanostructured scaffolds of CS and chondroitin sulphate could have potential use for partial thickness/chondral defects	[93]
Chitosan, β-sodium glycerophosphate and hydroxyethyl cellulose	Reconstruction of cartilage was performed *in vitro* by using temperature responsive CS hydrogels as scaffolds before transplantation *in vivo*	Bovine chondrocytes (BCH)	Chondrocytes survived in the reconstructed cartilage and retained their ability to secrete matrix when cultured *in vitro*. The reconstructions repaired cartilage defects upon *in vivo* transplantation	Chitosan-based hydrogel could support matrix accumulation of chondrocytes and could repair cartilage defects	[94]
Chitosan-graft-glycolic acid (GA) and phloretic acid (PA) (CS-GA/PA)	Injectable biodegradable hydrogels were prepared based on CS derivatives for cartilage regeneration	Bovine chondrocytes (BCH)	Hydrogel can be readily degraded by lysozyme. Chondrocytes were viable and retained their round shape on CS-GA/PA hydrogel	Water soluble CS derivative, CS-GA/PA has a high potential as an injectable biomaterial for cartilage tissue engineering	[95]

(*Continued*)

Table 11.2 Cont.

Components	Study	Cell lines used	Observation	Conclusions	Ref.
Chitosan/poly(3-caprolactone) (CS/PCL)	CS/PCL blend scaffold was prepared and characterized for cartilage tissue engineering	Bovine articular chondrocytes	This scaffold supported the most neo-cartilage formation, as demonstrated by an increase in glycosaminoglycan production	Chitosan/PCL blend three-dimensional fiber-mesh scaffolds has a potential for articular cartilage tissue repair	[83]
Methacrylated glycol CS (MeGC) and hyaluronic acid	Injectable CS composite hydrogels containing hyaluronic acid were explored for repair of cartilage defects	Auricular chondrocytes from rabbits	Incorporation of hyaluronic acid in MeGC hydrogels increases the proliferation and deposition of cartilaginous extracellular matrix by encapsulated chondrocytes	Composite hydrogels have high potential for use as a carrier of cells and bioactive molecules for the treatment of cartilage damage	[96]

irradiation; barrier formation; thermoregulation, and mechanical functions [101].

Chitin and CS polymers possess hemostasis properties which accelerate tissue regeneration and stimulate the collagen synthesis of fibroblasts. Due to this property CT and CS are widely explored as skin substitute materials in skin/dermal tissue engineering [102–104]. Agarwal *et al.*, developed gelatin-based carboxymethyl CS scaffolds for dermal tissue engineering. The work demonstrated the effect of carboxymethyl CS and gelatin ratio on variations in their physico-chemical and biological characteristics. SEM images revealed that the scaffolds were highly porous with pore size ranging between 90 μm and 170 μm. Equilibrium swelling and water retention experiments on prepared scaffold showed high water uptake (400–1100%) and water retention capacity (>300%). The degradation of the scaffolds was dependent on the amount of gelatin present in the formulation. The scaffold supported adhesion, spreading, growth, and proliferation of 3T3 mouse fibroblasts. The cells seeded on the scaffolds showed expression of collagen type I, HIF1α, and VEGF which confirms their potential in cell growth, proliferation along with angiogenesis during wound healing [105].

Recently, Anjum *et al.*, [106] developed antimicrobial and scar preventive wound dressings by coating of CS, polyethylene glycol and PVP on the cotton fabric. The SEM images of the coated fabric revealed porous structure with 54–70% porosity of the material and the pore size was in the range of 75–120 microns. The water vapor transmission rate was in the range of 2000–3500 g/m^2/day which may offer a suitable material for the wound dressing with moderate exudates absorption. The scaffold showed antimicrobial nature against both gram positive and gram negative bacteria. Fast healing was observed in drug loaded dressing which treated wounds with minimum scarring, as compared to the other groups in wound healing and tissue compatibility studies on full-thickness skin wounds.

11.5.4 Hepatic Tissue

Liver is one of the most important and complex organs involved in many metabolic processes and also serve several essential functions in the body. Liver transplantation is the first choice of treatment for liver failure. Due to the shortage of suitable donors and costly surgical procedure, liver transplantation was limited in its application. An alternative method for the treatment for liver failure is by using a bioartificial liver composing of circulating isolated hepatocytes and polymeric scaffolds.

Chitin and CS are promising biomaterials which can be applied in liver tissue engineering due to its various properties. The reason for selecting CT and CS polymers for hepatic regeneration is due to the presence of GAGs in CT and CS structure, which are main components of the liver ECM [107, 108]. Chitosan and its complexes with GAGs can modulate the activities of vascular ECs and smooth muscle cells *in vitro* and *in vivo*, hence have a significant potential for the design of new biologically active biomaterials [109].

Self-assembled homogenous and porous silk fibroin/CS scaffolds was developed which supported attachment of hepatocytes [110]. Similarly a water-stable and porous silk/fibroin/CS composite scaffold was developed which had controlled mechanical properties. The pore size of scaffold was in the range of 100–150 μm and showed significant hepatocellular carcinoma (HepG2) cellular viability [111]. Chitosan/collagen/heparin scaffold was fabricated and showed superior blood compatibility in implantable bioartificial liver applications [112]. Chitosan/collagen matrix (CCM) was prepared by cross-linking agent, that is, 1-Ethyl-3-(3-dimethylaminopropyl) carbodiimide/*N*-Hydroxysuccinimide and this matrix showed good hepatocyte compatibility, moderate mechanical strength, as well as excellent blood compatibility [113].

Another strategy involved in hepatic tissue engineering is modification of CS with galactose (GC) residues to enhance hepatocytes attachment and to maintain cell viability. Galactose residues allow to bind to the asialoglycoprotein receptor (ASGPR) expressed on the surface of hepatocytes. Galactose-containing polymers have been used to induce the selective adhesion of primary hepatocytes [114, 115]. Galactosylated CS was developed and hepatocyte attachment was achieved through specific interaction between asialoglycoprotein receptors of hepatocytes and galactose moieties of GC [116]. Similarly alginate-GC scaffolds have a potential ability to improve hepatocyte attachment [117]. Another ligand, that is, fructose, also one of the specific ligands of ASGPR in hepatocytes, was explored. Fructose-conjugated CS scaffold was developed for liver tissue engineering applications and induced the formation of cellular aggregates with enhanced liver specific metabolic activities and cell density [108, 118].

11.5.5 Cardiac Tissue

Myocardial infarction (MI) is one of the leading causes of death and disabilities in the world, which can cause irreversible damage to cardiac muscle [119]. Many reports have shown that injectable tissue engineering is an effective method to repair the damaged cardiac tissue [120, 121]. Effective vascularization is an essential requirement for the success of injectable cardiac tissue engineering [122]. The aim of cardiac tissue engineering is the development of three-dimensional heart muscle that can be utilized to increase the function of failing myocardium.

Thermosensitive CS-based hydrogels have been widely used in the field of injectable cardiac tissue engineering because of their unique biocompatibility and controlled biodegradability [123]. Chitosan-based hydrogels can protect the transplanted cells, reduce the size of infarct, and promote the angiogenesis as well as improve cardiac functions [124, 125]. Li *et al.*, produced thermosensitive hydrogels by conjugating CS chloride with glutathione. They showed excellent antioxidant capacity against H_2O_2 microenvironment and biocompatibility to favor the survival of cardiomyocytes [126]. Shu *et al.*, [127] developed a thermosensitive CS chloride-RoY (CSCl-RoY) hydrogel to improve angiogenesis. The effect of CSCl-RoY hydrogels on vascularization *in vitro* under hypoxia was investigated using human umbilical vein ECs. Results showed that CSCl-RoY hydrogels can promote the survival, proliferation, migration, and tube formation of human umbilical vein ECs under hypoxia compared with CSCl hydrogel. Further they investigated its therapeutic properties *in vivo* in MI rat model and results demonstrate that the introduction of RoY peptide can not only improve angiogenesis at MI region but also improve the cardiac functions.

In cardiac tissue engineering, several natural materials have also been tested with CS, including alginate [128, 129], gelatin [130], collagen [131], myocardial ECM [132], and silk fibroin [133]. Ceccaldi *et al.*, developed CS–alginate composite with MSCs and tested *in vivo* in a rat MI model. Results showed improved cardiac function, new blood vessel formation, cellular survival, and cell proliferation [128]. Pok *et al.*, [130] developed a three-dimensional biodegradable patch/scaffold composed of self-assembled polycaprolactone (PCL) inserted in a gelatin-CS hydrogel. Results showed that patch has a sufficient tensile strength for use as a cardiac patch. This allows migration of cardiac cells in a biomimetic environment.

11.5.6 Nerve Tissue

Chitin and CS are well studied for nerve regeneration applications [134]. They have sufficient biological properties which can serve as the main materials for artificial nerve tubes. Chitin and CS are blended with other biomaterials to create scaffolds that are more appropriate for directing the desired cell behaviors and to mechanically strengthen the tissue engineering of the nervous system [134]. Furthermore, to enhance their affinity with nerve cells, many different substrates and bioactive molecules have been added into CT-based scaffold [135].

Chitin and CS have been studied extensively for nerve regeneration. Yuan *et al.*, showed that CS fibers supported the adhesion, migration and proliferation of Schwann cells (SCs), which guide for regeneration of the axons in the nervous system [136]. Haipeng *et al.*, also reported that CS membrane cultured with neurons can grow well and that CS scaffold can considerably elevate the repair of the peripheral nervous system [137].

One strategy for nerve tissue regeneration has drawn interest of researchers is by growing neural stem cells (NSCs) because of their potential for neural regeneration [138]. Scanga *et al.*, screened four amine functionalized hydrogels *in vitro* for adult murine NSCs viability, migration, and differentiation. Chitosan has more surface amine content and the lowest equilibrium water content, which likely contribute to the greater NSCs viability. They also found that only CS hydrogel supported survival of multipotent stem cells and differentiation of progenitors into neurons, astrocytes, and oligodendrocytes. Chitosan hydrogel showed greater cell migration than other hydrogels after plating with intact colonies of NSCs [139].

In another strategy, HA-coated CS scaffold including laminin-1 or laminin peptides was developed as scaffolds for peripheral nerve reconstruction. They showed enhanced growth of regenerating axons. In their study, CS tubes were obtained and surface modified with HA, to avoid excessive swelling and to improve their mechanical strength during *in vivo* implantation. Laminin-1, IKVAV (Ile-Lys-Val-Ala-Val), and YIGSR (Tyr-Ile-Gly-Ser-Arg) containing peptides were adsorbed on the tubes. Laminin and laminin-1 peptides improved the growth of regenerating axons, when the bridging of a defect in the sciatic nerve was evaluated [140]. Furthermore, laminin peptide was immobilized on molecularly aligned CS by covalent bonding for neural regeneration [135].

Cao *et al.*, studied the physical, mechanical, and degradation properties of CS films. The affinity of these films with SCs was also studied. Three different types of CS films were prepared by cross-linking with hexamethylene diiso-cyanate (HDI), epichlorohydrin (ECH), and glutaraldehyde (GA). Cross-linking increased their hydrophilicity and elastic modulus, whereas the swelling degree and the degradation rate of the CS films were decreased. Between the cross-linked films, HDI cross-linked CS films showed enhanced attachment, spreading and proliferation of SCs in comparison to the other two cross-linked films [141]. In another study, CT hydrogel tubes were fabricated from CS solutions using acylation chemistry. Chitosan tubes were also obtained by alkaline hydrolysis of CT tubes and controlled the resulting amine content. *In vitro*, CT and CS supported adhesion and differentiation of ganglionic neurons. Chitosan films showed significantly enhanced neural outgrowth compared to CT films, indicating the dependence of nerve cell affinity on the amine content [142]. Mingyu *et al.*, prepared CS blend with a peptide to make the mechanical properties of scaffolds more similar to those of nerve tissues and to enhance nerve cell attachment. This blend also showed an improved attachment, differentiation, and growth of the cells when compared to cells cultured on CS membranes [143].

11.5.7 Vascular Tissue

Tissue engineering of blood vessels is an optimal alternative for blood vessel substitution in vascular diseases, such as blood vessel damage, atherosclerosis, and aneurysms. A biocompatible polymer with suitable degradation rate and biological properties that interact favorably with blood and cells is suitable for vascular tissue engineering.

Reports with respect to the applications of CT and CS as a scaffold for tissue engineering of blood vessels were few. A biocompatible tubular scaffold was produced in which collagen has been cross-linked with CS in order to mimic the morphological and mechanical properties of blood vessels. This scaffold showed desirable porosity, enhanced cell adhesion, proliferation, and ECM production [144, 145]. Similarly, CS/collagen blended scaffolds have also been employed in adipose tissue regeneration for vascular applications where adipocytes were seeded on scaffold. The cytocompatibility and biocompatibility of scaffolds were confirmed *in vitro* and *in vivo*, respectively [146].

Another strategy for vascular tissue engineering is to form a complex of CS/heparin to obtain good haemocompatibility. The CS/heparin scaffolds showed excellent biocompatibility as revealed by the reduced activation of coagulation, complement, and blood cells [147–149]. Chitosan-heparin scaffolds *in vivo* were observed to stimulate cell proliferation and the formation of a thick, dense, and highly vascularized granulation layer [109]. Similarly, Madihally *et al.*, fabricated CS scaffold, including heparin modified porous tubes, which had the potential for application in vascular tissue engineering [19].

11.5.8 Corneal Tissue

Cornea is the most important and naked part of an eye ball. It is subjected to damage easily as it is directly exposed to external environment. If the damaged corneal epithelia is not healed soon, many corneal diseases follow, that is, infection, ulcer, scar formation, etc. These diseases affect the transparency of cornea and may cause blindness. To overcome these issues and to maintain good vision, promotion of fast and complete wound healing as well as restoration of function of corneal epithelial cells (CEpCs) is very important [150, 151].

Tissue engineering methods for the treatment of corneal diseases are an area of intensive research. Scaffolds seeded with corneal cells on the polymeric membranes have been used for corneal tissue engineering [152]. Even though these scaffolds have shown good biocompatibility, their transplantation onto the corneal stroma has caused a range of problems [153, 154]. Chitosan is widely used for preparation of corneal tissue engineering due to its good optical transmittance; its transparency is above 85% at 400 nm [155].

Reports for corneal tissue engineering using CT and CS as a scaffolding material are available. Liang *et al.*, developed a water soluble hydroxyethyl CS (HECTS) membrane and examined its application as a scaffold in repair of mechanically damaged corneal epithelia. The results showed that HECTS significantly promoted growth of CEpCs *in vitro.* The corneal epithelium got repaired faster compared with control after transplanting the CEpCs seeded membranes onto the damaged corneal epithelium [156].

Chen *et al.*, investigated the biocompatibility of collagen–CS–sodium hyaluronate (Col–CS–NaHA) complexes on corneal tissues using cultivating rabbit corneal cells. This study evaluated the feasibility of Col–CS–NaHA complexes as substrates for cultivating rabbit corneal cells. The results showed that cells seeded on Col–CS–NaHA were allowed to proliferate and partly form confluent monolayer after 9 days in culture. Study confirmed that Col–CS–NaHA complex may serve as a suitable scaffold for corneal tissue engineering [157]. Wang *et al.*, fabricated hyaluronic acid immobilized chitosan (CS-HYA) films to improve hydrophilicity and biocompatibility of CS. They were characterized transparency, water absorption percentage and contact angle of the films. Results showed that there was no significant variation in transparency before and after immobilization. Also the human corneal epithelial cells (HCECs) grew better on CS-HYA films than on CS films and CS-HYA films have shown higher cell viability than the CS films [155].

11.5.9 Other Tissues

Apart from above mentioned uses also chitin and CS as scaffolding material they also find application in periodontal tissue engineering [158, 159], disc tissue engineering [160], tendon [161] and ligament [162], dental tissue, spinal cord tissue, esophageal tissue and breast tissue, and also for organ-specific tissue regeneration.

11.6 Chitin – and CS-Based Scaffolds for Drug and Growth Factors Delivery

Delivery of therapeutics has been a very active area, especially for CT, CS, and their derivatives as a carrier for various active agents [163, 164]. These polymers have been effectively used in drug delivery as a hydrogel system, drug conjugate, and biodegradable release system. These drug delivery systems can be used in both implantable as well as injectable forms through oral, nasal, and ocular routes. The aim for delivery of drug and growth factors is to cure, restore or replace damaged body parts or lost organs by transplanting supportive scaffolds with appropriate active agent in combination with biomolecules to generate new tissue.

Tissue regeneration demands complex biological process engaging chemotaxis, division of cells, neovascularization, and synthesis of ECM. To begin and sustain the phases of tissue repair, growth factors such as basic fibroblast growth factor (bFGF) and TGF-β play an important role [165]. Growth factors typically act as signaling molecules between cells that bind to specific receptors on the surface of their target cells and provide their effects [166]. Triggering of the receptors eventually leads to the stimulation of a number of processes, including those involved in tissue repair [167]. The application potential of CT – and CS-based membranes and scaffolds for drug and growth factors delivery was briefly summarizes in Tables 11.3 and 11.4, respectively.

11.7 Chitin – and CS-Based Scaffolds/ Dressings in Market

Based on the biocompatibility, biodegradability, regenerative, and hemostatic properties of CT, CS, and their derivatives commercial CT – and CS-based scaffolds/dressings have flooded the market. Brief description and functions of marketed scaffolds/dressings have been outlined in the Table 11.5.

11.8 Conclusions

Tissue engineering is regarded as an alternative approach to autograft and allograft for the regeneration of various types of tissues and organs. Recent years have witnessed intensified efforts in developing CT and CS scaffolds and their application in the field of tissue engineering due to their minimal foreign body response, regenerative nature, biocompatibility,

Table 11.3 Chitin – and CS-based membranes and scaffolds in drug delivery.

Components	Drug	Study	Remarks	Ref.
Chitosan	5-Fluorouracil (5-FU)	5-FU loaded scaffolds composed of CS fibers were prepared and characterized for the drug loading and release	• Drug loading was not very high may be due to high solubility of 5-FU in water • The drug release was fast at the initial period and almost remained constant at the later time period	[168]
Chitosan	Dexamethasone	CS scaffolds were loaded with dexamethasone for tissue engineering applications	• Chitosan scaffolds were successfully impregnated with dexamethasone • The release of dexamethasone from CS scaffolds was sustained	[169]
Chitosan–pectin	Pentoxifylline	Chitosan was cross-linked with pectin and made into porous scaffold	• Cross-linking of the scaffold reduces the drug release • Drug release rate altered its effectiveness in inhibiting TNF-α and IL-6 release from activated macrophage cells	[170]
n-Hydroxy-apatite-CS (*n*-HA/CS)	Ampicillin	Three-dimensional porous scaffold carrying drug loaded microspheres has been developed to regenerate tissue	• The *n*-HA/CS porous scaffold carrying drug-loaded microspheres has high connectivity and slow drug release rate • Mechanical property of scaffold enhanced when alginate microspheres were coated with CS	[171]
Chitosan	Amikacin and vancomycin	The release of inhibitory levels of amikacin and vancomycin from CS sponge was assessed by turbidity assay	• Activity of released amikacin and vancomycin indicated inhibitory levels of elution from the CS sponge	[172]
Chitosan-g-β-cyclodextrin (CS-*g*-β-CD)	Ketoprofen	Biodegradable CS-*g*-β-CD scaffolds were prepared with controlled drug release for treatment of tissue damages	• The drug dissolution profile showed that CS-*g*-β-CD scaffolds provided a slower release of the entrapped ketoprofen than CS scaffold • No obvious cytotoxicity of CS-*g*-β-CD scaffolds as shown by MTT assay	[173]

Table 11.4 Chitin – and CS-based membranes and scaffolds in delivery of growth factors.

Components	Growth factor	Cell line/animal model	Study	Remarks	Ref.
Brushite–CS	VEGF and PDGF	Rabbit bone model	Brushite–CS composite scaffold loaded with PDGF or a combination of PDGF and VEGF was examined in a bone defect	• Brushite–CS system is capable of controlling the release rate and localization of both GFs within a bone defect • The bone formation was considerably enhanced with PDGF loaded brushite–CS scaffolds as well as with the PDGF/VEGF combination	[174]
Chitosan	Insulin	Prechondrogenic ATDC5 cells	The effect of insulin loaded CS particle aggregated scaffolds was studied in chondrogenic differentiation	• Insulin loaded scaffolds upregulated SOX-9 and aggrecan expression in culture of prechondrogenic ATDC5 cells	[175]
Chitosan-HA	bFGF	–	*In vitro* release of bFGF from CS-HA Scaffolds was studied	• Sustained release of bFGF was achieved up to 7 days because of the presence of HA in composite scaffold	[176]

Table 11.5 Chitin – and CS-based marketed scaffolds/dressings [177–181].

Material	Product	Manufacturer	Description	Function/use
Chitin and its derivatives	Chitipack P®	Eisai Co.	Dispersed and swollen CT supported on poly(ethylene therephthalate)	Favors early granulation and suitable for defects which are difficult to suture
	Syvek-Patch®	Marine Polymer Technologies	Made up of CT microfibrils	It agglutinates red blood cells, activates platelets and promotes fibrin gel formation within the patch thus acting in a redundant way even on heparinized patients
	Beschitin®	Unitika Co.	Nonwoven material manufactured from CT	Favors early granulation, no retractive scar formation
Chitosan and its derivatives	HemCon Bandage™	HemCon	Freeze-dried CS acetate salt	Use to stop bleeding in emergency
	BST-CarGel®	Biosyntech company	Chitosan-glycerophosphate hydrogels	A biodegradable gel for cartilage repair
	Tegasorb®	3M	It contains CS particles that swell while absorbing exudate and produce a soft gel	For leg ulcers, sacral wounds, chronic wounds
	Chitoseal®	Abbott	Based on CS, backed with cellulose coating	For bleeding wounds
	TraumaStat®	Ore-Medix	Freeze-dried CS containing highly porous silica	To control moderate to severe bleeding
	Chitoflex®	HemCon	Chitosan-based antibacterial, biocompatible wound dressing was designed to be stuffed into a wound	To control moderate to severe bleeding

biodegradability, and ease of processing. However, efforts to improve the mechanical properties of CT – and CS-based composite biomaterials are essential for tissue engineering application, especially for the construction of complex tissue and organs. In this chapter, we have mainly concentrated on the selection of CT and CS as scaffolding material and looked at different types of scaffold fabrication techniques as well as the CT – and CS-based scaffolds for delivery of drug and growth factors. We also discussed some marketed scaffolds fabricated by using CT and CS polymers and their composites. It is hope that this chapter will update researchers for the use of CT and CS as a scaffold for the regeneration of various types of tissues and organs.

11.9 Future perspectives

Advances in scaffold design and biomaterial engineering have resulted in a context of *in vitro* and *in vivo* studies for tissue engineering. At present, CT and CS are the most promising biopolymers for tissue engineering and biomedical applications. Especially, the possibility to ease of processing and generate structures with predictable pore sizes and degradation rates make CT and CS a suitable material for scaffold. However, their mechanical properties need to be improved for making them ideal biopolymers for tissue engineering.

Abbreviations

CT	chitin
CS	chitosan
ECM	extra cellular matrix
MSC	mesenchymal stem cells
GAGs	glycosaminoglycans
GFs	growth factors
TGFβ	transforming growth factor beta
PDGF	platelet-derived growth factor
VEGF	vascular endothelial growth factor
bFGF	basic fibroblast growth factor
EGF	endothelial growth factor
IGF	insulin growth factor
BMP	bone morphogenetic protein
PRP	platelet-rich plasma

PRF	platelet-rich fibrin
HCF	heparin-conjugated fibrin
ECs	endothelial cells
HA	hydroxyapatite
ATMSCs	adipose tissue-derived mesenchymal stem cells
CEpCs	corneal epithelial cells
HDI	hexamethylene diisocyanate
GC	galactose
NSCs	neural stem cells
SCs	Schwann cells
GA	glycolic acid
PA	phloretic acid
PLLA	poly (L-lactic) acid
PLGA	poly (lactic-co-glycolic) acid
BCH	Bovine chondrocytes
hBMSCs	human bone marrow mesenchymal stem cells
CP	calcium phosphate
β-TCP	beta-tricalcium phosphate
ESCs	embryonic stem cells
PCL	polycaprolactone

References

1. Langer, R., Biodegradable polymer scaffolds for tissue engineering. *Nat. Biotechnol*, 12, 689–693, 1994.
2. Flanagan, T.C., Wilkins, B., Black, A., A collagen-glycosaminoglycan co-culture model for heart valve tissue engineering applications. *Biomaterials*, 27(10), 2233–2246, 2006.
3. Solchaga, L.A., Goldberg, V.M., Caplan, A.I., Cartilage regeneration using principles of tissue engineering. *Clin. Orthop. Relat. Res.*, 391, 161–170, 2001.
4. Hollister, S.J., Porous scaffold design for tissue engineering. *Nat. Mater.*, 4(7), 518–524, 2005.
5. Hutmacher, D.W., Scaffold design and fabrication technologies for engineering tissues—state of the art and future perspectives. *J. Biomater. Sci., Polym. Ed.*, 12(1), 107–124, 2001.
6. Karageorgiou, V., Kaplan, D., Porosity of 3D biomaterial scaffolds and osteogenesis. *Biomaterials*, 26(27), 5474–5491, 2005.
7. Ma, P.X., Langer, R., Fabrication of biodegradable polymer foams for cell transplantation and tissue engineering. *Tissue Eng. Methods Protocols*, 18, 47–56, 1999.
8. Cima, L., Vacanti, J., Vacanti, C., Tissue engineering by cell transplantation using degradable polymer substrates. *J. Biomech. Eng.*, 113(2), 143–151, 1991.

9. Kweon, H., Yoo, M.K., Park, I.K., A novel degradable polycaprolactone networks for tissue engineering. *Biomaterials*, 24(5), 801–808, 2003.
10. Mandal, B.B., Kundu, S.C., Non-bioengineered silk gland fibroin protein: characterization and evaluation of matrices for potential tissue engineering applications. *Biotechnol. Bioeng.*, 100(6), 1237–1250, 2008.
11. Mandal, B.B. and S.C. Kundu, Cell proliferation and migration in silk fibroin 3D scaffolds. *Biomaterials*, 30(15), 2956–2965, 2009.
12. Mandal, B.B., Kundu, S.C., Osteogenic and adipogenic differentiation of rat bone marrow cells on non-mulberry and mulberry silk gland fibroin 3D scaffolds. *Biomaterials*, 30(28), 5019–5030, 2009.
13. Ma, P.X., Biomimetic materials for tissue engineering. *Adv. Drug Deliv. Rev.*, 60(2), 184–198, 2008.
14. Li, J., Shi, R., Fabrication of patterned multi-walled poly-l-lactic acid conduits for nerve regeneration. *J Neurosci. Methods*, 165(2), 257–264, 2007.
15. Lee, C., Grodzinsky, A., Spector, M., The effects of cross-linking of collagen-glycosaminoglycan scaffolds on compressive stiffness, chondrocyte-mediated contraction, proliferation and biosynthesis. *Biomaterials*, 22(23), 3145–3154, 2001.
16. Khor, E., Lim, L.Y., Implantable applications of chitin and chitosan. *Biomaterials*, 24(13), 2339–2349, 2003.
17. Chandy, T., Sharma, C.P., Chitosan-as a biomaterial. *Biomater. Artif. Cells Artif. Org.*, 18(1), 1–24, 1990.
18. Paul, W., Sharma, C., Chitosan, a drug carrier for the 21st century: a review. *STP Pharma. Sci.*, 10(1), 5–22, 2000.
19. Madihally, S.V., Matthew, H.W., Porous chitosan scaffolds for tissue engineering. *Biomaterials*, 20(12), 1133–1142, 1999.
20. Salem, A., Stevens, R., Pearson, R., Interactions of 3T3 fibroblasts and endothelial cells with defined pore features. *J. Biomed. Mater. Res.*, 61(2), 212–217, 2002.
21. Freyman, T., Yannas, I., Gibson, L., Cellular materials as porous scaffolds for tissue engineering. *Prog. Mater. Sci.*, 46(3), 273–282, 2001.
22. Murphy, C.M., Haugh, M.G., O'Brien, F.J., The effect of mean pore size on cell attachment, proliferation and migration in collagen–glycosaminoglycan scaffolds for bone tissue engineering. *Biomaterials*, 31(3), 461–466, 2010.
23. Chapekar, M.S., Tissue engineering: challenges and opportunities. *J. Biomed. Mater. Res* 53(6), 617–620, 2000.
24. Kumar, J.P., Lakshmi, L., Jyothsna, V., Synthesis and characterization of diopside particles and their suitability along with chitosan matrix for bone tissue engineering *in vitro* and *in vivo*. *J. Biomed. Nanotechnol.*, 10(6), 970–981, 2014.
25. Peter, M., Binulal, N., Nair, S., Novel biodegradable chitosan–gelatin/nano-bioactive glass ceramic composite scaffolds for alveolar bone tissue engineering. *Chem. Eng. J.*, 158(2), 353–361, 2010.
26. O'brien, F.J., Biomaterials & scaffolds for tissue engineering. *Mater. Today*, 14(3), 88–95, 2011.
27. Babensee, J.E., Anderson, J.M., McIntire, L.V., Host response to tissue engineered devices. *Adv. Drug Deliv. Rev.*, 33(1), 111–139, 1998.

28. Tomihata, K., Ikada, Y., *In vitro* and *in vivo* degradation of films of chitin and its deacetylated derivatives. *Biomaterials*, 18(7), 567–575, 1997.
29. Li, J., Du, Y., Liang, H., Influence of molecular parameters on the degradation of chitosan by a commercial enzyme. *Polym. Degrad. Stab.*, 92(3), 515–524, 2007.
30. Zhang, Y., Ni, M., Zhang, M., Calcium phosphate-chitosan composite scaffolds for bone tissue engineering. *Tissue Eng.*, 9(2), 337–345, 2003.
31. Sowjanya, J., Singh, J., Mohita, T., Biocomposite scaffolds containing chitosan/alginate/nano-silica for bone tissue engineering. *Colloids Surf. B Biointerf.*, 109, 294–300, 2013.
32. Nolsoe, H., Imer, S., Hultin, H.O., Study of how phase separation by filtration instead of centrifugation affects protein yield and gel quality during an alkaline solubilisation process–different surimi-processing methods. *Int. J. Food Sci. Tech.*, 42(2), 139–147, 2007.
33. Nam, Y.S., Park, T.G., Biodegradable polymeric microcellular foams by modified thermally induced phase separation method. *Biomaterials*, 20(19), 1783–1790, 1999.
34. Sachlos, E., Czernuszka, J., Making tissue engineering scaffold work: review on the application of SFF technology to the production of tissue engineering scaffolds. *Eur. Cell Mater.*, 5(1), 29–40, 2003.
35. Smith, L.A., Beck, J.A., Ma, P.X., Nanofibrous scaffolds and their biological effects. *Nanotechnol. Life Sci.*, 9, 188–215, 2006.
36. Lim, J.I., Im, H., Lee, W.-K., Fabrication of porous chitosan-polyvinyl pyrrolidone scaffolds from a quaternary system via phase separation. *J. Biomater. Sci., Polym. Ed.*, 26(1), 32–41, 2015.
37. Li, W.J., Laurencin, C.T., Caterson, E.J., Electrospun nanofibrous structure: a novel scaffold for tissue engineering. *J. Biomed. Mater. Res.*, 60(4), 613–621, 2002.
38. Deepthi, S., Sundaram, M.N., Kadavan, J.D., Layered chitosan-collagen hydrogel/aligned PLLA nanofiber construct for flexor tendon regeneration. *Carbohydr. Polym.*, 153, 492–500, 2016.
39. Plikk, P., Malberg, S., Albertsson, A.-C., Design of resorbable porous tubular copolyester scaffolds for use in nerve regeneration. *Biomacromolecules*, 10(5), 1259–1264, 2009.
40. Lim, J.I., Lee, Y.-K., Shin, J.-S., Preparation of interconnected porous chitosan scaffolds by sodium acetate particulate leaching. *J. Biomater. Sci., Polym. Ed.*, 22(10), 1319–1329, 2011.
41. Schoof, H., Apel, J., Heschel, I., Control of pore structure and size in freeze-dried collagen sponges. *J. Biomed. Mater. Res.*, 58(4), 352–357, 2001.
42. Boland, E.D., Espy, P.G., Bowlin, G.L., *Tissue engineering scaffolds,* In *Encyclopedia of biomaterials and biomedical engineering*.pp. 1630–1638, 2004.
43. Pattnaik, S., Nethala, S., Tripathi, A., Chitosan scaffolds containing silicon dioxide and zirconia nano particles for bone tissue engineering. *Int. J. Biol. Macromolec.*, 49(5), 1167–1172, 2011.

44. Irmak, G., Demirtaş, T.T., Çetin Altındal, D., Sustained release of 17β-estradiol stimulates osteogenic differentiation of adipose tissue-derived mesenchymal stem cells on chitosan-hydroxyapatite scaffolds. *Cells Tissues Organs*, 199(1), 37–50, 2014.
45. Thomson, R., *et al.*, Biodegradable polymer scaffolds to regenerate organs, in *Biopolymers Ii*. 1995, Springer. p. 245–274.
46. Costa-Pinto, A.R., *et al.*, Adhesion, proliferation, and osteogenic differentiation of a mouse mesenchymal stem cell line (BMC9) seeded on novel melt-based chitosan/polyester 3D porous scaffolds. *Tissue Eng. Part A*, 14(6), 1049–1057, 2008.
47. Lam, C.X.F., *et al.*, Scaffold development using 3D printing with a starch-based polymer. *Mater. Sci. Eng.: C*, 20(1), 49–56, 2002.
48. Sachlos, E., Czernuszka, J., Making tissue engineering scaffolds work. Review: the application of solid freeform fabrication technology to the production of tissue engineering scaffolds. *Eur. Cell Mater.*, 5(29), 39–40, 2003.
49. Melchels, F., *et al.*, CAD/CAM-assisted breast reconstruction. *Biofabrication*, 3(3), 1–8, 2011.
50. Kufelt, O., *et al.*, Water-soluble photopolymerizable chitosan hydrogels for biofabrication via two-photon polymerization. *Acta Biomater.*, 18186–195, 2015.
51. Kim, I.-Y., *et al.*, Chitosan and its derivatives for tissue engineering applications. *Biotechnol. Adv.*, 26(1), 1–21, 2008.
52. Putnam, A.J., Mooney, D.J., Tissue engineering using synthetic extracellular matrices. *Nat. Med.*, 2(7), 824, 1996.
53. Mohamed, K.R., El-Rashidy, Z.M., Salama, A.A., *In vitro* properties of nano-hydroxyapatite/chitosan biocomposites. *Ceram. Int.*, 37(8), 3265–3271, 2011.
54. Xianmiao, C., *et al.*, Properties and *in vitro* biological evaluation of nano-hydroxyapatite/chitosan membranes for bone guided regeneration. *Mater. Sci. Eng.: C*, 29(1), 29–35, 2009.
55. Zo, S.M., *et al.*, Chitosan-hydroxyapatite macroporous matrix for bone tissue engineering. *Curr. Sci. (Bangalore)*, 103(12), 1438–1446, 2012.
56. Weir, M.D., Xu, H.H., Osteoblastic induction on calcium phosphate cement–chitosan constructs for bone tissue engineering. *J. Biomed. Mater. Res.: Part A*, 94(1), 223–233, 2010.
57. Dahlan, K., *et al.*, Synthesis and characterization of calcium phosphate/chitosan composites. *Int. J. Basic Appl. Sci.*, 12(1), 50–57, 2012.
58. Lai, G.-J., *et al.*, Composite chitosan/silk fibroin nanofibers for modulation of osteogenic differentiation and proliferation of human mesenchymal stem cells. *Carbohydr. Polym.*, 111, 288–297, 2014.
59. Lee, J.H., Jeong, B.O., The effect of hyaluronate-carboxymethyl cellulose on bone graft substitute healing in a rat spinal fusion model. *J. Korean Neurosurg. Soc.*, 50(5), 409–414, 2011.
60. Lee, M., *et al.*, Biomimetic apatite-coated alginate/chitosan microparticles as osteogenic protein carriers. *Biomaterials*, 30(30), 6094–6101, 2009.

61. Anitha, A., *et al.*, Chitin and chitosan in selected biomedical applications. *Prog. Polym. Sci.*, 39(9), 1644–1667, 2014.
62. Seol, Y.-J., *et al.*, Chitosan sponges as tissue engineering scaffolds for bone formation. *Biotechnol. Lett.*, 26(13), 1037–1041, 2004.
63. Zhang, Y., Zhang, M., Synthesis and characterization of macroporous chitosan/calcium phosphate composite scaffolds for tissue engineering. *J. Biomed. Mater. Res.*, 55(3), 304–312, 2001.
64. Zhang, Y., Zhang, M., Calcium phosphate/chitosan composite scaffolds for controlled *in vitro* antibiotic drug release. *J. Biomed. Mater. Res.*, 62(3), 378–386, 2002.
65. Sendemir-Urkmez, A., Jamison, R.D., The addition of biphasic calcium phosphate to porous chitosan scaffolds enhances bone tissue development *in vitro. J. Biomed. Mater. Res.: Part A*, 81(3), 624–633, 2007.
66. Zhang, Y., Zhang, M., Three-dimensional macroporous calcium phosphate bioceramics with nested chitosan sponges for load-bearing bone implants. *J. Biomed. Mater. Res.*, 61(1), 1–8, 2002.
67. Thein-Han, W., Misra, R., Biomimetic chitosan–nanohydroxyapatite composite scaffolds for bone tissue engineering. *Acta Biomater.*, 5(4), 1182–1197, 2009.
68. Chesnutt, B.M., *et al.*, Design and characterization of a novel chitosan/nanocrystalline calcium phosphate composite scaffold for bone regeneration. *J. Biomed. Mater. Res.: Part A*, 88(2), 491–502, 2009.
69. Zhao, F., *et al.*, Effects of hydroxyapatite in 3-D chitosan–gelatin polymer network on human mesenchymal stem cell construct development. *Biomaterials*, 27(9), 1859–1867, 2006.
70. Peter, M., *et al.*, Preparation and characterization of chitosan–gelatin/nanohydroxyapatite composite scaffolds for tissue engineering applications. *Carbohydr. Polym.*, 80(3), 687–694, 2010.
71. Hu, Q., *et al.*, Preparation and characterization of biodegradable chitosan/hydroxyapatite nanocomposite rods via *in situ* hybridization: a potential material as internal fixation of bone fracture. *Biomaterials*, 25(5), 779–785, 2004.
72. Chen, J.D., Wang, Y., Chen, X., *In situ* fabrication of nano-hydroxyapatite in a macroporous chitosan scaffold for tissue engineering. *J. Biomater. Sci., Polym. Ed.*, 20(11), 1555–1565, 2009.
73. Ge, Z., *et al.*, Hydroxyapatite–chitin materials as potential tissue engineered bone substitutes. *Biomaterials*, 25(6), 1049–1058, 2004.
74. Kawakami, T., *et al.*, Experimental study on osteoconductive properties of a chitosan-bonded hydroxyapatite self-hardening paste. *Biomaterials*, 13(11), 759–763, 1992.
75. Beskardes, I.G., *et al.*, Microwave-assisted fabrication of chitosan–hydroxyapatite superporous hydrogel composites as bone scaffolds. *J. Tissue Eng. Regen. Med.*, 9(11), 1233–1246, 2015.
76. Thein-Han, W., *et al.*, Chitosan–gelatin scaffolds for tissue engineering: physico-chemical properties and biological response of buffalo embryonic

stem cells and transfectant of GFP–buffalo embryonic stem cells. *Acta Biomater.*, 5(9), 3453–3466, 2009.

77. Costa-Pinto, A., *et al.*, Chitosan–poly (butylene succinate) scaffolds and human bone marrow stromal cells induce bone repair in a mouse calvaria model. *J. Tissue Eng. Regen. Med.*, 6(1), 21–28, 2012.
78. Jayakumar, R., *et al.*, Bioactive and osteoblast cell attachment studies of novel α-and β-chitin membranes for tissue-engineering applications. *Int. J. Biol. Macromolec.*, 45(3), 260–264, 2009.
79. Florczyk, S.J., *et al.*, Evaluation of three-dimensional porous chitosan–alginate scaffolds in rat calvarial defects for bone regeneration applications. *J. Biomed. Mater. Res.: Part A*, 101(10), 2974–2983, 2013.
80. Li, L., *et al.*, *In vitro* bioactivity of bioresorbable porous polymeric scaffolds incorporating hydroxyapatite microspheres. *Acta Biomater.*, 6(7), 2525–2531, 2010.
81. Kavya, K., *et al.*, Fabrication and characterization of chitosan/gelatin/nSiO_2 composite scaffold for bone tissue engineering. *Int. J. Biol. Macromolec.*, 59, 255–263, 2013.
82. Grande, D., *et al.*, Evaluation of matrix scaffolds for tissue engineering of articular cartilage grafts. *J. Biomed. Mater. Res.*, 34(2), 211–220, 1997.
83. Neves, S.C., *et al.*, Chitosan/Poly (ε-caprolactone) blend scaffolds for cartilage repair. *Biomaterials*, 32(4), 1068–1079, 2011.
84. Zhu, Y., *et al.*, Designed composites for mimicking compressive mechanical properties of articular cartilage matrix. *J. Mech. Behav. Biomed. Mater.*, 36, 32–46, 2014.
85. Li, C., *et al.*, A viscoelastic chitosan-modified three-dimensional porous poly (L-lactide-co-ε-caprolactone) scaffold for cartilage tissue engineering. *J. Biomater. Sci., Polym. Ed.*, 23(1), 405–424, 2012.
86. Mirahmadi, F., *et al.*, Enhanced mechanical properties of thermosensitive chitosan hydrogel by silk fibers for cartilage tissue engineering. *Mater. Sci. Eng.: C*, 33(8), 4786–4794, 2013.
87. Yan, L.P., *et al.*, Genipin-cross-linked collagen/chitosan biomimetic scaffolds for articular cartilage tissue engineering applications. *J. Biomed. Mater. Res.: Part A*, 95(2), 465–475, 2010.
88. Kuo, C.-Y., *et al.*, Incorporation of chitosan in biomimetic gelatin/chondroitin-6-sulfate/hyaluronan cryogel for cartilage tissue engineering. *Carbohydr. Polym.*, 117, 722–730, 2015.
89. da Silva, M.A., *et al.*, Chitosan/polyester-based scaffolds for cartilage tissue engineering: assessment of extracellular matrix formation. *Acta Biomater.*, 6(3), 1149–1157, 2010.
90. Kosher, R.A., Church, R.L., Stimulation of *in vitro* somite chondrogenesis by procollagen and collagen. *Nature*, 258, 327–330, 1975.
91. Kosher, R.A., Lash, J.W., Minor, R.R., Environmental enhancement of *in vitro* chondrogenesis: IV. Stimulation of somite chondrogenesis by exogenous chondromucoprotein. *Dev. Biol.*, 35(2), 210–220, 1973.

92. Choi, B., *et al.*, Visible-light-initiated hydrogels preserving cartilage extracellular signaling for inducing chondrogenesis of mesenchymal stem cells. *Acta Biomater.*, 12, 30–41, 2015.
93. Silva, J.M., *et al.*, Nanostructured 3D constructs based on chitosan and chondroitin sulphate multilayers for cartilage tissue engineering. *PloS One*, 8(2), e55451, 2013.
94. Hao, T., *et al.*, The support of matrix accumulation and the promotion of sheep articular cartilage defects repair *in vivo* by chitosan hydrogels. *Osteoarthr. Cartil.*, 18(2), 257–265, 2010.
95. Jin, R., *et al.*, Injectable chitosan-based hydrogels for cartilage tissue engineering. *Biomaterials*, 30(13), 2544–2551, 2009.
96. Park, H., *et al.*, Injectable chitosan hyaluronic acid hydrogels for cartilage tissue engineering. *Acta Biomater.*, 9(1), 4779–4786, 2013.
97. Hasırcı, V., *et al.*, Versatility of biodegradable biopolymers: degradability and an *in vivo* application. *J. Biotechnol.*, 86(2), 135–150, 2001.
98. Marler, J.J., *et al.*, Transplantation of cells in matrices for tissue regeneration. *Adv. Drug Deliv. Rev.*, 33(1), 165–182, 1998.
99. Chen, G., *et al.*, Culturing of skin fibroblasts in a thin PLGA–collagen hybrid mesh. *Biomaterials*, 26(15), 2559–2566, 2005.
100. Ma, L., *et al.*, Collagen/chitosan porous scaffolds with improved biostability for skin tissue engineering. *Biomaterials*, 24(26), 4833–4841, 2003.
101. Metcalfe, A.D., Ferguson, M.W., Tissue engineering of replacement skin: the crossroads of biomaterials, wound healing, embryonic development, stem cells and regeneration. *J. R. Soc. Interf.*, 4(14), 413–437, 2007.
102. Taravel, M., Domard, A., Collagen and its interaction with chitosan: II. Influence of the physicochemical characteristics of collagen. *Biomaterials*, 16(11), 865–871, 1995.
103. Cho, Y.-W., *et al.*, Water-soluble chitin as a wound healing accelerator. *Biomaterials*, 20(22), 2139–2145, 1999.
104. Ma, J., *et al.*, A preliminary *in vitro* study on the fabrication and tissue engineering applications of a novel chitosan bilayer material as a scaffold of human neofetal dermal fibroblasts. *Biomaterials*, 22(4), 331–336, 2001.
105. Agarwal, T., *et al.*, Gelatin/Carboxymethyl chitosan based scaffolds for dermal tissue engineering applications. *Int. J. Biol. Macromolec.*, 9, 358–369, 2016.
106. Anjum, S., *et al.*, Development of antimicrobial and scar preventive chitosan hydrogel wound dressings. *Int. J. Pharm.*, 508(1), 92–101, 2016.
107. Lindahl, U., Hook, M., Glycosaminoglycans and their binding to biological macromolecules. *Ann. Rev. Biochem.*, 47(1), 385–417, 1978.
108. Li, J., *et al.*, Culture of hepatocytes on fructose-modified chitosan scaffolds. *Biomaterials*, 24(13), 2317–2322, 2003.
109. Chupa, J.M., *et al.*, Vascular cell responses to polysaccharide materials: *in vitro* and *in vivo* evaluations. *Biomaterials*, 21(22), 2315–2322, 2000.
110. She, Z., Liu, W., Feng, Q., Self-assembly model, hepatocytes attachment and inflammatory response for silk fibroin/chitosan scaffolds. *Biomed. Mater.*, 4(4), 045014, 2009.

111. She, Z., *et al.*, Silk fibroin/chitosan scaffold: preparation, characterization, and culture with HepG2 cell. *J. Mater. Sci. Mater. Med.*, 19(12), 3545–3553, 2008.
112. Wang, X., *et al.*, Preparation and characterization of a collagen/chitosan/heparin matrix for an implantable bioartificial liver. *J. Biomater. Sci., Polym. Ed.*, 16(9), 1063–1080, 2005.
113. Wang, X., *et al.*, Cross-linked collagen/chitosan matrix for artificial livers. *Biomaterials*, 24(19), 3213–3220, 2003.
114. Weigel, P.H., Rat hepatocytes bind to synthetic galactoside surfaces via a patch of asialoglycoprotein receptors. *J. Cell Biol.*, 87(3), 855–861, 1980.
115. Kobayashi, A., *et al.*, Enhanced adhesion and survival efficiency of liver cells in culture dishes coated with a lactose-carrying styrene homopolymer. *Die Makromolekulare Chemie., Rapid Commun.*, 7(10), 645–650, 1986.
116. Park, I.-K., *et al.*, Galactosylated chitosan as a synthetic extracellular matrix for hepatocytes attachment. *Biomaterials*, 24(13), 2331–2337, 2003.
117. Chung, T.W., *et al.*, *Preparation of alginate/galactosylated chitosan scaffold for hepatocyte attachment.* Biomaterials, 23(14), 2827–2834, 2002.
118. Li, J., *et al.*, Culture of primary rat hepatocytes within porous chitosan scaffolds. *J. Biomed. Mater. Res.: Part A*, 67(3), 938–943, 2003.
119. Wang, F., Guan, J., Cellular cardiomyoplasty and cardiac tissue engineering for myocardial therapy. *Adv. Drug Deliv. Rev.*, 62(7), 784–797, 2010.
120. Iyer, R.K., *et al.*, Engineered cardiac tissues. *Curr. Opin. Biotechnol.*, 22(5), 706–714, 2011.
121. Galvez-Monton, C., *et al.*, Cardiac tissue engineering and the bioartificial heart. *Revista Espanola de Cardiología (English Edition)*, 66(5), 391–399, 2013.
122. Christman, K.L., Lee, R.J., Biomaterials for the treatment of myocardial infarction. *J. Am. Coll. Cardiol.*, 48(5), 907–913, 2006.
123. Kumar, M.N.R., A review of chitin and chitosan applications. *React. Funct. Polym.*, 46(1), 1–27, 2000.
124. Liu, Z., *et al.*, The influence of chitosan hydrogel on stem cell engraftment, survival and homing in the ischemic myocardial microenvironment. *Biomaterials*, 33(11), 3093–3106, 2012.
125. Lu, W.-N., *et al.*, Functional improvement of infarcted heart by co-injection of embryonic stem cells with temperature-responsive chitosan hydrogel. *Tissue Eng.: Part A*, 15(6), 1437–1447, 2008.
126. Li, J., *et al.*, A chitosan–glutathione based injectable hydrogel for suppression of oxidative stress damage in cardiomyocytes. *Biomaterials*, 34(36), 9071–9081, 2013.
127. Shu, Y., *et al.*, RoY peptide-modified chitosan-based hydrogel to improve angiogenesis and cardiac repair under hypoxia. *ACS Appl. Mater. Interf.*, 7(12), 6505–6517, 2015.
128. Ceccaldi, C., *et al.*, Evaluation of polyelectrolyte complex-based scaffolds for mesenchymal stem cell therapy in cardiac ischemia treatment. *Acta Biomater.*, 10(2), 901–911, 2014.

129. Deng, B., *et al.*, Delivery of alginate-chitosan hydrogel promotes endogenous repair and preserves cardiac function in rats with myocardial infarction. *J. Biomed. Mater. Res.: Part A*, 103(3), 907–918, 2015.
130. Pok, S., *et al.*, A multilayered scaffold of a chitosan and gelatin hydrogel supported by a PCL core for cardiac tissue engineering. *Acta Biomater.*, 9(3), 5630–5642, 2013.
131. Chiu, L., Janic, K., Radisic, M., Engineering of oriented myocardium on three-dimensional micropatterned collagen-chitosan hydrogel. *Int. J. Artif. Organs*, 35(4), 237–250, 2012.
132. Pok, S., *et al.*, Use of myocardial matrix in a chitosan-based full-thickness heart patch. *Tissue Eng.: Part A*, 20(13), 1877–1887, 2014.
133. Yang, M.-C., *et al.*, The cardiomyogenic differentiation of rat mesenchymal stem cells on silk fibroin–polysaccharide cardiac patches *in vitro*. *Biomaterials*, 30(22), 3757–3765, 2009.
134. Lin, S.-J., *et al.*, Formation of melanocyte spheroids on the chitosan-coated surface. *Biomaterials*, 26(12), 1413–1422, 2005.
135. Matsuda, A., *et al.*, Immobilization of laminin peptide in molecularly aligned chitosan by covalent bonding. *Biomaterials*, 26(15), 2273–2279, 2005.
136. Yuan, Y., *et al.*, The interaction of Schwann cells with chitosan membranes and fibers *in vitro*. *Biomaterials*, 25(18), 4273–4278, 2004.
137. Haipeng, G., *et al.*, Studies on nerve cell affinity of chitosan-derived materials. *J. Biomed. Mater. Res.*, 52(2), 285–295, 2000.
138. Ashton, R.S., *et al.*, Scaffolds based on degradable alginate hydrogels and poly (lactide-co-glycolide) microspheres for stem cell culture. *Biomaterials*, 28(36), 5518–5525, 2007.
139. Scanga, V.I., *et al.*, Biomaterials for neural-tissue engineering-Chitosan supports the survival, migration, and differentiation of adult-derived neural stem and progenitor cells. *Can. J. Chem.*, 88(3), 277–287, 2010.
140. Itoh, S., *et al.*, Hydroxyapatite-coated tendon chitosan tubes with adsorbed laminin peptides facilitate nerve regeneration *in vivo*. *Brain Res.*, 993(1), 111–123, 2003.
141. Cao, W., *et al.*, Physical, mechanical and degradation properties, and Schwann cell affinity of cross-linked chitosan films. *J. Biomater. Sci., Polym. Ed.*, 16(6), 791–807, 2005.
142. Freier, T., *et al.*, Chitin-based tubes for tissue engineering in the nervous system. *Biomaterials*, 2005. 26(22), 4624–4632.
143. Mingyu, C., *et al.*, Surface modification and characterization of chitosan film blended with poly-L-lysine. *J. Biomater. Appl.*, 19(1), 59–75, 2004.
144. Zhu, C., *et al.*, Initial investigation of novel human-like collagen/chitosan scaffold for vascular tissue engineering. *J. Biomed. Mater. Res.: Part A*, 89(3), 829–840, 2009.
145. Shi, H., *et al.*, Enhanced angiogenesis in porous collagen–chitosan scaffolds loaded with angiogenin. *Tissue Eng.: Part A*, 14(11), 1775–1785, 2008.
146. Barbosa, M. A., A.P.P., Amaral, I.F., *"Chitosan," in comprehensive biomaterials*, Elsevier, Oxford, UK. pp. 221–237, 2011.

147. Fukutomi, M., *et al.*, Changes in Platelet, Granulocyte, and Complement Activation During Cardiopulmonary Bypass Using Heparin-coated Equipment. *Artif. Organs*, 20(7), 767–776, 1996.
148. Belboul, A., Al-Khaja, N., Does heparin coating improve biocompatibility? A study on complement, blood cells and postoperative morbidity during cardiac surgery. *Perfusion*, 12(6), 385–391, 1997.
149. Svenmarker, S., *et al.*, Clinical effects of the heparin coated surface in cardiopulmonary bypass. *Eur. J. Cardiothorac. Surg.*, 11(5), 957–964, 1997.
150. Sheardown, H., Cheng, Y.-L., Mechanisms of corneal epithelial wound healing. *Chem. Eng. Sci.*, 51(19), 4517–4529, 1996.
151. Qu, L., *et al.*, Reconstruction of corneal epithelium with cryopreserved corneal limbal stem cells in a rabbit model. *Vet. J.*, 179(3), 392–400, 2009.
152. Gao, X., *et al.*, Preparation and properties of a chitosan-based carrier of corneal endothelial cells. *J. Mater. Sci. Mater. Med.*, 19(12), 3611–3619, 2008.
153. Lloyd, A.W., Faragher, R.G., Denyer, S.P., Ocular biomaterials and implants. *Biomaterials*, 22(8), 769–785, 2001.
154. Hadlock, T., *et al.*, Ocular cell monolayers cultured on biodegradable substrates. *Tissue Eng.*, 5(3), 187–196, 1999.
155. Wang, Y., *et al.*, A study on the performance of hyaluronic acid immobilized chitosan film. *Biomed Mater.*, 4(3), 035009, 2009.
156. Liang, Y., *et al.*, Tissue-engineered membrane based on chitosan for repair of mechanically damaged corneal epithelium. *J. Mater. Sci.: Mater. Med.*, 25(9), 2163–2171, 2014.
157. Chen, J., *et al.*, Study on biocompatibility of complexes of collagen–chitosan–sodium hyaluronate and cornea. *Artif. Organs*, 29(2), 104–113, 2005.
158. Peng, L., *et al.*, Preparation and evaluation of porous chitosan/collagen scaffolds for periodontal tissue engineering. *J. Bioact. Compat. Polym.*, 21(3), 207–220, 2006.
159. Zhang, Y., *et al.*, Combination of scaffold and adenovirus vectors expressing bone morphogenetic protein-7 for alveolar bone regeneration at dental implant defects. *Biomaterials*, 28(31), 4635–4642, 2007.
160. Shao, X., Hunter, C.J., Developing an alginate/chitosan hybrid fiber scaffold for annulus fibrosus cells. *J. Biomed. Mater. Res.: Part A*, 82(3), 701–710, 2007.
161. Funakoshi, T., *et al.*, Application of tissue engineering techniques for rotator cuff regeneration using a chitosan-based hyaluronan hybrid fiber scaffold. *Am. J. Sports Med.*, 33(8), 1193–1201, 2005.
162. Hayami, J.W., *et al.*, Design and characterization of a biodegradable composite scaffold for ligament tissue engineering. *J. Biomed. Mater. Res.: Part A*, 92(4), 1407–1420, 2010.
163. de la Fuente, M., *et al.*, Chitosan-based nanostructures: a delivery platform for ocular therapeutics. *Adv. Drug Deliv. Rev.*, 62(1), 100–117, 2010.
164. Saber, A., Strand, S.P., Ulfendahl, m., Use of the biodegradable polymer chitosan as a vehicle for applying drugs to the inner ear. *Eur. J. Pharm. Sci.*, 39(1), 110–115, 2010.

165. Giannoni, P., Hunziker, E.B., Release kinetics of transforming growth factor-β1 from fibrin clots. *Biotechnol. Bioeng.*, 83(1), 121–123, 2003.
166. Chua, C.C., *et al.*, Heparan sulfate proteoglycans function as receptors for fibroblast growth factor-2 activation of extracellular signal–regulated kinases 1 and 2. *Circ. Res.*, 94(3), 316–323, 2004.
167. Robson, M.C., Mustoe, T.A., Hunt, T.K., The future of recombinant growth factors in wound healing. *Am. J. Surg.*, 176(2), 80S–82S, 1998.
168. Denkbas, E., Seyyal, M., Piskin, E., Implantable 5-fluorouracil loaded chitosan scaffolds prepared by wet spinning. *J. Membr. Sci.*, 172(1), 33–38, 2000.
169. Duarte, A.R.C., Mano, J.F., Reis, R.L., Preparation of chitosan scaffolds loaded with dexamethasone for tissue engineering applications using supercritical fluid technology. *Eur. Polym. J.*, 45(1), 141–148, 2009.
170. Lin, H.-Y., Yeh, C.-T., Controlled release of pentoxifylline from porous chitosan-pectin scaffolds. *Drug Deliv.*, 17(5), 313–321, 2010.
171. Shi, P.J., *et al.* Fabrication of n-HA/CS porous scaffold carrying ampicillin-loaded microspheres. In *Mater. Sci. Trans. Tech. Publ.*, 544, 941–944, 2007.
172. Noel, S.P., *et al.*, Chitosan sponges to locally deliver amikacin and vancomycin: a pilot *in vitro* evaluation. *Clin. Orthop. Relat. Res.*, 468(8), 2074–2080, 2010.
173. Prabaharan, M., Jayakumar, R., Chitosan-graft-β-cyclodextrin scaffolds with controlled drug release capability for tissue engineering applications. *Int. J. Biol. Macromolec.*, 44(4), 320–325, 2009.
174. De la Riva, B., *et al.*, Local controlled release of VEGF and PDGF from a combined brushite–chitosan system enhances bone regeneration. *J. Contr. Rel.*, 143(1), 45–52, 2010.
175. Malafaya, P.B., Oliveira, J.T., Reis, R.L., The Effect of Insulin-Loaded Chitosan Particle–Aggregated Scaffolds in Chondrogenic Differentiation. *Tissue Eng.: Part A*, 16(2), 735-747, 2009.
176. Tıglı, R.S., *et al.*, *In vitro* release of dexamethasone or bFGF from chitosan/hydroxyapatite scaffolds. *J. Biomater. Sci., Polym. Ed.*, 20(13), 1899–1914, 2009.
177. Iliescu, M., *et al.*, Ultrastructure of hybrid chitosan–glycerol phosphate blood clots by environmental scanning electron microscopy. *Microsc. Res. Tech.*, 71(3), 236–247, 2008.
178. Buschmann, M.D., *et al.*, Cartilage repair with chitosan-glycerol phosphate-stabilized blood clots, in *Cartilage repair strategies*, Springer, pp. 85–104, 2007.
179. Hirsch, J.A., *et al.*, Non-invasive hemostatic closure devices:"patches and pads". *Tech. Vasc. Interv. Radiol.*, 6(2), 92–95, 2003.
180. Muzzarelli, R.A., Chitins and chitosans for the repair of wounded skin, nerve, cartilage and bone. *Carbohydr. Polym.*, 76(2), 167–182, 2009.
181. Brown, M.A., M.R. Daya, and J.A. Worley, Experience with chitosan dressings in a civilian EMS system. *J. Emerg. Med.*, 37(1), 1–7, 2009.

12

Nanopolymer Chitosan in Cancer and Alzheimer Biomedical Application

Athika Darumas Putri, Bayu Tri Murti, Myalowenkosi Sabela*, Suvardhan Kanchi* and Krishna Bisetty

Department of Chemistry, Durban University of Technology, Durban, South Africa

Abstract

This work focuses on the biomedical applications of chitosan, the most used natural polymer developed due to its biodegradable properties and structural benefits. Chitosan is a cationic polymer containing hydroxyl/amine-groups on its entire structure which makes it favorable to be functionalized. The use of chitosan nanopolymer in cancer and Alzheimer's disease (AD) fields has achieved several advancements in early screening stage of cancer, sophisticated enhancement of diagnosis, increasing the rate of drug delivery and overcoming the therapeutic and diagnostic problem of AD. Similarly, its excellence in biocompatibility, biodegradability, and the unique ability to cross blood brain barrier (BBB), chitosan biopolymer also opens up the use of chitosan for biomedical applications. In this chapter, we discuss the chemistry of chitosan and its advanced applications on theranostics areas of cancer and Alzheimer's.

Keywords: Chitosan, chitosan nanopolymer, cancer, Alzheimer's disease, theranostics

12.1 Introduction

Cancer is the most devastating disease in the world with no effective cure due to the challenging issues of its management from diagnosis to treatment therapy. However, exorbitant costs and inadequate results of the recent diagnostic techniques provoke further investigations in cancer management [1]. While there is progress, cancer treatment based on radio/chemotherapy

**Corresponding authors*: myalosabela@gmail.com; ksuvardhan@gmail.com

Shakeel Ahmed and Saiqa Ikram (eds.) Chitosan, (311–360) © 2017 Scrivener Publishing LLC

still has drawbacks with selectivity as it is commonly not able to distinguish the sick-cells among the normal cells. This brings unexpected adverse side effects that cannot be easily cured instead of a pleasing with "excess" drug therapy [2]. In addition, drug therapy for cancer is mostly deprived of sensitivity to target cells due to the depletion of bioavailability and low half-life time, therefore response to achieve higher dose administration and consequently toxicity issues persist [3]. Alzheimer's corresponds to a conformational disorder which is characterized by various pathological markers including a large amount of amyloid-β (Aβ) plaques, vascular damage from wide-ranging plaques deposition, and neuronal cell loss. They remain incurable, irreversible, and a fatal disease. The lack of reliable results on the Alzheimer's etiology is a mysterious challenge to be unraveled.

Nanotechnology has contributed many benefits to many important research fields due to its varied applications and controllable properties [4, 5]. It also blends its functionality with various materials, including polymers. Today, nanopolymers offer great advantages as biofunctional material in the field of biomedical applications. In this particular area, cancer and Alzheimer's management emerging nanopolymer have opened up a reliable avenue in recent decades. Most importantly, the often used polymer type is biodegradable in nature which qualifies them for the wide range of applications pertaining to the cancer and Alzheimer's treatment [6, 7]. This work focuses on the biomedical applications of chitosan, the most used natural polymer developed due to its biodegradable properties and structural benefits. Chitosan is a cationic polymer containing hydroxyl/amine-groups on its entire structure, which makes it favorable to be functionalized [8, 9]. The use of chitosan nanopolymer in cancer and Alzheimer's disease (AD) fields has achieved several advancements in early screening stage of cancer, sophisticated enhancement of diagnosis, increasing the rate of drug delivery and overcoming the therapeutic and diagnostic problem of AD. Similarly, its excellence in biocompatibility, biodegradability, and the unique ability to cross blood brain barrier (BBB), chitosan biopolymer also opens up the use of chitosan for biomedical applications. Thus in this chapter, we discuss thoroughly the chemistry of chitosan and its advanced applications on theranostics areas of cancer and Alzheimer's as the most researched neurodegenerative disease.

12.1.1 Chemistry of Chitosan

Chitosan is not naturally available but, hierarchically, it comes from deacetylation process of chitin, such a cellulose extract group, which is abundantly a part of organic composite in insects which belongs to the Crustaceans family. The deacetylation of chitins occurs through strong basic solution

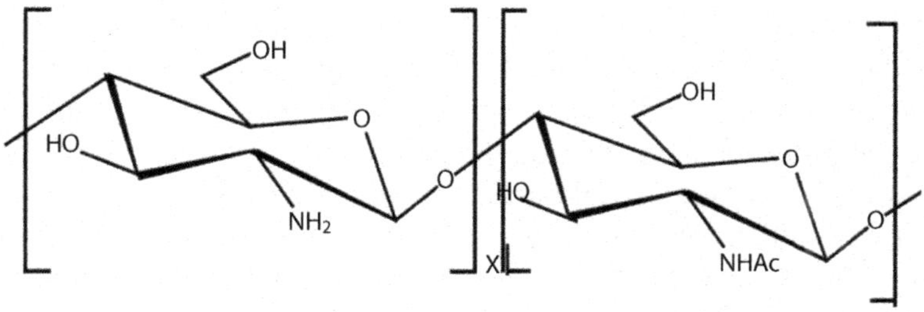

Figure 12.1 Basic chemical structure of chitosan (adopted from [12]).

and thus has been characterized by the degree of *N*-deacetylation process. Chitosan is structurally distinguishable from mutual cellulose since it has linear monosaccharides linked to 2-amino-2-deoxy-β-D-glucan. The less composition of *N*-acetyl-D-glucosamine in β-D-(1→4) glycosides differs chitosan from the chitin, which has degree of acetylation below 50 % of chitin [10, 11]. In Figure 12.1, it is depicted that chitosan structure has partially positive charges and the known chemical properties includes linear polyamine, reactive amino groups, reactive hydroxyl groups, ability to chelate with many transitional metal ions [12].

Basically, chitosan corresponds to pseudonatural cationic polymer containing aminopolysaccharides which are derived from continuous and progressive deacetylation of chitin [13, 14]. Initially, the acetylated degree is a labeled structural parameter to confer the nature of physicochemical properties and biocompatibility among the elemental structures.

Solubility property is one of the foremost issues of biomaterial applications in biological functions and chitosan has poor solubility in aqueous media or in organic solvents. Nevertheless, several attempts have been conducted in order to modify the solubility properties of chitosan. According to Riva *et al.*, the most favorable way was chemical modification of the chain via either alteration of functional groups with the grafted polymer chain or small molecules onto the chitosan backbone or by quaternization of the amino group [15]. Indeed, primary amine and hydroxyl groups of chitosan are two favorable sites to modify chitosan to enhance the solubility property [15].

Today, demand for chitosan is significantly of interest in therapeutic and diagnostic (theranostic) applications due to its unique chemical structures, multidimensional properties, and greatly sophisticated functions [16]. The key consideration is the biocompatibility of chitosan in human body and does not provoke any known adverse effect in contact with human cells [14]. Due to its sturdy biocompatible element and renewable resources, chitosan has been interesting to be designed in various experimental approaches for years which brought about tremendous benefits in many vital fields [17–20], particularly in pharmaceutical drug composing [21–23], cancer therapy and treatment [24, 25], and medical diagnostic [26–29]. The use of chitosan in medical and pharmaceutical issues has been growing up rapidly in the past decades. However, there is a special attention of chitosan applications in the neurodegenerative diseases due to their great distinctive capabilities in crossing two physiological barriers, that is, BBB and blood-cerebrospinal barrier [30]. Chitosan's which were notable in various pharmaceutical applications involves *N*-carboxymethyl, hydrophobic, methoxyphenyl, tyrosine glucan, highly cationic, polyurethane-type and hydroxyalkyl [14]. Traditionally, chitosan is produced from naturally occurring sources, which are the crustacean shells, exoskeleton of insects, fungal mycelia [31], and vegetables [32]. However, the animal source displays some limitations as seasonal supplies and the variability of physicochemical properties [33]. It also reflects non-eco-friendly basis due to the abundant limit of the animal sources. The mushroom source adds many advantages in terms of reproducibility of resulting chitosan [34]. From the allergenic point of view, vegetables-based chitosan is generally preferred, especially in applications at biomedical and healthcare issue [15]. Recently, several breakthroughs have been notably made by industries developing renewable, non-genetically modified organism (GMO), and nonsynthetic manipulation technique based on enzymatic reaction to replace traditional chitosan production.

12.1.2 Brief Overview on Cancer

As a progressive disease, cancer starts from repetitive exposure to the potential chemical compounds, which alters the normal pathway of cell metabolism. Moreover, genetic changes in different species characterize the different physiological changes of cancer which are difficult to detect. The mutative evidence and rapid proliferation give impact to tedious and unstoppable cancer spreading [35]. Thus nowadays, the destination in cancer drugs therapy is to breakthrough early prognosis and therapeutic surveillance [36]. However, cancer diagnostic and general cancer therapies including drugs (anticancer agent) administration and treatments are still facing a lack of effectiveness due to unspecific site targeting, unreached cellular uptake, severe plasma solubility, cytotoxicity issue of the unaffected cells, and short half-life time [37, 38].

Numerous biomedical researches on cancer are being attempted to overcome the severe diagnosis and deficient drug dose delivery of cancer therapy because it is one of the most dangerous diseases in the world [39–41]. Several efforts are formulated towards tackling these challenges, initiated from the stage of detection until drug therapy strategies. Recently, nanopolymerization has paved a new improvement in the potential therapeutic strategy of molecular targeting and drugs delivery, in which, the use of it has successfully minimized the shortcomings in recognizing the target cancer cells among the normal cells [42]. Several studies showed that the nanodelivery strategies are favorable in drug encapsulation in order to increase the solubility in plasma and drug surface intact, to avoid the gastrointestinal disruption, and to transport the drug to be amassed in the target cells only [43, 44]. Other significant studies suggested that the nanoparticulated chitosan can be one of those strategies due to its positively charged structure enabling it to be easily accumulated in the target membrane cells [45, 46]. This is evidence in recent reports on the use of chitosan in cancer-related studies shown in Figure 12.2.

The miraculous biocompatibility, biodegradability, nontoxic, mucoadhesive, and nonimmunogenic properties make chitosan as the preferable escorting agent in drug delivery system than other delivery vehicles. It is molecularly designable, easy to formulate, and inexpensive to mass production. In cancer cell therapy, nanochitosan is mostly used in cancer gene therapy via intravenous administration by tailoring the small interfering-RNA to get more efficacies against the cancer gene [47]. In cancer diagnostic, the use of conductive nanopolymer chitosan via synergistic immobilization has also been considered due to its potency in increasing surface contact area of electrode, thus transducing more electricity and

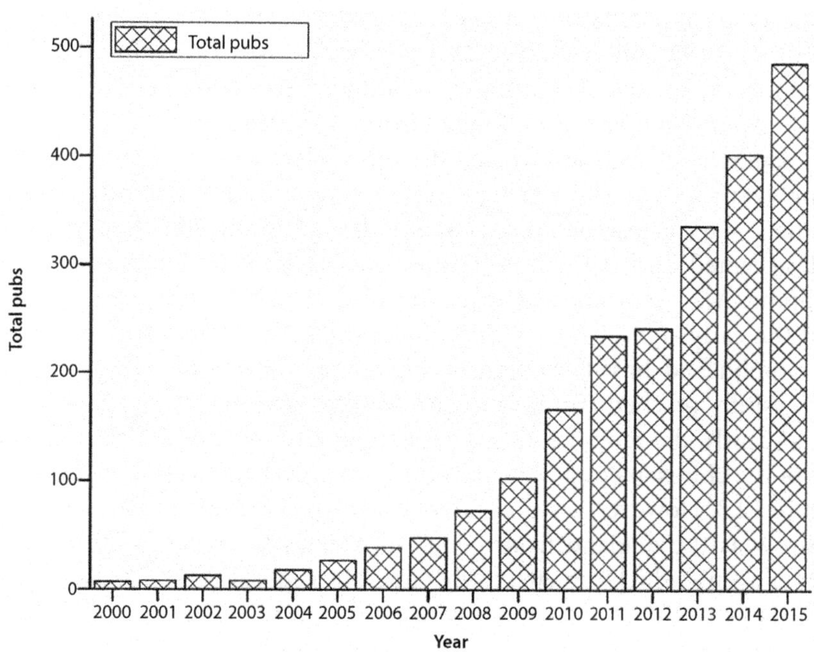

Figure 12.2 Trends in recent reports on the use of chitosan in cancer-related studies (Pubs: Publications). (Information obtained from ISI Web of Knowledge. Accessed on 12/08/2016.)

enhancing the sensor selectivity [48–50]. More futuristic strategies have also been established to complete a major piece of cancer therapeutic puzzle. Each will be comprehended in different sections to discover exhausted chitosan in its collaboration along with anticancer agents and biomarkers in prospective cancer diagnostic and therapy.

12.1.3 Brief Overview on Alzheimer's Disease

Alzheimer's Disease was firstly introduced by Dr. Aloysius (Alois) Alzheimer, a German psychiatrist and neuropathologist in the year 1906 at 37th Conference of South-West German Psychiatrists in Tübingen [51, 52]. Alzheimer's is a brain-related disease with ultimately devastating, irreversible and incurable disorder and can slowly cause serious brain problems which involves decrease of memory, thinking capability, and neuronal behavior. Many people prostheses which AD is apart as normal aging but that is not true [53]. AD is a disorder that can be prevented, even though the remedy to block the initiation and progression is yet to be discovered [54]. As the most probable prevalence of neurodegenerative disease, AD has spread out around the world extensively. In 2015, 46.8 million people are known to be suffering from dementia, whereas in 2030 and 2050 these statistical figures can reach to 74.7 million and 131.5 million, respectively [55]. In the United States of America, prevalence was estimated at over 5.3 million in 2015, and as many as 13.8 million (an increase of almost 3-fold) are projected to have AD by 2050, excluding the development of curative breakthroughs either prevention or curing the disease [56].

Scientist and clinicians argued the most effective way of curing AD is to prevent it. Researchers believe that early detection or diagnosis should be the key to preventing, reducing and discontinuing AD. This has been well-accepted during the last decades since there is no effective cure for stopping AD. As a consequence, the pivotal part to prevent AD is to find the early sign of AD or early diagnostics. Tremendous reports have been published based on specific biomarkers and some of them have successfully reached clinical test, while some are already in the market. Significantly, the remarkable progress along year by year and the percentage of the chitosan employment in the biomedical application is shown in Figure 12.3. However, research in treatment and therapy of AD continues to grow as the demand from people continues to escalate. The need of proper agents which can effectively target the cause of AD such as amyloid β (Aβ) plaques, blocking cholinesterase enzyme, reducing tau protein, etc., has also been raised as the extensive increment of Alzheimer's prevalence number in the recent years.

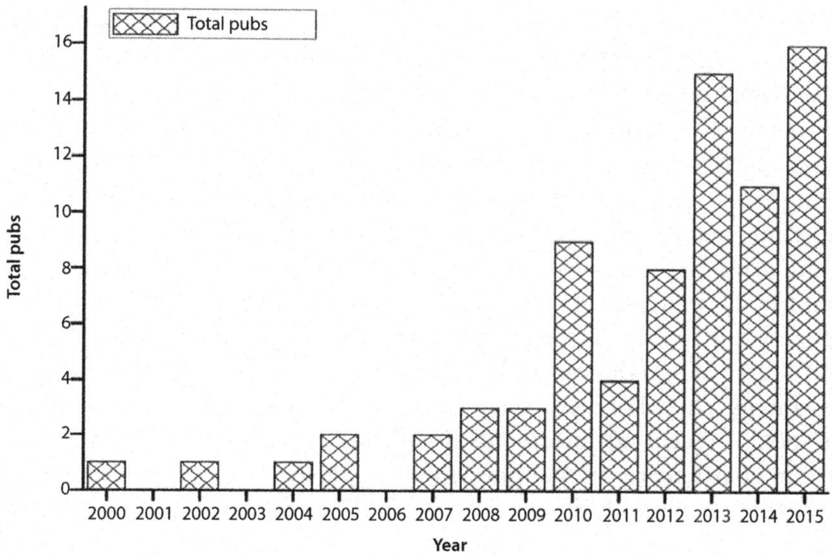

Figure 12.3 Trends in recent reports on the use of chitosan in Alzheimer's studies (Pubs: Publications). (Information obtained from ISI Web of Knowledge. Accessed on 12/08/2016.)

During the last decade, applications of chitosan in Alzheimer's were mostly highlighted to the drug delivery system and pharmaceutics area. Tremendous researches have demonstrated the use of chitosan as potential drug nanocarriers and pharmaceutical excipients. Interestingly, FDA has approved chitosan as one of the pharmaceutical excipient and listed in *Handbook of Pharmaceutical Excipient* [57] and *United State Pharmacopeia and National Formulary* [58]. In addition to aforementioned applications, herein, we also emphasized other remarkable applications of chitosan in Alzheimer's involved treatment, therapy, sensing and diagnosis.

12.2 Chitosan Applications in Cancer

12.2.1 Drug Delivery System

Increasing bioavailability of the anticancer agent mainly reflects to the bioadhesive properties in the drug composed materials. Bioadhesive characteristics are important since they affect the drug molecules contact either in mucous membrane or epithelial walls of tumor cells. Chitosan is a biodegradable glucose-polymer with a high degree of acetylation and has abundance of amino groups, making it preferable to be a bioadhesive component in drug formulation [21, 59–62]. Chitosan's biodegradable properties refer to its positively charged composition which is very suitable to enhance mucoadhesive contact and even safe administration in drug delivery [63–65]. Most chitosan used for drug delivery is a base type as the D-glucosamine residues have *pKa* of 6.2–7.0 that cannot be dissolved in high and neutral pH solutions, but is highly soluble and forms salts in inorganic and organic solvents, such as glutamic acid, hydrochloric acid, acetic acid. The solubility mainly depends on the degree of deacetylation [66, 67]. Nowadays, nanoparticle technology is known to benefit in drug delivery due to its particle size-efficiency and considerable affectivity [68]. The degree of acetylation of chitosan can turn the chitosan into the nano-size range, instead of controlling its molecular weight [11]. In this section, we look at recent advancements of chitosan advantages in drug delivery system, particularly as nanopolymer. In Table 12.1, a summary of the utmost used methods of emerging the nanoparticulated-chitosan in drug delivery system, which is generally familiar with chemically cross-linking, diffusion as emulsifying solvent, ionotropic gelation, and complex coacervation was reported [67].

Vecchione *et al.* reported an advanced formulation of oil in water (O/W) nanoemulsion for delivering curcumin into the target colon cancer cells by increasing its bioavailability in an *in vivo* approach [69]. Since the chitosan and thiol combination provided a highly encapsulated degree to the curcumin, investigations of size ratio with codelivery of piperine, nanoemulsion size, and degree of thiolation were studied. The results obtained from anti-inflammatory studies revealed that this formulation exhibited the cell toxicity of target colon cancer cells successfully without any impact to the normal cells [69]. Another use of micelle–chitosan has been conducted by Nguyen *et al.*, for the detection of SK-BR3 protein markers over-expressed HER-2 cell [70]. The micelle formation enabled the encapsulation of Paclitaxel more efficiently (83.28 ± 0.13%) with loading capacity of 9.12 ± 0.34%. The drug release also increased up to 85–93% in pH 7.5 after 12 h, while the IC_{50} was very efficient for highly loaded nanoparticles (NPs) with apoptosis of 95% in 6 h [70].

Encapsulated doxorubicin-integrated nanoscale metal organic framework nanoparticles (IRMOF-3) have been developed by Chowdhuri and colleagues for optical cancer imaging and drug delivery strategies [71]. *O*-carboxymethyl chitosan (OCMC) was merged into super paramagnetic iron oxide (Fe_3O_4) NPs and acted as pH responsive component in order to increase drug release and induce cell apoptosis. This technique has benefitted Fe_3O_4 NPs to act as the T2 weight-MRI (magnetic resonance imaging) contrast agent. Folic acid was loaded to IRMOF-3 and assembled on the

Table 12.1 Applications of chitosan for cancer drug delivery system.

No.	Chitosan Name (Derivatives)	Nanocomposites/ Biocomposites	Carried Molecule/ Target	Remarkable Functions	References
1	CS	Curcumin/CS-thiol nanoemulsion	Curcumin/colon cancer cells	Chitosan acted as emulsion agent providing stable dispersion of chitosan site to bind the curcumin.	[69]
2	OCMC	Fe_3O_4@OCMC@ IRMOF-3/ FA-DOX	DOX/folate HeLa cells	OCMC was pH responsive agent to increase drug release and induce apoptosis.	[71]
3	CS	MYR-CS micelles	MYR/glioblastoma cancer	Chitosan increased the positive charges density on the nanoparticles surface at biological pH which affects the drug uptake, dispersibility, and biodegradable properties.	[72]
4	CS	Vitamin E-catechol/ HRS-AgNPs/Cs	Vitamin E and catechol/MCF-7 cells	Chitosan conjugated to Ag-petal extract nanoparticles (HRS-AgNPs) enhanced drug encapsulation up to 76% thus improving antioxidant activity as compared to single element either of Vitamin E or catechol.	[73]
5	CS	CD33 protein on HL-60 (Acute myeloid leukemia cells)	MTX/FA/CMC NPs	Joint formulation of CD33 targeting and improved drug released strategy in pH dependence was enhanced by employing MTX and antibody on FA modified CMC.	[74]

(Continued)

Table 12.1 Cont.

No.	Chitosan name (derivatives)	Nanocomposites/ biocomposites	Carried molecule/ target	Remarkable functions	Reference
6	CS PTX/CS/ GMO	PTX/CS/GMO	PTX/ human breast cancer cells (MDA-MB-231)	Conjugation GMO and chitosan could effectively enhance sustain released properties of Paclitaxel due to positively charges increase and inward hydrophobic-outward hydrophilic of the nanoparticles complex.	[75]
7	CS	AuNRs/ mesoporous silica/V7-CMC capped-Gemcitabine	S2VP10 and MiaPaca2 cells	The multi-complex of chitosan capped gold nanorods-mesoporous silica along with Gemcitabine improved signaling specificity and cell toxicity towards the targeted cancer cells up to 97% and 96.5% of S2VP10 and MiaPaca2 cells, respectively.	[76]
8	CS	DOX/glycol chitosan nanobubbles	Anaplastic thyroid cancer	Glycol chitosan in nanobubbles formula harnessed in loaded-doxorubicin delivery by exhibiting anti-tumor activity and drug release in intracellular *in vivo*.	[78]
9	CS	PRV/CS NPs	PRV/HepG2 cells	Chitosan increased the positive charges density on the nanoparticles surface at biological pH which affects the drug uptake, dispersibility, and biodegradable properties.	[79]
10	CS	Curcumin/ALG-CS-Pluronic NPs	Curcumin/HeLa cells	More efficient drug delivery strategy was shown by increasing *in vitro* drug release profile and kinetic mechanism.	[80]

11	CS	5-fluorouracil@ CD-CS hydrogel	5-fluorouracil/ A549 cells	The complex of the hydrogel increased cancer cells detection capability, imaging visualization, and cellular uptake to initiate apoptosis in A549 cells.	[81]
12	CS	Tc-99m/MTX/CS NPs	MTX/MCF-7 cells	Radiolabeled (Technetium 99m) MTX loaded CS NPs increased the efficiency of encapsulation up to 64% and exhibit the drug uptake in target cells.	[82]
13	CS	RGD-HGC NPs	cRGDyk/B16F10 melanoma tumors	System of acidc RGDyk-loaded glycol chitosan modified with 5beta-cholanic (HGC) exhibited the encapsulation of cRGDyk peptide more than 85% and the drug sustained release within 1 week.	[83]

**Abbreviation*: CS = chitosan; CMC = carboxylmethyl chitosan; NPs = nanoparticles; OCMC = O-carboxymethyl chitosan; AgNPs = silver nanoparticles; AuNRs = gold nanorods; FA = folic acid; DOX = Doxorubicin; MYR = Myricetin; MTX = Methotrexate; PRV = Pravastatin; GMO = glyceryl monooleate; PTX = Paclitaxel; Tc-99m = Technetium 99m; RGD = Arg-Gly-Asp.

surface of OCMC functionalized Fe_3O_4NPs followed by conjugating the doxorubicin in nanoparticle complex. In order to induce fluorescent signal, carbon-dots were also employed at the outermost side of the complex. This targeted folate receptor increased more drug molecules internalization to the specific hepatocellular cancer (HeLa cells) as opposed to the normal cells [71].

In another report by Wang *et al.*, the improvement of drug delivery was effectively achieved by loading Myricetin (MYR) into chitosan-conjugated Pluoronic P 123/F68 micelles [72]. In addition, several approaches have been effectively deducted in order to tackle the BBB, acute toxicity, and bio-distribution. *In vitro* and *in vivo* studies revealed that MYR-micelles exhibited the cellular uptake and cytotoxicity on the target cells as compared to the performance of bare MYR. The results showed an increase of therapeutic index for glioblastoma cancer therapy [72]. A nanoformulation of loaded chitosan conjugated silver nanoparticles (AgNPs) from petal extract has been developed by Nayak *et al.* to deliver vitamin E and catechol to MCF-7 breast cancer cells [73]. The antioxidant potential, threatening metabolites, and antioxidant reductant assays were assessed to understand the biocompatibility and anticancer efficacy of the formulation strategy. The results showed increased encapsulation up to 76% with higher antioxidant activity than using each substance [73].

Methotrexate (MTX) has been effectively modified using folic acid and carboxymethyl chitosan (CMC) in NPs system via self-assembling [74]. Folic acid conjugated CMC NPs acted as hydrophobic region containing pteroyl group to surface the loaded drugs, while the amphiphilic group was self-assembled in the aqueous media. The specific antibody was emerged on the complex to recognize the target CD33 protein on HL-60 leukemia cells. The CD33 – nanoparticles binding rates and cells detection were found to be higher than bare MTX to CD33. In this study, it was observed that the drug release rate increased in certain acidic condition due to the presence of folic acid. This particular finding has enhanced the pH sensitive drug distribution to target cells and increased the drug release [74]. Another report by Tricker *et al.* on sustained release of Paclitaxel improved by chitosan and glyceryl monooleate (GMO) NPs has been established. This combination addressed the shortage of more positive charges into the complex by taking advantage of the properties of inward hydrophobic-outward hydrophilic nature of the NPs complex [75]. The use of CMC as an enhancer in targeting nanomedicine of Gemcitabine has been observed by Zeiderman *et al.*, to target the pancreatic tumors [76]. The drug contains gold nanorods and mesoporous silica covering variant-7 pHL1P peptide assembled to CMC to escalate the cytotoxicity of pancreatic tumors.

Employing multispectral optoacoustic tomography, this approach has gained better output up to 96.5% of cell death as compared to the single Gemcitabine [76].

Quantum dots "turn on-off" assembly model shown in Figure 12.4, has also been developed by combining multi-components in order to achieve a moderate and highly imaging signal in ovarian cancer (OVCAR3) therapy-diagnostic [77].

Activation of quantum dots complex, Mn^{2+} embedded CdS and ZnS quantum dots (CdS/ZnS QDs), was designed by releasing dopamine as the fluorescence quencher through electronic transfer of disulfide bond reduction. While in "turn off" mode, QDs were in the condition of complete complex form with neither any particular drug release nor recognition. Fundamentally, to obtain an optimal performance from any degradation or agglomeration, QDs were hydrothermally cross-linked with chitosan in nanocomposite gel format thus generating nontoxic and multifunctional probe. Folic acid and polyethylene-glycol were assembled onto the

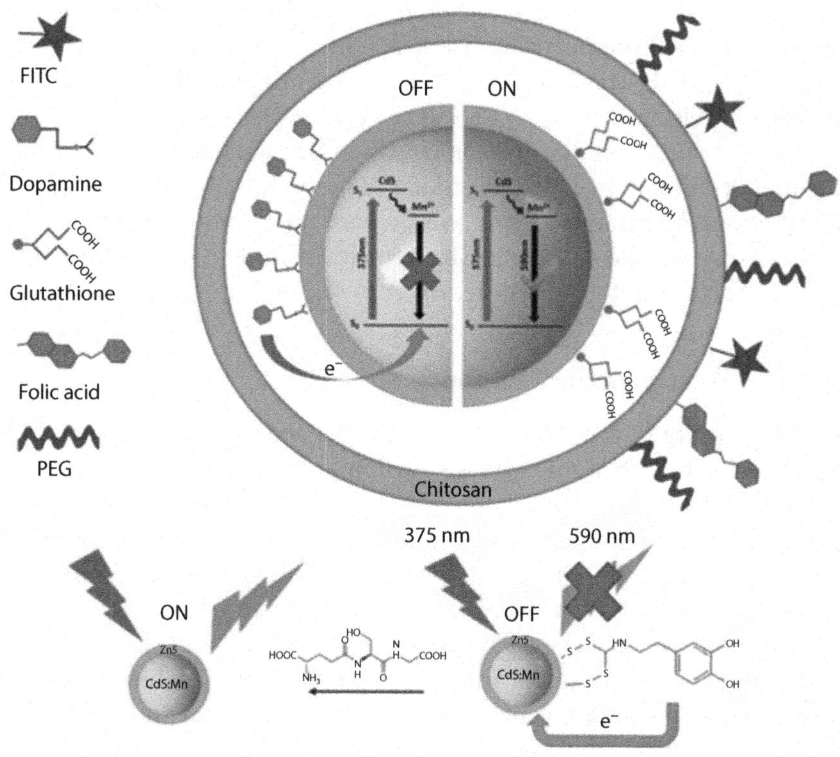

Figure 12.4 The schematic model of "turn on-off" complex of chitosan functionalized QDs as noncytotoxic platform for cancer therapy-diagnostic (reproduced from [77]).

chitosan complex as biorecognition and dispersing agent, respectively. In the final step, fluorescein isothiocyanate (FITC) dye was enclosed to comprehend the visual reading of QDs complex. The results determined that the QDs signal was concentrated in targeted cells (OVCAR3) as compared to the both control macrophages and cells assessed [77].

A combination strategy to exhibit antitumor activities has been established by loading Doxorubicin into glycol chitosan nanobubbles [78]. The targeted cells of anaplastic thyroid cancer was treated by glycol chitosan nanobubbles with shocked wave conjugated to fluorescent label to envision the antitumor activity in target cell cancer, instead of increasing the formulation stability. A distinctive data obtained was more efficient as compared to the free Doxorubicin alone, that by composing low dosage of Doxorubicin, the highest cytotoxicity effect was achieved with reduction of 40% [78]. In another work, a novel strategy to treat HepG2 hepatocellular cancer has been reported by Badran and colleagues. Assembling the Pravastatin (PRV) anticancer drug into chitosan NPs enhanced the efficiency of the drug loading entrapment up to the range between 49.05% and 72.04%. This drug design increased the PRV release from 52% to 92% after 6 h and 48 h, respectively. These results were also comparable to the free drug loaded, where the modified loaded drug enabled inhibition of the HepG2 cells growth up to 51% as opposed to 38% of free PRV after 72 h [79]. From these results, one can conclude that the use of biodegradable polymer of chitosan in drug suspending can substantially enhance the drug distribution on the target cancer cells. In particular, minimization of chitosan NPs size into the nanorange size benefits in enhanced efficiency of the delivered drug into the target site, either by *in vitro* or *in vivo* approaches.

12.2.2 Treatment and Therapy

The bullet point in cancer therapy is to detect and kill the targeted cancer cells selectively with less or without any impact on normal cells. This effort particularly related to the quality of human life patients after having the drug therapy or treatment. In turn, any dosage of therapy administered within convinced period of time probably could induce consequence of uncomfortable patient or organ failure and leads to the therapy deprivation [25, 84]. Due to this, any drug therapy or treatment should accomplish the primary purposes including: good biodegradable composing elements, biocompatibility, nontoxic, nonimmunogenic, controllable drug release, low-priced, and safe in preparation. Chitosan is such a selected polysaccharide that fulfills those criteria as compared to single element, such as liposome in therapy [24]. Optical imaging technology is one of noninvasive and sensitive method used in cancer therapy to monitor various disease and therapy progression. This technique is based on the photon transmission to specific tissues with fluorescent signals amplification produced by fluorescent dye and bioluminescent protein. However, some current treatments are still lacking sensitivity in cells-tissues targeting which leads to reduce the visualization performance, thus hampering an improved cancer treatment [85]. Nanoparticles are able to exhibit the sensitivity of imaging signal intensity up to cellular level since they have a wide dispersion and photo-stable dye molecules in nanosize range. Also, there are tremendous modifications that can offer such as flexible functionalization of functional groups, ligands, or conjugation with other supporting molecules (Table 12.2). In this section, recently used dyes employing fluorescent NPs are discussed.

Glycol chitosan functionalized 5β-cholanic acid (CA) and gadolinium (Gd(III)) NPs have exhibited the fluorescent signal of optical imaging near-infrared fluorescence with Cy5.5 dye. *In vivo* study suggested that this complex was successfully delivered and distributed via cytoplasm. The result showed that there was a signal enhancement by the site of targeted tumor cells without any toxicity effect to biological circumstance [86]. Cisplatin, as antineoplastic drug of various cancer types, has been attempted to increase the cancer treatment by loading it into the fibrin gels, which were conjugated into hyaluronate modified with *N*-trimethyl chitosan and NPs hybrid. The use of fibrin gels and hyaluronate acted as a local-delivery agent and receptor-specific substrate of CD44 and RHAMM on the target cancer cells. The selected chitosan derivative such as *N*-trimethyl chitosan, was used for the effectively loading of Cisplatin drug supported by hyaluronate. Therefore, it provides biocompatible and efficient surface to Cisplatin. The results obtained from *in vivo* experiments suggested that there was an increase up to 8% for the complex in antiproliferative activity studies in high fibrinogen concentration of 22 mg/ml which leads to efficient drug release and less toxicity [87].

A modified drug controlled delivery system conjugated to the NPs has been established by Liu *et al.* to maintain the drug release in different pHs [88]. Mesoporous silica NPs were utilized to provide a large surface area to load ibuprofen drug molecules via sol-gel method in cetyltrimethyl ammonium bromide (CTAB). In this study, chitosan was employed as an outermost layer to minimize the toxicity potential of highly loaded silica nanoparticles. Additionally, an abundant property of amino groups benefitted as chitosan was subjected to drug-nanoparticle complex to maintain the pH stability through the hydrogen linkage. As a result, the complex showed a better drug release than that of chitosan-unwrapped complex

Table 12.2 Chitosan in cancer therapy applications.

No.	Chitosan name (derivatives)/ Nanocomposite	Target	Type of cancer	Remarkable functions	References
1	Cy5.5-CS NPs-Gd(III)-CA	-	Murine squamous carcinoma cells (SCC7)	Glycol chitosan functionalized 5β-cholanic acid (CA) and gadolinium (Gd(III)) nanoparticles has exhibited the fluorescent signal of optical imaging near-infrared fluorescence with Cy5.5 dye.	[86]
2	*N*-trimethyl chitosan	-	Melanoma	To generate an effectively loading of nanoparticles for cisplatin complexed hyaluronate thus increase the efficiency of drug control release to target cancer cells up to more than 48 hrs.	[87]
3	Chitosan	-	Breast cancer	As pH-responsive drug nanocarrier together with mesoporous silica nanoparticles.	[88]
4	Chitosan	-	Pancreatic cancer	The quercetin and 5-fluorouracyl-loaded into chitosan carrier exhibited significant toxicity towards pancreatic cancer cells both in the 2D as well as in the 3D cultures.	[89]
5	Hydroxypropyl chitosan	-	-	Synergistic anti-cancer efficacy through the regulation of apoptosis-related genes in Janus kinase/signal transducer and activator of transcription pathways and mitochondrial pathways in xenograft tumor-bearing mice.	[90]

6	Chitosan	ICG	-	The CS-N-Arg/ICG complex prevented ICG from aggregation and precipitation, thus it could serve as a theranostic nanomaterial for image-guided photothermal cancer therapy.	[91]
7	Chitosan	-	Bladder tumor	Provide a significantly improved perspective in intravesical immunotherapy of bladder tumors.	[92]
8	Hydroxypatite/CS nanofilm	-	EpCAM, CK, and vimentin of circulating tumor cells (CTCs) in blood system	Combination between HA-CTS nanofilm and QDs nanoparticles increased efficiency captured cells up to more than 80% (EpCAM (+) CTCs) and typical QDs used in this study has exhibited imaging signal as compared to the common dye.	[93]

**Abbreviations:* CS = chitosan; CS-N-Arg = chitosan-arginine conjugate; ICG = indocyanine green; Cy5.5 = optical imaging near-infrared (NIRF) dye; CS-NPs = chitosan nanoparticles; Gd(III) = chelate gadolinium; CA = cholanic acids; HA-CTS = hydroxyapatite-chitosan; CTCs= circulating tumor cells; QDs = quantum dots.

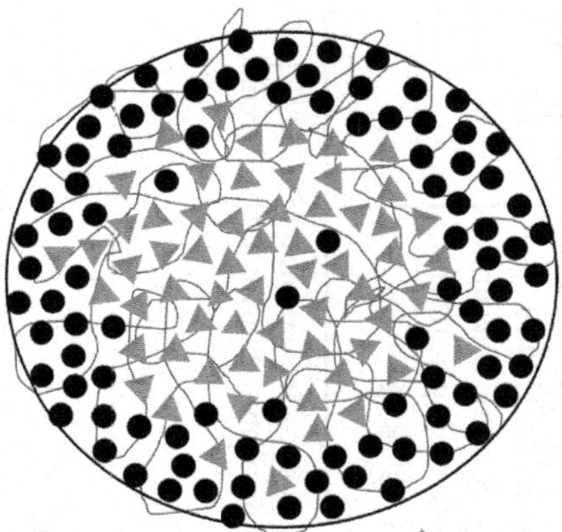

Figure 12.5 The schematic model of chitosan functionalized dual drugs in cancer drug delivery strategies (reproduced from [89]).

in which the amount of drug release was more concentrated in cancer cells [88].

Multidrug loading technique has been harnessed in an advanced treatment for cancer, in which, a synergetic encouragement of combining some prospered drugs can effectively exhibit drug half-time release, cancerous cell apoptosis, and efficiency of dosages. David *et al.* have revealed this unique function for quercetin and 5-fluorouracil in pancreatic cancer treatment [89]. This method was designed using ionic gelation method in which chitosan NPs were engaged to provide an efficient drug loading while preserving a good biocompatibility. Interestingly, chitosan also attracted the drugs uptake toward the cells due to a bundle of amino groups. The characterization study of NPs has successfully shown that most of the quercetin was concentrated in the center of the gel, while 5-fluorouracil was likely to concentrate in the periphery site as illustrated in Figure 12.5. A good release profile of drug complex in different media *in vitro* has been achieved with ratio of 3:1:2 for chitosan, quercetin, and 5-fluorouracil respectively, as well as drugs efficiency up to 95% of quercetin and 75% of 5-fluorouracil [89].

12.2.3 Diagnostic Strategy

In recent years, biosensors have successfully changed the sight of medical technology, for an example, they are proficient for monitoring blood glucose levels in diabetics. The most challenging part from cancer biosensors todays is to set the goal to achieve a precise medical prognosis from the spot results. Also, driving into the successful treatment is kind of advantage to be derived to detect the cancer in the earlier stages. Numerous biorecognition elements in various techniques have been developed by utilizing specific ligands, such as antibodies, immobilized on supporting layer of microfluidics; thereby making a good affinity binding to the abnormal cells extracted from huge volume of samples [94–98]. Using whole-cell as biomarker is also taken into account since the biological cell surface signature is one of the advantages for utilizing this detection strategy. It also aims directly at the target cell and overcomes protein expression and purification work which are likely less effective [99–102]. Meanwhile, the biorecognition elements are mostly not applied as such in stand-alone mode, considering the stability and system life time. However, they are applicable to be immobilized, the backing supports need to be decisive in order to overcome the critical construction for the next steps of sample application and signal reading. Some polymers offer their abilities to conduct the electrical properties as well as being a preferable platform for biorecognition binding [103]. Considering the physico-chemical and kinetic properties, chitosan has been a remarkable element in many pivotal fields due to its biodegradability to harmless polymer, biocompatibility, nontoxicity, metal ion chelation, high binding to protein, hydrophilicity, and abundant resource. The most benefitted part of it is the amino functional group which is accessible to be modified or functionalized via intra and inter-molecular hydrogen bonding [104]. With respect to this, nowadays chitosan has been valued as a good conducting polymer, redox initiator, sol-gel and hydrogel material in biosensor purpose as can be seen in Table 12.3 [103, 105, 106].

Liu *et al.* have assembled chitosan in biosensor fabrication integrated with $CuInS_2$ QDs and AuNPs to exhibit the fluorescent signal in lectin detection [107]. Lectin, which is involved in cancer malignancies, has natural bonding between its carbohydrate regions with protein group of CMC. In the case of signal quenching, the opulent functional groups, that is, amino, carboxyl, and hydroxyl groups within CMC, bridged the favorable interaction of quantum dots via electrostatic and hydrogen bonds thus; greatly increasing electronic transfer from AuNPs amplifying the fluorescent signals by the time lectin is introduced. As a result, a satisfactory detection limit has been achieved (0.08 nmol/l) [107].

An integrated-*sandwich* electrochemical technique has been proposed by Wang and colleagues emerging chitosan modified nanocubes and AuNPs to provide an excellent hallmark in the detection of multiple cancer proteins, carcinoembryonic antigen (CEA) and α-fetoprotein in one-time sensing [108]. The main functions of chitosan in this biosensor are to increase the consistency and biocompatibility of cadmium hexacyanocobaltates nanocubes as a signal amplifier to the bound proteins

Table 12.3 Biosensing applications of chitosan biomedical diagnosis.

No.	Techniques	Cancer targeting	Nanomaterials/nanocomposite	LOD	Linear range (R^2)	Reference
1	QCM	Folic acid (in MCF-7)	water soluble and low molecular weight chitosan	430 cells/mL	4.5×10^2–1.01×10^5 cells/mL (0.9984)	[26]
2	Fluorescence	Lectin (any cancer)	CMC/$CuInS_2$-QDs/AuNPs	0.08 nM	0.2–192.5 nM (0.999)	[107]
3	SWV	CEA and alpha-fetoprotein (cancer markers)	GCE/CS-AuNPs/Ab1/CEA-AFP/Ab2/CdNCs-CuNCs/AuNPs	CEA: 0.0175 ng/mL; alpha-fetoprotein: 0.0109 ng/mL	0.025–250 ng/mL (for CEA and alpha-feto-protein) (0.998)	[108]
4	EIS	K562 cells	CS/AuNPs	8.71 cells/mL	1.34×10^4–1.34×10^8 cells/mL (0.9986)	[113]
5	EIS	Folic acid (in HL-60 cells)	CMC/rGO	5.0×10^2 cells/mL	5.0×10^2–5.0×10^6 cells/mL (0.995)	[114]
6	ECL	PSA	CS/$EuPO_4$/anti-PSA antibody	177.33 fg/mL	0.0005–80 ng/mL (0.975)	[115]
7	DPV	PSA	CNTs(COOH)/CS/Apt-PSA	0.75 ng/mL	12.5—500 ng/mL (0.9973)	[116]
8	a. DPV b. EIS	PSA	Au-PAMAM/aptamer-HRP	10 fg/mL (DPV) 5 pg/mL (EIS)	a. 0.1 pg/mL–90 ng/mL (0.9831) (DPV) b. 0.05–35 ng/mL (0.9941) (EIS)	[117]

9	EIS	H_2O_2 in HepG2 (liver cancer cells) and LO2 (liver normal cells)	PtNPs/rGO-CS-Fc	20 nM	0.02–3.0 μM	[118]
10	CV	CEA	Fe_3O_4/AuNPs/Ab2/CEA/Ab1/GO/CS–Fc/GCE	0.39 pg/mL	0.001–30 ng/mL	[119]
11	SWV	Pancreatic cancer markers (CEA, CA199, CA125, CA242)	GCE/CS/Ab1/CEA-CA199-CA125-CA24/Ab2/CS-PAA/GA/ionic metals NSs	0.02 ng/mL (CEA), 0.4 U/mL (CA199), 0.3 U/mL (CA125), and 0.4 U/mL (CA242)	0.1–100 ng/mL; (0.993) (CEA), 1–150 U/mL; (0.991) (CA199), (0.993) (CA125), and (0.998) (CA242)	[120]
12	SWV	Squamous cell carcinoma antigens	TH/Ag@KIT-6/CMC/ILs-Ab2/SCCA/BSA/Ab1/Ag@APTES-GS/GCE	17 fg/mL	10^{-5}–10^{2} ng/mL	[121]
13	Fluorescence	Folate receptor of HeLa cells	Carbon dots/CMC/Fe_3O_4 MNPs/FA/Cu/cells	15 μg/mL	-	[122]
14	CV	EpCAM	AgNPs/CS	2.7 pg/mL	(0.998)	[123]

**Abbreviation*: CS = Chitosan; CMC = carboxylmethyl chitosan; EIS = electrochemical impedance spectroscopy; SWV = square wave voltammetry; CV = cyclic voltammetry; DPV = differential pulse voltammetry; ELISA = enzyme-linked immunosorbent assay; Ab1 = primary antibody; Ab2 = secondary antibody; Apt = aptamer; PSA = prostate specific antigen; TH = thiol; CNF = carbon nanofiber; rGO = reduced graphene oxide; GCE = glassy carbon electrode; AuNPs = gold nanoparticles; AgNPs = silver nanoparticles; CNTs = carbon nanotubes; PtNPs = platinum nanoparticles; FA = folic acid, CdNCs-CuNCs = cadmium nanoclusters-cuprum nanoclusters; Pt = platinum; HRP = horse radish peroxide; ILs = interleukins; SCCA = squamous cell carcinoma antigens; APTES = 3-aminopropyltriethoxysilane; GS = graphene sheets; BSA = bovine serum albumin; PAMAM = polyamidoamine; Fc = ferrocene.

and bridging the AuNPs with the GCE surface via amino chains. This also benefits in increasing the surface area of AuNPs to the antiprotein antibodies. The linear range observed by square wave voltammetry (SWV) was 0.025 to 250 ng/ml and resulted a satisfactory detection limits of CEA and α-fetoprotein of 0.0175 ng/ml and 0.0109 ng/ml, respectively [108].

Acoustic wave biosensor (AWB) has revealed tremendous benefits for its high sensitivity, simplicity, and effective cost in cancer cells diagnostic due to its ultrasensitivity for the mass recording [109]. Quartz crystal microbalance (QCM), an upgraded feature of the AWB, exhibits the reading of mass signal difference with no invasive defect and label-free [110]. The system works based on the change on the mass properties loaded on its piezoelectrode surface and generates microwave signal to be amplified and thus recorded as resonant frequency (f_r), which is inversely proportional to the mass stimulus [111, 112]. Zhang *et al.*, studied an attractive QCM biosensor for the *in situ* detection of folate receptor on the breast cancer cell line (MCF-7) [26]. The use of chitosan in this study is attributed to covalently linkage between amine group and folic acid thus prior to recyclable functionalization of reusable sensor. On the other hand, the used chitosan-water soluble low molecular weight type encouraged good hydrophilicity properties through carboxyl and hydroxyl bonding within folic acid and chitosan to promote the *in situ* cell adhesion. Low detection limit for this reusable biosensor was achieved at 430 cells/ml within the range of 4.5×10^2–1.01×10^5 cells/ml [26].

Ding *et al.* investigated nanocomposite, AuNPs modified chitosan gel assimilated with reduced tetrachloroauric acid via *in situ* method to detect K562 cells adhesion, proliferation, and apoptosis [113]. The nano-gels has benefitted not only in the immobilization process through adsorption but also in the characterization of cells behavior. The concentration profile was characterized using electrochemical impedance spectroscopy (EIS) method with an increase of electron transfer resistance in higher cells concentration. The linear range observed was in the range of 1.34×10^4–1.34×10^8 cells/ml with a LOD of 8.71 cells/ml. This study suggested that the use of nanocomposites based chitosan gels improves cells entrapment capacity and preserves good biocompatibility toward the cells [113].

Yang and *et al.* reported the use of carboxylmethyl chitosan and polyethyleneimine (PEI) modified reduced graphene oxide (rGO) and indium tin-oxide on the modified-GCE for detecting the human promyelocytic leukemia cells (HL-60 cells) [114]. The group principally conducted layer-by-layer technique for those aforementioned materials to provide well-defined biocompatibility and stability biosensor. Folic acid was used as a biorecognition element against the folate receptor overexpressed on the cancer cells surface. The use of carboxylmethyl chitosan abridged the molecular groups between its carboxylic acid and beneficial groups of rGO, instead of supporting the binding rigidity and dispersion of folic acid. From this study, the linear range obtained was found to be from 5.0 x 10^2 to 5.0 x 10^6 cells/ml, with a sensitivity of 5.0×10^2 cells/ml [114].

A sandwich format biosensor has also been developed for the prostate cancer antigen (PSA) detection in prostate cancer cells [115]. Chitosan was applied as dispersion solvent with $EuPO_4$ and to bridge the covalent linkage between the basis surfaces with recognition antibody. The binding PSA on the electrode surface hindered the electron-chemiluminescent transduction thus resulting different signals in different concentrations. The corresponding LODs obtained as low as 177.33 fg/ml and linearity was in between 0.0005 to 80 ng/ml which were found to be satisfactory [115]. Another report for the detection of PSA has been presented by Tahmasebi and Noorbakhsh in which chitosan platform with functionalized carboxylic to carbon nanotubes was used [116]. The idea was to increase the sensitivity of the immobilized PSA-aptamers with amino functionalization. The electrochemical signals was well produced and resulted in the achievement of low detection limits (0.75 ng/ml) [116].

An ultrasensitive immunosensor using triple electrochemical signal amplifications has been developed for the detection of PSA [117]. By employing GO/chitosan film assembled on the GCE, the monoclonal antibody of PSA was immobilized and arranged in the sandwich format with poly(amidoamine)-AuNPs labeled aptamer to improve the sensitivity. Under the optimized condition, a wide linear range was produced from 0.1 pg/ml to 90 ng/ml with a limit of detection of 10 fg/ml. Additionally, ELISA was conducted to confirm the proposed method. Lower detection limit with EIS was observed at 5 pg/ml with satisfactory value obtained using ELISA method [117].

A development of nonenzymatic biosensor has been proposed by Bai *et al.* to detect the peroxide product from catalytic reaction of the various living cells, including cancer cells [118]. Platinum and GO NPs were combined as a complex with the chitosan and ferrocene carboxylic acid nanohybrid. The results from this study by electrochemical techniques attained good reproducibility and highly sensitivity with S/N of 3. Linear correlation and detection limit obtained were satisfied and found to be 0.20 to 3.0 pM and 20 nM, respectively [118]. Another report on the chitosan in sandwich-electrochemical biosensor has been conducted by Peng *et al.* to detect CEA in cancerous cells [119]. In this report, primary-secondary antibodies bounded to the CEA specifically in sandwich format. GO/

chitosan-ferrocene (GO/CS-Fc) was grafted as an advanced basis surface, while the labelling properties used secondary antibody-conjugated Fe_3O_4/AuNPs complex. A catalytic redox signal was recorded from the redox cycle of *p*-nitrophenol to be *p*-aminophenol in the presence of Fe_3O_4/AuNPs, the signal was ten times amplified. Sensitive results were found in the range of 1.0 to $3.0–10^4$ pg/ml with a LODs of 0.39 pg/ml [119].

Rong *et al.* have reported ionic metals assembled nanospheres complex as biosensor for pancreatic cancer markers [120]. Several dopants contributing in amplified electrochemical signals are Cu, Cd, Pb, and Zn. These dopants were aligned with the chitosan and poly-acrylic acid in nanosphere and enhances the current signals significantly. The molecules of aforementioned markers have specific antibodies that were immobilized on the sensor surface for sandwich immunoassay approach. Using chitosan as a surface to support immobilization of primary and secondary antibodies offers additional stability to the nanosphere complex. Consequently, this construction enhanced good signal interpretation using square wave voltammetry (SWV) with linear range from 0.1 ng/ml to 100 ng/ml of CEA and 1.0 U/ml to 150 U/ml of CA199, CA125, and CA242, with corresponding LODs of 0.02 ng/ml, 0.4 U/ml, 0.3 U/ml, and 0.4 U/ml, respectively [120].

Another study in electrochemical immunosensor has been reported by Li *et al.* to recognize the squamous cell carcinoma antigens [121]. The primary antibody was grafted on the graphene surface hybridized AgNPs and 3-aminopropyltriethoxysilane (Ag@APTES-GS). The presence of CMC benefitted in facilitation of the electron transfer during the electrochemical detection with a LODs of 1.7×10^{-5} ng/ml [121]. A fluorescent magnetic nanoparticles-based biosensor has been developed by Kumar *et al.* to detect excess Cu^{2+} and Hela cells for diagnostic and imaging purposes [122]. This investigation was advanced by the use of combined visual instruments including FTIR, XRD, VSM, FESEM, TEM, DLS, and photochemiluminescence. The folic acid assembled to the complex acted as a bio-recognition toward the folate receptor on the cancer cell surface. It was observed that at 15 μg/ml concentration of the complex, there was more enhanced fluorescent image as compared to nonfolic acid molecule with the same complex [122]. Microfluidic biosensor based on the ultrasensitive sandwich immunoassay format has been approached by Ortega and colleagues for the detection of EpCAM antigen [123]. The antigen targeted by anti-EpCAM antibody was covalently linked to central channel of microfluidic on the AgNPs-chitosan. The chitosan-encapsulated nanoparticles acted as a stable and specific recognition agent to target the antigens with a LODs of 2.7 pg/ml in contrast to ELISA method which gave an LOD of 13.9 pg/ml [123].

12.3 Chitosan Applications in Alzheimer's

12.3.1 Pharmaceutical Formulation and Drug Delivery Application (Nanocarrier)

The success of chemotherapeutic treatment on AD depends on the efficacy outcome of the drug itself. One of the main factors to ensure that efficacy is by delivering drug particles into the target site of the expected receptor. Many drugs have low-profile solubility and permeability, therefore the need for an inert excipient which can effectively deliver drug into receptor has grown extensively. Moreover, brain region is the most important site of the body and thus, is naturally protected from any foreign materials or xenobiotics through the presence of BBB and blood-cerebrospinal fluid. This obstacle is foremost issue in drug delivery into the brain region; most of the drugs fail in crossing these specialized membranes.

Biopolymers have been used in drug delivery systems of neurodegenerative diseases such as Alzheimer's. Today, drug delivery crossing the BBB remains among the most common applications of chitosan in AD as can be seen in Table 12.4. Drug delivery application appears to be the topic of interest along with better understanding of chitosan's chemical modifications, biodegradation distribution to various body organs, mucoadhesion, chitosan association with inorganic compounds, and advanced technological transformations [14]. Herein, we review the applications of chitosan in drug delivery system based on the AD biomarker such as amyloid, acetylcholinesterase etc., as the drug target.

12.3.1.1 Amyloid-β-Targeted Nanocarier

The abundance of amyloid-β (Aβ) in the brain is one of the major features of people affected with AD [148]. It has been widely known as the most hypothetically caused, due to the initiation and progression stages of Alzheimer's occurrence. Aβ is cleaved from large amyloid-precursor protein (APP) by β-site APP cleavage enzyme-1 (BACE-1) resulting 42-amino-acid peptide [Aβ (42)] through amyloidogenic pathways [149]. On the basis of the amyloid hypothesis, Aβ accumulation in the brain is a predominant influence toward the pathogenesis of AD [150]. Some subfragments of Aβ are known to be protective to neurons and antibodies seems to be cytotoxic [151, 152]. Chitosan has been employed to carry Aβ sub-fragments crossing BBB due to its low toxicity, mucoadhesive, and immune stimulation properties [124]. As we discussed in previous section whether BBB remains foremost issue in drug delivery system of brain

Table 12.4 Pharmaceutical formulation and drug delivery applications of chitosan in Alzheimer's disease.

No	Chitosan name (derivatives)	Nanocomposite/ biocomposite	Carried molecule/ target	Remarkable functions	Reference
1	Chitosan	NP-IF-A	Aβ	Aβ nanocarrier to cross BBB (increase the permeability of Aβ)	[124]
2	Chitosan	CPLGANP, 6CMR- CPLGANP	Anti-Aβ antibody/ Aβ	As nanocarrier, to enhance the stability, permeability of anti-Aβ antibody across the BBB and targeting Aβ proteins	[125]
3	Chitosan	FITC-BSA loaded SNVs	Anti-Aβ antibody/ Aβ	As smart nanovehicle capable to permeate BBB and targeting Aβ in brain	[126]
4	Chitosan	TNVs	Cerebrovascular amyloid	As theranostics drug nanovehicle	[127]
5	Carboxymethyl chitosan	GAL-Ce@HAp/CMC	Galantamine (AChE inhibitor)	Selective nanocarrier of Galantamine drug	[128]
6	Chitosan	Tac-NP, Tac-Np+Ps 80	Tacrine (AChE inhibitor)	Nanocarrier of Tacrine drug	[129]
7	Low and medium molecular weight chitosan	NSO-pDNA-chitosan-PLGA	pDNA and NSO	Gene delivery and NSO nanocarrier	[130]
8	Chitosan	GH-CX-NPs	Galantamine hydrobromide (AChE inhibitor)	Nanocarrier of Galantamine hydrobromide via intranasal administration	[131]
9	Chitosan	Donepezil loaded chitosan NPs	Donepezil (AChE inhibitor)	Nanocarrier of Donepezil via intranasal administration	[132]

10	Chitosan	CS-RHT NPs	Rivastigmine (AChE inhibitor)	Nanocarrier of Rivastigmine drug via intranasal administration	[133]
11	Chitosan	Donepezil loaded chitosan nanosuspension	Donepezil (AChE inhibitor)	Donepezil drug carrier by forming nanosuspension (via intranasal)	[134]
12	Chitosan	Tacrine loaded CS-NPs	Tacrine (AChE inhibitor)	Excipient in formulation of hydrophilic Tacrine loaded sustained release nanoparticles	[135]
13	Chitosan	CTS/PL/β-CD	Phospholipid	Used as excipient in phospholipids microspheres formulation	[136]
14	Chitosan	PIP-NPs	Piperine	Nanocarrier of Piperine via intranasal administration	[137]
15	Chitosan	C/P/T	Tacrine hydrochloride (AChE inhibitor)	Used as excipient of powder based nasal formulation of Tacrine HCl	[138]
16	Chitosan	Rivastigmine loaded chitosan ME and MEs	Rivastigmine hydrogen tartrate (AChE inhibitor)	Used as excipient of formulation of microemulsion and mucoadhesive microemulsions of Rivastigmine via intranasal route	[139]
17	Chitosan	cRTNP	Rivastigmine (AChE inhibitor)	Used as Rivastigmine nanocarrier into the brain	[140]
18	Chitosan	Tacrine loaded chitosan	Tacrine (AChE inhibitor)	As nanocarrier of Tacrine drug	[141]
19	Chitosan	Chitosan-TPP NPs	Tacrine (AChE inhibitor)	As cross-linking agent of Tacrine delivery (in mixture with sodium tripolyphosphate)	[142]

(Continued)

Table 12.4 Cont.

No	Chitosan name (derivatives)	Nanocomposite/ biocomposite	Carried molecule/ target	Remarkable functions	Reference
20	Chitosan	Chitosan/pre-miR-29b	pre-miR-29b/ N2a695 cells	As nanocarrier of recombinant pre-miR-29b using polyplexes	[143]
21	Trimethylated chitosan	RVG-peptide-linked TMC-PEG	SiRNA	As potential carrier of small interfering RNA into brain region	[144]
22	Magnctic Chitosan	Tacrine loaded mag-netic chitosan microparticles	Tacrine (AChE inhibitor)	Increase the concentration of Tacrine in the brain	[145]
23	Chitosan	Ch/Tacrine	Tacrine (AChE inhibitor)	Used as nanocarrier of Tacrine into brain region	[146]
24	Chitosan	Chitosan coated insulin loaded PEGylated NPs	Insulin	Used as nanocarrier of Insulin into brain region	[147]

**Abbreviations*: NP-IF-A = fluorescently labelled chitosan nanoparticle-intramembranous fragments of Aβ; CPLGANP = chitosan adsorbed to poly lactic-co-glycolic acid NPs; 6CMR-CPLGANP = chitosan-coated PLGANP loaded with 6-Coumarin; FITC-BSA loaded SNVs = fluorescein-isothiocyanate bovine serum albumin loaded smart nanovehicles containing chitosan polymeric core; TNVs = polymeric nanocore made from Magnevist® (MRI contrast agent) conjugated chitosan; GAL@Ce-HAp/CMC = Cerium oxide containing carboxymethyl chitosan-coated hydroxyapatite; Tac-Np = Tacrine bound to chitosan NPs; Tac-Np+Ps 80 = Tacrine bound to chitosan NPs coated with 1% Polysorbate 80; NSO-pDNA-chitosan-PLGA = PLGA NPs loaded with Nigella sativa oil and pDNA, either by adsorption on chitosan-modified particles or by encapsulation within PLGA matrix after pre-complexation with CTAB; GH-CX-NPs = Galantamine hydrobromide/chitosan complex NPs; CS-RHT NPs = Rivastigmine (RHT) loaded chitosan NPs; CTS/PL/β-CD = chitosan/phospholipid/β-cyclodextrin; PIP-NPs = Piperine loaded chitosan NPs; ME = microemulsion; MEs = mucoadhesive microemulsions; cRTNP = Rivastigmine loaded chitosan NPs; Chitosan-TPP NPs = Chitosan-tripolyphosphate NPs; Chitosan/pre-miR-29b = pre-miR-29b loaded chitosan; RVG-peptide-linked TMC-PEG = rabies virus glycoprotein siRNA/trimethylated chitosan; Ch/Tacrine = Tacrine loaded chitosan.

disease. Songjiang and Lixiang have successfully revealed the potency of chitosan NPs to permeate BBB and determined the immune response produced after intraperitoneal injection into mice. In their work, interesting chitosan nanoparticles were labeled fluorescently using IF-A. The findings anticipated that such nanodelivery system could be used to carry Aβ efficiently to permeate BBB with significant immunogenicity [124].

Chitosan is also capable to carry payload of proteins conjugated with specific antibody. It was prepared from ionic gelation with tripolyphosphate (TPP) to be applied as a smart nanovehicle to carry bovine serum albumin (BSA) coated with antiamyloid antibody, polyamine modified F(ab') portion of IgG4.1, across the BBB and target Aβ deposits. This model was examined *in vivo* through testing on the animal model and it showed promising capability to carry the payload protein across the BBB [126]. Further studies emphasized its potential role as a diagnostic and delivering agent across the main barrier of the brain. For instance, Agyare *et al.* recently reported theranostic nanovehicle (TNVs) with chitosan conjugation designed perform triple functions. These include targeting cerebrovascular amyloid, providing magnetic resonance imaging contrast for the detection of Cerebral Amyloid Angiopathy (CAA) in early stage, and lastly, treating cerebrovascular inflammation resulting from CAA. The TNVs nanocomposite consists of a chitosan polymeric nanocore made from Magnevist® (a MRI contrast agent) loaded with cyclophosphamide, an immunosuppressant employed to decrease the cerebrovascular inflammation in CAA. Putrescine modified F(ab')2 fragment of antiamyloid antibody, IgG4.1 was attached to the surface of the nanocore to target Aβ in the cerebrovascular site. Interestingly, the results from *in vivo* and *in vitro* experiments revealed therapeutic, diagnosis as well as delivery effects as expected [127].

12.3.1.2 Cholinesterase-Targeted Nanocarrier

Cholinergic imbalance has been found to be one of the foremost hallmark in AD. Alzheimer's is generated by deficit of cholinergic neurotransmitter e.g. acetylcholine resulting the impairment of signal transmission of the cholinergic system [153]. This reducing event is affected over catalysis process of acetylcholinesterase and buthyrilcholinesterase (BuChE) enzymes. Therefore, there is a need to discover an effective agent to inhibit/halting these enzymes.

A nanocomposite based on chitosan conjugation with cholinesterase inhibitor also seems to be more predominant in drug delivery applications. A chitosan derivative, CMC, was used as coating agent of hydroxyapatite along with galantamine and cerium oxide (ceria) due to its ability to exhibit prominent biocompatibility, biomedical actions, and biodegradability. This new delivery and therapeutic strategy of Alzheimer's was proposed by Wahba and co-workers [128]. The combination of three elements, in which hydroxyapatite particles termed to upload nanoceria as a new antioxidant and galantamine as AChE inhibitor has emerged as a highly efficient anti-Alzheimer's agent. The nanoceria and hydroxyapatite showed remarkable effects to deliver Galantamine across BBB and carrying into target site, scavenging the hazardous reactive oxygen species, repairing the degenerated nerve cells, and disposal the toxic Aβ plaques [128]. The CMC played a role in encouraging the interaction between these three compounds and enhancing the delivery capability of the proposed nanocomposite.

Tacrine, a first of the AChE inhibitor approved in 1993 [154], is used to improve the biocognition in AD patients [155]. Tacrine integrated with chitosan in order to increase its uptake across BBB. Magnetic chitosan was successfully employed to target Tacrine to the brain region by crossing main obstacle, BBB. The study on the magnetic chitosan as drug carrier was explored by Wilson and co-workers. Earlier, they employed magnetic chitosan NPs to carry anti-Alzheimer's drug known as Tacrine, whereby chitosan-NPs were prepared via emulsion cross-linking method. The obtained results suggested the good capability of magnetic chitosan in transporting Tacrine drug agent by increasing the concentration of the drug in the brain as to control the rat's brain. The developed formulations could decrease the total dose required for the Alzheimer's chemotherapy with simultaneous reduction in dose related to the toxicity [145]. The same research group further extended the study to observe the effect of Polysorbate 80 on its bio-distribution in animal model (rats). The purpose of this study was to form extended release profile of Tacrine. Indeed, the polysorbate 80 was successfully shown as profound candidate to reduce the drug release via coating system [129].

In vitro study of Tacrine loaded chitosan was reported by Tamilsevan and Raghavan. The objective was to use chitosan couple with sodium-TPP to formulate a hydrophilic drug loaded with sustained release of NPs. This study employed ionic gelation method to load Tacrine and produce the expected particle size. The safety of constructed nanocomposites via cytotoxicity test suggested the potential application of chitosan and TPP as an novel biopolymer to deliver hydrophilic drug to target the brain [135]. Hassani *et al.* employed TPP as cross-linking agent of chitosan via ionic gelation method. Typically, this report proposed a novel gelation method using a stirred cell with microengineered membrane as a suitable approach to well prepare chitosan-TPP NPs. The encapsulation efficiency was found to be 66.1% under the optimum conditions [142]. In 2013,

Elmizadeh *et al.* investigated the use of chitosan and magnetic chitosan produced by emulsification approach which was used to carry Tacrine. This study successfully resulted to the better percentage yield and suitable for Tacrine loading capability into chitosan NPs and magnetic chitosan biopolymer [141].

Recently, an attractive molecular dynamic (MD) study was made to investigate the overall compatibility of Tacrine drug with chitosan and poly(n-butylcyanoacrylate) (PBCA). MD is used as a major tool to study molecular systems. The computation parameters were calculated in this study involves interaction energy of the polymer/Tacrine systems, the radius of gyration of the chitosan and PBCA in the function of time, solubility and Flory-Huggins interaction. As a result, the Tacrine molecule was found to exhibit higher compatibility with PBCA rather than chitosan. Moreover, the chain length of polymer seemed to influence the interaction between the Tacrine molecules and polymers. Interaction between Tacrine and PBCA NPs remained stronger by increasing the length of polymer chain while it was not observed as a consistent trend for chitosan/Tacrine systems [146]. This study was particularly interesting because of employment of computational study to explore the proper polymer which is able to be conjugated with Alzheimer's drug in order to yield appropriate drug delivery into specific target.

Intranasal delivery is favorable route of the Alzheimer's drug to bypass BBB and administer into brain region. Compared to another route of administration, intranasal delivery has the following advantages: reducing the systemic adverse effects, prevention of first pass metabolism, easy to use by self-administration, capability of dose adjustment, non-invasive route, and improved patient compliance [156]. Chitosan seems to be foremost biopolymer in term of Alzheimer's drug delivery via intranasal route. An *in vivo* drug delivery study of chitosan was reported to develop a new complexation technique for loading Galantamine hydrobromide into chitosan and deliver in the brain. The study showed that complexation was a promising technique to entrap cationic drug, Galantamine hydrobromide, into cationic polymer, chitosan, through successful assessment of delivery profile in rat [131].

Another study on *nose-to-brain* delivery route of drug loaded chitosan was successfully developed by Bhavna and colleagues. Donepezil is well-known as one of the most potent AChE inhibitor that is available in the market to cure AD. In this report, the intranasal delivery was replaced with the common gastrointestinal route for Donepezil and yielded a better efficacy in rat model (as previously discussed). The nanocomposite of chitosan and Donepezil was formed through ionic cross linking method and resulted sophisticated delivery of tested drug through intranasal pathways [134]. Instead, Bhavna *et al.* continued their study with the basis of ionic gelation method to form direct composite of Donepezil and chitosan biopolymer. In this report, they suggested as well as emphasized the effectiveness of intranasal administration route to deliver Donepezil into brain region and useful for better treatment of AD [132].

Mucoadhesive microparticles based on chitosan and pectin polyelectrolyte complexes were used to carry Tacrine into brain region via nasal administration. Combinative method of direct spray-drying followed by lyophilization approaches was chosen to prepare the microparticles encapsulated Tacrine hydrochloride. The study observed the influence of different loading methods and composition of chitosan/pectin to the complex physical properties, ability to hydrate, interaction with mucin, and enhancement of drug permeation capability [138].

Rivastigmine is a potential inhibitor of both AChE and BuChE enzymes and remains 4–17 times more specific to inhibit AChE in brain as compared to heart and blood. The delivery of Rivastigmine into the brain is quite challenging because of poor penetration and lesser concentration in the brain and therefore chitosan was employed to enhance its drug delivery capability. Fazil and colleagues developed Rivastigmine loaded chitosan via ionic gelation method to increase the bioavailability and enhance the uptake of Rivastigmine to the brain via intranasal route. These results suggested that the constructed biocomposites has better brain targeting efficiency as compared to other formulations. The direct transport percentage of 71.80 ± 6.71% was achieved from this study [133]. The similar study was reported by Shah and co-workers, wherein Rivastigmine loaded chitosan was formulated through microemulsion (ME) and mucoadhesive microemulsion (MME) via *nose-to-brain* delivery. The formulated complex within polymer and drug was evaluated by *in vitro* and *ex vivo* study and yielded good results. The MME remained more capable to carry Rivastigmine into brain region via both systems as compared to ME [139]. Another study on the use of chitosan loaded with Rivastigmine was conducted to improve the therapeutic and safety profile of the drug. The author used ionotropic gelation method to produce the chitosan and drug complex. The formulation of complex was proved to enhance the therapeutic efficacy and profile of drug tolerance [140].

The added advantage of chitosan in drug delivery is that its properties could be enhanced in the presence of nanomaterials. The nanocomposites containing typical galantamine/cerium oxide/CMC-coated hydroxyappatite in order to upraises delivery and efficacy of galantamine drug [128]. Galantamine is used as Alzheimer's therapy due to the prohibit action of the AChE from hydrolyzing the neurotransmitter acetylcholine, which plays a pivotal role in enhancing memory and cognition, thereby alleviates the potential cause for the presence of Aβ peptides in AD [157]. In this study comparing the nanocomposite with and without chitosan to investigate its effect in delivery profile of selected drugs was investigated. It successfully proved the omnipotency of rod-like hydroxyapatite particles for selective transport of galantamine and nanoceria to AD targeted as well as affected brain regions [128]. Intranasal study on direct conjugation of chitosan with galantamine has been proposed by Hanafy *et al.* The complexes were prepared by ionic gelation method and parameters affecting their physicochemical properties were assessed. The bare complexes were distributed via intranasal route into different brain regions [131].

12.3.1.3 Other Functional Drug Carrier

Chitosan seems to be not always associated with the chemical drug molecules while it was functionalized for the brain delivery purpose. Some biofunctional molecules have been reported along with chitosan conjugation such as piperine, phospholipids, microRNAs, siRNA, and *p*DNA. Phospholipids play an important role in memory and learning abilities. Instead, phospholipids act as a choline source in acetylcholine synthesis. A particular study was conducted to prepare and optimize the formulation of biocomposites containing chitosan/phospholipids/β-cyclodextrin microspheres which can improve the cognitive impairment. The microspheres were produced using spray-drying method and yielded the improvement of learning and memory abilities in the tested rats. Indeed, the microspheres reduced the expression of protein kinase C-δ and inhibited the microglias activation. These results suggested that the optimized microspheres were proved to be a well candidate for treating AD [136]. A monodisperse intranasal chitosan NPs was elaborated with piperine to target brain region. Piperine is a phytopharmaceutical agent having neuroprotective effect. Piperine–chitosan composites could significantly alleviate piperine nasal irritation and showed nontoxic behavior in the brain. The composites were administered via intranasal route. Further, the resulting mucoadhesive chitosan NPs were successfully tailored for effective, safe, and noninvasive piperine delivery [137].

Instead of carrying payload protein, chitosan was able to be used as gene delivery for the Alzheimer's treatment. The applications of chitosan in this particular area were previously reviewed [14]. Functional gene molecules which could be adhered with chitosan includes microRNAs, small interfering RNA (siRNA) and *p*DNA. These molecules belong to nucleotide-based gene silencing therapeutic for Alzheimer's. Recent advances toward an effective therapy for AD intend to employ microRNAs (miR-29) to overwhelm human BACE1 expression and subsequent Aβ peptide while siRNA is able to effectively silence target genes with high specificity. Moreover, gene-based therapy using siRNA remains a powerful tool for AD [143, 144]. Further, chitosan loaded gene was also investigated by Doolanea and colleagues. Polylactic-co-glycolic acid loaded chitosan nanoparticles along with *Nigella sativa* oil (NSO) and *p*DNA were able to exhibit gene expression as well as neuroregeneration. Encapsulated NSO acted as the therapeutic agent for promoting neurite outgrowth which is essential for cell neuroregeneration stage while *p*DNA transfected N2a cell. The particle size and zeta potential of constructed nanocomposites remain highly dependent on the medium conditions. This study suggested the effectiveness of chitosan-PLGA NPs system as a novel gene delivery carrier [130].

12.3.2 Treatment and Therapy

Several studies have been reported on the chitosan which correspond to its oligosaccharide structure to directly reduce the monomeric Aβ. Chitosan oligosaccharide (COS) is a hydrolysis products of chitosan which composed of linear polymers of β-1-4-linked D-glucosamine [158]. COS has markedly prevented monomeric Aβ aggregation, dissembling preformed amyloid fibrils and Aβ-mediated neurotoxicity [159]. Previous studies, presented the potential of COS to reduce Aβ (42)-induced neurotoxicity through repression of oxidative stress and hindering Aβ-mediated phosphorylation of c-Jun N-terminal kinase [160]. The usage of chitosan to directly target the specific protein like Aβ is such an interesting aspect that can be explored due to its safety and other sophisticated properties. Recently, the first *in vivo* investigation on modulating effects of COS on cognitive deficits and their underlying molecular mechanisms was reported by Jia *et al.* This study emphasized the beneficial application of COS on cognitive impairment in Aβ (42) of animal model induced AD [161]. Taken together, COS remains as a potential candidate for further Alzheimer's therapy via reducing monomeric Aβ, inhibition on the oxidative stress, and cognitive impairment in Aβ (42). The study on toxicity effect might be considered in order to observe the accumulation effect of COS in the brain region.

More detailed study on the capability of chitosan as a positively charged biopolymer like chitosan and its quaternary derivative *N*-trimethyl chitosan was presented. Liu and co-workers investigated the possibility of these two polysaccharides to inhibit the aggregation process of Aβ (40) peptide. The systematic studies on pH-dependent experiments described an attractive electrostatic interaction between the positive charge of moieties in chitosan/*N*-trimethyl chitosan and the negative charge of residues in Aβ (40) played a dominating role in the inhibiting activity. Molecular docking study also revealed the possible interaction between chitosan/*N*-trimethyl chitosan and Aβ (40). The depiction of illustrative interaction between tested polymers and Aβ (40) and the plausible interaction within both are shown at Figure 12.6.

In this study, it was found that *N*-trimethyl chitosan exerted stronger binding interaction with Aβ (40) in comparison to the chitosan. Furthermore, these findings anticipated that the positively charged polysaccharides (such as chitosan and *N*-trimethyl chitosan) could serve as an useful bioactive polymer for the mediating mechanistic aggregation of Aβ (40) and other aggregation-prone protein [162].

Targeting Aβ, several structural analogs of sialic acid (*N*-acetylneuramic acid) decorated chitosan backbone was investigated to observe the possible attenuating effect of Aβ pathology. In this regard, the study employed some sub-structures of sialic acid which might be sequester Aβ or interference on Aβ binding event with cells. The result showed that sialic acid could have on Aβ binding and allowed more detailed and specific improvement of its substructures as a potential therapeutic agent against Alzheimer's [163]. Instead of Aβ, chitosan could also be used to inhibit AChE activity as one of the well-accepted Alzheimer's target. Gallic acid grafted chitosan was reported as a beneficial materials in the treatment and therapy of AD. This material complex was able to exhibit potent AChE inhibitory effect in dose dependent manner and moreover, it did not exert any toxicity effect on PC12. The inhibition mode was found to be non-competitive. This material complex could also helpful in preventing the AD by elevating cholinergic transmission [164]. The positive properties of chitosan would also help in permeating main obstacle, BBB, before distributing materials into brain site.

In the treatment of AD, chitosan might be having inflammatory effect while the chronic inflammatory response associated with Aβ and interleukin-1β (IL-1β) is known to be responsible for the AD pathology. A remarkable report was proposed on the use of water soluble chitosan as a novel inflammatory agent. This study examined cytotoxicity effect, production of pro-inflammatory cytokines and inducible nitric-oxide synthase on the human astrocytoma cell stimulated with IL-1β and Aβ (25–35) in order to observe the biological effect of water soluble chitosan. As a result, this chitosan was found to have no effect on the cell viability

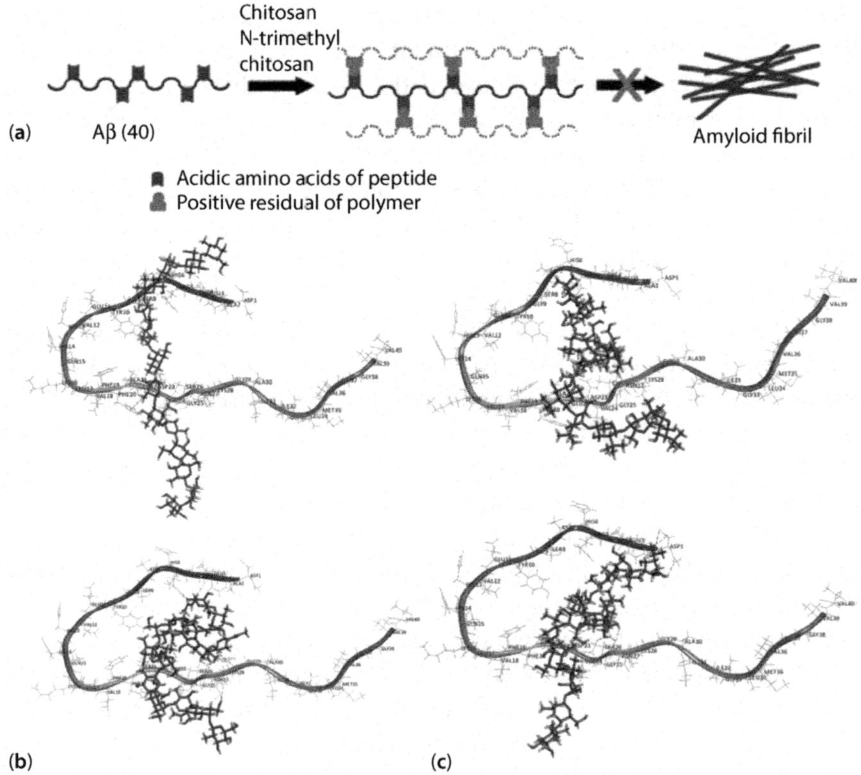

Figure 12.6 Pictorial depiction of (a) blocking interaction between chitosan/*N*-trimethyl chitosan and $A\beta$ (40), binding modes of (b) deacetylated chitosan, and (c) quartenarized *N*-trimethyl chitosan with full length of $A\beta$ (40) (adopted from [162]).

on human astrocytoma cells. Furthermore, the secretion and expression of pro-inflammatory cytokines, tumor necrosis factor-α and Interleukin-6, were significantly inhibited by pre-treatment with water soluble chitosan in the human astrocytoma cells. The expression of inducible nitric-oxide synthase was induced by IL-1β and $A\beta$ (25–35) and was partially inhibited by treatment with water soluble chitosan [165]. The water soluble chitosan had inflammatory effect which might be useful to decrease and delay pathologic stages in AD.

12.3.3 Sensing and Diagnosis

Chitosan was particularly employed as a composite substance of a fabricated biosensor platform. Since biosensors corresponds to bio-analytical sensor which utilized for detecting or quantifying the biological analyte/

target and assembled by molecular recognition entity integrated with a physicochemical transducer. Biomarkers play a central role to ensure the accuracy and reliability of diagnostic sensor device. It is the most critical part of the biosensor construction because of its important role in performing specific and selective analyte associated with the AD. Chitosan has been employed to a sensor device for detecting acetylcholine as one of the biomarker of Alzheimer's. The plasma concentration of acetylcholine is very important in order to study such neurodegenerative diseases and develop their remedies [166].

An amperometric biosensor was constructed by co-immobilizing covalently of dual enzyme, AChE and choline oxidase onto nanocomposite of chitosan/gold (Au) coated ferric oxide nanoparticles electrodeposited onto the surface of Au electrode as a working electrode. Chitosan was chosen due to the biocompatibility, nontoxicity, and ability to enhance the biosensor performance via conjugation with Au working electrode. The optimum performance of biosensor was reached within 3 s at +0.2 V, pH 7.0 at 30 °C. The linear working range and detection limit were found to be 0.005–400 mM and 0.005 mM, respectively. Interestingly, the constructed biosensor was tested on normal and suffered volunteers suffering with AD and yielded high specificity and good reproducibility as well as stability [166]. Therefore, chitosan-based sensor can amplify the sensor performance and thus, could be used as an efficient device in future diagnosis of early stage Alzheimer's.

12.4 Conclusion

The growth on biopolymer usage in biomedical technology has been noted over the past decade. Chitosan, as low-cost material, has successfully emerged plentiful benefits in high-definitive biomedical applications of cancer and Alzheimer's as drug nanocarrier, treatment and therapeutic agent as well as sensing and diagnosis nanocomposite. The nanorange size, wide surface area, oxygen-rich functional groups, and flexibility toward other excipients increases the demand of chitosan as a vital part in pharmaceutical compounding for biomedical applications. For instance, by tailoring chitosan in cancer drug delivery can increase drug molecule loading, enhance targeting selectivity, control the efficiency of drug release, and quicken the cancer cells apoptosis. When Chitosan is integrated with other nanomaterials such as QDs, it can enable their main function which is to trigger stability in activation of "on-off" imaging system. In terms of Alzheimer's, use of chitosan in drug delivery system appears to be the major application among others. This is due to the excellent features of chitosan, that is, biocompatibility, biodegradability, and unique capability to cross BBB membrane and therefore its being used in others neurodegenerative diseases as well. Apart from their functionalities, several issues of chitosan nanoparticles application still persist in the medical treatment and use in pharmaceutical excipient due to the concerns about accumulated toxicity toward living cells in neurology system. However, along with beneficial features, nanopolymer chitosan is capable to be emerged as promising biomaterials in the biomedical applications of proliferatic and neuronal-related disease such as cancer and Alzheimer's.

References

1. Siddiqui, M., Rajkumar, S.V., The high cost of cancer drugs and what we can do about it. *Mayo Clin. Proc.*, 87, 935, 2012.
2. Fonseca, A.C., Serra, A.C., Coelho, J.F.J., Bioabsorbable polymers in cancer therapy: Latest developments. *EPMA J.*, 6, 1, 2015.
3. Coelho, S.C., Pereira, M.C., Juzeniene, A., Juzenas, P., Coelho, M.A.N., Supramolecular nanoscale assemblies for cancer diagnosis and therapy. *J. Contr. Rel.*, 213, 152, 2015.
4. Cai, W., Gao, T., Hong, H., Sun, J., Applications of gold nanoparticles in cancer nanotechnology. *Nanotechnol. Sci. Appl.*, 2008, 2008.
5. Jabir, N.R., Tabrez, S., Ashraf, G.M., Shakil, S., Damanhouri, G.A., Kamal, M.A., Nanotechnology-based approaches in anticancer research. *Int. J. Nanomed.*, 7, 4391, 2012.
6. Tan, M.L., Choong, P.F.M., Dass, C.R., Recent developments in liposomes, microparticles and nanoparticles for protein and peptide drug delivery. *Peptides*, 31, 184, 2010.
7. Lai, P., Daear, W., Löbenberg, R., Prenner, E.J., Overview of the preparation of organic polymeric nanoparticles for drug delivery based on gelatine, chitosan, poly(d,l-lactide-co-glycolic acid) and polyalkylcyanoacrylate. *Colloid. Surf., B*, 118, 154, 2014.
8. Lattime, E.C., Gerson, S.L. (Ed.), *Gene therapy of cancer: translational approaches from preclinical studies to clinical implementation*, Academic Press, USA, 2013.
9. Ahmed, T.A., Aljaeid, B.M., Preparation, characterization, and potential application of chitosan, chitosan derivatives, and chitosan metal nanoparticles in pharmaceutical drug delivery. *Drug Des. Dev. Ther.*, 10, 483, 2016.
10. Muzzarelli, R.A.A., Chitins and chitosans as immunoadjuvants and non-allergenic drug carriers. *Mar. Drugs*, 8, 292, 2010.
11. Agnihotri, S.A., Mallikarjuna, N.N., Aminabhavi, T.M., Recent advances on chitosan-based micro- and nanoparticles in drug delivery. *J. Contr. Rel.*, 100, 5, 2004.

12. Dutta, P.K., Dutta, J., Tripathi, V., Chitin and chitosan: Chemistry, properties and applications. *J. Sci. Ind. Res.*, 63, 20, 2004.
13. Rinaudo, M., Chitin and chitosan: properties and applications. *Prog. Polym. Sci.*, 31, 603, 2006.
14. Kumar, M.N., Muzzarelli, R.A., Muzzarelli, C., Sashiwa, H., Domb, A.J., Chitosan chemistry and pharmaceutical perspectives. *Chem. Rev.*, 104, 6017, 2004.
15. Riva, R., Ragelle, H., *et al.*, Chitosan and chitosan derivatives in drug delivery and tissue engineering, In: *Chitosan for biomaterials II*, R. Jayakumar, M. Prabaharan, R.A.A. Muzzarelli, Eds., pp. 19–44, Springer, Germany, 2011.
16. Pillai, C.K.S., Paul, W., Sharma, C.P., Chitin and chitosan polymers: chemistry, solubility and fiber formation. *Prog. Polym. Sci.*, 34, 641, 2009.
17. Hirano, S., Application of chitin and chitosan in the ecological and environmental fields, In: *Applications of chitin and chitosan*, M.F.A. Goosen, Ed., pp. 31–54, CRC Press, USA, 1996.
18. Chen, R.H., Domard, A., Muzzarelli, R.A.A., Tokura, S., Wang, D.M., Advances in chitin/chitosan science and their applications. *Carbohydr. Polym.*, 84, 695, 2011.
19. Jimtaisong, A., Saewan, N., Utilization of carboxymethyl chitosan in cosmetics. *Int. J. Cosmet. Sci.*, 36, 12, 2014.
20. Kanatt, S.R., Chander, R., Sharma, A., Chitosan glucose complex - a novel food preservative. *Food Chem.*, 106, 521, 2008.
21. Bernkop-Schnürch, A., Dünnhaupt, S., Chitosan-based drug delivery systems. *Eur. J. Pharm. Biopharm.*, 81, 463, 2012.
22. Prabaharan, M., Mano, J.F., Chitosan-based particles as controlled drug delivery systems. *Drug Deliv.*, 12, 41, 2004.
23. Denkbas, E.B., Ottenbrite, R.M., Perspectives on: Chitosan drug delivery systems based on their geometries. *J. Bioact. Compat. Polym.*, 21, 351, 2006.
24. Dass, C.R., Choong, P.F., The use of chitosan formulations in cancer therapy. *J. Microencapsul.*, 25, 275, 2008.
25. Feng, S.S., Chien, S., Chemotherapeutic engineering: Application and further development of chemical engineering principles for chemotherapy of cancer and other diseases. *Chem. Eng. Sci.*, 58, 4087, 2003.
26. Zhang, S., Bai, H., Luo, J., Yang, P., Cai, J., A recyclable chitosan-based QCM biosensor for sensitive and selective detection of breast cancer cells in real time. *The Analyst*, 139, 6529, 2014.
27. Feng, K.J., Yang, Y.H., Wang, Z.J., Jiang, J.H., Shen, G.L., Yu, R.Q., A nanoporous CeO_2/chitosan composite film as the immobilization matrix for colorectal cancer DNA sequence-selective electrochemical biosensor. *Talanta*, 70, 561, 2006.
28. Tiwari, A., Gong, S., Electrochemical detection of a breast cancer susceptible gene using cDNA immobilized chitosan-co-polyaniline electrode. *Talanta*, 77, 1217, 2009.

29. Yu, H., Yan, F., Dai, Z., Ju, H., A disposable amperometric immunosensor for α-1-fetoprotein based on enzyme-labeled antibody/chitosan-membrane-modified screen-printed carbon electrode. *Anal. Biochem.*, 331, 98, 2004.
30. Nagpal, K., Singh, S.K., Mishra, D.N., Drug targeting to brain: A systematic approach to study the factors, parameters and approaches for prediction of permeability of drugs across BBB. *Exp. Opin. Drug Deliv.*, 10, 927, 2013.
31. Hirano, S., Seino, H., *et al.*, Chitosan: a biocompatible material for oral and intravenous administrations, In: *Progress in biomedical polymers*, C.G. Gebelin, R.L. Dunn, Eds., pp. 283–290, Springer, Germany, 1990.
32. Croisier, F., Jérôme, C., Chitosan-based biomaterials for tissue engineering. *Eur. Polym. J.*, 49, 780, 2013.
33. Kannan, M., Nesakumari, M., Rajarathinam, K., Singh, A., Production and characterization of mushroom chitosan under solid-state fermentation conditions. *Adv. Biol. Res.*, 4, 10, 2010.
34. Wu, T., Zivanovic, S., Draughon, F.A., Sams, C.E., Chitin and chitosan value-added products from mushroom waste. *J. Agric. Food Chem.*, 52, 7905, 2004.
35. Skuse, A., Wombs, worms and wolves: Constructing cancer in early modern england. *SSHM,* 27, 632, 2014.
36. Byrne, J.D., Betancourt, T., Brannon-Peppas, L., Active targeting schemes for nanoparticle systems in cancer therapeutics. *Adv. Drug Deliv. Rev.*, 60, 1615, 2008.
37. Chan, A., Orme, R.P., Fricker, R.A., Roach, P., Remote and local control of stimuli responsive materials for therapeutic applications. *Adv. Drug Deliv. Rev.*, 65, 497, 2013.
38. Pulkkinen, M., Pikkarainen, J., Wirth, T., Tarvainen, T., Haapa-aho, V., Korhonen, H., Seppala, J., Jarvinen, K., Three-step tumor targeting of paclitaxel using biotinylated PLA-PEG nanoparticles and avidin-biotin technology: formulation development and *in vitro* anticancer activity. *Eur. J. Pharm. Biopharm.*, 70, 66, 2008.
39. Kunii, T., Ogura, S., Mie, M., Kobatake, E., Selection of DNA aptamers recognizing small cell lung cancer using living cell-SELEX. *The Analyst,* 136, 1310, 2011.
40. Cukierman, E., Khan, D.R., The benefits and challenges associated with the use of drug delivery systems in cancer therapy. *Biochem. Pharmacol.*, 80, 762, 2010.
41. Sriraman, S.K., Aryasomayajula, B., Torchilin, V.P., Barriers to drug delivery in solid tumors. *Tissue Barriers,* 2, e29528, 2014.
42. Torchilin, V.P., Targeted polymeric micelles for delivery of poorly soluble drugs. *Cell. Mol. Life Sci.*, 61, 2549, 2004.
43. Patel, N.R., Pattni, B.S., Abouzeid, A.H., Torchilin, V.P., Nanopreparations to overcome multidrug resistance in cancer. *Adv. Drug Deliv. Rev.*, 65, 1748, 2013.

44. Rejinold, N.S., Sreerekha, P.R., Chennazhi, K.P., Nair, S.V., Jayakumar, R., Biocompatible, biodegradable and thermo-sensitive chitosan-g-poly (N-isopropylacrylamide) nanocarrier for curcumin drug delivery. *Int. J. Biol. Macromol.*, 49, 161, 2011.
45. Duceppe, N., Tabrizian, M., Advances in using chitosan-based nanoparticles for *in vitro* and *in vivo* drug and gene delivery. *Exp. Opin. Drug Deliv.*, 7, 1191, 2010.
46. Rai, P., Mallidi, S., Zheng, X., Rahmanzadeh, R., Mir, Y., Elrington, S., Khurshid, A., Hasan, T., Development and applications of photo-triggered theranostic agents. *Adv. Drug Deliv. Rev.*, 62, 1094, 2010.
47. Pille, J.Y., Li, H., Blot, E., Bertrand, J.R., Pritchard, L.L., Opolon, P., Maksimenko, A., Lu, H., Vannier, J.P., Soria, J., Malvy, C., Soria, C., Intravenous delivery of anti-RhoA small interfering RNA loaded in nanoparticles of chitosan in mice: safety and efficacy in xenografted aggressive breast cancer. *Hum. Gene. Ther.*, 17, 1019, 2006.
48. Hernandez-Ibanez, N., Garcia-Cruz, L., Montiel, V., Foster, C.W., Banks, C.E., Iniesta, J., Electrochemical lactate biosensor based upon chitosan/carbon nanotubes modified screen-printed graphite electrodes for the determination of lactate in embryonic cell cultures. *Biosens. Bioelectron.*, 77, 1168, 2016.
49. Liu, S., Kang, M., Yan, F., Peng, D., Yang, Y., He, L., Wang, M., Fang, S., Zhang, Z., Electrochemical DNA biosensor based on microspheres of cuprous oxide and nano-chitosan for Hg(II) detection. *Electrochim. Acta*, 160, 64, 2015.
50. Ibrahim, M., Mahmoud, A.A., Osman, O., Refaat, A., El-Sayed el, S.M., Molecular spectroscopic analysis of nano-chitosan blend as biosensor. *Spectrochim. Acta A Mol. Biomol. Spectrosc.*, 77, 802, 2010.
51. Korolev, I.O., Alzheimer's disease: a clinical and basic science review. *MSRJ*, 4, 24, 2014.
52. Alzheimer, A., Stelzmann, R.A., Schnitzlein, H.N., Murtagh, F.R., An english translation of Alzheimer's 1907 paper, "Uber eine eigenartige Erkankung der Hirnrinde". *Clin. Anat.*, 8, 429, 1995.
53. Alzheimer's Association (Ed.), *Basics of Alzheimer's disease: what it is and what you can do*, Alzheimer's Association, USA, 2015.
54. de la Torre, J.C., Alzheimer's disease is incurable but preventable. *J. Alzheimers Dis.*, 20, 861, 2010.
55. Azheimer's Disease International (Ed.), *World Alzheimer's Report 2015: The Global Impact of Dementia: Analysis of Prevalence, Incidence, Cost and Trends*, Azheimer's Disease International, UK, 2015.
56. Alzheimer's Association, 2015 Alzheimer's disease facts and figures. *Alzheimers Dement.*, 11, 332, 2015.
57. Rowe, R.C., Sheskey, P.J., Weller, P.J. (Ed.), *Handbook of pharmaceutical excipients*, Pharmaceutical Press London, UK, 2006.
58. Pharmacopeia, U. (Ed.), *United States Pharmacopeia and National Formulary (USP 37–NF 32)*, United States Pharmacopeial Convention, USA, 2014.

59. Bernkop-Schnürch, A., Weithaler, A., Albrecht, K., Greimel, A., Thiomers: Preparation and *in vitro* evaluation of a mucoadhesive nanoparticulate drug delivery system. *Int. J. Pharm.*, 317, 76, 2006.
60. Zheng, F., Shi, X.W., Yang, G.F., Gong, L.L., Yuan, H.Y., Cui, Y.J., Wang, Y., Du, Y.M., Li, Y., Chitosan nanoparticle as gene therapy vector via gastrointestinal mucosa administration: Results of an *in vitro* and *in vivo* study. *Life Sci.*, 80, 388, 2007.
61. De Campos, A.M., Sánchez, A., Alonso, M.a.J., Chitosan nanoparticles: a new vehicle for the improvement of the delivery of drugs to the ocular surface. Application to cyclosporin A. *Int. J. Pharm.*, 224, 159, 2001.
62. Sandri, G., Bonferoni, M.C., Rossi, S., Ferrari, F., Gibin, S., Zambito, Y., Di Colo, G., Caramella, C., Nanoparticles based on N-trimethylchitosan: Evaluation of absorption properties using *in vitro* (Caco-2 cells) and *ex vivo* (excised rat jejunum) models. *Eur. J. Pharm. Biopharm.*, 65, 68, 2007.
63. Hirano, S., Noishiki, Y., The blood compatibility of chitosan and N-acylchitosans. *J. Biomed. Mater. Res.*, 19, 413, 1985.
64. Baldrick, P., Pharmaceutical excipient development: The need for preclinical guidance. *Regul. Toxicol. Pharmacol.*, 32, 210, 2000.
65. Carreño-Gómez, B., Duncan, R., Evaluation of the biological properties of soluble chitosan and chitosan microspheres. *Int. J. Pharm.*, 148, 231, 1997.
66. Thanou, M., Verhoef, J.C., Junginger, H.E., Chitosan and its derivatives as intestinal absorption enhancers. *Adv. Drug Deliv. Rev*, 50, S91, 2001.
67. Tan, M.L., Choong, P.F., Dass, C.R., Review: doxorubicin delivery systems based on chitosan for cancer therapy. *J. Pharm. Pharmacol.*, 61, 131, 2009.
68. Wang, J.J., Zeng, Z.W., Xiao, R.Z., Xie, T., Zhou, G.L., Zhan, X.R., Wang, S.L., Recent advances of chitosan nanoparticles as drug carriers. *Int. J. Nanomed.*, 6, 765, 2011.
69. Vecchione, R., Quagliariello, V., Calabria, D., Calcagno, V., De Luca, E., Iaffaioli, R.V., Netti, P.A., Curcumin bioavailability from oil in water nano-emulsions: *In vitro* and *in vivo* study on the dimensional, compositional and interactional dependence. *J. Contr. Rel.*, 233, 88, 2016.
70. Nguyen, K.T., Le, D.V., Do, D.H., Le, Q.H., Development of chitosan graft pluronic (R) F127 copolymer nanoparticles containing DNA aptamer for paclitaxel delivery to treat breast cancer cells. *Adv. Nat. Sci. Nanosci. Nanotechnol.*, 7, 8, 2016.
71. Chowdhuri, A.R., Singh, T., Ghosh, S.K., Sahu, S.K., Carbon dots embedded magnetic nanoparticles @chitosan @metal organic framework as a nanoprobe for pH sensitive targeted anticancer drug delivery. *ACS Appl. Mater. Interfaces*, 8, 16573, 2016.
72. Wang, G., Wang, J.J., Tang, X.J., Du, L., Li, F., *In vitro* and *in vivo* evaluation of functionalized chitosan-Pluronic micelles loaded with myricetin on glioblastoma cancer. *Nanomedicine*, 12, 1263, 2016.
73. Nayak, D., Minz, A.P., Ashe, S., Rauta, P.R., Kumari, M., Chopra, P., Nayak, B., Synergistic combination of antioxidants, silver nanoparticles and chitosan in

a nanoparticle based formulation: Characterization and cytotoxic effect on MCF-7 breast cancer cell lines. *J. Colloid. Interf. Sci.*, 470, 142, 2016.

74. Hu, Z.Y., Zheng, H., Li, D., Xiong, X., Tan, M.Y., Huang, D., Guo, X., Zhang, X.Q., Yan, H., Self-assembled nanoparticles based on folic acid modified carboxymethyl chitosan conjugated with targeting antibody. *J. Wuhan Univ. Technol. Mat. Sci. Edit.*, 31, 446, 2016.
75. Trickler, W.J., Nagvekar, A.A., Dash, A.K., A novel nanoparticle formulation for sustained Paclitaxel delivery. *AAPS Pharm. Sci. Tech.*, 9, 486, 2008.
76. Zeiderman, M.R., Morgan, D.E., Christein, J.D., Grizzle, W.E., McMasters, K.M., McNally, L.R., Acidic pH-targeted chitosan-capped mesoporous silica coated gold nanorods facilitate detection of pancreatic tumors via multispectral optoacoustic tomography. *ACS Biomater. Sci. Eng.*, 2, 1108, 2016.
77. Maxwell, T., Banu, T., Price, E., Tharkur, J., Campos, M., Gesquiere, A., Santra, S., Non-cytotoxic quantum dot–chitosan nanogel biosensing probe for potential cancer targeting agent. *Nanomaterials,* 5, 2359, 2015.
78. Marano, F., Argenziano, M., Frairia, R., Adamini, A., Bosco, O., Rinella, L., Fortunati, N., Cavalli, R., Catalano, M.G., Doxorubicin-loaded nanobubbles combined with extracorporeal shock waves: Basis for a new drug delivery tool in anaplastic thyroid cancer. *Thyroid,* 26, 705, 2016.
79. Badran, M.M., Harisa, G.I., AlQahtani, S.A., Alanazi, F.K., Zoheir, K.M.A., Pravastatin-loaded chitosan nanoparticles: Formulation, characterization and cytotoxicity studies. *J. Drug Deliv. Sci. Technol.*, 32, 1, 2016.
80. Das, R.K., Kasoju, N., Bora, U., Encapsulation of curcumin in alginate-chitosan-pluronic composite nanoparticles for delivery to cancer cells. *Nanomedicine,* 6, 153, 2010.
81. Sachdev, A., Matai, I., Gopinath, P., Carbon dots incorporated polymeric hydrogels as multifunctional platform for imaging and induction of apoptosis in lung cancer cells. *Colloid. Surf., B,* 141, 242, 2016.
82. Ekinci, M., Ilem-Ozdemir, D., Gundogdu, E., Asikoglu, M., Methotrexate loaded chitosan nanoparticles: Preparation, radiolabeling and *in vitro* evaluation for breast cancer diagnosis. *J. Drug Deliv. Sci. Technol.*, 30, 107, 2015.
83. Kim, J.H., Kim, Y.S., Park, K., Kang, E., Lee, S., Nam, H.Y., Kim, K., Park, J.H., Chi, D.Y., Park, R.W., Kim, I.S., Choi, K., Chan Kwon, I., Self-assembled glycol chitosan nanoparticles for the sustained and prolonged delivery of anti-angiogenic small peptide drugs in cancer therapy. *Biomaterials,* 29, 1920, 2008.
84. Brannon-Peppas, L., Blanchette, J.O., Nanoparticle and targeted systems for cancer therapy. 64, Supplement, 2012.
85. Jiang, S., Gnanasammandhan, M.K., Zhang, Y., Optical imaging-guided cancer therapy with fluorescent nanoparticles. *J. R. Soc. Interface,* 7, 3, 2010.
86. Nam, T., Park, S., Lee, S.Y., Park, K., Choi, K., Song, I.C., Han, M.H., Leary, J.J., Yuk, S.A., Kwon, I.C., Kim, K., Jeong, S.Y., Tumor targeting chitosan nanoparticles for dual-modality optical/MR cancer imaging. *Bioconjug. Chem.*, 21, 578, 2010.

87. Viale, M., Rossi, M., Russo, E., Cilli, M., Aprile, A., Profumo, A., Santi, P., Fenoglio, C., Cafaggi, S., Rocco, M., Fibrin gels loaded with cisplatin and cisplatin-hyaluronate complexes tested in a subcutaneous human melanoma model. *Invest. New Drugs,* 33, 1151, 2015.
88. Liu, W.T., Yang, Y., Shen, P.H., Gao, X.J., He, S.Q., Liu, H., Zhu, C.S., Facile and simple preparation of pH-sensitive chitosan-mesoporous silica nanoparticles for future breast cancer treatment. *Exp. Polym. Lett.,* 9, 1068, 2015.
89. David, K.I., Jaidev, L.R., Sethuraman, S., Krishnan, U.M., Dual drug loaded chitosan nanoparticles-sugar-coated arsenal against pancreatic cancer. *Colloid. Surf., B,* 135, 689, 2015.
90. Wu, X., He, C., Wu, Y., Chen, X., Cheng, J., Nanogel-incorporated physical and chemical hybrid gels for highly effective chemo–protein combination therapy. *Adv. Funct. Mater.,* 25, 6744, 2015.
91. Jheng, P.R., Lu, K.Y., Yu, S.H., Mi, F.L., Free DOX and chitosan-N-arginine conjugate stabilized indocyanine green nanoparticles for combined chemo-photothermal therapy. *Colloid. Surf., B.,* 136, 402, 2015.
92. Erdogar, N., Iskit, A.B., Eroglu, H., Sargon, M.F., Mungan, N.A., Bilensoy, E., Antitumor efficacy of Bacillus Calmette-Guerin loaded cationic nanoparticles for intravesical immunotherapy of bladder tumor induced rat model. *J. Nanosci. Nanotechnol.,* 15, 10156, 2015.
93. Chen, Y.Y., Cheng, B.R., He, Z.B., Wang, S.Y., Wang, Z.M., Sun, M., Song, H.B., Fang, Y., Chen, F.F., Xiong, B., Capture and identification of heterogeneous circulating tumor cells using transparent nanomaterials and quantum dots-based multiplexed imaging. *J. Cancer,* 7, 69, 2016.
94. Wang, K., Marshall, M.K., Garza, G., Pappas, D., Open-tubular capillary cell affinity chromatography: Single and tandem blood cell separation. *Anal. Chem.,* 80, 2118, 2008.
95. Plouffe, B.D., Radisic, M., Murthy, S.K., Microfluidic depletion of endothelial cells, smooth muscle cells, and fibroblasts from heterogeneous suspensions. *Lab Chip,* 8, 462, 2008.
96. Maheswaran, S., Sequist, L.V., Nagrath, S., Ulkus, L., Brannigan, B., Collura, C.V., Inserra, E., Diederichs, S., Iafrate, A.J., Bell, D.W., Digumarthy, S., Muzikansky, A., Irimia, D., Settleman, J., Tompkins, R.G., Lynch, T.J., Toner, M., Haber, D.A., Detection of mutations in EGFR in circulating lung-cancer cells. *N. Engl. J. Med.,* 359, 366, 2008.
97. Phillips, J.A., Xu, Y., Xia, Z., Fan, Z.H., Tan, W., Enrichment of cancer cells using aptamers immobilized on a microfluidic channel. *Anal. Chem.,* 81, 1033, 2009.
98. Huang, Z., King, M.R., An immobilized nanoparticle-based platform for efficient gene knockdown of targeted cells in the circulation. *Gene Ther.,* 16, 1271, 2009.
99. Shangguan, D., Meng, L., Cao, Z.C., Xiao, Z., Fang, X., Li, Y., Cardona, D., Witek, R.P., Liu, C., Tan, W., Identification of liver cancer-specific aptamers using whole live cells. *Anal. Chem.,* 80, 721, 2008.

100. Fang, X., Tan, W., Aptamers generated from cell-SELEX for molecular medicine: a chemical biology approach. *Acc. Chem. Res.*, 43, 48, 2010.
101. Liu, R., Zhang, X., Pu, Y., Yin, L., Li, Y., Zhang, X., Liang, G., Li, X., Zhang, J., Small-sized titanium dioxide nanoparticles mediate immune toxicity in rat pulmonary alveolar macrophages *in vivo*. *J. Nanosci. Nanotechnol.*, 10, 5161, 2010.
102. Cohen, J., Wilson, A., Manzolillo, K., Clinical and economic challenges facing pharmacogenomics. *Pharmacogenomics J.*, 13, 378, 2013.
103. Wang, X., Uchiyama, S., Polymers for biosensors construction, In: *State of the art in biosensors-general aspects*, T. Rinken, Ed., pp. 67–86, INTECH Open Access Publisher, Japan, 2013.
104. Zhang, J., Xia, W., Liu, P., Cheng, Q., Tahirou, T., Gu, W., Li, B., Chitosan modification and pharmaceutical/biomedical applications. *Mar. Drugs*, 8, 1962, 2010.
105. Higgins, M.J., Molino, P.J., Yue, Z., Wallace, G.G., Organic conducting polymer–protein interactions. *Chem. Mater.*, 24, 828, 2012.
106. Vidal, J.C., Garcia Ruiz, E., Castillo, J.R., Recent advances in electropolymerized conducting polymers in amperometric biosensors. *Microchim. Acta*, 143, 93, 2003.
107. Liu, Z., Liu, H., Wang, L., Su, X., A label-free fluorescence biosensor for highly sensitive detection of lectin based on carboxymethyl chitosan-quantum dots and gold nanoparticles. *Anal. Chim. Acta*, 932, 88, 2016.
108. Wang, Z., Chen, X., Ma, Z., Chitosan coated copper and cadmium hexacyanocobaltate nanocubes as immunosensing probes for the construction of multiple analytes platform. *Biosens. Bioelectron.*, 61, 562, 2014.
109. Lange, K., Rapp, B.E., Rapp, M., Surface acoustic wave biosensors: a review. *Anal. Bioanal. Chem.*, 391, 1509, 2008.
110. Lee, T.Y., Song, J.T., Detection of carcinoembryonic antigen using AlN FBAR. *Thin Solid Films*, 518, 6630, 2010.
111. Tukkiniemi, K., Rantala, A., Nirschl, M., Pitzer, D., Huber, T., Schreiter, M., Fully integrated FBAR sensor matrix for mass detection. *Procedia Chem.*, 1, 1051, 2009.
112. Nirschl, M., Rantala, A., Tukkiniemi, K., Auer, S., Hellgren, A.-C., Pitzer, D., Schreiter, M., Vikholm-Lundin, I., CMOS-integrated film bulk acoustic resonators for label-free biosensing. *Sensors*, 10, 4180, 2010.
113. Ding, L., Hao, C., Xue, Y., Ju, H., A bio-inspired support of gold nanoparticles-chitosan nanocomposites gel for immobilization and electrochemical study of K562 leukemia cells. *Biomacromolecules*, 8, 1341, 2007.
114. Yang, G.H., Cao, J.T., Li, L.L., Rana, R.K., Zhu, J.J., Carboxymethyl chitosan-functionalized graphene for label-free electrochemical cytosensing. *Carbon*, 51, 124, 2013.
115. Ma, H.M., Zhou, J., Li, Y., Han, T.Q., Zhang, Y., Hu, L.H., Du, B., Wei, Q., A label-free electrochemiluminescence immunosensor based on $EuPO_4$ nanowire for the ultrasensitive detection of prostate specific antigen. *Biosens. Bioelectron.*, 80, 352, 2016.

116. Tahmasebi, F., Noorbakhsh, A., Sensitive electrochemical prostate specific antigen aptasensor: Effect of carboxylic acid functionalized carbon nanotube and glutaraldehyde linker. *Electroanalysis,* 28, 1134, 2016.
117. Kavosi, B., Salimi, A., Hallaj, R., Moradi, F., Ultrasensitive electrochemical immunosensor for PSA biomarker detection in prostate cancer cells using gold nanoparticles/PAMAM dendrimer loaded with enzyme linked aptamer as integrated triple signal amplification strategy. *Biosens. Bioelectron.,* 74, 915, 2015.
118. Bai, Z., Li, G., Liang, J., Su, J., Zhang, Y., Chen, H., Huang, Y., Sui, W., Zhao, Y., Non-enzymatic electrochemical biosensor based on Pt NPs/RGO-CS-Fc nano-hybrids for the detection of hydrogen peroxide in living cells. *Biosens. Bioelectron.,* 82, 185, 2016.
119. Peng, D., Liang, R.-P., Huang, H., Qiu, J.-D., Electrochemical immunosensor for carcinoembryonic antigen based on signal amplification strategy of graphene and Fe_3O_4/Au NPs. *J. Electroanal. Chem.,* 761, 112, 2016.
120. Rong, Q., Feng, F., Ma, Z., Metal ions doped chitosan–poly(acrylic acid) nanospheres: Synthesis and their application in simultaneously electrochemical detection of four markers of pancreatic cancer. *Biosens. Bioelectron.,* 75, 148, 2016.
121. Li, Y.Y., Zhang, Y.H., Jiang, L.P., Chu, P.K., Dong, Y.H., Wang, P., An electrochemical immunosensor comprising thionin/silver nanoparticles decorated KIT-6 for ultrasensitive detection of squamous cell carcinoma antigen. *RSC Adv.,* 6, 6932, 2016.
122. Kumar, A., Chowdhuri, A.R., Laha, D., Chandra, S., Karmakar, P., Sahu, S.K., One-pot synthesis of carbon dot-entrenched chitosan-modified magnetic nanoparticles for fluorescence-based Cu^{2+} ion sensing and cell imaging. *RSC Adv.,* 6, 58979, 2016.
123. Ortega, F.G., Fernandez-Baldo, M.A., Serrano, M.J., Messina, G.A., Lorente, J.A., Epithelial cancer biomarker EpCAM determination in peripheral blood samples using a microfluidic immunosensor based in silver nanoparticles as platform. *Sens. Actuator B-Chem.,* 221, 248, 2015.
124. Songjiang, Z., Lixiang, W., Amyloid-beta associated with chitosan nanocarrier has favorable immunogenicity and permeates the BBB. *AAPS Pharm. Sci. Tech.,* 10, 900, 2009.
125. Jaruszewski, K.M., Ramakrishnan, S., Poduslo, J.F., Kandimalla, K.K., Chitosan enhances the stability and targeting of immuno-nanovehicles to cerebro-vascular deposits of Alzheimer's disease amyloid protein. *Nanomedicine,* 8, 250, 2012.
126. Agyare, E.K., Curran, G.L., Ramakrishnan, M., Yu, C.C., Poduslo, J.F., Kandimalla, K.K., Development of a smart nano-vehicle to target cerebrovascular amyloid deposits and brain parenchymal plaques observed in Alzheimer's disease and cerebral amyloid angiopathy. *Pharm. Res.,* 25, 2674, 2008.
127. Agyare, E.K., Jaruszewski, K.M., Curran, G.L., Rosenberg, J.T., Grant, S.C., Lowe, V.J., Ramakrishnan, S., Paravastu, A.K., Poduslo, J.F., Kandimalla, K.K., Engineering theranostic nanovehicles capable of targeting cerebrovascular amyloid deposits. *J. Contr. Rel.,* 185, 121, 2014.

128. Wahba, S.M., Darwish, A.S., Kamal, S.M., Ceria-containing uncoated and coated hydroxyapatite-based galantamine nanocomposites for formidable treatment of Alzheimer's disease in ovariectomized albino-rat model. *Mater. Sci. Eng. C. Mater. Biol. Appl.*, 65, 151, 2016.
129. Wilson, B., Samanta, M.K., Santhi, K., Kumar, K.S., Ramasamy, M., Suresh, B., Chitosan nanoparticles as a new delivery system for the anti-Alzheimer drug tacrine. *Nanomedicine*, 6, 144, 2010.
130. Doolaanea, A.A., Mansor, N., Mohd Nor, N.H., Mohamed, F., Co-encapsulation of Nigella sativa oil and plasmid DNA for enhanced gene therapy of Alzheimer's disease. *J. Microencapsul.*, 33, 114, 2016.
131. Hanafy, A.S., Farid, R.M., ElGamal, S.S., Complexation as an approach to entrap cationic drugs into cationic nanoparticles administered intranasally for Alzheimer's disease management: preparation and detection in rat brain. *Drug Dev. Ind. Pharm.*, 41, 2055, 2015.
132. Bhavna, Md, S., Ali, M., Bhatnagar, A., Baboota, S., Sahni, J.K., Ali, J., Design, development, optimization and characterization of Donepezil loaded chitosan nanoparticles for brain targeting to treat Alzheimer's disease. *Sci. Adv. Mater.*, 6, 720, 2014.
133. Fazil, M., Md, S., Haque, S., Kumar, M., Baboota, S., Sahni, J.K., Ali, J., Development and evaluation of rivastigmine loaded chitosan nanoparticles for brain targeting. *Eur. J. Pharm. Sci.*, 47, 6, 2012.
134. Bhavna, Md, S., Ali, M., Ali, R., Bhatnagar, A., Baboota, S., Ali, J., Donepezil nanosuspension intended for nose to brain targeting: *in vitro* and *in vivo* safety evaluation. *Int. J. Biol. Macromol.*, 67, 418, 2014.
135. Tamilselvan, N., Raghavan, C.V., Formulation and characterization of anti-alzheimer's drug loaded chitosan nanoparticles and its *in vitro* biological evaluation. *J. Young Pharm.*, 7, 29, 2015.
136. Shan, L., Tao, E.X., Meng, Q.H., Hou, W.X., Liu, K., Shang, H.C., Tang, J.B., Zhang, W.F., Formulation, optimization, and pharmacodynamic evaluation of chitosan/phospholipid/beta-cyclodextrin microspheres. *Drug Des. Devel. Ther.*, 10, 417, 2016.
137. Elnaggar, Y.S.R., Etman, S.M., Abdelmonsif, D.A., Abdallah, O.Y., Intranasal piperine-loaded chitosan nanoparticles as brain-targeted therapy in Alzheimer's disease: optimization, biological efficacy, and potential toxicity. *J. Pharm. Sci.*, 104, 3544, 2015.
138. Saladini, B., Bigucci, F., Cerchiara, T., Gallucci, M.C., Luppi, B., Microparticles based on chitosan/pectin polyelectrolyte complexes for nasal delivery of tacrine hydrochloride. *Drug Deliv. Transl. Res.*, 3, 33, 2013.
139. Shah, B.M., Misra, M., Shishoo, C.J., Padh, H., Nose to brain microemulsion-based drug delivery system of rivastigmine: formulation and *ex-vivo* characterization. *Drug Deliv.*, 22, 918, 2015.
140. Nagpal, K., Singh, S.K., Mishra, D.N., Optimization of brain targeted chitosan nanoparticles of Rivastigmine for improved efficacy and safety. *Int. J. Biol. Macromol.*, 59, 72, 2013.

141. Elmizadeh, H., Khanmohammadi, M., Ghasemi, K., Hassanzadeh, G., Nassiri-Asl, M., Garmarudi, A.B., Preparation and optimization of chitosan nanoparticles and magnetic chitosan nanoparticles as delivery systems using Box-Behnken statistical design. *J. Pharm. Biomed. Anal.*, 80, 141, 2013.
142. Hassani, S., Laouini, A., Fessi, H., Charcosset, C., Preparation of chitosan–TPP nanoparticles using microengineered membranes – Effect of parameters and encapsulation of tacrine. *Colloid. Surf., A*, 482, 34, 2015.
143. Pereira, P.A., Tomas, J.F., Queiroz, J.A., Figueiras, A.R., Sousa, F., Recombinant pre-miR-29b for Alzheimer's disease therapeutics. *Sci. Rep.*, 6, 19946, 2016.
144. Gao, Y., Wang, Z.Y., Zhang, J., Zhang, Y., Huo, H., Wang, T., Jiang, T., Wang, S., RVG-peptide-linked trimethylated chitosan for delivery of siRNA to the brain. *Biomacromolecules*, 15, 1010, 2014.
145. Wilson, B., Samanta, M.K., Santhi, K., Sampath Kumar, K.P., Ramasamy, M., Suresh, B., Significant delivery of tacrine into the brain using magnetic chitosan microparticles for treating Alzheimer's disease. *J. Neurosci. Methods*, 177, 427, 2009.
146. Eslami, M., Nikkhah, S.J., Hashemianzadeh, S.M., Sajadi, S.A., The compatibility of tacrine molecule with poly(n-butylcyanoacrylate) and chitosan as efficient carriers for drug delivery: a molecular dynamics study. *Eur. J. Pharm. Sci.*, 82, 79, 2016.
147. Pankaj, D., Rakesh Kumar, T., Narendra Kumar, J., Nanoparticulate carrier mediated intranasal delivery of insulin for the restoration of memory signaling in Alzheimer's disease. *Curr. Nanosci.*, 9, 46, 2013.
148. Selkoe, D.J., Physiological production of the beta-amyloid protein and the mechanism of Alzheimer's disease. *Trends Neurosci.*, 16, 403, 1993.
149. Humpel, C., Identifying and validating biomarkers for Alzheimer's disease. *Trends Biotechnol.*, 29, 26, 2011.
150. Hardy, J., Selkoe, D.J., The amyloid hypothesis of Alzheimer's disease: progress and problems on the road to therapeutics. *Science (New York, N.Y.)*, 297, 353, 2002.
151. Morgan, D., Diamond, D.M., Gottschall, P.E., Ugen, K.E., Dickey, C., Hardy, J., Duff, K., Jantzen, P., DiCarlo, G., Wilcock, D., Connor, K., Hatcher, J., Hope, C., Gordon, M., Arendash, G.W., A beta peptide vaccination prevents memory loss in an animal model of Alzheimer's disease. *Nature*, 408, 982, 2000.
152. Fradinger, E.A., Monien, B.H., Urbanc, B., Lomakin, A., Tan, M., Li, H., Spring, S.M., Condron, M.M., Cruz, L., Xie, C.W., Benedek, G.B., Bitan, G., C-terminal peptides coassemble into Abeta42 oligomers and protect neurons against Abeta42-induced neurotoxicity. *Proc. Natl. Acad. Sci. U.S.A.*, 105, 14175, 2008.
153. Kihara, T., Shimohama, S., Alzheimer's disease and acetylcholine receptors. *Acta Neurobiol. Exp. (Wars)*, 64, 99, 2004.
154. Čolović, M.B., Krstić, D.Z., Lazarević-Pašti, T.D., Bondžić, A.M., Vasić, V.M., Acetylcholinesterase inhibitors: pharmacology and toxicology. *Curr. Neuropharmacol.*, 11, 315, 2013.

155. Scarpini, E., Scheltens, P., Feldman, H., Treatment of Alzheimer's disease: current status and new perspectives. *Lancet Neurol.*, 2, 539, 2003.
156. Lochhead, J.J., Thorne, R.G., Intranasal delivery of biologics to the central nervous system. *Adv. Drug Deliv. Rev.*, 64, 614, 2012.
157. Matharu, B., Gibson, G., Parsons, R., Huckerby, T.N., Moore, S.A., Cooper, L.J., Millichamp, R., Allsop, D., Austen, B., Galantamine inhibits β-amyloid aggregation and cytotoxicity. *J. Neurol. Sci.*, 280, 49, 2009.
158. Kato, Y., Onishi, H., Machida, Y., Application of chitin and chitosan derivatives in the pharmaceutical field. *Curr. Pharm. Biotechnol.*, 4, 303, 2003.
159. Dai, X., Hou, W., Sun, Y., Gao, Z., Zhu, S., Jiang, Z., Chitosan oligosaccharides inhibit/disaggregate fibrils and attenuate amyloid β-mediated neurotoxicity. *Int. J. Mol. Sci.*, 16, 10526, 2015.
160. Dai, X., Chang, P., Zhu, Q., Liu, W., Sun, Y., Zhu, S., Jiang, Z., Chitosan oligosaccharides protect rat primary hippocampal neurons from oligomeric beta-amyloid 1–42-induced neurotoxicity. *Neurosci. Lett.*, 554, 64, 2013.
161. Jia, S., Lu, Z., Gao, Z., An, J., Wu, X., Li, X., Dai, X., Zheng, Q., Sun, Y., Chitosan oligosaccharides alleviate cognitive deficits in an amyloid-beta1–42-induced rat model of Alzheimer's disease. *Int. J. Biol. Macromol.*, 83, 416, 2016.
162. Liu, H., Ojha, B., Morris, C., Jiang, M., Wojcikiewicz, E.P., Rao, P.P., Du, D., Positively charged chitosan and N-trimethyl chitosan inhibit Abeta40 fibrillogenesis. *Biomacromolecules*, 16, 2363, 2015.
163. Dhavale, D., Henry, J.E., Evaluation of sialic acid-analogs for the attenuation of amyloid-beta toxicity. *Biochim. Biophys. Acta*, 1820, 1475, 2012.
164. Cho, Y.S., Kim, S.K., Ahn, C.B., Je, J.Y., Inhibition of acetylcholinesterase by gallic acid-grafted-chitosans. *Carbohydr. Polym.*, 84, 690, 2011.
165. Kim, M.S., Sung, M.J., Seo, S.B., Yoo, S.J., Lim, W.K., Kim, H.M., Water-soluble chitosan inhibits the production of pro-inflammatory cytokine in human astrocytoma cells activated by amyloid beta peptide and interleukin-1beta. *Neurosci. Lett.*, 321, 105, 2002.
166. Chauhan, N., Pundir, C.S., Amperometric determination of acetylcholine-A neurotransmitter, by chitosan/gold-coated ferric oxide nanoparticles modified gold electrode. *Biosens. Bioelectron.*, 61, 1, 2014.

13

Biomedical Significance of Chitin- and Chitosan-Based Nanocomposites

Aasim Majeed*, Raoof Ahmad Najar, Shruti Choudhary, Sapna Thakur, Amandeep Singh, and Pankaj Bhardwaj

Molecular Genetics Laboratory, Central University of Punjab, Bathinda, India

Abstract

Chitin forms a primary structural component of exoskeletons of crustacea, arthropods, fungal and yeast cell walls. Structurally, it resembles with plant cellulose. The residue or waste of the marine food industry that includes crab, shrimp shell, etc., is the commercial sources of chitin. Among the three different isomorphs, α-chitin is most abundant isomorph mostly present in fungal and yeast cell walls, lobsters, crab tendons, shrimp shells, and insect cuticle while β-chitin is reported in squid pens. Chitin is insoluble in common solvents so it is relatively difficult to process it. Consequently, deacetylation of chitin to chitosan renders it more soluble. Due to higher solubility of chitosan in common solvents, it is rendered more suitable as a biopolymer. Both chitin and chitosan are considered as biocompatible, biodegradable, and nontoxic besides showing adhesiveness, bone formation ability, antitumor ability, central nervous system depressant, and immunoadjuvant property. Both chitin and chitosan can be processed into beads, gels, membranes, micro and nanofibres/particles/films, or sponges. Apart from industrial utilities, chitin and chitosan nanocomposites have extended their applications as an implantable biomaterial in several biomedical applications. Delivery of drugs to the targeted site with relative prolonged release, management of wounded tissues, and tissue engineering is among the prime applications of chitin and chitosan nanocomposites in the biomedical field. Besides, the use of these nanocomposites in dentistry and combating microbes is well established. This chapter focuses on biomedical significance of chitin and chitosan nanocomposites in a simplified manner.

Keywords: Tissue engineering, wound healing, drug delivery, nanocomposites

**Corresponding author*: majeedaasim@gmail.com

Shakeel Ahmed and Saiqa Ikram (eds.) Chitosan, (361–384) © 2017 Scrivener Publishing LLC

13.1 Introduction

Chitin and chitosan are multipurpose biomaterials. Chitin is a copolymer of poly β-(1,4)-2-acetamido-2-deoxy-D-glucopyranose containing low percentage of 2-amino-2-deoxy-β-glucopyranose residual units [1,2]. Chitin structure resembles to cellulose of plants and also to murein (a structural polymer in cell wall of bacteria, made up of *N*-acetyl-D-glucosamine but has different properties than cellulose). Chitin is the main ingredient in the exoskeletons of crustacea, arthropods, and fungal cell walls due to its highly ordered crystalline microfibrils with *N*-acetylglucosamine as the monomer linked with β-1,4 glycosidic linkages [3]. The residue or waste of the marine food industry that includes crab, shrimp shell, etc., is the commercial sources of chitin. It exhibits three polymorphic forms such as α-, β- and γ-forms. Among them, α-chitin is most abundant isomorph mostly present in fungal and yeast cell walls, lobsters, crab tendons, shrimp shells, and insect cuticle while β-chitin is reported in squid pens [4] and in tubes synthesized by pogonophoron and vestimetiferan worms [5]. Chitin possesses several hydrogen bonds which determine its reactivity and solubility. Also the linear chains are organized into sheets tightly packed by hydrogen bonds. IR spectroscopy, NMR, X-ray scattering, microscopy, and sorption techniques reveal that chitin possesses intra and intermolecular hydrogen bonds with high degree of crystallinity and polymorphism and ordered fibrillar structure [6].

Chitosan is a common constituent of fungal cell walls, particularly in the Zygomycete class [7]. Chitin is insoluble in common solvents so it is relatively difficult to process it. Consequently, deacetylation of chitin to chitosan renders it more soluble. Due to a higher solubility of chitosan in common solvents, it is rendered more suitable as a biopolymer [8]. Chitosan is derived by the process of deacetylation of chitin in which alkaline treatment is given to chitin to remove acetyl groups to form chitosan. As a natural product it is found in shells of crustaceans. It is a copolymer composed of *N*-acetyl-d-glucosamine and d-glucosamine units, available in different grades depending upon the degree of acetylated moieties [9]. Reports are also available according to which chitosan is copolymer composed of *N*-acetyl-2-amino-2-deoxy-d-glucopyranose and 2-amino-2-deoxy-d-glucopyranose, where the two types of repeating units are linked by (1 →4)-glycosidic bonds [10]. The relative proportions *N*-acetyl-d-glucosamine and d-glucosamine residues provide specific structural changes and are responsible for main structural differences in chitosan. Chitosan is available in two grades low and high molecular weight. Low molecular weight chitosan grade is characterized by molecular weight ranging between 20 kDa and 190 kDa with degree of deacetylation (DD) < 75% where as high grade chitosan has molecular weight between 190 kDa and 375 kDa with degree of deacetylation (DD) > 75% [11]. It is obtained from carapaces of marine crustaceans such as crabs and shrimps [12]. It exhibits hydrophilic and polycationic properties because amino groups and hydroxyl groups are present in backbone of chitosan.

Due to their origin from natural sources, chitin and chitosan are considered biocompatible, biodegradable, and nontoxic [13]. Mostly natural polysaccharides are acidic in nature like cellulose but chitin and chitosan are basic polysaccharides. They exhibit diverse properties like polyelectrolyte behavior, metal chelation, ability to form films, solubility in various media, formation of polyoxysalt, and viscosity [14]. Chitin is hydrophobic in nature and dissolves in organic solvents like hexafuoroisopropanol, hexfluoroacetone, and chloroalcohols in conjunction with aqueous solution of mineral acids [15] and dimethylacetamide (DMAc) containing 5% lithium chloride (LiCl) [16]. Reactive amino groups and hydroxyl groups determine chemical properties of chitosan. Among the biological properties of chitosan biocompatibility, biodegradability, adhesiveness, bone formation ability, antitumor ability, central nervous system depressant, and immunoadjuvant property are well known [6]. Over the last decades, chitosan has proved to be safe excipient in drug formulations [17]. The flexibility of the polymer allows its processing into beads, gels, membranes, micro and nanofibres/particles/films, or sponges [18]. Both chitin and chitosan have extended their applications as an implantable biomaterial in several biomedical applications. There has been extensive research in this context from Asia itself.

13.2 Biomedical Applications

The medical field has witnessed the incorporation of diverse materials from living as well as non-living origin to solve different problems. Biomaterials are preferred over others because of their less toxicity, immunogenicity, and easy availability. Moreover, they require less input for their production. Cellulose, starch, carrageenan, gelatin, lipids, proteins, and others are the biomaterials of choice that have been employed to serve not only the industrial purposes but also medical field to a great extent. Chitin and chitosan possess special properties to be equally used as excellent biomaterial in the field of medicine. Their use in medicine is favored due to their property of being fabricated into gels, films, nanoparticles, and nanocomposites. Both physical and chemical properties of chitosan allow it to be

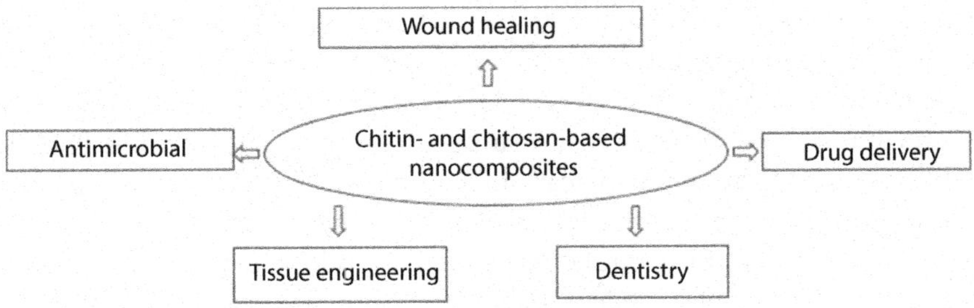

Figure 13.1 Biomedical applications of chitin- and chitosan-based nanocomposites.

incorporated with compounds of medical significance to effectively treat the ailment. Chitosan nanocomposites have been employed in various medical (Figure 13.1) applications described ahead.

13.2.1 Wound Healing

For utilization of a polymer in wound healing applications, a biomaterial should possess favorable features that can promote rapid skin regeneration. In addition, the wound care materials used for injuries should be (i) nontoxic, (ii) able to prevent bleeding, allergy, or infection, (iii) retain moisture, (iv) absorb excessive exudates, (v) permeable to gases, and (vi) fasten the healing process. Ranging from simpler wound covers to complex artificial skin constructs, water-soluble chitin, fluid-absorbing chitin, chitosan-cotton and keratin-chitosan films/fibres/hydrogels/sponges have been demonstrated to provide an accelerated healing rate. To further enhance these properties, the chitin and chitosan biomaterials are associated with different types of polymers (alginate, hyaluronic acid, polyethylene glycol, polyvinylalcohol, polyglutamic acid, 2-hydroxyethyl methacrylate). Hyaluronic acid is an extracellular molecule that holds *N*-acetylglucosamine and imparts wound repairing properties to chitin. Similar to the natural extracellular matrix, these composites can provide more surface area and are suitable as substrates for cell adhesion and proliferation [19]. Chitosan can provide favorable environment for tissue formation followed by angiogenesis and regular deposition of thin collagen fibres, leading to a precise repair of dermo-epidermal wounds. Overall, both chitin- and chitosan-based materials support polymorphonuclear cell proliferation, fibroblast activation, cytokine production, cell migration, stimulation of interleukins, collagen synthesis, promote granulation, vascularization, etc. [20]. During skin loss in case of severe burns, an extensive fluid loss may also accompany microbial infections. β-chitin nanocomposite hydrogels provide good tensile strength, more flexibility, tear proof features, moisture-retaining ability, gaseous permeability, and easy removal after healing. These hydrogels confer high moisture content and enzymatic degradability further increasing wound contraction. Water-soluble composites are also highly recommended as they are able to prevent excessive scar formation and cell death induced by serum deprivation. In the same manner, multilayered chitin dressings are designed for treating mustard-burns. Chitosan hydrogels can promote cell adhesion and proliferation. The positive charge allows for interactions with glycosaminoglycans which can attract growth factors. Chitin- and chitosan-based biomaterials retain their integrity even after sterilization procedures that are necessary for maintaining aseptic conditions. Taking into account the overall benefits viz. human compatibility, antimicrobial feature, regenerating environment offered by chitin- and chitosan-based materials, their potential as an effective wound management component is promising [20].

Controlling the microflora at the site of injury is another important consideration that can otherwise impede the healing. Encapsulating antibiotics within chitosan sponge or fibril is an innovative design which can act as an ideal material for better wound healing. In addition, there are incidences of antibiotic resistance at the site of injury that can pose a challenge in the form of infection and delayed healing. Blood loss at wound site is another cause of death after trauma. Considering the biomedical applications of chitin, multifunctional nanocomposite enclosing some essential antibacterial drugs is fabricated to improve wound curing [21]. The cross-linking density determines the swelling ability of a polymer which can thereafter control the absorption rate, antibiotic release or diffusion and hence, it controls over bleeding and infection. Furthermore, chitin can also indirectly stimulate complement-mediated chemotaxis and certain interleukins. Chitosan has also been found with intrinsic antimicrobial properties which in turn is largely influenced by its molecular weight, ionic strength, and pH of the surrounding environment. It is possible that positively charged chitosan molecules interact with the negatively charged microbial cell membranes leading to membrane damage, and leakage of cellular components or cause agglutination [22].

Silver (Ag) has been used in skin ointments for a long time. With the advent of the Ag nanoparticles, more medical applications have come up. Ag nanoparticles coupled to chitin and its derivatives are designed to enhance the antimicrobial properties of nanocomposites. Nanocrystalline chitin whiskers obtained from crab shells [23] were found to deliver antimicrobial compounds to the site of infection. Chitin-nano Ag composite scaffolds and porous chitosan-Ag nanocomposite displayed good inhibition against bacteria, higher healing rate, and effective clotting ability [20, 24]. Similarly, nano-titanium oxide-chitosan composite with collagen artificial skin demonstrated a unique bactericidal action in rat models and also shows faster recovery and immune-stimulatory effects in the experimental organisms [25]. A concentartion-dependent antifungal activity and higher epithelialization rate were revealed by chitosan formulations under varying conditions both *in vivo* and *in vitro.* Chitosan promoted an ordered organization in cutaneous tissue in patients who underwent plastic surgery. It was suggested to improve adherence, haemostasis, and tissue architecture during healing process. Polypropylene-collagen-chitosan composite could be used as an automatic releaser of wound dressing material. Again, silver and polyphosphate (as a procoagulant) were found to enhance clotting rate, platelet adhesion, thrombin generation, and absorbing rate. Polyelectrolyte chitosan complexes, gold/colloidal scaffolds could accelerate cell proliferation and wound healing. Copper nanoparticles enhance wound healing by modulating cytokines and growth factors. Chitosan possesses antibacterial and antifungal properties. Combining these properties of copper nanoparticles and chitosan in a single nanocomposite results in better and faster wound healing in open excision wounded Wistar rats. The nanocomposite promoted angiogenesis, deposition of collagen, and proliferation of fibroblasts as revealed by upregulation of VEGF and TGF-β1 [26]. Cotton gauze has gained prime importance as wound dressing material. Imparting antimicrobial and moisture-retaining ability to cotton gauzes will render it more advantageous in wound management. To achieve this, a nanocomposite of chitosan, zinc oxide, and silver is impregnated over the gauze. Moisture-retaining ability of the resulting blend increased upto 38% while effectiveness against *Escherichi coli* and *Staphylococcus aureus* is enhanced [27]. The matrix of castor oil after reinforcement with chitosan modified zinc oxide nanoparticles resulted into a nanocomposite having increased ability in (i) moisture retaining, (ii) porosity, (iii) hydrophilicity, (iv) heat stability, (v) permeability to O_2, (vi) biodegradability, (vii)Tensile strength, and (viii) flexibility. Further, the antibacterial property of the nanocomposite is enhanced against *E. coli, S. aureus, M. luteus.* The bactericidal action depends on chitosan-ZnO content. Cytocompatibility of this nanocomposite is promising and wound healing is rapid [28]. Chitosan films reinforced with chitin nanofibers also possess good wound healing property [29]. Further, the nanocomposite of chitosan and fibrin exhibits excellent wound healing properties as complete healing of the open excised wound results on 14th day in contrast to 22 days in control in albino rats. Moreover, this composite efficiently works against *E. coli* and *S. aureus* [30].

13.2.2 Drug Delivery

A competent drug delivery system must be able to achieve a targeted delivery along with controlled rate of drug release. Both the features will not only improve the therapeutic effectiveness but also reduce the associated side effects. Biopolymers have high stability as drug carrier. Additional loading capacities and regulation over several physical/chemical attributes render them more potent in drug delivery systems [31]. The possibility of generating nanoscale dimensions in biodegradable polymers has further opened new realms in the field of therapeutics and pharmaceutics. The natural polymers are promising drug transporters with reduced side effects on sites other than target and have allowed to cross the biological barriers especially in case of poorly absorbed drugs [32]. Specially designed drug delivery systems are required for drug transport only at the affected sites for maximum effect. Those with poor solubility will have limited tissue availability and a lesser absorption rate. This will lead to a poor kinetics profile lowering its medical applications. Various formulations are under investigations to overpower these limitations. However, the major concern with the use of polymeric nanoparticles is that they do not have an exact binding site where the drug molecule can interact. An alternate to this is to employ those polymers that have hydro/electrophilic domains. These nanocomposites can not only improve the target specificity but also augment the availability of ample drug dose to the tissue and reduce side effects. *In silico* modeling strategies are exploited for optimizing several features of a nanocarrier in order to develop an ideal delivery systems. Computational-based approach has simplified both structure and ligand-based drug discovery and designing. Owing to the stability, flexibility in surface, degradability, compatibility with the biological systems, the nanocomposite materials can be easily modified chemically or mechanically to encapsulate proteins/peptides, polysachharide, or other therapeutic drugs. Nanocomposites can be designed as per need by altering the method of preparation to obtain different properties suiting to various applications. They can even deliver biomolecules like DNA (vectors) or peptides by either enclosing them inside or attaching to their exterior surface. Owing to their size, these particles can pass the tiniest spaces like capillaries, cells, and tissue gaps to reach at target site. Avoiding rapid clearance from macrophages prolongs their duration in blood stream. The load release can be controlled by maintaining pH, ion, temperature, and availability of materials which increase drug utility and reduce toxic effects.

A nanocomposite generated from chitosan and organic rectorite after fabricating into films was tested for controlled drug release and antibacterial activity. The results revealed slow drug release profile of the nanocomposite which enhances further on increasing the content of the organic rectorite. Further, the films also exhibit antimicrobial activity along with excellent water barrier and UV protective property [33]. Recently, nanogels have gained importance as effective drug carriers. A nanogel composite of chitosan, sodium alginate, and hydroxyapatite was generated through gamma-rays-mediated cross-linking. Delivery kinetics of doxorubicin, an anticancer drug, revealed a pH sensitive release profile with maximum release at pH 5. This nanocomposite gel is claimed to be effective for liver cancer [34]. Chitosan bears positive charge so can interact with negatively charged molecules thereby forming sol-gel transition state. This property makes chitosan practically feasible for nanoencapsultion purposes [35]. Due to this property, encapsulation of water-insoluble anticancer drug, camptothecin into the nanoparticles generated from modified chitosan results into the formation of composite nanoparticles having the loading capacity of above 80% and sustained release of the drug upto 1 week in human breast cancer xenografts implanted subcutaneously in nude mice. The nanoparticles show enhanced transport through blood circulation and increased accumulation in the cancer tissues to counter the tumor cells effectively. This nanocomposite holds great promise for cancer therapy. Synthesis of nanocomposite beads from zinc oxide nanoparticles and chitosan hydrogel beads after testing for their potential to release ibuprofen, revealed extended drug release profile of the nanocomposite [36]. Upto 96 h sustained release of methotrexate, an anticancer drug is observed when encapsulated in the nanocomposite of chitosan and fibrin. Further, the composite exhibits excellent wound healing properties as complete healing of the open excised wound results on 14th day in contrast to 22 days in control in albino rats. Moreover, this composite efficiently works against *E. coli* and *S. aureus* [30]. A nanocomposite formulated from cationic chitosan and anionic gold shows enhanced ascorbic acid release [37]. Similarly, a pH-dependent sustained release of ibuprofen is achieved by using the nanocomposites of chitosan-montmorillonite [38]. Likewise a pH- and matrix-dependent controlled release of curcumin occurs after encapsulation of the drug in starch-chitosan nanocomposite [39]. Moreover to detect a target certain biomarkers can be helpful, for example, lactobionic acid can bind to receptors on liver. A delivery system for DOX designed from carboxymethyl chitosan and graphene oxide with lactobionic acid, show pH-based release of anticancer drug [40]. *In situ* polymerization of chitosan with acrylic acid was suggested for pH-dependent drug release in gut.

Dispersion of montmorillonite particles in the solution of chitosan yields a nanocomposite which is biocompatible and enhances cell proliferation *in vitro*. Further, the mucoadhesive property of this nanocomposite reveals its potential as a new candidate for drug delivery [35]. Similarly, a prolonged drug release was observed by using the nanocomposite of ZnO nanoparticles and chitosan [40]. Through solvent evaporation method, a nanocomposite was prepared through blending of chitosan and sodium alginate with a matrix of cloisite 30B solution which also behaves as co-emulsifier. The nanocomposite upon testing for its drug delivery potential under varying pH, time duration, and curcumin drug concentration shows pronounced drug release in basic medium [41]. Similar blending of starch and chitosan with cloisite yields a nanocomposite that shows pH- and matrix-dependent drug release [42]. In order to ascertain the effect of chitosan on the release profile of the Glucantime drug, clay nanoparticles were blended with the hydrogel of poly (*N*-2-vinil-pirrolidone) containing chitosan to generate a nanocomposite through gamma irradiation-mediated cross-linking. The nanocomposite shows promising drug release for treating skin wounds leishmaniasis [43]. A nanocomplex of cationic nanoparticles of chitosan and anionic gold colloids was prepared to generate a chitosan-gold nanocomposite which meets the requirements for drug carrier and delivery systems [44]. A nanocomposite generated from chitosan-polyvinyl alcohol and cloisite 30B is reported to be a good candidate for delivery of the anticancer drug, curcumin [45].

In order to achieve minimal toxicity, the modern drug delivery systems aim to minimize the systemic distribution of the drug in addition to maximizing their target accumulation. This is specially designed for tumor cells. Using the modern sophisticated nanocarriers or delivery systems, an anticancer drug can be specifically targeted to cancer cells either through stimuli-based approach or through a specific receptor-mediated targeting. The former approach depends on the physiochemical properties of the drug which are released in changed pH, enzyme system, or other factor while the later approach relies on designing specific structural moieties over the drugs that can act as legends for specific receptor on the target cells. Magnetic nanoparticles are one of the materials that can used to enhance the target accumulation of the drug with minimum systemic effects. Magnetic particles of magnetite and chitosan have been designed through coacervation approach which is responsive to magnetic gradients. Using these magnetic nanoparticles, gemcitabine which is an anticancer drug has been introduced intravenously. The nanoparticles exhibit good loading ability as well as slow drug release profile. Under magnetic gradient, these nanoparticles show increased accumulation of the drug at the target site as revealed by the Prussian blue staining. This magnetically active nanomedicine can be used as an effective approach for cancer therapy [46].

Though chitin- and chitosan-based nanocomposites have been under extensive research, there are lacunae in the field of *in vivo* biomolecule transport via these materials. Further improvements can be done in getting the composites available for oral or other routes of administration in humans based on more and more clinical trials. For instance, nasal route of drug administration is an alternative route for intravenous injection which can be largely assisted by nanocarrier. However, these drug-laden carriers can get easily washed away by the mucous/ciliary movements causing a reduction in availability of the therapeutic agent [47]. Combining the magnetic and stimuli responsiveness as well as fluorescent imaging in a single nanocomposite of chitosan has made it an innovative delivery system. The nanoparticles exhibit pH sensitive release kinetics and show extended delivery of the drug over longer periods. This strategy not only allows us to increase target drug accumulation through magnetic gradient application but also allows to moniter the transport through fluorescence imaging. Therefore this nanocomposite holds great promise in smart drug delivery systems [48]. A nanocomposite complex of magnetic particles, chitosan and cis-aconitic anhydride loaded with doxorubicin. Cis-aconitite anhydride acts as a pH sensitive linker of the drug with the nanocomposite. The nanocomposite shows good releasing property at pH 5 [49]. Cancer is charecterized by high acidic local environment due to Warburg effect, consequently pH-based delivery systems prove useful in smart delivery strategies. Chitosan-based magnetic nanoparticles loaded with anticancer drugs are capable to release their load depending upon pH. In a study, *N*-naphthyl-*O*-dimethymaleoyl chitosan-based drug-loaded magnetic nanoparticles (*N* Chitosan-DMNPs) were fabricated which exhibit greater stability at higher pH and release rapidly the loaded drug, doxorubicin under low pH. This nanocomposite is believed to possess highly potent ability to diagnose and monitor the disease thereby revealing its significance in theragnosis for cancer therapy [50].

13.2.3 Tissue Engineering

Tissue reconstruction involves reviving the function of an injured tissue by seeding cells onto a matrix. The principle behind the tissue engineering states that healthy stem cell progenitors, when recruited at the injured site, can help to regenerate the damaged tissue. Traditionally, the metallic plates used during bone healing surgeries have a tendency to deform under mechanical pressure. Hence, the composite-based tissue engineering came into the picture to resist the failures by imitating the tissue structure and function. Progenitor cells added along with growth factor onto the surface of biopolymer will proliferate and differentiate into the desired cell type. The research pursues to substitute the diseased/damaged tissues by creating a suitable environment for proper proliferation and differentiation of cells. Bone and cartilage systems have specialized structure and composition to support body mechanics, and only a few plastic tissue products were approved for clinical use. The main reason behind the failure of most of the implants is a lack of nutrient and gaseous transport that culminate into the death of implanted cells. Under natural conditions, tissues face many stresses that the engineered constructs will not be able to tolerate. The wholesome elasticity, strength, and other properties of natural tissue may vary under different development and physiological conditions of an organism. So, a good temporary matrix is the major limiting factor in this context.

Tissue engineering recruits the living cells modified either genetically or under the influence of their extra-cellular environment to design bio-implants of organs or tissues. The aim is to sustain the function of a particular system by providing a natural substitute. For this purpose, the scaffold should possess typical features to prevent the entire framework from collapsing including certain other characteristics of high porosity with an appropriate pore size, more surface area, biodegradability with the rate of degradation equal to that of new tissue formation, and nontoxicity to living cells [51]. Biocomposites should be able to mold in different shapes, sizes with sufficient porosity, permitting host cell migration, and tissue growth. They should be biodegradable, bioactive, water soluble, complex forming, and biocompatible. As the biomaterial will be directly interacting with the cells to promote their adhesion, division, migration, and differentiation, therefore it is mandatory for the polymer to maintain the standards prior to their use in such biomedical implications [52]. Several biological active molecules are known to interact with the cells at nanoscales in a similar manner as the extracellular matrix does. The chitin-based nanocomposites have been serving as a replacement for the artificial ones in various biomedical applications. Nano-assemblies are designed to facilitate stem-cell differentiation or attachment of a particular cell type to promote the regeneration. Three-dimensional porous polymeric matrix further made robust and flexible by adding some minerals, for example, calcium which will increase its osteogenic potential. The acidic residues on the surface of carboxymethyl chitin will attract calcium ions that in turn can attract phosphate ions laying the foundation for calcium apatite layer. Higher chitin content and higher porosity promote higher calcification. To further improve the mechanical properties of chitin and chitosan, organic and inorganic compounds viz. cellulose, PEG, calcium phosphate, bioactive glass ceramic (BGC), hydroxyapatite (HAP), etc., can be supplemented with these polymers. For chitin to be used for bone repair and reconstruction, its mechanical property needs to be enhanced by adding other biomaterials like HAP and BGC. Bioglass is a biomaterial widely employed in orthopedics and dentistry to repair bone defects. Titanium coated with bioglass is recommended over HAP for their integrity as bone regenerating material. BGC can bind to both hard and soft tissues due to the formation of carbonated apatite layer on the coated surface contributed by silicon dioxide (SiO_2). It can also promote osteoblast and bone marrow stromal cell proliferation and differentiation [18]. It has been proposed that high pH and polymerization of SiO_2 tend to aggregate calcium oxide and P_2O_5 *in vivo* as a nucleation process leading to the growth of apatite layer. Silica chelates are necessary for mineralization of hard tissues.

Chitosan modified with silanol and calcium, chitosan-silica xerogel, chitin-nano silica hydrogel showed apatite-forming properties and was assured suitable for bone engineering. The sol-gel technique can be used to produce BGC nanoparticles. Chitin or chitosan/nano-BGC composites developed by lyophilization exhibited high porosity, nontoxicity, and high swelling capacity. This nanocomposite scaffold is active in alveolar bone regeneration [53]. Gelatin can be blended with nano-HAP as well as with nano-BGC to make these scaffolds comparatively highly porous [54]. Mammalian fibroblast cells and mesenchymal stem cells were able to proliferate and differentiate on the chitin HAP matrix. β-Chitn nano-HAP hydrogels can be easily manufactured and provide a close contact with surrounding tissues. β-Chitn hydrogel and nano-HAP are mixed to generate scaffolds that have been proved better than either of the biomaterial when used alone. Since silica exhibits similar properties to chitin, nano-silica chitin hydrogel with its bioactive and biocompatible properties is useful in bone reconstruction. These scaffolding materials can restore and maintain human tissues as they form a good cell anchorage and proliferation matrix [55]. As for the cartilage, the tissue undergoes negligible wearing that makes it rather complex both in structure and function, and delivers unique properties. The fixed negative charge due to a high content of aggrecan and proteoglycan in the extracellular matrix support an increased internal swelling pressures and uneven residual stresses within the cartilage. Chitin can maintain the morphology of the chondrocytes without affecting the cell-specific properties or phenotype. Pure β-Chitin scaffolds can support the cartilage cells advocating its potential in cartilage tissue engineering. Chitosan/HAP/genipin-cross-linked nanofibrous scaffolds induce cell differentiation and are reported with enhanced mechanical and osteoconductive properties [56].

Technological advances have enabled to produce sophisticated nanostructures to support the growth of artificial organs/tissues from diverse cell types. Nanotechnology combined with freeze-drying, particle leaching, electrospinning, soft lithography, and photolithography can build the foundation that mimics the natural occurring matrices for an overall cell growth and differentiation. Electrospinning led to the production of carboxymethyl chitin and PVA-based nanofibers for artificial tissue engineering. Glutaraldehyde vapors followed by heat rendered them aqueous insoluble. Cells could attach and spread onto these nanofibrous scaffolds [57]. Chitosan/HAP and chitosan/calcium phosphate composites prepared by lyophilization technique can have potential use in tissue engineering. These scaffolds are more porous and biocompatible than pure chitosan. HAP reduces water absorption of chitosan thereby increasing the tensile strength under humid conditions. These composites are biologically well accepted and osteoconductive in nature, a prerequisite for biomedical application. A chitosan-based novel montmorillonite (MMT) and HAP nanocomposite display an improved nanomechanical strength, thermal property and provide better cell proliferation. The exfoliated MMT can restrict the degradation of chitosan and addition of HAP confers additional benefits to chitosan. The nanomaterials can have potent applications in bone engineering [58]. Chitin can arrange itself into a particular hierarchy; the long-range packaging can be exploited to construct three-dimensional patterns or liquid crystalline structures that are reported either in gel suspensions or biomineralization. Chitin flakes hydrolyzed in the presence of acid form crystalline whiskers. The degradation rate of chitosan is inversely related to the degree of the crystal formation. Electrospun chitosan whiskers were created using polyvinyl alcohol that were found compatible with osteoblasts and promoted their proliferation [59]. A different porous chitin- and chitosan-based nano diopside composite designed using freeze drying is reported with better swelling, degradation, bioactive, cell adherence, and biocompatible abilities with the human gingival fibroblasts [60].

Thermoreversible nanocomposite hydrogels designed from chitosan could encapsulate the cells and other materials required for bone tissue formation. Chitin and its derivatives are known to strengthen these gel-like structures. *In situ* cross-linking and rapid gelatinization *in vivo* enable them to serve as an injectable biomaterial for cartilage regeneration. Likewise, pH sensitive hydrogels can be prepared where pH affects the cross-linking, stiffness, and swelling rate of the gel. An injectable hydrogel produced by a combined ionic and covalent bonding with a thermosensitive character showed a long-lasting gelation [61, 62] suitable for cartilage tissue genesis. Electrospun silk nanofibers added to chitosan/glycerophosphate are tougher as regenerating material [63]. Incorporating PEG in the *N*-succinyl chitosan-based hydrogels could decrease the degradation of the biomaterial by enzymes without affecting cell adhesion and migration properties. Such a gel can withhold human adipose stem cells up to one week [64] and would favor chondrogenesis and osteogenesis in repair strategies when supplemented with some growth factors [65]. Photosensitive chitosan-gelatin scaffolds are created using carbodiimide as a cross-linking agent. Injectable pluronic/chitosan/keratin nanohydrogels armored with Laponite proved highly tough, thermoresponsive, and biocompatible and were proposed as an excellent matrix for cartilage tissue formation [66].

Nanoparticles of zirconium oxide (ZrO_2), when incorporated to chitin-chitosan, can promote bone formation. The nanocomposite scaffolds designed through lyophilization were better in swelling, nontoxic, show uniform cell adhesion and provided a controlled degradation [67]. Numerous studies with chitin and chitosan as bone reconstruction substitutes have further shown improved properties through the use of nanophase ceramics like titania, zirconia, alumina, etc. Zirconia (crystalline zirconium dioxide) is used extensively in dentistry and orthopedic implants, and constructs prosthetic devices. Its biocompatibility allows cells to divide and proliferate without any side effect. It is inert as titanium and does not display any physical/chemical bonding. In addition to this, ZrO_2 exhibits excellent osteointegration that makes it an ideal material for bone implant [68]. Fucoidan, an anionic polysaccharide, comprises of sulfate ester and fucose groups essential for cellular activities due to which the biomolecule is being targeted in tissue remodeling. It can stimulate collagen matrix formation, both *in vitro* and *in vivo* angiogenesis and other osteogenic genes such as bone morphogenetic protein-2, osteocalcin, alkaline phosphatase, collagen 1, osteonectin, etc., to support artificial bone development. Fucoidan augmented with the chitosan nano-HAP scaffold was incorporated into a three-dimensional scaffold using the freeze-drying method. The resulting structure demonstrated a high rate of differentiation in mesenchymal progenitors to produce bone minerals [69]. The properties of chitosan and its derivatives for bone tissue engineering applications have been extensively reviewed [70]. Despite the developments in the field of regenerating bone/cartilage through chitin- and chitosan-based nanomaterials, there are negligible reports for skin or other tissues due to inadequate success rate in this context.

Cardiac tissue consists of electrically responsive cells which are capable of transmitting the electrical signals responsible for pumping action of the heart. However adult and damaged cardiomyocytes have poor regenerative potential [71]. There is a burning interest to regenerate cardiac tissue by tissue regeneration technology. This requires electrically active material for implantation to regain cardiac tissue functionality. Different experiments have been carried out to address this issue [72–75] and diverse conductive materials have been developed over time [76–78]. Diverse reports mention the biocompatibility, nontoxicity, less immunogenicity of chitosan. Besides, chitosan possesses haemostatic properties [79, 80] and is antithrombogenic [81]. Chitosan in itself is nonconductive to electrical signals so in order to make it effective in cardiac tissue engineering, it is blended with electrically active materials. In one such study, the doping material used was carbon nanofibers to generate a chitosan-based nanocomposite that exhibited excellent conductivity. The scaffolds of this composite on implanting with neonatal rat heart cells and culturing for two weeks under nonelectrical stimulation result into the enhanced proliferation of cardiac cells which filled the porous scaffold rapidly as compared to chitosan scaffold alone. This could be attributed to transmission of electrical signals through the composite scaffold. Such nanocomposites pave way to refine and improve the biomaterials for cardiac tissue engineering [82]. Through ionic cross-linking, chitosan and sodium tripolyphosphate were blended in presence of magnetite and plasminogen activator to generate chitosan-based magnetic nanoparticles encapsulating plasminogen activator. *In vitro* assay shows that these nanoparticles are noncytotoxic and nonhaemolytic. The nanoparticles are sensitive to magnetic gradient and release the plasminogen activator in serum. In an experiment involving rat having a clot in the left iliac artery, the nanoparticles were guided to the clot through the application of the magnetic field. After about 50 minutes, the blood flow in the blocked artery restored to normal thereby validating the thrombolytic activity of the nanocomposite [83].

13.2.4 Dental Application

Growth of microbes in the mouth results in tooth decay. The decaying away of hard materials of teeth and demineralization are characteristic features of most dental problems. Bacterial growth result into plaque formation over the teeth causing the black rust like teeth. Pain, sensitivity to high and cold temperatures plus gum swelling are among the fruits of dental disorders and may also need the removal of entire teeth. Dental caries are common and mainly treated by filling. Since the growth of microbes is the main reason of the dental problems, designing effective and sophisticated biomaterials to tackle the microbial growth will solve the problem to a great extent. In order to prevent demineralization and gingival problems in teeth caused by enhanced microbial growth, use of chitosan and zinc oxide nanoparticles as antimicrobial agents in the composite dental resins resulted in inhibition of both function as well as number of microbes. Such nanocomposite is thought to be valuable in preventing film and plaque formation by microbes which protects the teeth damage [84]. One of the diseases that result into tooth loss is periodontitis in which the tooth supporting tissues are progressively destroyed [85]. The disease is characterized by intensive growth of *Aggregatibacter actinomycetemcomitans*. In an attempt to design an effective therapy for the disease, a nanocomposite of chitosan gel and copper nano particles was generated which showed sustained release of copper ions resulting into long term inhibition of the pathogen. In order to combat periodontal gum disease, chitosan-TPP/oligonucleotide nanoparticle nanocomposite shows slow release of the nucleotide. This nanocomposite has promising impact on periodontal disease.

13.2.5 Antimicrobial Applications

Chitin and chitosan nanocrystals manufactured though acid hydrolysis have high tensile strength, low density and biodegradability. The chitin domains are modified during the process to yield nanoscale crystalline structures. Being produced from a renewable source, they have attracted immense interest in pharmaceutical industry. Chitosan can form complexes with different sources like with organic and inorganic compounds, gelatin-drug films or as mucoadhesive membranes, nanospheres, silver nanocomposites, etc. Chitosan-silver nanocomposite hydrogel beads possess comparatively more swelling capacity and a higher antibacterial activity than hydrogel alone, depending upon the content of Ag nanoparticles [86]. ZnO has an excellent photocatalytic activity, high thermal conductivity along with other properties and is considered safe for consumption. Zinc complex with chitosan-polyacrylamide was examined with antimicrobial properties and controlled olfloxacin release. Similarly, polyacrylamide/cupric sulphate composite employed for olfloxacin delivery exhibited antibacterial activity [87]. Chitosan-silver chloride nanoparticle nanocomposite shows complete healing of the burn wounds in rat models upto 14 days. Collagen regeneration at the wound site increases. The composite also exhibited antibacterial properties to combat infection at the wounded sites [88]. Hydrogel made from nanocomposite, chitosan/PVA show strong inhibition against *E. coli* and *S. aureus*. Further, the composite possesses enhanced mechanical stability, prevents biofilm formation, and also acts as a scaffold for dermal fibroblast cell growth [89]. Gamma irradiation synthesized nanocomposite membranes of chitosan, polyvinyl alcohol and silver exhibit antibacterial, non-thrombogenic and -haemolytic property. There occurs significant reduction of the microbes due to this composite [90]. With a view to synthesize an antimicrobial topical cream, chitosan, montmorillonite clay and nylon 6 in the ratio of 2:1:1 served as an effective composite to act against *E. coli, Pseudomonas* sp. *A. niger, Candida albicans and C. tropicalis.* However the nanocomposite did not show any inhibitory effect against *S. aureus* [91]. Using sodium tripolyphosphate as a cross-linker, a nanocomposite was synthesized from chitosan hydrogel beads and CuO nanoparticles. The nanocomposite exhibits a unique swelling behavior which depends on pH and possesses good antibacterial action against *E. coli* and *S. aureus* [92]. Incorporation of ZnO nanorods into the matrix of carboxymethyl chitosan generates an antibacterial nanocomposite hydrogel which shows bactericidal action against *E. coli* and *S. aureus* thereby validating its potential in the medical field [93]. A similar nanocomposite of chitosan with silver nano particles is effective against *S. aureus, Bacillus subtilis, Streptococcus faecalis, E. coli, P. aeruginosa and Neisseria gonorrhoeae* along with the fungal species, *C. albicans* [94].

13.3 Conclusion

Chitin and chitosan have natural source-based origin. They are biocompatible, biodegradable, and nontoxic, less allergic and less immune reactive. Both chitin and chitosan are excellent biomaterials and are serving mankind for decades. They have revolutionized the tissue engineering and drug delivery realms of the medical field. Since both of them have intrinsic antimicrobial property, extensive research needs to be carried out to use this property to generate effective and sophisticated materials for dental applications. Besides, production of systems for combating infections in different organ systems will practically broaden its antimicrobial application for medical purposes. Though chitin- and chitosan-based nanocomposites have been under extensive research, there are lacunae in the field of *in vivo* biomolecule transport and drug delivery via these materials. Further improvements can be done in getting the composites available for oral or other routes of administration in humans based on more and more clinical trials. Despite the developments in the field of regenerating bone/cartilage through chitin- and chitosan-based nanomaterials, there are negligible reports for skin or other tissues due to inadequate success rate in this context. Efforts need to be intensified in this regard.

References

1. Struszczyk, H., Pośpieszny, H., Gamzazade, A., Chitin and chitosan. *Pol. -Russian monogr.*, 1, 2002.
2. Agrawal, A., Rajamani, V., Reddy, V.S., Mukherjee, S.K., and Bhatnagar, R.K. Transgenic plants over-expressing insect-specific microRNA acquire insecticidal activity against Helicoverpa armigera: an alternative to Bt-toxin technology. *Trans. Res.*, 24(5), 791–801, 2015.
3. Peniston, Q.P., Johnson, E.L., Process for the manufacture of chitosan. Google Patents, 1980.
4. Nisha, S., Seenivasan, A., Vasanth, D. Chitin and its derivatives: structure, production, and their applications. International conference on Signal Processing, Communication, Power and Embedded System (SCOPES), 2016.
5. Blackwell, J., Parker, K., Rudall, K. Chitin in pogonophore tubes. *J. Mar. Biol. Assoc. U. K.*, 45(3), 659–661, 1965.
6. Dutta, P.K.,Dutta, J., Tripathi, V., Chitin and chitosan: Chemistry, properties and applications. *J. Sci. Ind. Res.*, 63(1), 20–31, 2004.
7. Fai, A.E.C., Stamford, T., Stamford-Arnaud, T.M., Santa-Cruz, P.D., Silva, M.C., Campos-Takaki, G.M., Stamford, T.L., Physico-chemical characteristics and functional properties of chitin and chitosan produced by Mucor circinelloides using yam bean as substrate. *Molecules,* 16(8), 7143–7154, 2011.
8. Croisier, F., Jerome, C., Chitosan-based biomaterials for tissue engineering. *Eur. Polym. J.* , 49(4), 780–792, 2013.
9. Hoppe-Seiler, F., Chitin and chitosan. *Ber. Dtsch. Chem. Ges.*, 27, 3329–3331, 1994.
10. Roberts, G.A., Structure of chitin and chitosan. In: *Chitin chemistry,* 1–53, Springer, 1992
11. Dash, M., Chiellini, F., Ottenbrite, R., Chiellini, E., Chitosan—A versatile semi-synthetic polymer in biomedical applications. *Prog. Polym. Sci.*, 36(8), 981–1014, 2011.
12. Shahidi, F., Arachchi, J.K.V., Jeon, Y.J., Food applications of chitin and chitosans. *Trends Food Sci. Technol.*, 10(2), 37–51, 1999.
13. Muzzarelli, R., Muzzarelli, C., Chitosan chemistry: relevance to the biomedical sciences. *Polysaccharides I,* Springer, pp. 151–209, 2005.
14. Domard, A., Smidsrød, O., *Advances in chitin science.* Norges teknisk-naturvitenskapelige universitet Institutt for bioteknologi, 2003.
15. Dutta, P.K., Ravikumar, M., Dutta, J., Chitin and chitosan for versatile applications. *J. Macromol. Sci., Part C: Polym. Rev.*, 42(3), 307–354, 2002.
16. Dutta, P., Khatua, M., Dutta, J., Prasad, R., Use of Chitosan-DMAc/LiCl gel as drug carriers. *Int. J. Chem. Sci.,* 1: p. 93, 2003.
17. Felt, O., Buri, P., Gurny, R., Chitosan: a unique polysaccharide for drug delivery. *Drug Dev. Ind. Pharma.*, 24(11), 979–993, 1998.

18. Jayakumar, R., Menon, D., Manzoor, K., Nair, S., Tamura, H., Biomedical applications of chitin and chitosan based nanomaterials—A short review. *Carbohydr. Polym.*, 82(2), 227–232, 2010.
19. Naseri, N., Algan, C., Jacobs, V., John, M., Oksman, K., Mathew, A. P., Electrospun chitosan-based nanocomposite mats reinforced with chitin nanocrystals for wound dressing. *Carbohydr. Polym.*, 109, 7–15, 2014.
20. Azuma, K., Izumi, R., Osaki, T., Ifuku, S., Morimoto, M., Saimoto, H., Okamoto, Y., Chitin, chitosan, and its derivatives for wound healing: old and new materials. *J. Func. Biomater.*, 6(1), 104–142, 2015.
21. Rădulescu, M., Holban, A.M., Mogoantă, L., Bălşeanu, T.-A., Mogoşanu, G.D., Savu, D., Bezirtzoglou, E., Fabrication, characterization, and evaluation of bionanocomposites based on natural polymers and antibiotics for wound healing applications. *Molecules*, 21(6), p. 761, 2016.
22. El-Sherbiny, I.M., El-Baz, N.M., A review on bionanocomposites based on chitosan and its derivatives for biomedical applications. *Eco-friendly Polymer Nanocomposites*, Springer, pp. 173–208, 2015.
23. Gopalan Nair, K., Dufresne, A., Crab shell chitin whisker reinforced natural rubber nanocomposites. 1. Processing and swelling behavior. *Biomacromolecules*, 4(3), 657–665, 2003.
24. Madhumathi, K., Kumar, P.S., Abhilash, S., Sreeja, V., Tamura, H., Manzoor, K., Jayakumar, R., Development of novel chitin/nanosilver composite scaffolds for wound dressing applications. *J. Mat. Sci.: Mater. Med.*, 21(2), 807–813, 2010.
25. Peng, C.C., Yang, M.H., Chiu, W.T., Chiu, C.H., Yang, C.S., Chen, Y.W., Peng, R.Y., Composite nano-titanium oxide–chitosan artificial skin exhibits strong wound-healing effect—an approach with anti-inflammatory and bactericidal kinetics. *Macromol. Biosci.*, 8(4), 316–327, 2008.
26. Gopal, A., Kant, V., Gopalakrishnan, A., Tandan, S.K., Kumar, D., Chitosan-based copper nanocomposite accelerates healing in excision wound model in rats. *Eur. J. Pharma., 731*, 8–19, 2014.
27. Abbasipour, M., Mirjalili, M., Khajavi, R., Majidi, M.M., Coated cotton gauze with Ag/ZnO/chitosan nanocomposite as a modern wound dressing. *J. Eng. Fabr. Fibers (JEFF)*, 9(1), 2014.
28. Díez-Pascual, A.M., Díez-Vicente, A.L., Wound healing bionanocomposites based on castor oil polymeric films reinforced with chitosan-modified ZnO nanoparticles. *Biomacromolecules*, 16(9), 2631–2644, 2015.
29. Shelma, R., Paul, W., Sharma, C.P., Chitin nanofibre reinforced thin chitosan films for Wound healing application. *Trends Biomater. Artif. Organs*, 22(2), 111–115, 2008.
30. Vedakumari, W.S., Prabu, P., Sastry, T.P., Chitosan-fibrin nanocomposites as drug delivering and wound healing materials. *J. Biomed. Nanotechnol.*, 11(4), 657–667, 2015.
31. Petrak, K., Essential properties of drug-targeting delivery systems. *Drug Discov. Today*, 10(23), 1667–1673, 2005.

32. Pillai, O., Panchagnula, R., Polymers in drug delivery. *Curr. Opin. Chem. Biol.*, 5(4), 447–451, 2001.
33. Wang, X., Du, Y., Luo, J., Lin, B., Kennedy, J.F., Chitosan/organic rectorite nanocomposite films: Structure, characteristic and drug delivery behaviour. *Carbohydr. Polym.*, 69(1), 41–49, 2007.
34. Taleb, M.F.A., Alkahtani, A., Mohamed, S.K., Radiation synthesis and characterization of sodium alginate/chitosan/hydroxyapatite nanocomposite hydrogels: a drug delivery system for liver cancer. *Polym. Bullet.*, 72(4), 725–742, 2015.
35. Min, K.H., Park, K., Kim, Y.S., Bae, S.M., Lee, S., Jo, H.G., Kwon, I.C., Hydrophobically modified glycol chitosan nanoparticles-encapsulated camptothecin enhance the drug stability and tumor targeting in cancer therapy. *J. Control. Release*, 127(3), 208–218, 2008.
36. Yadollahi, M., Farhoudian, S., Barkhordari, S., Gholamali, I., Farhadnejad, H., Motasadizadeh, H., Facile synthesis of chitosan/ZnO bio-nanocomposite hydrogel beads as drug delivery systems. *Int. J. Biol. Macromol.*, 82, 273–278, 2016.
37. Lee, S.C., Lee, S.W., Kang, I.J.,.Preparation and characterization of chitosan–gold nanocomposites for drug delivery application. *Surf. Rev. Lett.*, 17(02), 165–172, 2010.
38. Abdeen, R., Salahuddin, N., Modified chitosan-clay nanocomposite as a drug delivery system intercalation and *in vitro* release of ibuprofen. *J. Chem.*, 2013.
39. Mohanty, D.P., Biswal, S.K., Nayak, L., Preparation of starch-chitosan nanocomposites for control drug release of curcumin. *Inter. J. Curr. Eng. Technol.*, 5, 336–331, 2015.
40. Pan, Q., Lv, Y., Williams, G.R., Tao, L., Yang, H., Li, H., Zhu, L., Lactobionic acid and carboxymethyl chitosan functionalized graphene oxide nanocomposites as targeted anticancer drug delivery systems. *Carbohydr. Polym.*, 2016.
41. Malesu, V.K., Sahoo, D., Nayak, P.L. Chitosan–sodium alginate nanocomposites blended with cloisite 30B as a novel drug delivery system for anticancer drug curcumin. *Int. J. Appl. B. Pharm. Tech.*, 2(3), 402–411, 2011.
42. Mohanty, D.P., Biswal, S.K., Nayak, L. Preparation of starch-chitosan nanocomposites for control drug release of curcumin. *Inter. J. Curr. Eng. Technol.*, 5, 336–331, 2015.
43. Oliveira, M.J., Amato, V.S., Lugão, A.B., Parra, D.F. Drug delivery glucantime in PVP/chitosan membranes. International Nuclear Atlantic Conference - INAC 2015 Sao Paulo, SP, Brazil, October 4–9, 2015.
44. Lee, S.C., Lee, S.W., Kang, I.J. Preparation and characterization of chitosan–gold nanocomposites for drug delivery application. *Surf. Rev. Lett.*, 17(02), 165–172, 2010.
45. Parida, U.K., Nayak, A.K., Binhani, B.K., Nayak, P.L. Synthesis and characterization of chitosan-polyvinyl alcohol blended with cloisite 30B for controlled release of the anticancer drug curcumin. *J. Biomat. Nanobiotechnol.*, 2(04), p. 414, 2011.

46. Arias, J.L., Reddy, L.H., Couvreur, P., Fe 3 O 4/chitosan nanocomposite for magnetic drug targeting to cancer. *J. Mat. Chem.*, 22(15), 7622–7632, 2012.
47. Salatin, S., Barar, J., Barzegar-Jalali, M., Adibkia, K., Milani, M.A., Jelvehgari, M., Hydrogel nanoparticles and nanocomposites for nasal drug/vaccine delivery. *Arch. Pharma. Res.*, 39(9), 1181–1192, 2016.
48. Li, L., Chen, D., Zhang, Y., Deng, Z., Ren, X., Meng, X., Zhang, L. Magnetic and fluorescent multifunctional chitosan nanoparticles as a smart drug delivery system. *Nanotechnol.*, *18*(40), p. 405102, 2007.
49. Xu, S.S., Wu, J., Jiang, W., Synthesis and characterisation of a pH-sensitive magnetic nanocomposite for controlled delivery of doxorubicin. *J. Microencapsul.*, 32(6), 533–537, 2015.
50. Lim, E.K., Sajomsang, W., Choi, Y., Jang, E., Lee, H., Kang, B., Huh, Y.M. Chitosan-based intelligent theragnosis nanocomposites enable pH-sensitive drug release with MR-guided imaging for cancer therapy. *Nanoscale Res. Lett.*, 8(1), p. 467, 2013.
51. Gloria, A., De Santis, R., Ambrosio, L., Polymer-based composite scaffolds for tissue engineering. *J. Appl. Biomater. Biomech.*, 8(2), 2010.
52. Tabata, Y., Biomaterial technology for tissue engineering applications. *J. R. Soc. Interface*, 6(Suppl 3), S311–S324, 2009.
53. Peter, M., Binulal, N.S., Soumya, S., Nair, S.V., Furuike, T., Tamura, H., Jayakumar, R., Nanocomposite scaffolds of bioactive glass ceramic nanoparticles disseminated chitosan matrix for tissue engineering applications. *Carbohydr. Polym.*, 79(2), 284–289, 2010.
54. Peter, M., Binulal, N.S., Soumya, S., Nair, S.V., Furuike, T., Tamura, H., Jayakumar, R., Nanaocomposites scaffolds of bioactive glass ceramics nanoparticles disseminated chitosan matrix for tissue engineering applications. *Carbohydr. pPlym.*, 79, 284–289, 2010.
55. Kumar, P.S., Srinivasan, S., Lakshmanan, V.K., Tamura, H., Nair, S.V., Jayakumar, R., β-Chitin hydrogel/nano hydroxyapatite composite scaffolds for tissue engineering applications. *Carbohydr. Polym.*, 85(3), 584–591, 2011.
56. Frohbergh, M.E., Katsman, A., Botta, G.P., Lazarovici, P., Schauer, C.L., Wegst, U.G., Lelkes, P.I., Electrospun hydroxyapatite-containing chitosan nanofibers cross-linked with genipin for bone tissue engineering. *Biomaterials*, 33(36), 9167–9178, 2012.
57. Shalumon, K., Binulal, N., Selvamurugan, N., Nair, S., Menon, D., Furuike, T., . . . Jayakumar, R., Electrospinning of carboxymethyl chitin/poly (vinyl alcohol) nanofibrous scaffolds for tissue engineering applications. *Carbohydr. Polym.*, 77(4), 863–869, 2009.
58. Katti, K.S., Katti, D.R., Dash, R., Synthesis and characterization of a novel chitosan/montmorillonite/hydroxyapatite nanocomposite for bone tissue engineering. *Biomed. Mater.*, 3(3), p. 034122, 2008.
59. Pangon, A., Saesoo, S., Saengkrit, N., Ruktanonchai, U., Intasanta, V., Hydroxyapatite-hybridized chitosan/chitin whisker bionanocomposite

fibers for bone tissue engineering applications. *Carbohydr. Polym.*, 144, 419–427, 2016.
60. Moatary, A., Teimouri, A., Bagherzadeh, M., Chermahini, A. N., & Razavizadeh, R. Design and fabrication of novel chitin hydrogel/chitosan/nano diopside composite scaffolds for tissue engineering. *Ceramics International,* 2016.
61. Boere, K.W., Soliman, B.G., Rijkers, D.T., Hennink, W.E., Vermonden, T., Thermoresponsive injectable hydrogels cross-linked by native chemical ligation. *Macromolecules*, 47(7), 2430–2438, 2014.
62. Moura, M.J., Faneca, H., Lima, M.P., Gil, M.H., Figueiredo, M.M., *In situ* forming chitosan hydrogels prepared via ionic/covalent co-cross-linking. *Biomacromolecules*, 12(9), 3275–3284, 2011.
63. Mirahmadi, F., Tafazzoli-Shadpour, M., Shokrgozar, M.A., Bonakdar, S., Enhanced mechanical properties of thermosensitive chitosan hydrogel by silk fibers for cartilage tissue engineering. *Mater/ Sci. Eng.: C*, 33(8), 4786–4794, 2013.
64. Hassan, W., Dong, Y., Wang, W., Encapsulation and 3D culture of human adipose-derived stem cells in an *in-situ* cross-linked hybrid hydrogel composed of PEG-based hyperbranched copolymer and hyaluronic acid. *Stem Cell Res. Ther.*, 4(2), 1, 2013.
65. Han, F., Zhou, F., Yang, X., Zhao, J., Zhao, Y., Yuan, X., A pilot study of conically graded chitosan–gelatin hydrogel/PLGA scaffold with dual-delivery of TGF-β1 and BMP-2 for regeneration of cartilage–bone interface. *J. Biomed. Mater. Res. Part B: Appl. Biomater.*, 103(7), 1344–1353, 2015.
66. Eslahi, N., Simchi, A., Mehrjoo, M., Shokrgozar, M.A., Bonakdar, S., Hybrid cross-linked hydrogels based on fibrous protein/block copolymers and layered silicate nanoparticles: tunable thermosensitivity, biodegradability and mechanical durability. *RSC Adv.*, 6(67), 62944–62957, 2016.
67. Jayakumar, R., Ramachandran, R., Kumar, P.S., Divyarani, V., Srinivasan, S., Chennazhi, K., Nair, S. Fabrication of chitin–chitosan/nano ZrO_2 composite scaffolds for tissue engineering applications. *Int. J. Biol. Macromol.*, 49(3), 274–28, 2011.
68. Deepthi, S., Venkatesan, J., Kim, S.-K., Bumgardner, J.D., Jayakumar, R., An overview of chitin or chitosan/nano ceramic composite scaffolds for bone tissue engineering. *Int. J. Biol. Macromol.,* 2016.
69. Lowe, B., Venkatesan, J., Anil, S., Shim, M. S., Kim, S.-K., Preparation and characterization of chitosan-natural nano hydroxyapatite-fucoidan nanocomposites for bone tissue engineering. *Int. J. Biol. Macromol.,* 2016.
70. Puvaneswary, S., Raghavendran, H.B., Talebian, S., Murali, M.R., Mahmod, S.A., Singh, S., Kamarul, T., Incorporation of Fucoidan in β-Tricalcium phosphate-Chitosan scaffold prompts the differentiation of human bone marrow stromal cells into osteogenic lineage. *Sci. Rep.*, 6, 2016.
71. Shin, M., Ishii, O., Sueda, T., Vacanti, J.P., Contractile cardiac grafts using a novel nanofibrous mesh. *Biomaterials*, 25, 3717–3723, 2004.

72. Yang, M.C., Wang, S.S., Chou, N.K., Chi, N.H., Huang, Y.Y., Chang, Y.L., Shieh, M.J., Chung, T.W. The cardiomyogenic differentiation of rat mesenchymal stem cells on silk fibroinpolysaccharide cardiac patches *in vitro*. *Biomaterials*, 30, 3757–3765, 2009.
73. Barsotti, M.C., Felice, F., Balbarini, A., Di Stefano, R. Fibrin as a scaffold for cardiac tissue engineering. *Biotechnol. Appl. Biochem.*, 58, 301–310, 2011.
74. Zhang, T., Wan, L.Q., Xiong, Z., Marsano, A., Maidhof, R., Park, M., Yan, Y.N.A., Vunjak-Novakovic, G., Channelled scaffolds for engineering myocardium with mechanical stimulation. *J. Tissue Eng. Regener. Med*, 6, 748–756, 2012.
75. Patra, C., Talukdar, S., Novoyatleva, T., Velagala, S.R., Muhlfeld, C., Kundu, B., Kundu, S.C., Engel, F.B., Silk protein fibroin from Antheraea mylitta for cardiac tissue engineering. *Biomaterials*, 33, 2673–2680, 2012.
76. Radisic, M., Park, H., Shing, H., Consi, T., Schoen, F.J., Langer, R., Freed, L.E., Vunjak-Novakovic, G. Functional assembly of engineered myocardium by electrical stimulation of cardiac myocytes cultured on scaffolds. *Proc. Natl. Acad. Sci. U.S.A.*, 101, 18129–18134, 2004.
77. Tandon, N., Marsano, A., Maidhof, R., Wan, L., Park, H., Vunjak-Novakovic, G., Optimization of electrical stimulation parameters for cardiac tissue engineering. *J. Tissue Eng. Regener. Med.*, 5, E115–E125, 2011.
78. Dvir, T., Timko, B.P., Brigham, M.D., Naik, S.R., Karajanagi, S.S., Levy, O., Jin, H.W., Parker, K.K., Langer, R., Kohane, D.S., Nanowired three-dimensional cardiac patches. *Nat. Nanotechnol.*, 6, 720–725, 2011.
79. Madihally, S.V., Matthew, H.W., Porous chitosan scaffolds for tissue engineering. *Biomaterials*, 20, 1133–1142, 1999.
80. Ueno, H., Mori, T., Fujinaga, T. Topical formulations and wound healing applications of chitosan. *Adv. Drug Deliv. Rev.*, 52, 105–115, 2001.
81. Hoekstra, A., Struszczyk, H., Kivekas, O. Percutaneous microcrystalline chitosan application for sealing arterial puncture sites. *Biomaterials*, 19, 1467–1471, 1998.
82. Martins, A.M., Eng, G., Caridade, S.G., Mano, J.F., Reis, R.L., Vunjak-Novakovic, G., Electrically conductive chitosan/carbon scaffolds for cardiac tissue engineering. *Biomacromolecules*, 15(2), 635–643, (2014).
83. Chen, J.P., Liu, C.H., Hsu, H.L., Wu, T., Lu, Y.J., Ma, Y.H. Magnetically controlled release of recombinant tissue plasminogen activator from chitosan nanocomposites for targeted thrombolysis. *J. Mater. Chem. B*, 4(15), 2578–2590, 2016.
84. Mirhashemi, A., Bahador, A., Kassaee, M., Daryakenari, G., Ahmad-Akhoundi, M., Sodagar, A., Antimicrobial Effect of Nano-Zinc Oxide and Nano-Chitosan Particles in Dental Composite Used in Orthodontics. *J. Med. Bacteriol.*, 2(3–4), 1–10, 2015.
85. Friedewald, V.E., Kornman, K.S., Beck, J.D., Genco, R., Goldfine, A., Libby, P., Roberts, W.C., The American Journal of Cardiology and Journal of Periodontology Editors' Consensus: Periodontitis and Atherosclerotic Cardiovascular Disease. *J. Periodontol.*, 80(7), 1021–1032, 2009.

86. Yadollahi, M., Farhoudian, S., Namazi, H., One-pot synthesis of antibacterial chitosan/silver bio-nanocomposite hydrogel beads as drug delivery systems. *Int. J. Biol. Macromol.*, 79, 37–43, 2015.
87. Pathania, D., Gupta, D., Agarwal, S., Asif, M., Gupta, V.K., Fabrication of chitosan-g-poly (acrylamide)/CuS nanocomposite for controlled drug delivery and antibacterial activity. *Mater. Sci. Eng.: C*, 64, 428–435, 2016.
88. Kang, Y.O., Jung, J.Y., Cho, D., Kwon, O.H., Cheon, J.Y., Park, W.H., Antimicrobial silver chloride nanoparticles stabilized with chitosan oligomer for the healing of burns. *Materials*, 9(4), 215, 2016.
89. Rinehart, S.J., Campbell, T.D., Burke, K.J., Garcia, B., Mlynarski, A., Brain, S.J., Keleher, J.J., Synthesis and characterization of a chitosan/PVA antimicrobial hydrogel nanocomposite for responsive wound management materials. *J. Microb. Biochem. Technol.*, 2016.
90. Elbarbary, A.M., El-Sawy, N.M., Radiation synthesis and characterization of polyvinyl alcohol/chitosan/silver nanocomposite membranes: antimicrobial and blood compatibility studies. *Polym. Bull.*, 1–18, 2016
91. Prakash, N., Rajkumar, E., Sudha, P.N., Udaya Prakash, N.K., Antimicrobial activity of binary ternary composites of chitosan amended with nylon 6 and montmorillonite clay. *Int. J. Pharm. Pharm. Sci.*, 6, 118, 2014.
92. Farhoudian, S., Yadollahi, M., Namazi, H. Facile synthesis of antibacterial chitosan/CuO bio-nanocomposite hydrogel beads. *Int. J. Biol. Macromol.*, 82, 837–843, 2016.
93. Wahid, F., Yin, J.J., Xue, D.D., Xue, H., Lu, Y.S., Zhong, C., Chu, L.Q. Synthesis and characterization of antibacterial carboxymethyl Chitosan/ZnO nanocomposite hydrogels. *Int. J. Biol. Macromol.*, 88, 273–279, 2016.
94. Mohamed, R.R., Sabaa, M.W. Synthesis and characterization of antimicrobial cross-linked carboxymethyl chitosan nanoparticles loaded with silver. *Int. J. Biol. Macromol.*, 69, 95–99, 2014.

14

Potential Biomedical Applications of Chitosan – and Chitosan-Based Nanomaterials

Gulshan Singh[1]*, Murli Manohar[2], Suresh Kumar Arya[3], Waseem Ahmad Siddiqui[2] and Thor Axel Stenström[1]

[1]*Institute for Water and Wastewater Technology (IWWT), Durban University of Technology, Durban, South Africa*
[2]*Department of Biochemistry, Jamia Hamdard (Hamdard University), New Delhi, India*
[3]*Department of Chemistry, Jawaharlal Nehru Rajkeeya Mahavidyalaya, Pondicherry University, Port Blair, India*

Abstract

Chitin is considered as the second most abundant biomaterial after cellulose in terms of its application in various fields including biomedical field. Chitin is a long-chain polymer of an *N*-acetyl glucosamine, a derivative of glucose, and present in structural element in the exoskeleton of several living organism such as insects, crustaceans (mainly shrimps, lobsters, and crabs) and cell walls of fungi. Chitin and its derivative chitosan have gained a lot of attention in the field of nanotechnology for the development of polymeric nanomaterials, nanofibers, and nanoparticles, due to its excellent film forming ability, biodegradability, high permeability, biocompatibility, nontoxic property, and reduced cost. The chitin – and chitosan-based nanomaterials have been extensively used in tissue engineering, wound dressing, drug delivery, cancer diagnosis, and biosensors. In addition, the novel applications of these natural polymers can be utilized in a variety of nanostructural forms and sizes are also gaining attention. In this chapter, different biomedical applications of chitosan – and chitosan-based nanomaterials have been discussed and the importance of this biosourced polymer and its derivatives in improving human health have been highlighted.

Keywords: Chitin, chitosan, biomedical applications, burns, tissue engineering, drug delivery

**Corresponding author*: gsingh.gulshan@gmail.com

Shakeel Ahmed and Saiqa Ikram (eds.) Chitosan, (385–408) © 2017 Scrivener Publishing LLC

14.1 Introduction

Chitin is a long-chain polymer of an *N*-acetylglucosamine, a derivative of glucose, and present in structural element in the exoskeleton of several living organism such as insects, crustaceans (mainly shrimps, lobsters, and crabs) and cell walls of fungi [1]. Chitin is considered as the second most abundant biomaterial after cellulose in terms of its application in various fields [2]. Chitosan is a derivative of chitin and can be obtained by deacetylation in the solid state under alkaline conditions (concentrated NaOH) or by enzymatic hydrolysis by using chitin deacetylase. Therefore, chitosan is also termed as "deacetylated chitin." Chitosan is composed of β-(1-4)-linked D-glucosamine and *N*-acetyl-D-glucosamine randomly distributed within the polymer (Figure 14.1). Chitosan was first discovered by Henri Braconnot in 1811 [3, 4]. Chitin is insoluble by most organic solvents and due to that, its applications are limited. Therefore, chitosan is more preferred over chitin in biomedical applications because chitosan is soluble in acidic, neutral, and alkaline solutions. Chitosan is a biocompatible, biodegradable, and nontoxic polymer, hence, its unique properties are used for several biomedical beneficial purposes such as tissue engineering, wound healing, drug delivery, gene delivery, nanofibers, microparticles, and nanoparticles [1, 5–8]. Nowadays, nanotechnology has become an emerging and promising area and has provided immense opportunities and applications in different fields including biomedical research. Chitosan has gained more attention in the field of nanotechnology for the development of polymeric nanomaterials, nanofibers, and nanoparticles, due to its excellent film forming ability, biodegradability, high permeability, biocompatibility, nontoxic property and reduced cost [9]. Chitosan-based nanomaterials have been found to have wide application in tissue engineering [10, 11], drug delivery [12] wound healing [13],

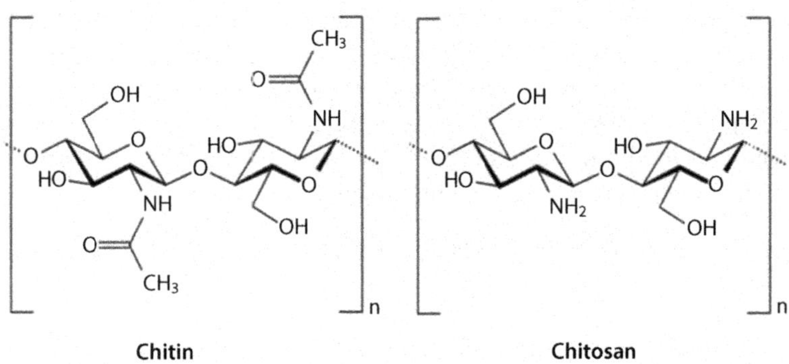

Figure 14.1 Structure of chitin and chitosan.

biosensors [14], and gene delivery [15]. Chitosan is easily incorporated with nanofibers [16], beads [17], microparticles [18], nanoparticles [19], and scaffolds [10]. Therefore, chitin and chitosan nanofibers find novel applications in biomedical fields due to their high surface area and porosity. In this chapter, different biomedical applications of chitin - and chitosan-based nanomaterials are described and the importance of this biosourced polymer and its derivatives in improving human health have been highlighted.

14.2 Production of Chitin and Chitosan

The cuticles of various crustaceans mainly crabs and shrimps serve as the prime sources of raw material for the production of chitin. Chitin is found as a constituent of a complex network with proteins onto which calcium carbonate deposits to form the rigid shell, crustaceans or more specifically shellfish and fish scales. There are three different steps involved in the preparation of chitosan from fish scales mainly, demineralization, decolorisation, and deacetylation as shown in Figure 14.2 [20].

Step 1: Raw fish scales are washed thoroughly with water, oven dried and soaked in 1% HCL solution for 36 h. Step 2: It is further washed, oven dried and kept in 2N NaOH solution for 36 h for demineralization. Step 3: The product of step 2 are transferred into a Potassium permanganate solution

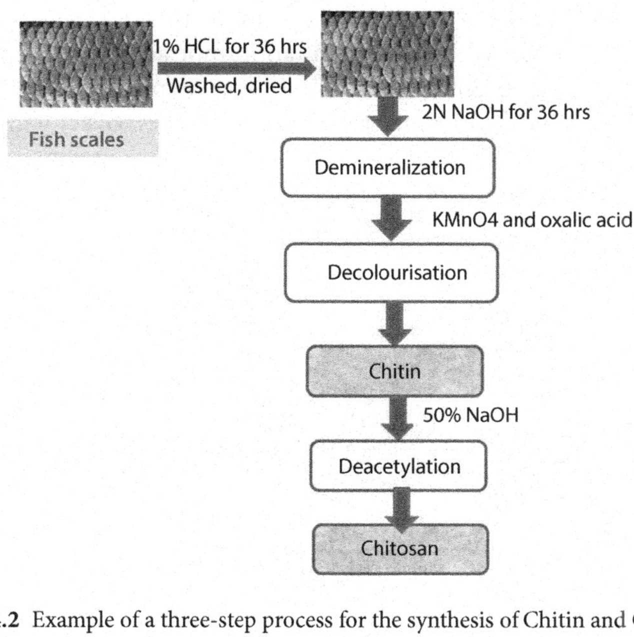

Figure 14.2 Example of a three-step process for the synthesis of Chitin and Chitosan from fish scales.

(1g of KMnO4 in 100 ml water) for 1 h, followed by an Oxalic acid step (1g of Oxalic acid in 100 ml water) for decolorization of the sample. These processes form chitin as a product which is further treated with 50% w/v NaOH for deacetylation, which results in chitosan as the end product.

14.3 Bioactivities of Chitin and Chitosan

Chitosan and its derivatives have several useful properties such as biocompatibility, biodegradability, safety and also interesting biological activities, therefore scientist have gained more interest to their applications especially in biomedical area. There are several important biological activities linked to chitosan, for example, antitumor, antibacterial, antifungal, antioxidant, and anti-HIV activities. These are briefly discussed below and illustrated in Figure 14.3.

14.3.1 Antitumor Activity

The peculiar properties of chitosan nanoparticles could provide a higher affinity for biological membranes (negatively charged) and offer site-specific targeting *in vivo*. Chitosan nanoparticles can elicit dose-dependent inhibitory effects on the proliferation of various tumor cell lines, while no toxicity against normal human cells are encountered. Several reports are available that address the important antitumor activities of chitin and chitosan [6, 21]. The problem with these two polymers is however their poor solubility (in comparison to their modified derivatives) which limits their application in biomedical applications [22]. In contrast, the hydrolyzed products of chitosan, for example, *N*-acetyl-D-glucosamine oligomers

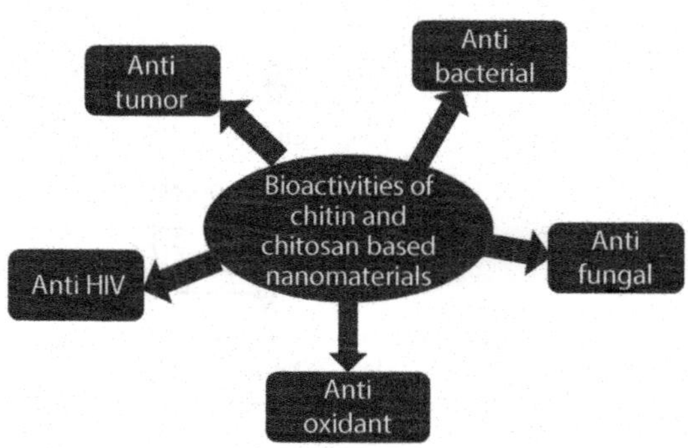

Figure 14.3 Various bioactivities of chitin – and chitosan-based nanomaterials.

(chitin oligosaccharide; NACOS) and D-glucosamine oligomers (chitosan oligosaccharide, COS) are readily soluble in water due to their short chain lengths [23]. Antitumor properties of chitosan and its derivatives and chitosan-based nanomaterials have been tested in several *in vitro* and *in vivo* models. An *in vitro* study has shown the chemo-preventive effects of COS in human colorectal adenocarcinoma cell line, HT-29 [24]. In a parallel study, COS pretreatment inhibited pro-inflammatory responses and invasiveness of HT-29 cancer cells [25]. The antitumor effect of Chitosan nanoparticles (CNP) on growth of human hepatocellular carcinoma (BEL7402) was evaluated [26].The study reported a strong antitumor effect of CNP on human hepatoma cell line *in vitro* and *in vivo*. These findings pave a way for the successful application of CNP in the treatment of hepatocellular carcinoma.

Furthermore, antiangiogenic activity of COS has been reported in human breast cancer cell line (i.e., MDA-MB-231) and revealed that, COS treatment to breast cancer cells reduces matrix metalloproteinase-9 (MMP-9) secretion that might be responsible for the invasiveness of breast cancer [27, 28]. Another *in-vitro* study revealed antimetastatic potential of COS in human hepatocellular carcinoma cell line, HepG2 and the authors found that COS has potential for inhibiting cellular proliferation and cell cycle [29]. The study further validated their *in-vitro* finding by using a mouse model and COS was found to reduce the growth of HepG2 xenografts in severe combined immune deficient (SCID) mice [29]. Chitin and chitosan oligosaccharides (COS) also have the ability to affect the production of several cytokines that might cause increased proliferation and differentiation of T-cells, which in turn increases the cytotoxicity of cancer cells [30, 31]. Recently, it has been shown that chitosan inhibit adhesion and proliferation of several melanoma cells, that is, A375, SKMEL28 and RPMI7951 cells via inhibition of specific caspases, alteration in the ratio of Bax and Bcl2 proteins that further turned on Fas ligand (FasL)-induced apoptosis signaling pathway in these cancerous cells [32].

The anticancer potential β-chitosan nanoparticles (synthesized from depolymerization of β-chitosan extracted from the gladius) against human hepatoma cells was analyzed [33]. The cell viability of HepG2 cells was assessed in the presence of different doses of β-chitosan nanoparticles (β-CNP). The inhibitory concentration of β-CNP was found to be 30 μg/ml. The results obtained were further supported by measuring different biochemical parameters such as TBARS and lipid hydroperoxides, enzymatic and non-enzymatic antioxidant (SOD, CAT, GPx and GSH). The anticancer of β-CNP in HepG2 cells suggests that β-CNP could be used as a potential drug for the efficient treatment of hepatocellular carcinoma in the future.

The cytotoxic effect of engineered chitosan nanoparticles (CS-NPs) on an *in vitro* human liver cancer cell model (HepG2) was analyzed by Loutfy *et al.*, [34]. The sulforhodamine B colorimetric assays and flow cytometric analysis was used for the cytotoxicity screening. The concentration of 100 μg/ml of engineered CS-NPs (controlled to a 150 nm size and charges of 40 mV) showed genotoxic potential on HepG2 after 48 h of cell exposure through intrinsic apoptotic caspase-independent mechanisms. The study further suggested the molecular and protein levels quantitative analysis to confirm the actual impact of chitosan size and time on the genotoxic effect.

The antitumor effect of chitosan oligosaccharides (COSs) against ten tumor cells, were evaluated [35]. The COS treated groups showed the cell death and necrosis and it was observed that the COS stimulated the mRNA expression of tumor necrosis factor α. In conclusion, the COS may be used as an antitumor agent and can be used as adjuvant in oncoptherapy (Figure 14.4).

These studies suggested that chitin – and chitosan-based nanomaterials possess antitumor effects *in vitro* and *in vivo*, leading to good opportunities for their application as a supplementary antitumor drug and drug carrier.

14.3.2 Antibacterial Activity

Chitosan and its derivatives have been investigated as antimicrobial agents against a wide range of target organisms including bacteria. Chitosan is

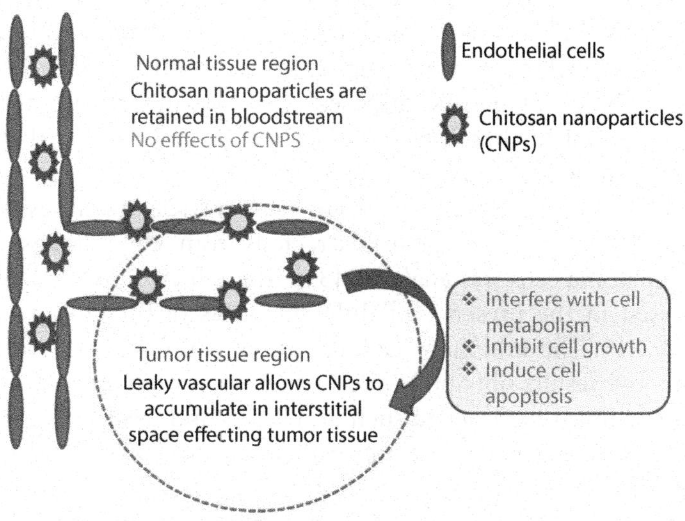

Figure 14.4 Antitumor activity of chitosan nanomaterials.

considered to be a bacteriocidal or bacteriostatic compound, but recent data in the literature tend toward characterizing chitosan as bacteriostatic rather than bactericidal. The exact mechanism is not fully understood and several other factors may contribute to the antibacterial action [36].The antimicrobial activity of chitosan was demonstrated against several bacterial species (*Bacillus cereus, Escherichia coli, Staphlyococcus aureus, Salmonella enterica*) [37]. The antimicrobial activity of chitosan can be explained by two main types of mechanisms. In the first mechanism the membrane permeability of bacteria is altered due to interaction of positively charged chitosan and negatively charged groups present on the surface of cells (Figure 14.5). This would block the entry of essential materials in to the cells or/and lead to the leakage of vital soluble compounds out of the cell [38]. The second mechanism is based on the the binding of chitosan to the cell DNA, which would result in the inhibition of the microbial RNA synthesis [39]. The combination of both mechanism might play role in the antimicrobial property of chitosan [39, 40]. This hypothesis was further supported by the studies from Muzzarelli *et al.*, [41].

The antimicrobial potential of chitosan and its derivative was evaluated against *Staphylococcus aureus* (gram positive bacteria) and *Escherichia coli* (gram negative bacteria) [37]. The results revealed that the growth of the gram-positive microorganism (*S. aureus*) was reduced, while growth of the *E. coli* (Gram-negative) strain was found to be less sensitive to both polymers. The antimicrobial effects of chitosan along with citrus extract for pathogens like *Escherichia coli* O157:H7 and *Salmonella enterica* on turkey

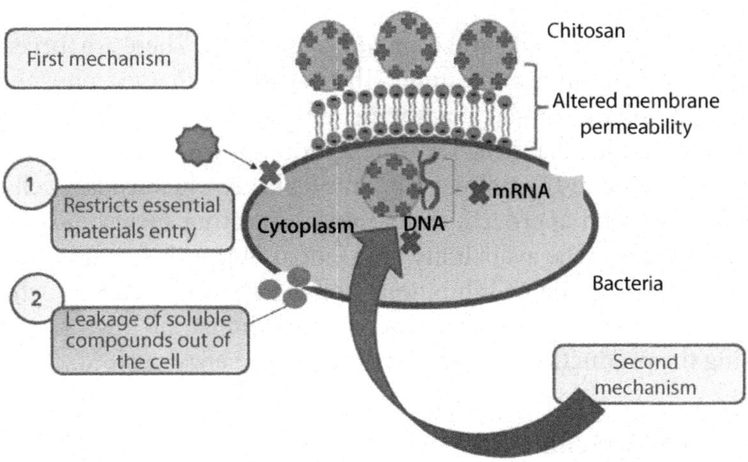

Figure 14.5 The schematic representation of antimicrobial mechanism of action of chitosan.

meat was evaluated [42]. It was shown that the chitosan alone or in combination with citrus extract was able to inhibit bacterial growth on meat, however, citrus extract alone didn't perform well. The study emphasizes that citrus extract combined with chitosan showed additive antibacterial effect against both pathogenic bacteria [42]. The effect of chitosan nanoparticles (CSNPs) on the efficiency of methylene blue (MB)-mediated antimicrobial photodynamic inactivation of *Staphylococcus aureus* and *Pseudomonas aeruginosa* biofilms was also evaluated [43]. APDI mediated by the mixture of MB and CSNPs showed significant antibiofilm photo-inactivation in comparison to MB-induced APDI proving the antibacterial potential of chitosan.

14.3.3 Antifungal

Chitosan has been reported to have antifungal activity, when exists in free form, against different fungi, for example, *Rhizopus oryzae*, *Aspergillus niger*, *Rhizopus stolonifera*, *Phomopsis asparagi* and *Alternaria alternate* [44–46]. The antifungal activity of chitosan is influenced by a number of factors like its molecular weight, concentration, degree of substitution, types of fungus, and also the types of functional groups in the chitosan derivatives chains [45, 47–49]. It has been inferred from different studies that the antifungal activity is contributed by the polycationic nature of the chitosan. Due to this chitosan does not need any chemical modification and possess natural antifungal activity [45, 48].

Chitosan is also effective against several molds (*Botrytis cinerea*, *Fusarium oxysporum*) [50, 51] and yeast (*Saccharomyces cerevisiae*, *Candida albicans*, *Candida glabrata* and *Candida krusei*) [52]. There are three mode of action of chitosan against fungi. The negatively charged phospholipid components of the fungal membrane interacts with positively charged chitosan in first mechanism. This interaction upsurge the membrane permeability leading to the cellular contents leakage, which causes cell death [53]. In the second approach, chitosan as a chelating agent binds to trace elements blocking the availability of essential nutrients for fungal growth [54]. In the last mode of inhibition, after penetrating cell wall, chitosan can bind to DNA of fungi and subsequently inhibits the mRNA synthesis affecting the production of essential proteins and enzymes[55].

14.3.4 Antioxidant

Reactive oxygen species (ROS) and reactive nitrogen species (RNS) are generated during the normal metabolism and ultimately leading to oxidative stress [56–58] and harmful effect such as cancer, aging and pancreatic inflammatory diseases [36, 57–59]. Antioxidants are well-known for their beneficial health effects by protecting our body against the harmful effect of ROS, which exert oxidative damage to membrane lipids, protein and DNA. Many studies have reported the antioxidant activity of chitosan and its derivatives during recent years [60]. The effect of chitin, chitosan, and their derivatives may be used as functional ingredients in food formulations to promote consumer health and to improve the shelf life of food products [60]. It can be attributed to *in vitro* and *in vivo* free radical-scavenging activities. An *in vitro* study showed that chitosan and its derivatives (chitosan oligosaccharides) had oxygen radicals scavenging activity. Among them, low molecular weight chitosans have higher potential of scavenging free radicals [61].

Antioxidant activity of chitosan can be improved by gamma ray irradiation and it was further reported that gamma-irradiated chitosan and its derivatives could act as potential natural antioxidant agents [62]. Another recent study also showed that molecular weight reduction by gamma irradiation increased antioxidant capacity of chitosan [63]. Interestingly, chitosan nanoparticles coupled with chlorogenic acid showed sustained release property, retained antioxidant activity and enhanced bioavailability [64].

14.3.5 Anti-HIV

Anti-HIV therapy, which includes different chemical drugs alone or in combination, has definitely improved the health of patients suffering from HIV/AIDS. Although this therapy has given a lot of hope for the betterment of human health still the side effects of the drugs and emergence of HIV drug resistance are the main roadblocks and lead to failure of anti-HIV therapy. Therefore, novel anti-HIV therapeutics from natural sources may supplement the treatment alternatives. There are some evidences in the literature showing the application of marine-derived sulphated chitosan derivatives as anti-HIV agents [65]. These chitosan derivatives were analyzed for their role in (i) inhibiting HIV-1 replication *in vitro* and (ii) inhibiting blood coagulation. The polysaccharide derivative of chitin, that is, *N*-carboxymethyl chitosan-*N*-*O*-sulfate (NCMCS) was found to impede the propagation of the human immunodeficiency virus type 1 (HIV-1) in human CD4+ cells and that of Rauscher murine leukemia virus (RLV) in murine fibroblasts [66]. Chitosan derivatives showed the inhibition of both virus at different concentrations in a dose-dependent manner. NCMCS was found to block the binding of HIV-1 to human CD4+ target cells and inhibited HIV-1 reverse transcriptase. However, the sites of sulfation also play major role in the inhibitory effect of chitin sulfates on HIV-1 infection [66]. Another study indicates that a regioselective sulfation of O-2 (C-2S) and/or O-3 (C-3S) groups of chitin showed a much higher inhibitory effect against AIDS virus *in vitro* than the 6-*O*-sulfonated (C-6S) derivatives [67]. Moreover, the product with sulfation at both 2 and 3 positions (C-2,3S) completely inhibited the infection of AIDS virus to T-lymphocytes at 0.28 μg/ml concentration with minimal cytotoxicity. These reports suggest that the specific interaction of these new types of chitin sulfates with GP120, that is, glycoprotein exposed on the surface of the HIV envelope of the AIDS virus depends significantly on the sites of sulphation rather than on the total degree of substitution on sugar residues [67]. A recent study evaluated the chitosan nano formulations as potent anti-HIV drugs and it was reported that the chitosan nanoparticles loaded with saquinavir (a drug that can treat HIV infection) has greater cell targeting efficiency and led to efficient control of the viral proliferation by targeting T-cells [68]. A parallel study demonstrated that tenofovir-loaded (an anti-HIV microbicide), chitosan-thioglycolic acid-conjugated nanoparticles exhibit superior biophysical properties and can potentially maximize the retention time of tenofovir, intended for the prevention of HIV transmission [69].

14.4 Biomedical Application of Chitin – and Chitosan-Based Nanomaterials

14.4.1 Wound Healing and Burns

Chitin and chitosan appear to have beneficial properties for wound healing and tissue engineering because of their unique structural, physicochemical and functional properties [70, 71]. Antibacterial resistance is one of the main hurdles in the field of wound care and management because it may cause infection and delayed wound healing. Chitosan derivatives with quaternary ammonium groups have high antimicrobial efficacy. It is also well known that these cationic polymers mainly target the cytoplasmic membrane of bacterial cells [72]. The best biomaterials for wound dressing should be biocompatible and promote the growth of dermis and epidermis layers. Further, it has been reported that composite nanofibrous membrane of chitosan/collagen was found to have promising effects on wound healing [73–75]. The membrane was found to promote wound healing and induce cell migration and proliferation. In another study, the nanofibrous membrane was found to be better than gauze and commercial collagen sponge in wound healing [76]. Chitosan and its derivatives when interact with RBCs (red blood cells) become sticky due to positive-negative charge interactions resulting in quick healing of wounds (Figure 14.6).

Several studies reported the use of chitosan scaffolds and membranes for the treatment of patients with deep burns and wounds. Madhumathi *et al.*, [77] have shown that the developed chitin /nanosilver composite scaffolds had excellent antibacterial activity against *S. aureus* and *E. coli* for wound healing applications.

Chitosan has been reported to be used for wound healing in patients undergoing plastic surgery [78], skin grafting [79, 80] and endoscopic sinus surgery [81]. Nowadays, a number of chitosan and its derivatives-based wound dressings are available in the market in the form of nonwovens, films, sponges, composites, and nanofibers. One of the chitosan-coated commercially available products is HemCon® hemostatic bandages, which have been used widely for the cure of external hemorrhage in cases of prehospital bleeding [82].

For the treatment of chronic wound infections caused by *Staphylococcus aureus*, an injectable hydrogel system, loaded with antibiotic encapsulated tigecycline nanoparticles was used [83]. It was reported that the developed hydrogel system proved to be an effective medium for frug delivery and can be applied on the bacterial infected site to treat wound infections [83].

14.4.2 Tissue Engineering

Currently, tissue engineering has become an important therapeutic strategy for the development of medical applications. Recently, biomaterial research has been directed toward the development of scaffolds for regenerative medicine, where chitosan is a suitable candidate as a natural polymer with

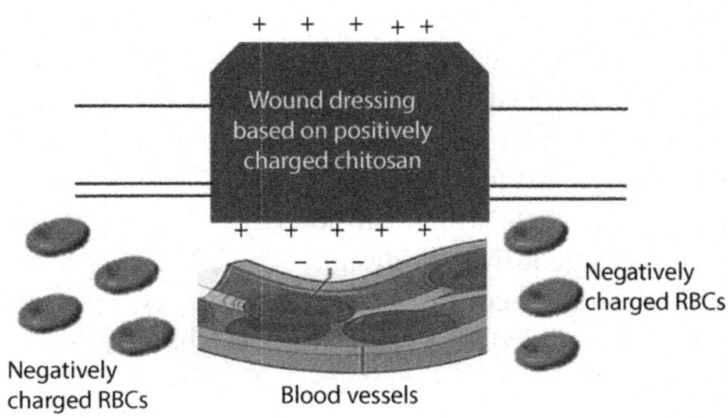

Figure 14.6 Wound dressing application of chitosan and their derivatives.

novel properties of biocompatibility, and biodegradability without toxicity. Additionally, chitosan and its derivatives are promising biomedical three-dimensional scaffolds for tissue engineering and two-dimensional scaffolds especially for wound healing purposes. The properties which facilitates its application in tissue engineering are their gel forming properties, ease of chemical modification, porous structure and high affinity to *in vivo* macromolecules [84]. The important properties that are required for material to qualify for the tissue engineering scaffolds are: (i) it should be biodegradable, (ii) It should not incite any acute or chronic reactions, (iii) it should exhibit surface characteristics that facilitates cell attachment, differentiation and proliferation, and (iv) mechanical strength and ability to be manufactured in different shapes [11]. Chitin - and chitosan-based nanomaterials serve as relevant candidates for the preparation of tissue scaffolds substituting the missing or injured tissue and organ [85]. The chitosan hydrogels and foams had been utilized to form three-dimensional scaffolds for tissue engineering with a promising outcome [11]. Hydrogels are frequently used as scaffolds for tissue engineering but also utilized in other biomedical applications like delivery of growth factor and drug to targeted tissue or organ [86].The high water content of hydrogels and their bendable property makes them well-suited with a majority of living tissues and with minimum damage to surrounding tissue throughout implantation process [10].

14.4.3 Drug Delivery

Chitosan and its derivatives have found application in drug release, *in situ* gelation, permeation enhancement, mucoadhesion, inhibition of efflux pump and transfection due to their cationic character [87]. The biodistribution profile and pharmacokinetic parameters of EGFR-targeted chitosan nanoparticles (TG CS nanoparticles) for siRNA/cisplatin combination therapy of lung cancer were explored [88]. The results suggested that TG CS nanoparticles exhibit a six fold higher tumor targeting efficiency compared with the nontargeted nanoparticles. The study concludes that the TG Chitosan nanoparticles offers an attractive drug delivery platform for RNAi therapeutics against lung cancer.

A chitosan nanoparticle which can load two drugs (Aspirin and probucol; PRO) simultaneously was designed [89].The properties of chitosan nanoparticles such as controlled release behaviors and encapsulation capacity of Aspirin and probucol were studied. The results indicated that the encapsulation efficiency, drug release behaviors and loading capacity were affected by several factors including chitosan molecular weight, pH, and drug concentrations.

Self-assembled chitosan-dispersed low density lipoprotein nanoparticles (CS-LDL NPs) were fabricated for drug delivery by Tian *et al.*, [90]. Doxorubicin hydrochloride (DOX) was then encapsulated in formulated CS-LDL NPs and it was observed that encapsulated drug showed less toxicity and significant cellular uptake in comparison to free drug when incubated with gastric cancer cells.

Another study by Abruzzo *et al.*, [91] attempted to prepare different formulations of chitosan-based nanoparticles that have capability to encapsulate lipophilic molecules and have potential for targeted drug delivery to cancerous cells. Different concentrations of chitosan, hyaluronic acid and sulphobutyl-ether-β-cyclodextrin with or without curcumin were used for the preparation of nanoparticles. It was observed that the nanoparticles with ratio 20:0:1 provided a lesser curcumin internalization in the Intestinal epithelial cell line hence offered less cytotoxicity whereas higher cellular uptake hence higher cytotoxicity in colorectal cancer (HT29) cell lines. Chitosan nanoparticles were encapsulated with pravastatin (PRV) to be used as drug carrier for the treatment of liver cancer [92]. The *in vitro* drug release and cytotoxicity showed that this formulation could be used as potential drug carrier for cancer treatment.

In another recent study, docetaxel–nicotinamide complex was loaded with chitosan nanoparticles to increase its bioavailability [93]. The study reported that there was increase in saturated solubility, drug release, cellular uptake and cytotoxicity in comparison to free docetaxel, proving the utility of this formulation in drug delivery systems. Lecithin/chitosan nanoparticles encapsulated with Amphotericin-B were formulated for prolonged ocular application [94]. Improved bioavailibilty, efficient mucoadhesive properties, and precorneal residence time make this formulation more effective than commercially available formulations [94]. Few examples of application of Chitosan based nanomaterials in drug delivery have been presented in Table 14.1.

14.4.4 Biosensors

The flexibility in chitin – and chitosan-based nanomaterial's fabrication have allowed their application for the development of different types of biosensors. The primary role of chitosan is to specifically immobilize and capture cells, biomolecules or nanoparticles. Different types of biosensors, for example, amperometric, potentiometric, and conductimetric sensors have been fabricated utilizing the unique properties of chitosan [106]. A quartz crystal microbalance (QCM) biosensor was developed using chitosan and folic acid conjugate selectively recognized by MCF-7 cancer cell over-expressed

Table 14.1 Application of chitosan based nanomaterials in drug delivery.

Polymers used*	Drug	Method of preparation	Application in treatment	References
Self-assembled Nps composed of chitosan(CS) and low density lipoprotein (LDL)	doxorubicin hydrochloride (DOX)	one-step green synthesis of mixing CS solution and LDL suspension	Gastric cancer	[90]
Chitosan-based Nps	curcumin	ionic gelation method	Colorectal cancer	[91]
EGFR-targeted chitosan Nps	Mad2 siRNA	electrostatic interaction	Lung cancer	[88]
Chitosan Nps	Pravastatin (PRV)	ionic gelation method	liver cancer treatment	[92]
Chitosan-lactate Nps	CD73-siRNA	ionic gelation	breast tumour	[95]
Chitosan Nps	Docetaxel(DTX)–nicotinamide	modified ionic gelation method	Potential for oral, instead of intravenous, administration of DTX	[93]
Chitosan Nps	Flurbiprofen-loaded deformable liposomes	modified ethanol injection method followed by coating	Ocular	[96]
Carboxymethyl dextran chitosan Nps	HMGA2 siRNA and doxorubicin	NR**	Colorectal cancer.	[97]
Carboxymethyl dextran Chitosan Nps	IL17RB siRNA and doxorubicin	NR**	Breast cancer	[98]
lecithin/chitosan Nps	Amphotericin-B	ionic gelation method	Ocular	[94]

Chitosan polyplex Nps	miR-145 (a tumor suppressor micro RNA)	fabricating method determined by response surface analysis and D-optimal design	breast cancer	[99]
Chitosan Nps	chlorogenic acid (CGA),	ionic gelation method.	enhanced bioavailability	[64]
Chitosan nanospheres	5-fluorouracil (5-FU)	coacervation and emulsion droplet coalescence methods	Inhibits tumor cell proliferation	[100]
Polylactic acid – grafted chitosan	Amphotericin B	Dialysis	Ocular	[101]
Chitosan and dextran sulphate	Flutamide	Poly Electrolyte complex	Prostate cancer	[102]
Succinate modified chitosan	5-aminolaevulinic acid	Ionic gelation	Oral Cancer	[103]
Chitosan Nps	Aspirin and probucol	gelation of chitosan with tripolyphosphate (TPP) by ionic cross-linking	restenosis	[89]
polylactic-co-glycolic acid (PLGA) nanospheres surface modified with chitosan (CS)	Mitoxantrone	solvent evaporation technique	anticancer drug carrier	[104]
Chitosan Nps	paclitaxel	solvent evaporation and emulsification cross-linking method	Cancer treatment	[105]

*Nps: Nanoparticles

**NR: Information not retrieved

folic acid receptors [107]. The concentration of the MCF-7 cell was analyzed and a wide linear range of 4.5×10^2–1.01×10^5 cells/ml with a detection limit of 430 cells/ml, was obtained. It was reported that this chitosan-based biosensors could prove to be a promising detection strategy for the efficient detection of MCF-7 cancer cell over-expressed folic acid receptors.

For the electrochemical detection of morphine (MO), a simple and efficient sensor formed on carbon paste electrode (CPE) modified by chitosan-coated magnetic nanoparticle (CMNP) was developed [108].Under optimized conditions, the sensor provides a detection limit of 3 nM. The applicability of developed sensor was validated in urine and serum samples for the monitoring of MO and results obtained were found to be satisfactory. In another study, Chitosan-based magnetite nanocomposites were synthesized having the potential application as electrochemical sensors [109].

The direct electrochemistry of hemoglobin immobilized on nanocomposites (composed of graphene, quantum dots and chitosan) for biosensing application was analyzed [110]. The biosensor showed good linear response to hydrogen peroxide over the concentration range 1.5–195 μM with a sensitivity of 0.53 $\mu A/\mu M/cm^2$ and limit of detection of 0.68 μM. Due to short response time, good stability and high sensitivity of the developed biosensor, it can be applied for the determination of H_2O_2 in urine samples.

14.5 Conclusion and Future Perspective

Chitin - and chitosan-based nanomaterials have been extensively used in different research areas due to their unique and bendable properties and this book chapter, we have briefly summarized the current biomedical applications of chitin - and chitosan-based nanomaterials in tissue engineering, wound dressing, drug delivery, cancer diagnosis, and biosensors. In addition the novel applications where these natural polymers can be utilized in a variety of nanostructural forms and sizes have been discussed. Many fundamental technological advances have offered improved capabilities of chitin - and chitosan-based nanomaterials for different application mainly in the field of drug delivery and biosensors. Nanostructured composite scaffolds can be developed as promising tissue engineered constructs or for wound healing. Multifunctional use of chitin - and chitosan-based nanomaterials proved to aid simultaneous cancer targeting and drug delivery. With new inspirations proposed by the ability of chitin - and chitosan-based nanomaterials and immense opportunities brought by nanotechnology, it is reasonable to expect the substantial developments that will further escalate their wide application in different sectors in the near future. These naturally obtained chitosan biomaterials have been an interesting area of research in recent years, but the mechanical pathways influencing the unique properties of these chitosans remain unclear. More detailed *in vitro* and *in vivo* studies are required to elucidate the abilities of chitosan derivatives. This will definitely help to expand the usage of this very competent polymer in clinical trials and eventually will lead to the betterment of human health.

Acknowledgement

The support from the South African Research Chair Initiative of the Department of Science and Technology and National Research Foundation of South Africa is sincerely acknowledged.

References

1. Azuma, K., Ifuku, S., Osaki, T., Okamoto, Y., Minami, S., Preparation and biomedical applications of chitin and chitosan nanofibers. *J. Biomed. Nanotechnol.,* 10, 2891, 2014.
2. Muzzarelli, R.A., M. El. Mehtedi., and M. Mattioli-Belmonte., Emerging biomedical applications of nano-chitins and nano-chitosans obtained via advanced eco-friendly technologies from marine resources. *Mar. Drugs.,* 12, 5468, 2014.
3. Dai, T., Tanaka, M., Huang, Y.Y., Hamblin, M.R., Chitosan preparations for wounds and burns: antimicrobial and wound-healing effects. *Exp. Rev. Anti. Infect. Ther.,* 9, 857, 2011.
4. Labrude, P., Becq, C., Pharmacist and chemist Henri Braconnot. *Rev. Hist. Pharm. (Paris).,* 51, 61, 2003.
5. Jayakumar, R., Prabaharan, M., Nair, S.V., Tamura, H., Novel chitin and chitosan nanofibers in biomedical applications. *Biotechnol. Adv.,* 28, 142, 2010.
6. Azuma, K., Osaki. T., Minami, S., Okamoto, Y., Anticancer and anti-inflammatory properties of chitin and chitosan oligosaccharides. *J. Funct. Biomater.,* 6, 33, 2015.
7. Cheung, R.C., Ng, T. B., Wong, J.H., Chan, W.Y., Chitosan: an update on potential biomedical and pharmaceutical applications. *Mar. Drugs.,*13, 5156, 2015.
8. Ahmadi, F., Oveisi, Z., Samani, S.M., Amoozgar, Z., Chitosan based hydrogels: characteristics and pharmaceutical applications. *Res. Pharm. Sci.,*10, 1, 2015.
9. Rinaudo, M., Chitin and chitosan: properties and applications. *Prog. Polym. Sci.,* 31, 603, 2006.
10. Rodríguez-Vázquez, M., Vega-Ruiz, B., Ramos-Zúñiga, R., Saldaña-Koppel, D.A., Quiñones-Olvera, L.F., Chitosan and its potential use as a scaffold for tissue engineering in regenerative medicine. *Biomed. Res. Int.,* 15, 2015.

11. Croisier, F., C, Jérôme., Chitosan-based biomaterials for tissue engineering. *Eur. Polym. J.*, 49, 780, 2013.
12. Islam, N., Ferro, V., Recent advances in chitosan-based nanoparticulate pulmonary drug delivery. *Nanoscale.*, 8, 14341, 2016.
13. Patrulea, V., Ostafe, V., Borchard, G., Jordan, O., Chitosan as a starting material for wound healing applications. *Eur. J. Pharm. Biopharm.*, 97, 417, 2015.
14. Zhang, W., Li, X., Zou, R., Wu, H., Shi, H., Yu, S., Liu, Y., Multifunctional glucose biosensors from Fe(3)O(4) nanoparticles modified chitosan/graphene nanocomposites. *Sci. Rep.*,5, 11129, 2015.
15. Saranya, N., Moorthi, A., Saravanan, S., Devi, M.P., Selvamurugan, N., Chitosan and its derivatives for gene delivery. *Int. J. Biol. Macromol.*, 48, 234, 2011.
16. Hadipour-Goudarzi, E., Montazer, M., Latifi, M., Aghaji, A.A.G., Electrospinning of chitosan/sericin/PVA nanofibers incorporated with *in situ* synthesis of nano silver. *Carbohydr. Polym.*, 113, 231, 2014.
17. Thakre, D., Jagtap, S., Sakhare, N., Labhsetwar, N., Meshram, S., Rayalu, S., Chitosan based mesoporous Ti–Al binary metal oxide supported beads for defluoridation of water. *Chem. Eng. J.*, 158, 315, 2010.
18. Gomaa, Y.A., El-Khordagu, L.K., Boraei, N.A., Darwish I.A., Chitosan microparticles incorporating a hydrophilic sunscreen agent. *Carbohydr. Polym.*, 81, 234, 2010.
19. Elgadir, M.A., Uddin, Md.S., Ferdosh, S., Adam, A., Chowdhury, A.J.K., Sarker, Md. Z. I., Impact of chitosan composites and chitosan nanoparticle composites on various drug delivery systems: a review. *J. Food. Drug. Anal.*, 23, 619, 2015.
20. Kumari, S., Rath, P.K., Extraction and characterization of chitin and chitosan from (Labeo rohit) fish scales. *Procedia. Mater. Sci.*, 6, 482, 2014.
21. Chien, R.C., M.T, Yen., J.L, Mau., Antimicrobial and antitumor activities of chitosan from shiitake stipes, compared to commercial chitosan from crab shells. *Carbohydr. Polym.*,138, 259, 2016.
22. Lodhi, G., Kim, Y.S., Hwang, J.W., Kim, S.K., Jeon, Y.J, Je, J.Y., Ahn, C.B., Moon, S,H., Jeon, B.T., Park, P.J., Chitooligosaccharide and its derivatives: preparation and biological applications. *Biomed. Res. Int.*, 654913, 2014.
23. Nam, K.S., M.K. Kim., Y.H. Shon., Chemopreventive effect of chitosan oligosaccharide against colon carcinogenesis. *J. Microbiol. Biotechnol.*, 17, 1546, 2007.
24. Qi, L., Xu, Z., Chen, M., *In vitro* and *in vivo* suppression of hepatocellular carcinoma growth by chitosan nanoparticles. *Eur. J. Cancer*, 43, 184, 2007.
25. Nam, K.S., Shon, Y.H., Suppression of metastasis of human breast cancer cells by chitosan oligosaccharides. *J. Microbiol. Biotechnol.*,19, 629, 2009.
26. Quan, H., Zhu, F., Han, X., Xu, Z., Zhao, Y., Miao, Z., Mechanism of anti-angiogenic activities of chitooligosaccharides may be through inhibiting heparanase activity. *Med. Hypotheses.*, 73, 205, 2009.
27. Shen, K.T., Chen, M.H., Chan, H.Y., Jeng, J.H., Wang, Y.J., Inhibitory effects of chito-oligosaccharides on tumor growth and metastasis. *Food. Chem. Toxicol.*, 47, 1864, 2009.

28. Tokoro, A., Kobayashi, M., Tatewaki, N., Suzuki, K., Okawa, Y., Mikami, T., Suzuki, S., Suzuki, M., Protective effect of N-acetyl chitohexaose on Listeria monocytogenes infection in mice. *Microbiol. Immunol.*, 33, 357, 1989.
29. Tokoro, A., Suzuki, K., Mikami, T., Suzuki, S., Suzuki, M.,Growth-inhibitory effect of hexa-N-acetylchitohexaose and chitohexaose against Meth-A solid tumor. *Chem. Pharm. Bull. (Tokyo).*, 36, 784, 1988.
30. Gibot, L., Bouhout, S., Bolduc, S., Auger, F.A., Moulin, V.J., Anticancer properties of chitosan on human melanoma are cell line dependent. *Int. J. Biol. Macromol.*, 72, 370, 2015.
31. Subhapradha, N., Shanmugam, A., Fabrication of beta-chitosan nanoparticles and its anticancer potential against human hepatoma cells. *Int. J. Biol. Macromol.*, 94, 194, 2017.
32. Loutfy, S.A., El-Din, H.M.A., Elberry, M.H.,Allam, N.G., Hasanin, M.T.M., Abdellah, A.M., Synthesis, characterization and cytotoxic evaluation of chitosan nanoparticles: *In vitro* liver cancer model. *Adv. Nat. Sci.: Nanosci. Nanotechnol.*,7, 035008, 2016.
33. Zou, P., Yang, X., Zhang, Y., Du, P., Yuan, S., Yang, D., Wang, J., Antitumor effects of orally and intraperitoneally administered chitosan oligosaccharides (COSs) on S180-bearing/residual mouse. *J. Food. Sci.*, 81, H3035, 2016.
34. Pillai, S., Oresajo, C., Hayward, J., Ultraviolet radiation and skin aging: roles of reactive oxygen species, inflammation and protease activation, and strategies for prevention of inflammation-induced matrix degradation – a review. *Int. J. Cosmet. Sci.*, 27, 17, 2005.
35. Goy, R.C., Morais, S.T.B., Assis, O.B.G., Evaluation of the antimicrobial activity of chitosan and its quaternized derivative on E. coli and S. aureus growth. *Rev. Bras. Farmacogn.*, 26(1), 122–127, 2016.
36. Zheng, L.Y., Zhu, J.F., Study on antimicrobial activity of chitosan with different molecular weights. *Carbohydr. Polym.*, 54, 527, 2003.
37. Sudarshan, N.R., Hoover, D.G., Knorr, D., Antibacterial action of chitosan. *Food. Biotechnol.*, 6, 257, 1992.
38. Chung, Y.C., Chen, C.Y., Antibacterial characteristics and activity of acid-soluble chitosan. *Bioresour. Technol.*, 99, 2806, 2008.
39. Muzzarelli, R., Tarsi, R., Filippini, O., Giovanetti, E., Biagini G., Varaldo, P.E., Antimicrobial properties of N-carboxybutyl chitosan. *Antimicrob. Agents. Chemother.*, 34, 2019, 1990.
40. Vardaka, V.D., Yehia, H.M., Savvaidis, I.N., Effects of Citrox and chitosan on the survival of Escherichia coli O157:H7 and Salmonella enterica in vacuum-packaged turkey meat. *Food, Microbiol.*, 58, 128, 2016.
41. Darabpour, E.N., Kashef., Mashayekhan, S., Chitosan nanoparticles enhance the efficiency of methylene blue-mediated antimicrobial photodynamic inactivation of bacterial biofilms: An *in vitro* study. Photodiagnosis. *Photodyn. Ther.*, 14, 211, 2016.
42. Zhong, Z., Chen, R., Xing, R., Chen, X., Liu, S., Guo, Z., Ji, X., Wang, L., Li, P., Synthesis and antifungal properties of sulfanilamide derivatives of chitosan. *Carbohydr. Res.*, 342, 2390, 2007.

43. Ziani, K., Fernández-Pan, I., Royo, M., Mate J.I., Antifungal activity of films and solutions based on chitosan against typical seed fungi. *Food Hydrocoll.*, 23, 2309, 2009.
44. Guerra-Sánchez, M.G., Vega-Pérez, J., Velázquez-del Valle , M.G., Hernández-Lauzardo , A.N., Antifungal activity and release of compounds on Rhizopus stolonifer (Ehrenb.:Fr.) Vuill. by effect of chitosan with different molecular weights. *Pestic. Biochem. Physiol.*, 93, 18, 2009.
45. Seyfarth, F., Schliemann, S., Elsner, P., Hipler, U.C., Antifungal effect of high – and low-molecular-weight chitosan hydrochloride, carboxymethyl chitosan, chitosan oligosaccharide and N-acetyl-D-glucosamine against Candida albicans, Candida krusei and Candida glabrata. *Int. J. Pharm.*, 353, 139, 2008.
46. Ing, L.Y., Zin, N.M., Sarwar, A., Katas, H., Antifungal Activity of Chitosan Nanoparticles and correlation with their physical properties. *Int. J. Biomater.*, 632698, 2012.
47. Tayel, A.A., Moussa, S., El-Tras, W.F., Knittel, D.,Opwis, K., Schollmeyer E. Anticandidal action of fungal chitosan against Candida albicans. *Int. J. Biol. Macromol.* 47, 454, 2010.
48. Munoz, Z., Moret, A., Sensitivity of Botrytis cinerea to chitosan and acibenzolar-S-methyl. *Pest. Manag. Sci.*, 9, 974, 2010.
49. Al-Hetar, M.Y., Zainal Abidin, M.A., Sariah, M., Wong, M.Y., Antifungal activity of chitosan against Fusarium oxysporum f. sp. cubense. *J. Appl. Polym. Sci.*, 120, 2434, 2011.
50. Alburquenque, C., Bucarey, S.A., Neira-Carrillo, A., Urzúa, B., Hermosilla, G., Tapia, C.V., Antifungal activity of low molecular weight chitosan against clinical isolates of Candida spp. *Med. Mycol.*, 48, 1018, 2010.
51. García-Rincón, J., Vega-Pérez , J., Guerra-Sánchez, M.G., Hernández-Lauzardo, A.N., Peña-Díaz, A., Velázquez-Del Valle, M.G., Effect of chitosan on growth and plasma membrane properties of Rhizopus stolonifer (Ehrenb.:Fr.) Vuill. *Pestic. Biochem. Physiol.*, 97, 275, 2010.
52. Roller, S., Covill, N., The antifungal properties of chitosan in laboratory media and apple juice. *Int. J. Food. Microbiol.*, 47, 67, 1999.
53. Kong, M., Chen, X.G., Xing, K., Park, H.J., Antimicrobial properties of chitosan and mode of action: a state of the art review. *Int. J. Food. Microbiol.*, 144, 51, 2010.
54. Manohar, M., Fatima, I., Saxena, R., Chandra, V., Sankhwar, P.L., Dwivedi, A., (–)-Epigallocatechin-3-gallate induces apoptosis in human endometrial adenocarcinoma cells via ROS generation and p38 MAP kinase activation. *J. Nutr. Biochem.*, 24, 940, 2013.
55. Park, B.K., Kim, M.M., Applications of chitin and its derivatives in biological medicine. *Int. J. Mol. Sci.*,11, 5152, 2010.
56. Leung, P.S., Chan, Y.C., Role of oxidative stress in pancreatic inflammation. *Antioxid. Redox. Signal.*, 11, 135, 2009.
57. Maynard, S., Schurman, S.H., Harboe, C., de Souza-Pinto, N.C., Bohr, V.A., Base excision repair of oxidative DNA damage and association with cancer and aging. *Carcinogenesis*, 30, 2, 2009.

58. Ngo, D.H., Kim, S.K., Antioxidant effects of chitin, chitosan, and their derivatives. *Adv. Food. Nutr. Res.*, 2014.
59. Park, P.J., Je, J.Y., Kim, S.K., Free radical scavenging activity of chitooligosaccharides by electron spin resonance spectrometry. *J. Agric. Food. Chem.*, 51, 4624, 2003.
60. Abd El-Rehim, H.A., El-Sawy, N.M., Hegazy, el-SA., Soliman, el-SA., Elbarbary, A.M., Improvement of antioxidant activity of chitosan by chemical treatment and ionizing radiation. *Int. J. Biol. Macromol.*, 50, 403, 2012.
61. García, M.A., Paz, Nilia de la., Castro, C., Rodríguez, M., Rapado, M., Zuluaga, R., Gañán, P., Casariego, A., Effect of molecular weight reduction by gamma irradiation on the antioxidant capacity of chitosan from lobster shells. *J. Radiat. Res. Appl. Sci.*, 8, 190, 2015.
62. Nallamuthu, I., Devi, A., Khanum, F., Chlorogenic acid loaded chitosan nanoparticles with sustained release property, retained antioxidant activity and enhanced bioavailability. *Asian. J. Pharm. Sc.*, 10, 203, 2015.
63. Vo, T.S. Kim, S.K., Potential anti-HIV agents from marine resources: an overview. *Mar. Drugs*,8, 2871, 2010.
64. Sosa, M.A., Fazely, F., Koch, J.A., Vercellotti, S.V., Ruprecht, R.M., N-carboxymethylchitosan-N,O-sulfate as an anti-HIV-1 agent. *Biochem. Biophys. Res. Commun.*, 174, 489, 1991.
65. Nishimura, S.I., Kai, H., Shinada, K.,Yoshida, T., Tokura, S., Kurita, K., Nakashima, H., Yamamoto, N., Uryu, T., Regioselective syntheses of sulfated polysaccharides: specific anti-HIV-1 activity of novel chitin sulfates. *Carbohydr. Res.*, 306, 427, 1998.
66. Ramana, L.N., Sharma, S., Sethuraman, S., Ranga, U., Krishnan, U.M., Evaluation of chitosan nanoformulations as potent anti-HIV therapeutic systems. *Biochim. Biophys. Acta.*, 1840, 476, 2014.
67. Meng, J., Zhang, T., Agrahari, V., Ezoulin, M.J., Youan, B.B., Comparative biophysical properties of tenofovir-loaded, thiolated and nonthiolated chitosan nanoparticles intended for HIV prevention. *Nanomedicine (Lond)*, 9, 1595, 2014.
68. Francesko, A., Tzanov, T., Chitin, chitosan and derivatives for wound healing and tissue engineering. *Adv. Biochem. Eng. Biotechnol.*, 125, 1, 2011.
69. Yang, T.L., Chitin-based materials in tissue engineering: applications in soft tissue and epithelial organ. *Int. J. Mol. Sci.*, 12, 1936, 2011.
70. Tashiro, T., Antibacterial and Bacterium Adsorbing Macromolecules. *Macromol. Mater. Eng.*, 286, 63, 2001.
71. Chen, Z., Mo, X., He C., Wang, H., Intermolecular interactions in electrospun collagen–chitosan complex nanofibers. *Carbohydr. Polym.* 72, 410, 2008.
72. Chen, J.-P., Chang, G.-Y., Chen, J.-K., Electrospun collagen/chitosan nanofibrous membrane as wound dressing. *Colloids. Surf. A. Physicochem. Eng. Asp.*, 313, 183, 2008.
73. Chen, F., Li, X., Mo, X., He, C., Wang, H., Ikada, Y., Electrospun chitosan-P(LLA-CL) nanofibers for biomimetic extracellular matrix. *J. Biomater. Sci .Polym. Ed.* 19, 677, 2008.

74. Jayakumar, R., Prabaharan, M., Nair, S.V., Tamura, H., Novel chitin and chitosan nanofibers in biomedical applications. *Biotechnol. Adv.*, 28, 142, 2010.
75. Madhumathi, K., Sudheesh Kumar, P.T., Abhilash, S., Sreeja, V., Tamura, H., Manzoor, K., Nair, S.V., Jayakumar, R., Development of novel chitin/nanosilver composite scaffolds for wound dressing applications. *J. Mater. Sci. Mater. Med.*, 21, 807, 2010.
76. Biagini, G., Bertani, A., Muzzarelli, R., Damadei, A., DiBenedetto, G., Belligolli, A., Riccotti, G., Zucchini, C., Rizzoli, C., Wound management with N-carboxybutyl chitosan. *Biomaterials,* 12, 281, 1991.
77. Stone, C.A., Wright, H., Clarke, T., Powell, R.,Devaraj, V.S., Healing at skin graft donor sites dressed with chitosan. *Br. J. Plast. Surg.,* 53, 601, 2000.
78. Azad, A.K., Sermsintham, N., Chandrkrachang, S., Stevens, W.F., Chitosan membrane as a wound-healing dressing: characterization and clinical application. *J. Biomed. Mater. Res. B. Appl. Biomater.,* 69, 216, 2004.
79. Valentine, R., Athanasiadis, T., Moratti, S., Hanton, L., Robinson, S., Wormald, P.J., The efficacy of a novel chitosan gel on hemostasis and wound healing after endoscopic sinus surgery. *Am. J. Rhinol. Allergy.,* 24, 70, 2010.
80. Bennett,B.L.,Littlejohn,L.F.,Kheirabadi,B.S.,Butler,F.K.,Kotwal,R.S.,Dubick, M.A., Bailey, J.A., Management of External Hemorrhage in Tactical Combat Casualty Care: Chitosan-Based Hemostatic Gauze Dressings–TCCC Guidelines-Change 13-05. *J. Spec. Oper. Med.,* 14, 40, 2014.
81. Nimal, T.R., Baranwal, G., Bavya, M,C., Biswas, R., Jayakumar, R., Anti-staphylococcal Activity of Injectable Nano Tigecycline/Chitosan-PRP Composite Hydrogel Using Drosophila melanogaster Model for Infectious Wounds. *ACS, Appl, Mater, Interfaces.,*8, 22074, 2016.
82. Kim, I.Y., Seo, S.J., Moon, H.S., Yoo, M.K., Park, I.Y., Kim, B.C., Cho, C.S., Chitosan and its derivatives for tissue engineering applications. *Biotechnol. Adv.,* 26, 1, 2008.
83. Dutta, P.K., Rinki, K., Dutta, J., Chitosan: a promising biomaterial for tissue engineering scaffolds, In: *Chitosan for biomaterials II*, R. Jayakumar, M. Prabaharan, R.A.A. Muzzarelli, Eds., 2011, Springer Berlin Heidelberg: Berlin, Heidelberg. pp. 45–79.
84. El-Sherbiny, I.M., Yacoub, M.H., Hydrogel scaffolds for tissue engineering: Progress and challenges. *Glob. Cardiol. Sci. Pract.,* 3, 316, 2013.
85. Dünnhaupt, S., Barthelmes, J., Thurner, C.C., Waldner, C., Sakloetsakun, D., Bernkop-Schnürch, A., S-protected thiolated chitosan: Synthesis and *in vitro* characterization. *Carbohydr. Polym.,* 90, 765, 2012.
86. Nascimento, A.V., Gattacceca, F., Singh, A., Bousbaa, H., Ferreira, D., Sarmento B1., Amiji, M.M., Biodistribution and pharmacokinetics of Mad2 siRNA-loaded EGFR-targeted chitosan nanoparticles in cisplatin sensitive and resistant lung cancer models. *Nanomedicine (Lond).,* 11, 767, 2016.
87. Ajun, W., Yan, S., Li, G., Huili, Li., Preparation of aspirin and probucol in combination loaded chitosan nanoparticles and *in vitro* release study. *Carbohydr, Polym.,*75, 566, 2009.

88. Tian, J., Xu, S., Deng, H., Song, X., Li, X., Chen, J., Cao, F., Li, B., Fabrication of self-assembled chitosan-dispersed LDL nanoparticles for drug delivery with a one-step green method. *Int. J. Pharm.*, 17, 25, 2016.
89. Abruzzo, A., Zuccheri, G., Belluti, F., Provenzano, S., Verardi, L., Bigucci, F., Cerchiara, T., Luppi, B., Calonghi, N., Chitosan nanoparticles for lipophilic anticancer drug delivery: Development, characterization and *in vitro* studies on HT29 cancer cells. *Colloids. Surf. B Biointerfaces.*,145, 362, 2016.
90. Badran, M.M., Harisa, G.I., AlQahtani, S.A., Alanazi,F.K., Zoheir, K.M.A., Pravastatin-loaded chitosan nanoparticles: Formulation, characterization and cytotoxicity studies. *J. Drug. Deliv. Sci. Technol.*, 32, 1, 2016.
91. Jing, Li., Wu, J., Zhang. J., Wang, Y., Fang, L., Shen., Qi., Oral bioavailability and evaluation of docetaxel-nicotinamide complex loaded chitosan nanoparticles. *RSC Adv.*, 6, 35354, 2016.
92. Chhonker, Y.S., Prasad, Y.D., Chandasana, H., Vishvkarma, A., Mitra, K., Shukla, P. K., Bhatta, R. S.,.Amphotericin-B entrapped lecithin/chitosan nanoparticles for prolonged ocular application. *Int. J. Biol. Macromol.*, 72, 1451, 2015.
93. Jadidi-Niaragh, F., Atyabi, F., Rastegari, A., Mollarazi, E., Kiani, M., Raza1., Yousefi, M., Kheshtchin, N., Downregulation of CD73 in 4T1 breast cancer cells through siRNA-loaded chitosan-lactate nanoparticles. *Tumour. Biol.* 37, 8403, 2016.
94. Chen, H., Pan, H., Li, P., Wang, H.3., Wang, X., Pan, W., Yuan, Y., The potential use of novel chitosan-coated deformable liposomes in an ocular drug delivery system. *Colloids .Surf. B. Biointerfaces.*, 143, 455, 2016.
95. Siahmansouri, H., Somi, M.H., Babaloo, Z., Baradaran, B., Jadidi-Niaragh, F., Atyabi, F., Mohammadi, H., Ahmadi, M., Yousefi, M., Effects of HMGA2 siRNA and doxorubicin dual delivery by chitosan nanoparticles on cytotoxicity and gene expression of HT-29 colorectal cancer cell line. *J. Pharm. Pharmacol.*, 68, 1119, 2016.
96. Alinejad, V., Hossein, Somi, M., Baradaran, B., Akbarzadeh, P., Atyabi, F., Kazerooni, H., Samadi, K.H., Aghebati, M.L., Siah, M. H., Yousefi, M., Co-delivery of IL17RB siRNA and doxorubicin by chitosan-based nanoparticles for enhanced anticancer efficacy in breast cancer cells. *Biomed Pharmacother.*, 83, 229, 2016.
97. Tekie, F.S., Atyabi, F., Soleimani, M., Arefian, E., Atashi, A., Kiani, M., Khoshayand, MR., Amini M., Dinarvand, R., Chitosan polyplex nanoparticle vector for miR-145 expression in MCF-7: Optimization by design of experiment. *Int. J. Biol. Macromol.*, 81, 828, 2015.
98. Cavalli, R., Leone, F., Minelli, R., Fantozzi, R., Dianzani, C., New chitosan nanospheres for the delivery of 5-fluorouracil: preparation, characterization and *in vitro* studies. *Curr. Drug. Deliv.*, 11, 270, 2014.
99. Zhou, W., Wang, Y., Jian, J., Song, S., Self-aggregated nanoparticles based on amphiphilic poly(lactic acid)-grafted-chitosan copolymer for ocular delivery of amphotericin B. *Int. J. Nanomed.*, 8, 3715, 2013.

100. Anitha, A., Uthaman, S., Nair, S.V., Jayakumar, R., Lakshmanan, V.K., Enhanced delivery system of flutamide loaded chitosan-dextran sulphate nanoparticles for prostate cancer. *J. Biomed. Nanotechnol.*, 9, 335, 2013.
101. Yang, S.-J., Lin, C,F., Kuo, M.L., Tan, C.T., Photodynamic Detection of Oral Cancers with High-Performance Chitosan-Based Nanoparticles. *Biomacromolecules,* 14, 3183, 2013.
102. Chen, H., Yang, W., Chen, H., Liu, L., Gao, F., Yang, X., Jiang, Q., Zhang, Q., Wang, Y., Surface modification of mitoxantrone-loaded PLGA nanospheres with chitosan. *Colloids. Surf. B. Biointerfaces.*,73, 212, 2009.
103. Li, F., Li, J., Wen, X., Zhou, S., Tong, X., Su., P. Li H., Shi, D., Anti-tumor activity of paclitaxel-loaded chitosan nanoparticles: An *in vitro* study. *Mater. Sci. Eng. C.*, 29, 2392, 2009.
104. Shukla, S,K., Mishra, A.K., Arotiba, O.A., Mamba, B.B., Chitosan-based nanomaterials: A state-of-the-art review. *Int. J. Biol .Macromol.*, 59, 46, 2013.
105. Zhang, S., Bai, H., Luo, J., Yang, P., Cai, J., A recyclable chitosan-based QCM biosensor for sensitive and selective detection of breast cancer cells in real time. *Analyst,* 139, 6259 2014.
106. Dehdashtian, S., Gholivand, M. B., Shamsipur, M., Kariminia, S., Construction of a sensitive and selective sensor for morphine using chitosan coated Fe3O4 magnetic nanoparticle as a modifier. *Mater. Sci. Eng. C. Mater. Biol. Appl.*, 58, 53, 2016.
107. Freire, T.M., Dutra, L.M., Queiroz, D.C., Ricardo, N.M., Barreto, K., Denardin, J.C., Wurm, F.R., Sousa C.P., Correia, A.N., de Lima-Neto, P., Fechine, P.B., Fast ultrasound assisted synthesis of chitosan-based magnetite nanocomposites as a modified electrode sensor. *Carbohydr. Polym.* 151, 760, 2016.
108. Mohammad-Rezei, R., Razmi, H., Preparation and Characterization of hemoglobin immobilized on graphene quantum dots-chitosan nanocomposite as a sensitive and stable hydrogen peroxide biosensor. *Sensor. Lett.*, 14, 685, 2016.

Section IV

CHITOSAN IN AGRICULTURE AND WATER TREATMENT

15

Practical and Plausible Implications of Chitin- and Chitosan-Based Nanocomposites in Agriculture

Aasim Majeed*, Raoof Ahmad Najar, Shruti Choudhary, Wahid Ul Rehman, Amandeep Singh, Sapna Thakur and Pankaj Bhardwaj

Molecular Genetics Laboratory, Central University of Punjab, Bathinda, India

Abstract

Chitin and chitosan have profound impact on diverse physiological processes of plants. They elicit immune response in plants, aid in plant protection, manipulate plant growth and flowering. Further they act to adsorb various heavy metals, pollutants, pesticides, and insecticides from soil or water. These properties of chitin and chitosan render them to be applicable for agricultural purposes. They can form nanocomposites and nanoparticles which are applied for controlled delivery and release of herbicides and pesticides for effective pest and weed management in a more cost effective manner. Further, their nanocomposites are being employed for controlled release of fertilizers and micronutrients for better nutrient management in crops. Biosensors generated from chitosan nanocomposites can aid in monitoring soil condition in real time to help farmers in deciding the appropriate agronomic practices to be employed. Moreover, the chitosan nanocomposite-based gene delivery system helps in effective gene manipulation in crops for enhanced genetic engineering. These nanocomposites also bear activity against diverse plant pathogens to help in protecting crops from disease. This chapter focuses on the agricultural applications of chitin- and chitosan-based nanocomposites.

Keywords: Chitosan nanocomposites, biosensors, immunomodulatory, controlled release of insecticides and pesticides

**Corresponding author*: majeedaasim@gmail.com

Shakeel Ahmed and Saiqa Ikram (eds.) Chitosan, (411–430) © 2017 Scrivener Publishing LLC

15.1 Introduction

Chitin is the second most abundant biomaterial on the planet earth after cellulose. It is a polymeric carbohydrate consisting of *N*-acetylglucosamine units. The latter is the derivative of glucose. Chitin forms the most important component of fungal cell walls, arthropod, crustacean, and insect exoskeletons. Structurally it resembles cellulose in having crystalline nature of nanofibrils and whisker. Functionally it is close to keratin. Although chitin is of widespread occurrence in nature but commercially most of the chitin is harvested from the shells of crabs and shrimps [1]. Two structural variants exist in nature for chitin viz α- and β-forms [2]. A γ-chitin has also been proposed as a third allomorph [3, 4] but this variant has been considered as a mere variation of the α-form [5]. The α-form is abundant while β-form is rarer in nature. In the crystal structure of chitin, there occurs fine organization of its chains into sheets held tightly through intrasheet hydrogen bonds in both forms. The α-form, however, contains additional intersheet hydrogen bonds rendering it more stable than β-form in swelling [6]. Chitin occurs in partial deacetylated form in nature [7]. It is insoluble in usual solvents which hurdles its processing [1]. Widely used solvents for chitin include DMAc/LiCl, hexafluoroisopropyl alcohol, methanol saturated $CaCl_2 \cdot 2H_2O$, and hexafluoracetone sesquihydrate [8] though concentrated phosphoric acid [9], lithium thiocyanate [10], and NaOH [11] have also been used. Partial deacetylation by alkaline environment or chitin deacetylase hydrolysis of chitin in solid state yields a chitin derivate known as chitosan [1]. Its β-form is more susceptible to deacetylation than α-form [12]. Carboxymethyl chitin, next most prominent derivative of chitin is obtained when chitin is treated in presence of concentrated NaOH with monochloroacetic acid. Addition of chemical constituents like fluorine, sulphur, diethylamine, phosphoryl, and carbamates to chitin yields their corresponding chitin derivatives with specific properties [1]. Partial acid degradation of chitin yields oligochitins [12].

Chitin plays its role in affinity chromatography [13], macrophage activation, tumor cell suppression and wound healing [14], biosensors and immobilization of enzymes for industrial purposes [15]. Processing of chitin into fibers and films is done at relative ease [16]. Blending of these fibers with cellulose [17] or silk [18] yields a composites that can be used for antimicrobial, moisture controlling, deodorization, and nonallergic applications [2]. Chitin derivatives are also used in making paper. In the field of medicine, the composites of chitin fibers and films are mainly used in wound healing [19], drug delivery [20], and bone filling [21]. Grafting of 5-flurouracil and muramyl l-alanyl-isoglutamate with carboxymethyl chitin produces antitumor drug conjugate that is effectively used to target cancer cells [1].

Deacetylation of chitin beyond 50% results in the formation of a water soluble product called chitosan. The solubility of chitosan renders it feasible for generation of hydrogels, films as well as composites. It bears semicrystalline structure in solid state. Chitosan forms complexes with different metals like copper [22], silver nitrate [23] and iron [24], surfactants like SDS [25], polymers like polyacrilic acid and carboxymethyl cellulose [26], carrageenan, pectin, dextran sulfate [27].

Chitin- and chitosan-based biomaterials have left a profound and landmark impact in the biomedical field. Both chitin and chitosan forms hydrogels for diverse purposes. The hydrogels of cellulose and chitin exhibit excellent gelling properties [28]. Chitin-methacrylate hydrogels show promising cell attaching and proliferating property which makes them suitable for tissue engineering. Further these hydrogels are noncytotoxic [29]. Chitosan-based hydrogels are being used for controlled drug delivery [30], neural tissue engineering [31], and delivery of cytokines intra tumoraly [32]. Composites and nanocomposites of chitosan play greater role in tissue engineering [33], antimicrobial activity [34], drug delivery [35], and cancer therapy [36].

15.2 Applications of Chitin and Chitosan Nanocomposite in Agriculture

Although chitin- and chitosan-based nanocomposites have a profound role in biomedicine, their importance and use in agriculture are still at the horizon and infancy. Chitin and chitosan play diverse physiological roles when applied to plants. Chitosan elicits immune response in plants has antibacterial, antiviral, and antifungal activity. So the development of suitable nanocomposites of chitosan will definitely play a vital role in crop protection. Chitosan can adsorb different heavy metals, pollutants, and pesticides from soil or water so can be thought of managing agriculture soils through its nanocomposites. It is evident from the literature that although chitosan nanocomposites have been used to some extent for different agriculture purposes (Figure 15.1), chitin-based nanocomposites are still to be developed for this purpose. The different agricultural aspects in which chitosan-based nanocomposites play their role are discussed in the following headings:

15.2.1 Controlled Pesticide, Insecticide, and Herbicide Delivery

Global population is growing exponentially. Food production has to grow at the same pace to sustain the growing population. Besides producing the improved and high yielding varieties of diverse food crops, strategic

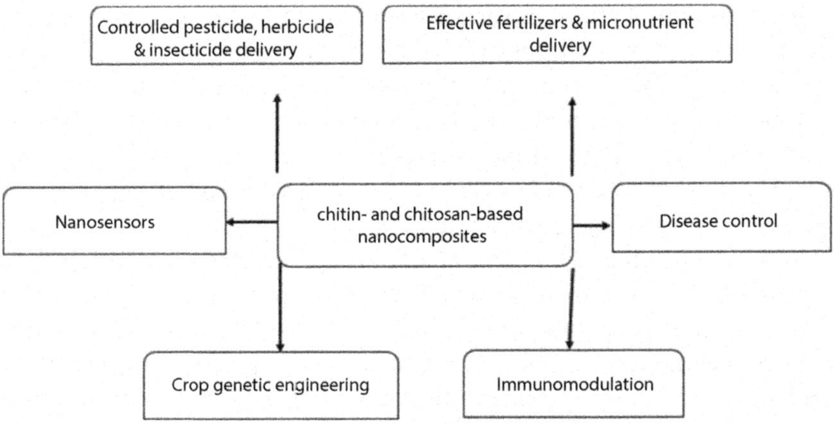

Figure 15.1 Agricultural implications of chitin- and chitosan-based nanocomposites.

and efficient management of pests and insects of food crops is required. Though many insect- and pest-resistant crop varieties have been produced and are currently under domestication, insects and pests are evolving in parallel to evade the host defence response against them. The traditional use of insecticides and pesticides for disease management in plants is still at its peak [37]. Insecticides and pesticides are primarily applied in the form of sprays at regular intervals. Few concerns are raised against their excessive use. First, their little temporal availability necessitates the need of multisprays at regular intervals which adds to the burden of financial crises to the farmers. Second, they are delivered poorly into the cell from the surface. For effective action, the chemicals sprayed over plants shall penetrate into the cell and also shall remain active for prolonged time to effectively counter the pests and insects. So controlled release technology is viewed as a promising solution to achieve these goals. Different materials are being used to deliver the insecticides and pesticides sustainably. Polyacrylamide hydrogels to deliver paraquat [38], modified starch copolymers prepared by gamma irradiation to deliver different insecticides [39], etc., have been applied for this purpose. Nanoparticles are also effectively used for this purpose. Controlled release of, tebuconazole chlorothalonil fungicides through polyvinyl pyridine nanoparticles [40], ivermectin and acetamiprid pesticides by polymeric nanocapsules [41] is among the fruits of controlled release technology. Nanoparticles of titanium oxide, silver oxide, iron oxide, and aluminium oxide have also been used.

Chitosan favors its use as an efficient carrier of chemical compounds because it acts as protective reservoir for the active elements of chemicals to be delivered, safeguards the chemicals from the external environment and facilitates their controlled release from the chitosan matrix. Chitosan possesses features that enable it to be used in controlled release technology of pesticides. It is biocompatible, biodegradable, and easily available and has been used to deliver different drugs [42–45]. Carbon nanotubes are porous, biocompatible, and biodegradable and have hollow structure. They exhibit fabulous absorptive properties. In an attempt to design and fabricate a novel material for controlled release of pesticides, a nanocomposite of gamma irradiated carbon nanotubes and chitosan was prepared and tested for the release profile of the Azinphos methyl (AZM), which is a widely used insecticide for protection of apple, lemon, and other fruit trees. The release profile as indicated by the HPLC shows that 2% release of AZM occurs after 5 min which gradually increases upto 100% after 4 h. Controlled release of AZM by the nanocomposite allows us to use minimum pesticide amount which could reduce the environmental hazards caused by these pesticides [42]. Diverse reports indicate that chitosan nanoparticles act as delivery matrix for sustained release of pesticides. The fabrication of chitosan and cashew tree gum into suitable microspheres renders them as efficient carriers of the *Lippia sidoides* oil which acts as an insecticide [46]. Similarly chitosan and alginate microcapsules bearing the encapsulation efficiency of about 82% and release time prolongation to about eight times, efficiently deliver immidacloprid chemical insecticide [47]. Likewise rotenone encapsulated *N*-(octadecanol-1-glycidyl ether)-*O*-sulfate chitosan nanoparticle micelles show sustained release of the insecticide over prolonged periods so can be effectively employed for agricultural use [48]. Nanocomposite particles of carboxymethyl chitosan and ricinoleic acid exhibit good carrier ability for azadirachitin, a biopesticide. The release prolongation is exceptionally greater (11 days) but the encapsulation efficiency of only around 56% is observed [49].

Weeds compete with the main crop for space, light, water, and nutrients and, therefore, deprive the main plants from essential growth requirements. Weeds are robust and grow fast to occupy the whole field in a little time. They rob the main crop from nutrients. A substantial amount of food production loss is attributed to the weed growth. So effective weed management needs utmost attention. Conventional hand plucking of weeds is labor intensive and time consuming. Consequently a myriad of herbicides are being developed to tackle the problem. Excessive use of these herbicides leads to environmental degradation. Efficient delivery of these weedicides and herbicides can solve the problem of their overuse which further would be cost effective and environmentally less hazardous. Different delivery systems have been developed for controlled and efficient delivery of weedicides as in case of pesticides. Nanocomposite particles based on alginate and chitosan prolong the release of paraquat and also alter its interaction with soil [50]. A nanocomposite of montmorillonite clay and chitosan acts as an excellent adsorbent for the herbicide clopyralid. It efficiently adsorbs the herbicide from soil or water which improves on further addition of chitosan content to the nanocomposite [51]. In an attempt to evaluate the impact of chitosan nanoparticles on the bioavailability of the herbicide, dichloroprop, in *Chlorella pyrenoidosa* (green alga), the results revealed that the chitosan nanoparticles modify and enhance its bioavailability to the green algae [52]. This proves its effectiveness as herbicide for agricultural purposes. Nanocomposites generated from chitosan/alginate and chitosan/tripolyphosphate show efficient encapsulation of the herbicides (imazapic and imazapyr) and their much reduced release profile. Moreover, the nanocomposite exhibits lesser cytotoxicity and genotoxicity thereby causing less damage to the underlying organisms. Further, the herbicidal activity of these compounds was improved when used as nanocomposites rather than the free compounds, so the composite can be used as an effective herbicide for better weed management [53]. Paraquat is commonly used non selective herbicide that is employed for weed control globally. Due to its water solubility as well as soil sorption, this compound confers toxicity to untargeted organisms. In order to produce an herbicide formulation with less toxicity, chitosan and tripolyphosphate were fabricated into a nanocomposite that could be used safely for better weed management. The nanocomposite possesses reduced release profile. Cytotoxicity and genotoxicity tests on *Zea mays* and *Brassica sp.* showed reduced toxicity [54]. Such nanocomposites would target only the unwanted weeds without causing damage to the main or other plants. Million tonnes of pesticides, weedicides, insecticides, and herbicides are being employed for agricultural purposes all over the world. Widespread use of pesticides and herbicides leads to their stagnation in various environmental spheres like aquatic habitats [55], ground water [56], etc. One of the commonly used insecticide for agricultural purposes is Malathion, an organophosphate compound. Strategic removal of these pesticides can provide friendly environment. In view of this, a nanocomposite of copper and chitosan efficiently adsorbs Malathion from agriculture run off due to the formation of dithionate complex between copper and Malathion. Thus this nanocomposite can be efficiently used to remove organophosphorus pesticides [57]. In a similar study, chitosan-silver oxide nanocomposites show successful removal of organochlorine pesticides [58]. Permethrin pesticide is efficiently adsorbed by chitosan/zinc oxide nanoparticles from water [59]. Extensive use of pesticides and insecticides has added to the problem of soil and water pollution. Use of controlled release formulations

(CRFs) will help in better management of pesticides and insecticides. A composite of carboxymethyl cellulose, chitosan, and bentonite was fabricated into a gel which exhibits good encapsulating properties due to carboxymethyl-chitosan polymer as well as excellent sorption ability due to bentonite for the pesticides, atrazine, and immidacloprid. Both encapsulation and sorption restrain the diffusion of these chemicals from the composite which was further enhanced by high hydrophobicity as well as low solubility of these pesticides. All these factors reduce the release profile of the pesticides. Further, the composite reduces the leaching behavior of the pesticides which is helpful in reducing the environmental pollution [60] Similarly to prevent excessive contamination of soil and water by anionic pesticides and herbicides, nanocomposite derived from montmorillonite and chitosan was investigated for the herbicide, clopyralid (3,6-dichloropyridine-2-carboxylic acid). The results revealed that the bionanocomposite acts as an excellent adsorbent for the anionic herbicide and can be used for the removal or immobilization of anionic pesticides and herbicides from both water and soil under acidic conditions [51].

15.2.2 Efficient Fertilizer and Micronutrient Delivery

A balanced plant growth requires appropriate nutrients and water. Plants absorb water and nutrients from soil which acts as a reservoir of mineral nutrients. Besides the naturally occurring minerals in the soil, external application of mineral nutrients to soil in the form of fertilizers facilitates rapid growth and development of plants. Consequently millions of tonnes of fertilizers are being applied worldwide to the agricultural crops for enhanced yield. However there arises a problem that a substantial part of this fertilizer input gets unused by the plant. According to a report, upto 70% of nitrogen, 90% of phosphorous, and 70% of potassium from applied fertilizers remain unabsorbed by the plant [61]. In order to combat this problem, use of fertilizers at regular intervals is needed. But this overuse of fertilizers induces nutrient burn to the plants which then start to die. Further the excessive use of fertilizers causes salt stress to the plant [62] which decreases the plant growth and development thereby yield. Moreover, additional use of fertilizers puts financial burden to the farmers. So a way has to be found out that would evacuate us from the whole scenario. One of the strategy to mitigate this problem is to develop the fertilizer delivery systems that maintain the optimum fertilizer content in the soil for prolonged periods. Consequently such slow release fertilizers can save losses due to the need for excessive fertilizer application and environmental degradation due to excessive stagnation of these fertilizers in soil and water. Different release mechanisms of fertilizer exist like polymer coating to the fertilizer granules where biodegradation of the polymer releases the encapsulated fertilizer for prolonged periods. The polymer coating immobilize the fertilizer to prevent its leaching [63]. The polymer also prevents the contact of the fertilizer with the external environment so as to protect it from degradation and chemical reaction [64]. A superabsorbent composite having an outer coating of urea, inner coating of chitosan, and a core of NPK fertilizer granule exhibits slow release property of N, P, and K along with enhancement in water-holding capacity. Besides, this compound fertilizer is nontoxic to soil [65]. Use of this composite fertilizer is a boon to agriculture. Similarly a composite fertilizer made from urea-kaolinite and chitosan shows decrease in nitrogen release which increases by increasing chitosan concentration [66]. As application urea is required in large quantities by the farmers, use of this composite fertilizer with controlled release ability will not only save the capital but also reduce the negative impact of overdosing. Also the chitosan nanoparticles containing NPK have been prepared by polymerizing methacrylic acid to generate an efficient slow release fertilizer [67]. Similar nanofertilizers prepared by employing chitosan nanoparticles will boost the agricultural productivity [68]. NPK fertilizer based on chitosan nanoparticles on application to the leaves is absorbed through stomata and translocated by phloem which increases the harvest index, crop index and mobilization index in wheat plants grown in sand. This enhanced plant growth upon foliar application of nanofertilizers opens the gate for innovative nutrient engineering in the field of agriculture [69]. Genipin cross-linked chitosan microspheres with urea show slow release profile of urea under experimental conditions and thus can be another addition to the controlled release fertilizers [70]. Moreover, the microspheres made from chitosan and montmorillonite encapsulating potassium form an efficient potassium fertilizer. Using time domain reflectometry, this composite fertilizer shows fertilizer release at two stages, the outer potassium is released upto days followed by internal release thereafter. Thus after initial increase in release upto there days, there occurs a stabilized and constant release of KNO_3. Carbon nanoparticles after penetrating successfully through the thick seed coat of tomato seeds induce water uptake by the seeds which ultimately enhances seed germination [71]. Zinc oxide nanoparticles enhance growth, yield as well as zinc accumulation in maize seeds [72]. Nanoparticles of silver-chitosan enhance the growth in chick pea. Over all fresh and dry weight as well as seed germination percentage and seedling length were greatly increased. Further the nanoparticle-treated plants show marked increase in chlorophyll content as compared to untreated plants. Biochemical investigations reveal increased activity of ascorbate peroxidase, catalase as well as α- and β-amylase. Further, the MDA content is also increased [73]. Foliar application of chitosan-based nanofertilizers containing NPK results in enhanced plant growth. The nanofertilizer was incorporated into the plant through stomata and transported via phloem. The foliar-treated plants showed significant reduction in their life cycle, with 40 days less than nontreated plants in addition to effect on increasing the yield [74].

Micronutrients like iron, copper, zinc, chlorine, molybdenum, boron, etc., play a vital role in plant growth and development. They form an essential component of many enzymes and coenzymes and are thus involved in key plant metabolic processes like photosynthesis, respiration, replication, transcription translation, DNA repair, etc. Micronutrient deficiencies lead to severe physiological symptoms that reduce growth, development, and reproduction of the plant. So a proper and sophisticated way of applying these micronutrients would enhance vigorous plant growth and increased yield. Foliar delivery systems are becoming familiar to deliver micronutrients to the plant. Nanoparticles loaded with micronutrients can be applied to leaves for better absorption and utilization of micronutrients by the plant [61]. Chemical conjugation of naphthyacelyl and chitosan generates a blend that enhances slow release of the plant hormone which continued upto 55 days [75]. Although chitosan-based nanocomposite materials have rarely been used to deliver micronutrients through foliar application, but they hold great promise and potential in this aspect.

15.2.3 Nanosensors

It would be fabulous if soil conditions like nutrient availability/deficiency and water availability/deficiency can be monitored in real time. Such systems would act as an alarm or a signal that catches the alterations in soil conditions. The farmer can then act accordingly to combat the issue. This would be a boon to the agriculture. The concept is still at its fancy but rapid progress and refinement is believed to be achieved in the near future. Detection of heavy metals in the soil through these nanosensors is already reported. This sensor is a film made from super nanoparticles of iron oxide in a matrix of chitosan. A coating of platinum to act as an electrode makes it sensitive to voltage changes. It is applied to detect arsenic, lead, and nickel from sewage water and human urine [76]. This nanocomposite can also be effectively employed to detect heavy metals in soil solutions in agricultural fields which would help in assessing the condition of soil before using it for cultivation of crop species. Similarly chitosan cross-linked carbon nanotubes hold the promise to detect cadmium (II) and mercury (II)

[77] from the agricultural soils. Organophosphorus pesticides and insecticides are primarily hydrolysed by organophosphorus hydrolases (OPH) in the soil. Thus the activity of these enzymes is an indication of the pesticide contamination in the soil. Detection of organophosphorus compounds can thus be achieved by employing OPH enzymes. In view of this, covalent embedding of OPH in chitosan matrix followed by electrostatic immobilization of this conjugate on negatively charged gold nanoparticles generates a novel nanobiosensor that could detect paraoxon, an organophosphorus pesticide [78]. Thus using this nano-biocomposite, health of agricultural soils can be assessed. Likewise a nano-biocomposite of chitosan and magnetic nanoparticles of iron oxide successfully detects urea level in any system [79]. After incorporation of urease and glutamate dehydrogenase on a nano-biocomposite film made from superparamagnetic iron oxide and chitosan followed by deposition of the resulting nanocomposite conjugate film on a glass plate coated with indium-tin oxide, detection of urea from the surroundings manages to assess urea level in system [80]. Such nanocomposites hold promise for agricultural purposes. They can be installed in the field soil to know about the condition of the soil and accordingly rectify the agronomic issue for precision farming.

15.2.4 Genetic Engineering in Crop Plants

Efficient delivery system is always important for a successful genetic engineering experiment. Based on the objective, different delivery methods like particle bombardment, electroporation, and polyethylene glycol mediated transfer exist for plants. One of most significant hindrances for gene transfer in plants is the occurrence of cell wall as an outer protective covering of plants. Nanotechnology has recently emerged as a new innovative gene delivery technology that can give new directions to crop improvement. Nanoparticles can be targeted for both mono and dicots and for any kind of plant tissue. Zinc oxide nanoparticles, carbon nanotubes [81], gold nanoparticles [82], and silica nanoparticles [83] have been already used as delivery vehicle in plants. Chitosan can be easily fabricated to nanoparticles. It can form complex with DNA and RNA through electrostatic interactions [84] and can protect DNA as well as RNA from nuclease degradation [85]. These properties of chitosan make it an interesting candidate as a delivery vehicle in gene transfer experiments. Though chitosan/DNA nanoparticles can be easily formed but they bear low transformation efficiency [86]. Recent reports indicate that chitosan nanoparticles can effectively deliver siRNA in gene silencing experiments with minimal toxicity [87–89]. Insect pest has an exoskeleton of chitin. They cause a substantial damage to the agriculture crops and reduce the yield. Proper pest management will enhance better plant growth and yield. One strategy to combat the insect attack is to design nanoparticles containing chitin synthase silencing constructs. This would knock-down the chitin synthase genes in the insects thereby causing failure to synthesize chitin. Under this scenario, insect growth and vigor would be severely hampered as chitin forms the main element of their exoskeleton. In an attempt to accomplish this, interference constructs targeting two genes viz AgCHS1 and AgCHS2 that are responsible for chitin synthesis were embedded on chitosan nanoparticles and the resulting complex was fed to the insect pests, *Anopheles gambiae* larvae with the result due to RNA interference by the DNA construct targeted for the chitin synthase genes, chitin synthesis severely dropped. This controlled the spread of the insect [90]. The success of this experiment opens a door for the use of chitosan-based nanoparticles in genetic engineering experiments for crop improvement, disease, and pest management.

15.2.5 Disease Control

Plants acquire different types of diseases both of fungal and bacterial origin as well as from viruses. Diseases cause huge losses in agricultural production each year. New varieties are developed that are resistant to diseases. However, this resistance is broken with the passage of time because pathogens continuously evolve to find way to evade the host defence mechanism. Chemical sprays hold promise but have negative impact on environment. So a new way has to be developed that is effective and environment friendly. Chitosan holds promise in this aspect. It is reported to possess antimicrobial property [91]. A group of gram-negative bacteria is responsible for wilts in crops like tomato, pumpkin, etc. The disease affects vascular tissues of the plant in which the pathogen multiplies in the xylem thereby causing xylem blockage. This promotes wilting. In an attempt to evaluate the role of bio nanocomposites in increasing the resistance against the bacterial wilt pathogen, a nanocomposite of silica and chitosan was generated to test its efficacy in tomato plants against *Ralstonia solanacearum*. The results revealed that the nanocomposite effectively promoted wilt resistance. Further, the shelf life and water-holding ability of the treated tomato plants were enhanced [92]. Anthracnose in mango is a fungal disease caused by *Colletotrichum gloeosporioides*. The disease is manifested on leaves, flowers, and fruits. Fruits bear dark spots on their skin and may even fall before plucking. This causes huge losses to the mango production. Chitosan possesses antifungal activity against the anthracnose pathogen [93]. To combat the disease, chitosan-silver nanocomposite significantly decreased the anthracnose in detached alphonso mango fruits to prevent the post-harvest decay of the fruits [94]. *Botrytis fabae* produces red brown spots on leaves, flowers, and stems of bean plants [95]. *Fusarium oxysporum* promotes plant invasion by secreting small proteins and several enzymes in the xylem vessels of the plant [96]. Both these fungal species are responsible for bean crop losses. Chitosan exhibits antimicrobial activity due to the interaction of its cationic amino groups with cellular components. Sulphur nanoparticles coated with rue, eucalyptus or thyme oils and stabilized in a matrix of chitosan generates a nanocomposite that shows effectiveness against *Fusarium oxysporum* at 40% rue or thyme concentration and *Botrytis fabae* at 60–80% concentrations. Eucalyptus oil shows inhibition at 20% concentration against both these pathogens [97]. The nanoparticles of Cu-chitosan exhibit excellent porosity as well as stability. Upon testing them on tomato plants for their growth promontory and antifungal effect, the results revealed that these nanoparticles enhance seed germination, biomass, and seedling length. Further the mycelial growth of *Alternaria solani* and *Fusarium oxysporum* was significantly reduced at 0.12% concentration of Cu-chitosan nanoparticles. Thus the nanoparticle holds a great promise to be used as an effective agrochemical [98]. Bacterial control agents (BCAs) from *Bacillus subtilis, Trichoderma viride*, and *Glomus mosseae* were centrifuged to generate a supernatant. Adsorption of BCAs on chitosan immobilized silica nanocomposites shows good sorption properties of the nanocomposite gel thereby its potential to be used as a carrier of BCAs. The efficacy of the nanocomposite-BCA complex upon testing on tomato plants reveals its inhibitory effect against *R. solanacearum.* The gel is found to form a thin film around the tomato root hairs. Promotion of wound healing is also observed in treated plants as compared to controls. In addition to great reduction in percentage of wilting caused by the pathogen, the seed germination and overall growth rate is enhanced [99]. Nanoparticles of copper and chitosan promote growth in tomato as revealed by enhanced seed germination, dry weigh, and seedling growth. Further, mycelial growth and spore germination in *A. solani* and *F. oxysporum* exhibited significant inhibition. The nanocomposite thus is effective against early blight and fusarium wilt in tomato [100]. In a similar study, copper-chitosan nanoparticles significantly increased germination percentage, root growth, dry weight, and shoot length in maize seedlings [101]. Likewise the chitosan nanoparticles and chitosan-pepper tree essential oil, nanocomposite exhibits antifungal effect on *Aspergillus parasiticus* spores and thus consequently reduces their viability [102].

15.2.6 Immunomodulation

Plants have evolved to acquire versatile defence mechanisms against biotic and abiotic stresses. Plants have learned to produce phytoelaxans, pathogenesis related proteins, antioxidant enzymes, and hypersensitive response to combat pathogen attack. Certain biomolecules act as elicitors to activate the plant immune response. The eliciting role of chitosan is well known to stimulate immune system [103]. In an attempt to assess the role of chitosan nanoparticles in eliciting the immune response in plants, treatment of chitosan nanoparticles on leaves induces the elevated expression of defence related enzymes, antioxidant enzymes as well as phenolics. Nitric oxide level is also increased. The immunomodulatory role of chitosan nanoparticles is hampered on inhibition of nitric oxide production. This suggests that chitosan nanoparticles exert their influence on immune system via nitric oxide signalling pathway. Such nanoparticles generate the plants with activated immune response and therefore can help in disease and pest management [104].

15.3 Conclusion

Although chitin- and chitosan-based nanocomposites have great promise in agriculture, limited practical application has been achieved in this aspect. Extensive research needs to be carried out to develop and generate newer, more sophisticated and better performing nanocomposites from both chitin and chitosan for soil monitoring along with controlled release delivery systems for pesticides, weedicides, fertilizers, and micronutrients. The development of nanocomposites for foliar application of these nanocomposites loaded with growth regulators as well as antipathogenic agents shall be given utmost care for better growth, performance, and protection of crops. In addition, the importance and effectiveness of nanocomposite-based approach in agriculture shall be highlighted among the farmers so that they feel safe and at relative ease in applying the nanotechnology in practical farming. It is surprizing that despite of having tremendous potential of these nanocomposite in monitoring soil conditions, no or negligible efforts have been made to utilize them practically for agricultural purposes.

References

1. Rinaudo, M., Main properties and current applications of some polysaccharides as biomaterials. *Polym. Int.*, 57(3), 397–430, 2008.

2. Rinaudo, M., Chitin and chitosan: properties and applications. *Prog. Polym. Sci.*, 31(7), 603–632, 2006.
3. Rudall, K.M., Kenchington, W., The chitin system. *Biol. Rev.*, 48(4), 597–633, 1973.
4. Rudall, K.M., Chitin and its association with other molecules. In: *Journal of Polymer Science Part C: Polymer Symposia* 28(1), 83–102, Wiley Subscription Services, Inc., A Wiley Company, 1969.
5. Atkins, E., Conformations in polysaccharides and complex carbohydrates. *J. Biosci.*, 8(1–2), 375–387, 1985.
6. Minke, R.A.M., Blackwell, J., The structure of α-chitin. *J. Mol. Biol.*, 120(2), 167–181, 1978.
7. Mathur, N.K., Narang, C.K., Chitin and chitosan, versatile polysaccharides from marine animals. *J. Chem. Educ*, 67(11), p. 938, 1990.
8. Tamura, H., Hamaguchi, T., Tokura, S., Destruction of rigid crystalline structure to prepare chitin solution. *Adv. Chitin Sci.*, 7, 84–87, 2004.
9. Vincendon, M., Regenerated chitin from phosphoric acid solutions. *Carbohydr. Polym.*, 32(3), 233–237, 1997.
10. Gagnaire, D., Saint-Germain, J., Vincendon, M., NMR studies of chitin and chitin derivatives. *Die Makromolekulare Chemie*, 183(3), 593–601, 1982.
11. Einbu, A., Naess, S.N., Elgsaeter, A., Varum, K.M., Solution properties of chitin in alkali. *Biomacromol.*, 5(5), 2048–2054, 2004.
12. Kurita, K., Tomita, K., Ishii, S., Nishimura, S.I., Shimoda, K., β-chitin as a convenient starting material for acetolysis for efficient preparation of N-acetylchitooligosaccharides. *J. Polym. Sci. Part A: Polym. Chem.*, 31(9), 2393–2395, 1993.
13. Datta, P.K., Basu, P.S., Datta, T.K., Isoiation and characterization of vicia FABA lectin affinity purified on chitin column. *Prep. Biochem. Biotechnol.*, 14(4), 373–387, 1984.
14. Hudson, S.M., Jenkins, D.W., Chitin and chitosan. *Encyclopedia of polymer science and technology*. New York: Wiley; 2001.
15. Krajewska, B., Application of chitin- and chitosan-based materials for enzyme immobilizations: a review. *Enzyme Microb. Technol.*, 35(2), 126–139, 2004.
16. Hirano, S., Wet-spinning and applications of functional fibers based on chitin and chitosan. In: *Macromolecular Symposia*, WILEY-VCH Verlag GmbH, 168(1), 21–30, 2001.
17. Hirano, S., Midorikawa, T., Novel method for the preparation of *N*-acylchitosan fiber and *N*-acylchitosan-cellulose fiber. *Biomaterials*, 19(1), 293–297, 1998.
18. Hirano, S., Nakahira, T., Nakagawa, M., Kim, S. K., The preparation and applications of functional fibres from crab shell chitin. *J. Biotechnol.*, 70(1), 373–377, 1999.
19. Yusof, N.L.B.M., Wee, A., Lim, L.Y., Khor, E., Flexible chitin films as potential wound-dressing materials: Wound model studies. *J. Biomed. Mat. Res. Part A*, 66(2), 224–232, 2003.

20. Kato, Y., Onishi, H., Machida, Y., Application of chitin and chitosan derivatives in the pharmaceutical field. *Curr. Pharm. Biotechnol.*, 4(5), 303–309, 2003.
21. Ito, M., Matahira, Y., Sakai, K., The application of chitin-chitosan to bone filling materials. *Kichin, Kitosan Kenkyu: Publ. Nippon Kichin, Kitosan Gakkai*, 4, 142–143, 1998.
22. Rhazi, M., Desbrieres, J., Tolaimate, A., Rinaudo, M., Vottero, P., Alagui, A., Contribution to the study of the complexation of copper by chitosan and oligomers. *Polymer*, 43(4), 1267–1276, 2002.
23. Peniche-Covas, C., Alvarez, L.W., Argüelles-Monal, W., The adsorption of mercuric ions by chitosan. *J. Appl. Polym. Sci.*, 46(7), 1147–1150, 1992.
24. Nieto, J.M., Peniche-Covas, C., Del Bosque, J., Preparation and characterization of a chitosan-Fe (III) complex. *Carbohydr. Polym.*, 18(3), 221–224, 1992.
25. Babak, V.G., Rinaudo, M., Physico-chemical properties of chitin-surfactant complexes. *Chitosan Pharm. Chem.*, 277–284, 2002.
26. Peniche, C., Argüelles-Monal, W., Chitosan based polyelectrolyte complexes. In: *Macromolecular Symposia*, WILEY-VCH Verlag GmbH, 168(1), 103–116, 2001.
27. Rusu-Balaita, L., Desbrieres, J., Rinaudo, M., Formation of a biocompatible polyelectrolyte complex: chitosan-hyaluronan complex stability. *Polym. Bull.*, 50(1–2), 91–98, 2003.
28. Abe, K., Ifuku, S., Kawata, M., Yano, H., Preparation of tough hydrogels based on β-chitin nanofibers via NaOH treatment. *Cellulose*, 21(1), 535–540, 2014.
29. Khor, E., Wu, H., Lim, L.Y., Guo, C.M., Chitin-methacrylate: Preparation, characterization and hydrogel formation. *Materials*, 4(10), 1728–1746, 2011.
30. Bhattarai, N., Gunn, J., Zhang, M., Chitosan-based hydrogels for controlled, localized drug delivery. *Adv. Drug Deliv. Rev.*, 62(1), 83–99, 2010.
31. Valmikinathan, C.M., Mukhatyar, V.J., Jain, A., Karumbaiah, L., Dasari, M., Bellamkonda, R.V., Photo-cross-linkable chitosan based hydrogels for neural tissue engineering. *Soft Matter*, 8(6), 1964–1976, 2012.
32. Lowry, E.D., *The investigation of chitosan-based hydrogels for intratumoral delivery of immunotherapeutic cytokines* (Doctoral dissertation), 2015.
33. Sultana, N., Mokhtar, M., Hassan, M.I., Jin, R.M., Roozbahani, F., Khan, T.H., Chitosan-based nanocomposite scaffolds for tissue engineering applications. *Mater. Manuf. Processes*, 30(3), 273–278, 2015.
34. Park, S.I., Daeschel, M.A., Zhao, Y., Functional properties of antimicrobial lysozyme-chitosan composite films. *J. Food Sci.*, 69(8), M215–M221, 2004.
35. Giri, T.K., Thakur, A., Alexander, A., Badwaik, H., Tripathi, D.K., Modified chitosan hydrogels as drug delivery and tissue engineering systems: present status and applications. *Acta Pharm. Sin. B*, 2(5), 439–449, 2012.
36. Lakshmanan, V.K., Snima, K.S., Bumgardner, J.D., Nair, S.V., Jayakumar, R., Chitosan-based nanoparticles in cancer therapy. In: *Chitosan for biomaterials*, Springer Berlin Heidelberg, I, 55–91, 2011.
37. Lichiheb, N., Personne, E., Bedos, C., Van den Berg, F., Barriuso, E., Implementation of the effects of physicochemical properties on the foliar

penetration of pesticides and its potential for estimating pesticide volatilization from plants. *Sci. Tot. Environ.*, 550, 1022–1031, 2016.

38. Aouada, F.A., de Moura, M.R., Orts, W.J., Mattoso, L.H., Polyacrylamide and methylcellulose hydrogel as delivery vehicle for the controlled release of paraquat pesticide. *J. Mat. Sci.*, 45(18), 4977–4985, 2010.
39. Abd El-Mohdy, H.L., Hegazy, E.A., El-Nesr, E.M., El-Wahab, M.A., Control release of some pesticides from starch/(ethylene glycol-co-methacrylic acid) copolymers prepared by γ-irradiation. *J. Appl. Polym. Sci.*, 122(3), 1500–1509, 2011.
40. Liu, Y., Yan, L., Heiden, P., Laks, P., Use of nanoparticles for controlled release of biocides in solid wood. *J. Appl. Polym. Sci.*, 79(3), 458–465, 2001.
41. Zheng, H., Shang, Q., Water suspension acetamiprid nanocapsule preparation and its repairing method. *Chem. Abs.*, 143, p. 73729, 2005.
42. Bibi, S., Nawaz, M., Yasin, T., Riaz, M., Chitosan/CNTs nanocomposite as green carrier material for pesticides controlled release. *J. Polym. Res.*, 23(8), p. 154, 2016.
43. Barchuk, M., Capková, P., Kolská, Z., Matoušek, J., Poustka, D., Splíchalová, L., Munzarová, M., Structure and surface properties of chitosan/PEO/gelatin nanofibrous membrane. *J. Polym. Res.*, 23(2), 1–7, 2016.
44. Kumar, S., Dutta, J., Dutta, P.K., Preparation and characterization of N-heterocyclic chitosan derivative based gels for biomedical applications. *Int. J. Biol. Macromol.*, 45(4), 330–337, 2009.
45. Islam, A., Yasin, T., Bano, I., Riaz, M., Controlled release of aspirin from pH-sensitive chitosan/poly (vinyl alcohol) hydrogel. *J. Appl. Polym. Sci.*, 124(5), 4184–4192, 2012.
46. Paula, H.C., Sombra, F.M., de Freitas Cavalcante, R., Abreu, F.O., de Paula, R.C., Preparation and characterization of chitosan/cashew gum beads loaded with Lippia sidoides essential oil. *Mater. Sci. Eng.: C*, 31(2), 173–178, 2011.
47. Guan, H., Chi, D., Yu, J., Li, X., A novel photodegradable insecticide: Preparation, characterization and properties evaluation of nano-Imidacloprid. *Pest. Biochem. Physiol.*, 92(2), 83–91, 2008.
48. Lao, S.B., Zhang, Z.X., Xu, H.H., Jiang, G.B., Novel amphiphilic chitosan derivatives: synthesis, characterization and micellar solubilization of rotenone. *Carbohydr. Polym.*, 82(4), 1136–1142, 2010.
49. Feng, B.H., Peng, L.F., Synthesis and characterization of carboxymethyl chitosan carrying ricinoleic functions as an emulsifier for azadirachtin. *Carbohydr. Polym.*, 88(2), 576–582, 2012.
50. dos Santos Silva, M., Cocenza, D.S., Grillo, R., de Melo, N.F.S., Tonello, P.S., de Oliveira, L.C., Fraceto, L.F., Paraquat-loaded alginate/chitosan nanoparticles: preparation, characterization and soil sorption studies. *J. Hazard. Mater.*, 190(1), 366–374, 2011.
51. Celis, R., Adelino, M.A., Hermosín, M.C., Cornejo, J., Montmorillonite-chitosan bionanocomposites as adsorbents of the herbicide clopyralid in

aqueous solution and soil/water suspensions. *J. Hazard. Mat.*, 209, 67–76, 2012.

52. Wen, Y., Chen, H., Yuan, Y., Xu, D., Kang, X., Enantioselective ecotoxicity of the herbicide dichlorprop and complexes formed with chitosan in two fresh water green algae. *J. Environ. Monit.*, 13(4), 879–885, 2011.
53. Maruyama, C.R., Guilger, M., Pascoli, M., Bileshy-José, N., Abhilash, P.C., Fraceto, L.F., de Lima, R., Nanoparticles based on chitosan as carriers for the combined herbicides imazapic and imazapyr. *Sci. Rep.*, 6, 2016.
54. Grillo, R., Pereira, A.E., Nishisaka, C.S., de Lima, R., Oehlke, K., Greiner, R., Fraceto, L.F., Chitosan/tripolyphosphate nanoparticles loaded with paraquat herbicide: an environmentally safer alternative for weed control. *J. Hazard. Mater.*, 278, 163–171, 2014.
55. Miliadis, G.E., Determination of pesticide residues in natural waters of Greece by solid phase extraction and gas chromatography. *Bull. Environ. Contam. Toxicol.*, 52(1), 25–30, 1994.
56. Sankararamakrishnan, N., Sharma, A.K., Sanghi, R., Organochlorine and organophosphorous pesticide residues in ground water and surface waters of Kanpur, Uttar Pradesh, India. *Environ. Int.*, 31(1), 113–120, 2005.
57. Jaiswal, M., Chauhan, D., Sankararamakrishnan, N., Copper chitosan nanocomposite: synthesis, characterization, and application in removal of organophosphorous pesticide from agricultural runoff. *Environ. Sci. Pollut. Res.*, 19(6), 2055–2062, 2012.
58. Rahmanifar, B., Dehaghi, S.M., Removal of organochlorine pesticides by chitosan loaded with silver oxide nanoparticles from water. *Clean Technol. Environ. Policy*, 16(8), 1781–1786, 2014.
59. Dehaghi, S.M., Rahmanifar, B., Moradi, A.M., Azar, P.A., Removal of permethrin pesticide from water by chitosan–zinc oxide nanoparticles composite as an adsorbent. *J. Saudi Chem. Soc.*, 18(4), 348–355, 2014.
60. Li, J., Yao, J., Li, Y., Shao, Y. Controlled release and retarded leaching of pesticides by encapsulating in carboxymethyl chitosan/bentonite composite gel. *J. Environ. Sci. Health, Part B*, 47(8), 795–803, 2012.
61. Kashyap, P.L., Xiang, X., Heiden, P., Chitosan nanoparticle based delivery systems for sustainable agriculture. *Int. J. Biol. Macromol.*, 77, 36–51, 2015.
62. Yaduvanshi, N.P.S., Swarup, A., Effect of continuous use of sodic irrigation water with and without gypsum, farmyard manure, pressmud and fertilizer on soil properties and yields of rice and wheat in a long term experiment. *Nutr. Cycl. Agroecosys.*, 73(2–3), 111–118, 2005.
63. Morita, A., Takano, H., Oota, M., Yoneyama, T., Nitrification and denitrification in an acidic soil of tea (Camellia sinensis L.) field estimated by δ15N values of leached nitrogen from the soil columns treated with ammonium nitrate in the presence or absence of a nitrification inhibitor and with slow-release fertilizers. *Soil Sci. Plant Nutr.*, 48(4), 585–593, 2002.
64. Tian, X. H., Saigusa, M., Response of tomato plants to a new application method of polyolefin-coated fertilizer. *Pedosphere,* 15(4), 491–498, 2005.

65. Wu, L., Liu, M., Liang, R., Preparation and properties of a double-coated slow-release NPK compound fertilizer with superabsorbent and water-retention. *Biores. Technol.*, 99(3), 547–554, 2008.
66. Roshanravan, B., Soltani, S.M., Rashid, S.A., Mahdavi, F., Yusop, M.K., Enhancement of nitrogen release properties of urea–kaolinite fertilizer with chitosan binder. *Chem. Spec. Bioavail.*, 27(1), 44–51, 2015.
67. Corradini, E., De Moura, M.R., Mattoso, L.H.C., A preliminary study of the incorparation of NPK fertilizer into chitosan nanoparticles. *Express Polym. Lett.*, 4(8), 509–515, 2010.
68. Hasaneen, M.N.A., Abdel-Azia, H.M.M., El-Baily, D.M.A., Omar, A.M., Prepration of chitosan nanoparticles for loading with NPK fertilizers. *J. Environ. Sci.*, 43(4), 557–564, 2014.
69. Abdel-Aziz, H.M., Hasaneen, M.N., Omer, A.M., Nano chitosan-NPK fertilizer enhances the growth and productivity of wheat plants grown in sandy soil. *Span. J. Agric. Res.*, 14(1), 0902, 2016.
70. Hussain, M.R., Devi, R.R., Maji, T.K. (2012). Controlled release of urea from chitosan microspheres prepared by emulsification and cross-linking method. *Iran. Polym. J.*, 21(8), 473–479.
71. Khodakovskaya, M., Dervishi, E., Mahmood, M., Xu, Y., Li, Z., Watanabe, F., Biris, A.S., Carbon nanotubes are able to penetrate plant seed coat and dramatically affect seed germination and plant growth. *ACS Nano*, 3(10), 3221–3227, 2009.
72. Subbaiah, L.V., Prasad, T.N.V.K.V., Krishna, T.G., Sudhakar, P., Reddy, B.R., Pradeep, T., Novel Effects of Nanoparticulate Delivery of Zinc on Growth, Productivity, and Zinc Biofortification in Maize (Zea mays L.). *J. Agric. Food Chem.*, 64(19), 3778–3788, 2016.
73. Anusuya, S., Banu, K.N. Silver-chitosan nanoparticles induced biochemical variations of chickpea (Cicer arietinum L.). *Biocatal. Agric. Biotechnol.*, *8*, 39–44, 2016.
74. Abdel-Aziz, H.M., Hasaneen, M.N., Omer, A.M. Nano chitosan-NPK fertilizer enhances the growth and productivity of wheat plants grown in sandy soil. *Span. J. Agric. Res.*, 14(1), 0902, 2016.
75. Tao, S., Pang, R., Chen, C., Ren, X., Hu, S., Synthesis, characterization and slow release properties of O-naphthylacetyl chitosan. *Carbohydr. Polym.*, 88(4), 1189–1194, 2012.
76. Ahmed, R.A., Fekry, A.M., Preparation and characterization of a nanoparticles modified chitosan sensor and its application for the determination of heavy metals from different aqueous media. *Int. J. Electrochem. Sci*, 8, 6692–6708, 2013.
77. Janegitz, B.C., Figueiredo-Filho, L.C.S., Marcolino-Junior, L.H., Souza, S.P., Pereira-Filho, E.R., Fatibello-Filho, O., Development of a carbon nanotubes paste electrode modified with cross-linked chitosan for cadmium (II) and mercury (II) determination. *J. Electroanalyt. Chem.*, 660(1), 209–216, 2011.

78. Karami, Rezvan, Afshin Mohsenifar, Seyede Maryam Mesbah Namini, Nahid Kamelipour, Tavoos Rahmani-Cherati, Taha Roodbar Shojaei, and Meisam Tabatabaei. A novel nanobiosensor for the detection of paraoxon using chitosan-embedded organophosphorus hydrolase immobilized on Au nanoparticles. *Preparative Biochemistry and Biotechnology* 46 (6), 559–566 2016.
79. Ali, A., AlSalhi, M.S., Atif, M., Ansari, A.A., Israr, M.Q., Sadaf, J.R., Willander, M., Potentiometric urea biosensor utilizing nanobiocomposite of chitosan-iron oxide magnetic nanoparticles. In: *Journal of Physics: Conference Series*, 414(1), 012024. IOP Publishing, 2013.
80. Kaushik, A., Solanki, P.R., Ansari, A.A., Sumana, G., Ahmad, S., Malhotra, B.D., Iron oxide-chitosan nanobiocomposite for urea sensor. *Sens. Actu. B: Chem.*, 138(2), 572–580, 2009.
81. Christou, P., McCabe, D.E., Swain, W.F., Stable transformation of soybean callus by DNA-coated gold particles. *Plant Physiol.*, 87(3), 671–674, 1988.
82. DeRosa, M.C., Monreal, C., Schnitzer, M., Walsh, R., Sultan, Y., Nanotechnology in fertilizers. *Nat. Nanotech.*, 5(2), 91, 2010.
83. Torney, F., Trewyn, B.G., Lin, V.S.Y., Wang, K., Mesoporous silica nanoparticles deliver DNA and chemicals into plants. *Nat. Nanotechnol.*, 2(5), 295–300, 2007.
84. Duceppe, N., Tabrizian, M., Advances in using chitosan-based nanoparticles for *in vitro* and *in vivo* drug and gene delivery. *Exp. Opin. Drug Deliv.*, 7(10), 1191–1207, 2010.
85. Borchard, G., Chitosans for gene delivery. *Adv. Drug Deliv. Rev.*, 52(2), 145–150, 2001.
86. Mao, H.Q., Roy, K., Troung-Le, V.L., Janes, K.A., Lin, K. Y., Wang, Y., Leong, K.W., Chitosan-DNA nanoparticles as gene carriers: synthesis, characterization and transfection efficiency. *J. Control. Release*, 70(3), 399–421, 2001.
87. Ragelle, H., Vandermeulen, G., Préat, V., Chitosan-based siRNA delivery systems. *J. Control. Release*, 172(1), 207–218, 2013.
88. Malmo, J., Sørgård, H., Vårum, K.M., Strand, S.P., siRNA delivery with chitosan nanoparticles: Molecular properties favoring efficient gene silencing. *J. Control. Release*, 158(2), 261–268, 2012.
89. Katas, H., Alpar, H.O., Development and characterisation of chitosan nanoparticles for siRNA delivery. *J. Control. Release*, 115(2), 216–225, 2006.
90. Zhang, X., Zhang, J., Zhu, K.Y., Chitosan/double-stranded RNA nanoparticle-mediated RNA interference to silence chitin synthase genes through larval feeding in the African malaria mosquito (Anopheles gambiae). *Insect Mol. Biol.*, 19(5), 683–693, 2010.
91. El Hadrami, A., Adam, L.R., El Hadrami, I., Daayf, F., Chitosan in plant protection. *Mar. Drugs*, 8(4), 968–987, 2010.
92. Gatahi, D.M., Wanyika, H., Kihurani, A.W., Ateka, E., Kavoo, A., Use of bio-nanocomposites in enhancing bacterial wilt plant resistance, tomato

production and water conservation in greenhouse farming. In: *Scientific Conference Proceedings*, 1, 2016.

93. Ali, A., Muhammad, M.T.M., Sijam, K., Siddiqui, Y., Potential of chitosan coating in delaying the postharvest anthracnose (Colletotrichum gloeosporioides Penz.) of Eksotika II papaya. *Int. J. Food Sci. Technol.*, 45(10), 2134–2140, 2010.
94. Chowdappa, P., Gowda, S., Chethana, C.S., Madhura, S., Antifungal activity of chitosan-silver nanoparticle composite against Colletotrichum gloeosporioides associated with mango anthracnose. *Afr. J. Microbiol. Res.*, 8(17), 1803–1812, 2014.
95. Purkayastha, R.P., Deverall, B.J., The growth of Botrytis fabae and B. cinerea into leaves of bean (Vicia faba L.). *Ann. Appl. Biol.*, 56(1), 139–147, 1965.
96. Houterman, P.M., Speijer, D., Dekker, H.L., de Koster, C.G., Cornelissen, B.J., Rep, M., The mixed xylem sap proteome of Fusarium oxysporum-infected tomato plants. *Mol. Plant Pathol.*, 8(2), 215–221, 2007.
97. Yela, A.V., Jimenez, V.J., Rodriguez, D.V., Quishpe, G.P., Evaluation of the antifungal activity of sulfur and chitosan nanocomposites with active ingredients of Ruta graveolens, Thymus vulgaris and Eucalyptus melliodora on the growth of Botrytis fabae and Fusarium oxysporum. *Biol. Med.*, 2016.
98. Saharan, V., Sharma, G., Yadav, M., Choudhary, M.K., Sharma, S.S., Pal, A., Biswas, P., Synthesis and *in vitro* antifungal efficacy of Cu–chitosan nanoparticles against pathogenic fungi of tomato. *Int. J. Biol. Macromol.*, 75, 346–353, 2015.
99. Gatahi, D.M., Wanyika, H., Kihurani, A.W., Ateka, E., Kavoo, A., Use of bionanocomposites in enhancing bacterial wilt plant resistance, tomato production and water conservation in greenhouse farming. In: *Scientific Conference Proceedings* (No. 1), 2016, April.
100. Saharan, V., Sharma, G., Yadav, M., Choudhary, M.K., Sharma, S.S., Pal, A., Biswas, P., Synthesis and *in vitro* antifungal efficacy of Cu–chitosan nanoparticles against pathogenic fungi of tomato. *Int. J. Biol. Macromol.*, 75, 346–353, 2015.
101. Saharan, V., Kumaraswamy, R.V., Choudhary, R.C., Kumari, S., Pal, A., Raliya, R., Biswas, P., Cu-chitosan nanoparticle mediated sustainable approach to enhance seedling growth in maize by mobilizing reserved food. *J. Agric. Food Chem.*, 64(31), 6148–6155, 2016.
102. Luque-Alcaraz, A.G., Cortez-Rocha, M.O., Velázquez-Contreras, C.A., Acosta-Silva, A.L., Santacruz-Ortega, H.D.C., Burgos-Hernández, A., Plascencia-Jatomea, M., Enhanced antifungal effect of chitosan/pepper tree (Schinus molle) essential oil bionanocomposites on the viability of Aspergillus parasiticus spores. *J. Nanomater.*, 2016, 38, 2016.
103. Katiyar, D., Hemantaranjan, A., Singh, B., Chitosan as a promising natural compound to enhance potential physiological responses in plant: a review. *Ind. J. Plant Physiol.*, 20(1), 1–9, 2015.
104. Chandra, S., Chakraborty, N., Dasgupta, A., Sarkar, J., Panda, K., Acharya, K., Chitosan nanoparticles: a positive modulator of innate immune responses in plants. *Sci. Rep.*, 5, 2015.

16

Scope of Electrospun Chitosan Nanofibrous Web for its Potential Application in Water Filtration

Rahul Gadkari*, Wazed Ali*, Apurba Das and R. Alagirusamy

Department of Textile Technology, Indian Institute of Technology, New Delhi, India

Abstract

Water should essentially be bereft of toxic chemicals and pathogens to ensure good health of all living beings as well as preserving the environment. Advances in nanoscale science and engineering have provided solution to many of the current problems involving water filtration. Utilization of specific nanoparticles, either embedded in membrane or in the form of nanofilament web membranes, or on other structure media, can effectively and economically filter contaminated water to consumable form. Considering the myriad properties of chitosan, this chapter focuses on the potential application of electrospun chitosan nanofibrous web for water filter/disinfection application.

Keywords: Chitosan, electrospun, water, filtration, nanofibrous

16.1 Introduction

Water is necessary for the survival of all living beings. However, the presence of various toxic impurities in drinking water generates various health risks, making it extremely important to protect the water from being contaminated. Water should essentially be bereft of toxic chemicals and pathogens to ensure good health of all living beings as well as preserving the environment. Existing supplies of fresh water have decreased due to

**Corresponding authors*: rahulgadkari91@gmail.com; wazed@iitd.ac.in

ʰakeel Ahmed and Saiqa Ikram (eds.) Chitosan, (431–452) © 2017 Scrivener Publishing LLC extended droughts, increasing population and competing demands from a variety of users. Consequently, it has become difficult to meet rising demands of clean drinkable water. Therefore, the need of the hour is to avail the society with decontaminated water. Conventional water purification/disinfection techniques, such as chemical treatment, purify the water via the carcinogen disinfection by-products generated due to the addition of biocides, e.g., chlorine-based components [1]. Another popular method of filtration is removal of microorganisms by conventional membranes wherein the quality of filtered water is considerably reduced due to the formation of biofilms [2]. At present, there are various technologies available in the market for water purification. However, majority of these technologies are difficult to put to use because of their high level of complexity, as well as higher operating and maintenance cost. In such circumstances, simple techniques with low level of mechanization should be used. The custom of sand filtration and chlorine disinfection, coagulation, sedimentation marked the conclusion of waterborne epidemics in the developing as well developed countries. However, outbreaks of waterborne diseases continue to occur at unexpected high levels. Thus, efficient disinfection or filtration without production of harmful by-product is a key environmental concern [3–4].

Advances in nanoscale science and engineering have provided solution to many of the current problems involving water filtration. Use of nanosorbants, nanocatalysts, bioactive nanoparticles, nanostructured catalytic membranes, and nanoparticle has been reported to enhance the water purification efficiency [5]. Utilization of specific nanoparticles, either embedded in membrane or in the form of nanofilaments web membranes, or on other structure media, can effectively and economically filter contaminated water to consumable form. Nanofibrous membranes have drawn significant attention in fluid filtration systems due to their capability to withstand fluid flux while removing micro- and nano-sized particulates from solution [6].

Nanomaterials act as superb adsorbents, catalysts, and sensors owing to their large surface area and higher reactivity. The advanced research in the stream of polymer science presented several natural and engineered nanomaterials like chitosan, silver nanoparticles, photocatalytic TiO_2, aqueous fullerene nanoparticles, and carbon nanotubes which can be effectively used in water filtration systems [5]. Chitosan is a polysaccharide having excellent biocompatibility with noticeable antimicrobial activity along with low level of immunogenicity and toxicity [7–10]. There are various natural sources of chitosan viz. lobsters, crabs, shrimps shells, prawn shell, crayfish shells, etc. [11, 12], from which chitosan polymer can be obtained by simple chemical treatment. Owing to its remarkable combination of properties, chitosan is highly potent of being used in several applications like water engineering, cosmetics, biomedical, photography, ophthalmology, etc. [13, 14].

Electrospinning is a useful method to spin polymers into continuous fibrous form, with diameters of fibres ranging from a few micrometers to a few nanometers. This technique results in the production of endless ultrafine fibres getting laid in random, as well as oriented alignment. This electrospun nanofibrous web possesses numerous properties like random orientation of fibrous bundles, high surface area to volume ratio, high aspect ratio, high air and water permeability, small pore size, high porosity, better fibre interconnectivity, and fibre integration [15]. All these properties are favorable for developing an efficient media for water purification.

Many researchers have attempted to produce nanofibrous web from various polymers by use of electrospinning and have been successful [16]. Also, number of studies have been conducted to understand the process control parameters required for the production of fibres with optimized physical properties and web orientation [17, 18]. Considering the myriad properties of both chitosan and electrospun nanofibrous webs, this chapter focuses on the potential of electrospun chitosan nanofibrous web for water filter/disinfection application.

16.2 Chitosan as an Efficient Material for Water Purification/Disinfection

16.2.1 Structure of Chitosan

Chitosan is a promising polysaccharide source. It is a linear biopolymer having β-1-4 linked units of poly(D-glucosamine) (80%) and poly(*N*-acetyl-D-glucosamine) (20%) (Figure 16.1). It is obtained from the extraction of chitin by elimination of acetate moiety through hydrolysis (deacetylation)

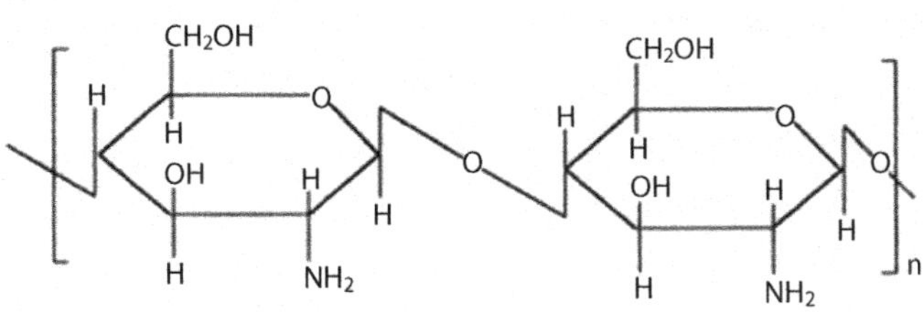

Figure 16.1 Structure of chitosan [13].

under strong alkaline conditions. Chitosan is the *N*-deacetylated derivative of chitin, but *N*-deacetylation reaction is incomplete (Figure 16.2). The available glucosamine and acetyle glucosamine units in it depend upon the degree of deacetylation of chitin. The degree of deacetylation (number of $-NH_2$ and $-CH_3CONH_2$ groups) controls properties such as solubility, acid base behavior, etc.

16.2.2 Physical, Chemical, and Biological Properties of Chitosan

Naturally occurring polysaccharides like cellulose, dextran; pectin, alginic acid; agar, agarose, and carrageenan's, are neutral or acidic in nature. Chitosan is highly basic polysaccharide, due to the presence of protonated amine group (NH_3+) in chitosan molecule [19, 20]. Distinctive properties like film forming ability, ability to bind with microbial cell, metal chelation ions, optical structural characteristics, polyoxysalt formation, viscosity and solubility in various media are due to polycationic nature of chitosan [13, 21]. Presence of one $-NH_2$ group and two -OH groups in each glycosidic unit of chitosan helps it to reflect useful chemical and biological properties (Figure 16.3).

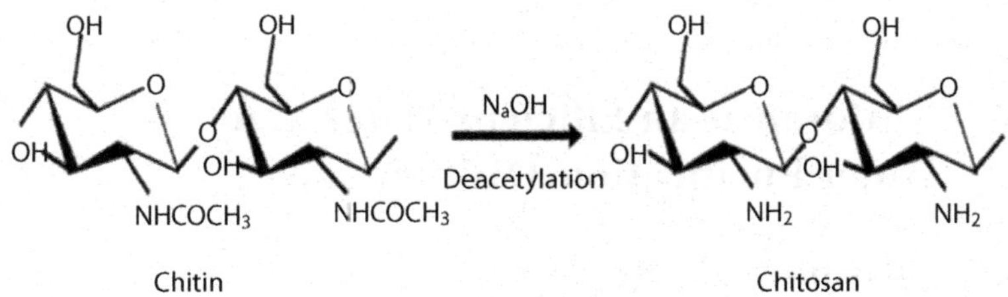

Figure 16.2 Deacetylation process by which chitosan is obtained from chitin [20].

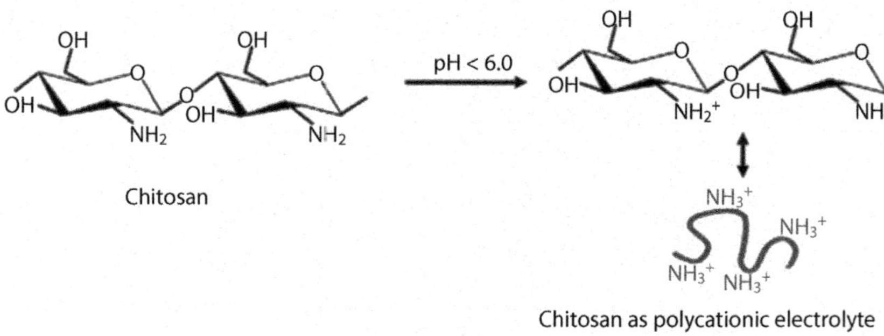

Figure 16.3 Conversion of chitosan to a polycationic electrolyte at pH < 6.0 [20].

Chitosan is biocompatible, biodegradable, safe, nontoxic, binds to mammalian and microbial cells aggressively, has a regenerative effect on connective gum tissue, is fungi static, spermicidal, and antibacterial [22]. It has been found that nanoscale chitosan and its derivatives show an antimicrobial property for bacteria, viruses, and fungi. It is reported that these are more active for control of fungi and viruses rather than bacteria. Chitosan shows higher antimicrobial activity for gram-positive bacteria compared to gram-negative bacteria [23].

16.2.3 Chitosan for Water Purification/Disinfection

16.2.3.1 Chitosan as a Coagulant/Flocculant

The water filtration process begins with the removal of suspended particles from raw water. Coagulation/flocculation process has been used to remove turbidity, color, and natural organic matter from raw water in order to obtain clean drinkable water. Aluminium sulphate, polyaluminum chloride (PACl), ferric chloride, and calcium carbonate are used to remove impurities as well as colloids from raw water. Aluminium salts have found increased acceptance as a coagulant, owing to their reasonable cost, high effectiveness, and ease of handling. However, the produced sludge contains aluminium salts which create a disposal problem, for example, an aluminium accumulation in the environment is difficult to dispose [24]. Chitosan is a natural coagulant/flocculant agent which provides advantages such as low sludge creation, ease of handling due to its biodegradability, and nontoxicity. Along with this, chitosan coagulation leads to faster settling velocity and creation of flocks of larger size [25]. Figure 16.4 shows the mechanism of flocculation of particles using chitosan. Chitosan carries positive charge along its polymeric chain, so the negatively charged colloidal materials and particulates are efficiently coagulated. The agglomeration/flocculation of the particles is a result of presence of chitosan. These particles are rapidly settled down in water, which can then be easily filtered out.

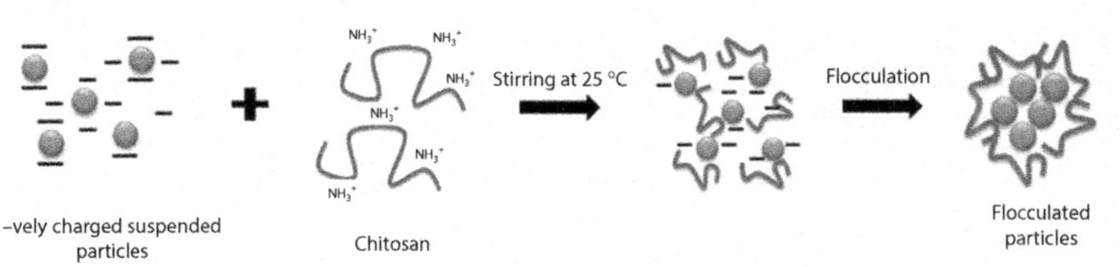

Figure 16.4 Mechanism of flocculation of particles using chitosan.

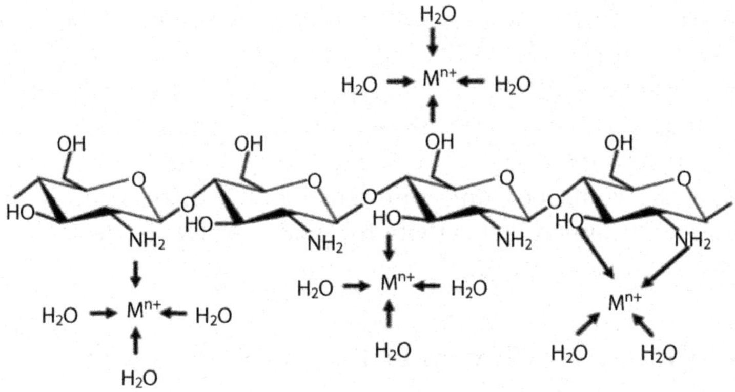

Figure 16.5 Removal of metal ions by chitosan [20].

16.2.3.2 Removal of Heavy Metals

In case of wastewater treatment, a perfect adsorbent for heavy metal ions should possess the properties like large surface area, appropriate pore size, pore volume, high adsorption capacity, better mechanical stability, good compatibility, ease of availability, ease of reinforcement, cost efficacy and eco-friendliness. Chitosan is a very good bio-sorbent for heavy metal ion removal and clears utmost of the above criteria. However, it shows an inferior adsorption because of its low acid stability, low thermal stability, lower porosity, inadequate mechanical properties, and resistance for mass transfer. Physical modification of chitosan can eliminate these drawbacks. Studies have reflected that converting chitosan powder into film, membranes, beads, nanoparticles, nanofibres or nanofibrous web form improve its mechanical and diffusing properties, as a result of polymer chain extension providing decrease in crystallinity and enhancement in chitosan swelling [26]. The amine groups of chitosan are strongly reactive with metal ions. Higher degree of deacetylation provides more number of amine groups, increasing metal binding efficiency of chitosan. After donation of free electrons which are present on nitrogen and oxygen in the amine and hydroxyl groups to the vacant orbitals of the metal, the amine groups form coordinate bonds with the metal ions (Figure 16.5).

Equation 16.1 shows the reaction of chitosan with metal ions.

$$M^{N+} \ldots + RNH_2 \leftrightarrow M(RNH_2)^{n+} \ldots \tag{16.1}$$

Mohanasrinivasan *et al.*, [27] studied heavy metal removal efficiency of chitosan (Table 16.1) using atonic absorption spectrophotometer. They

Table 16.1 Copper (Cu), Chromium (Cr), Iron (Fe), Zinc (Zn) removal percentage [27]

Metal	Removal (%)
Cu(II)	98.97
Cr(IV)	37.51
Fe(II)	65.2
Zn(II)	86.15

reported chitosan to be an effective polymer in removing metal ions Cu(II), Zn(II), Fe(II), and Cr(IV) from industrial effluent.

16.2.3.3 Removal of Organic Compounds

The presence of amino and hydroxyl groups in the molecular chain of chitosan provides possible adsorption interfaces with many of the organic pollutants like dyes, phenols, pesticides, herbicides, polychlorinated biphenyls and surfactants. Chlorine has been used for disinfecting the drinking water. It gives chlorinated hydrocarbons by-products like trichloromethane, dichlorobenzene, and chlorophenol which are carcinogenic in nature. Trichloromethane from wastewater can be effectively removed by chitosan, due to the presence of amine groups in it which integrates with halides in water. The modified chitosan, as well as the chitosan-based composites are found to be highly suitable for removal of most of the organic compounds, dyes from wastewater [28].

16.2.3.4 Removal of Bacteria

Chitosan is a natural antibacterial agent. Chitosan derivatives, nanoparticles, and nanofibres show an antibacterial activity toward fungi, viruses, and bacteria. These are promising for low-cost and low-end disinfection applications [23]. Positively charged chitosan particles on interacting with negatively charged bacteria cell membranes, result in increasing membrane permeability followed by eventual rupture and leakage of intracellular components (Figure 16.6). Chitosan inhibits the multiplication of bacteriophages in bacteria, and tempts to resist viral diseases [23]. Chitosan nanoparticles, nanofibres in the form of fibres membranes, ultrafiltration membranes, sponges and nanocoatings can be used for water filtration/disinfection applications.

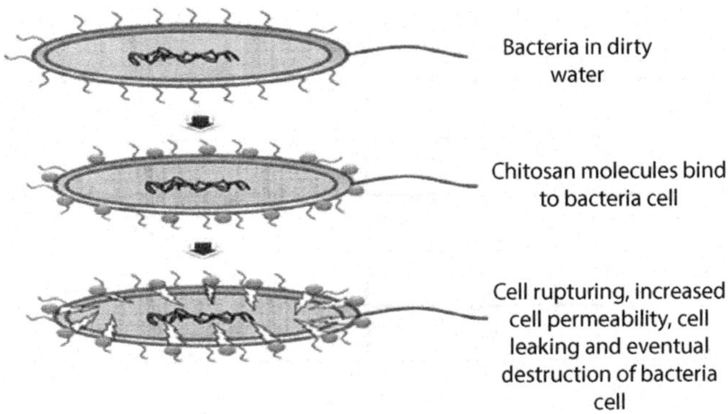

Figure 16.6 Mechanism of killing of the bacteria present in water by chitosan.

Table 16.2 Relative effectiveness of various antimicrobial agents [23].

Disinfectant	**CT[a] (mg-min/L)**
Chitosan	7.5–144[b]
Aqueous fullerene nanoparticles (nC_{60})	100
Chloramine	95–180
Silver ion	0.0075[c]–26
Free chlorine	0.03–0.05
Ozone	0.0007–0.02

[a]All CT values are for 99% disinfection of *Escherichia coli* unless otherwise specified.

[b]Minimum inhibition concentration multiplied by exposure time.

[c]CT for 99% disinfection of Pseudomonas aeruginosa.

Chitosan has advantages over other disinfectants (Table 16.2) as it possesses higher antibacterial activity toward wider range of bacteria, viruses and fungi. Lower toxicity towards higher order animals and humans make them more suitable for the same. However, chitosan effectiveness for disinfection is only at acidic pH because of its solubility and the availability of charged amino groups, but preparation of water-soluble derivatives of chitosan may eventually overcome this limitation.

16.3 Electrospinning Process

Electrospinning is a useful and efficient method for the production of continuous nanofibers ranging from 5 nm to 500 nm diameter by using a high

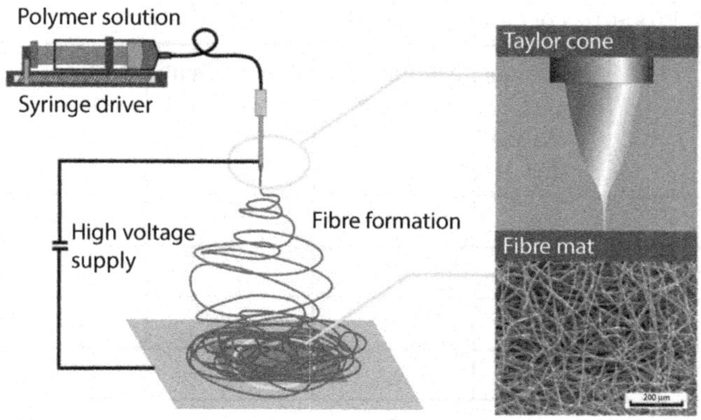

Figure 16.7 Principle of electrospinning process [43].

potential electric field. It is capable of processing synthetic, natural polymers, metals, ceramics, and composite materials. These nanofibers form nonwoven textile mats or webs and even three-dimensional-structured scaffolds which possess large surface area and high porosity [29].

In electrospinning technique, most of the polymers are dissolved in a solvent, forming a viscoelastic polymer solution. This polymer fluid is then fed to syringe driver for electrospinning. The electric field is applied between the syringe needle end and metallic collecting surface. When the intensity of the applied electric field is increased, the hemispherical surface of the solution at the tip of the capillary elongates and forms a conical shape known as Taylor cone (Figure 16.7). The electric field reaches a critical value where the repulsive electric force overcomes the surface tension force. At this critical value, a charged jet of the solution is ejected from the tip of the Taylor cone. The solvent evaporates as the jet travels in air. The charged polymer fibre is randomly deposited over a metallic collector [30].

Molecular weight, viscosity, conductivity, and surface tension of the polymer solution are important parameters for the electrospinning process. Along with this, the processing parameters like operating voltage, tip-to-target distance, temperature, pressure, and flow rate have an influence on the quality of electrospun fibre [16, 30].

16.3.1 Electrospinning of Chitosan to Obtain Nanofibres

Electrospinning of chitosan is a challenging task. Chitosan shows poor solubility in common organic solvents due to rigid D-glucosamine repeating unit, internal hydrogen bonding between hydroxyl, amine group, and high level of crystallinity [31, 32]. Therefore, several researchers have blended

Table 16.3 Properties of chitosan and PEO solutions [33].

Chitosan (g)	PEO (g)	Molecular mass of PEO (kDa)	Amount of solvent* (g)	Conductivity (ms/cm)	Surface tension (mN/m)	Viscosity (Pa.s)
1	1	2300	98	2.25	58.26	0.08
2	2	2300	96	3.50	58.61	0.69
3	3	2300	94	4.70	55.83	2.55
4	4	2300	92	5.60	49.63	6.52

*The solvent is aqueous 2 wt% acetic acid solution with a conductivity of 0.95 ms/cm.

the chitosan with other polymers those possess good electrospinning ability, like poly(vinyl alcohol) (PVA), poly(ethylene oxide)(PEO) in order to improve the electrospinability of the chitosan solution (Table 16.3) [31].

Duan *et al.*, [33] reported that ultrafine fibres could not be electrospun from chitosan solution in aqueous acetic acid, but formed when a certain amount of PEO is added. Ultrafine fibres of 80–180 nm diameter were produced after addition of PEO in 2:1 or 1:1 mass ratios of chitosan to PEO from 4–6 wt% chitosan/PEO solutions at 15 kV voltage, 20 cm capillary–collector distance, and flow rate 0.1 mL/h.

Desai *et al.*, [34] demonstrated the applicability of chitosan-based nanofibrous filter media to effectively filter out the heavy metal ions, pathogenic micro-organisms and contaminant particulate media from both air and water media. They produced the chitosan/PEO nanofibers using 90% chitosan in blend solution with fibre diameters as low as 80 nm, claiming that the fibre formation improved with addition of PEO. Filtration efficiency of the nanofibre mats is strongly related to the size of the fibres and its surface chitosan content. The nanofibrous filter media with 0.5 gm^{-2} showed hexavalent chromium binding capacity up to 35 mg chromium/g of chitosan.

Siavash Adibzadeh *et al.*, [31] fabricated chitosan/ PVA electrospun nanofibrous membranes containing silver nanoparticles (AgNPs) in order to improve antibacterial effectiveness for water filtration application. They prepared the chitosan solution by dissolving chitosan powder in 90% acetic acid/water solvent followed by extensive stirring at 80 °C for 4 h. In order to produce a homogeneous solution, PVA powder was dissolved in double-distilled water under gentle stirring conditions at 80 °C for 4 h. $AgNO_3$ was added to chitosan/PVA solution with a 70/30 mass ratio to prepare blend solution. Then, this blend solution was inserted into a 60 mL syringe with a metallic needle which was

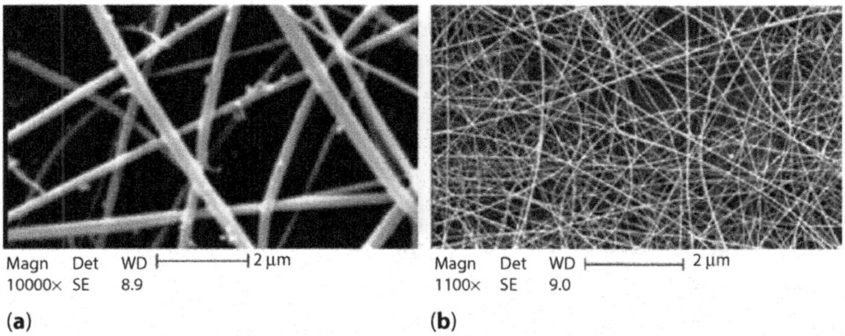

Figure 16.8 SEM images of nanofibers of chitosan hydrolysed for 48 h in aqueous 80% acetic acid 7 wt% [(a) magnification: ×10,000, (b) magnification: ×1100) [35].

connected to a high voltage power supply. The voltage was adjusted to 18 kV and the solution flow rate was set to 0.1 mL/h using a syringe pump. A rotational steel collector, covered with polyester substrate, collected nanofibers onto its surface during electrospinning. Needle tip to collector distance was adjusted at 15 cm.

The solution properties such as viscosity, ionic conductivity, surface tension, solvent type, blend weight ratio, and polymer concentration are some of the important factors which have direct effect on the electrospinability of the chitosan solution [32]. It has been stated that surface tension determines lower and upper limits of the electrospinning if other factors are kept constant. However, it has also been demonstrated that increasing the chitosan solution acidity leads to a slight increase in solution viscosity and decrease in the surface tension of chitosan solution [31, 32]. Therefore, use of concentrated acetic acid as a solvent can enhance the electrospinability of chitosan.

Homa Homayoni *et al.*, [35] reported that the problem associated with chitosan electrospinning due to its high viscosity can be resolved through application of an alkali treatment which hydrolyses chitosan chains, resulting in decreased molecular weight of chitosan. Solution of the treated chitosan in aqueous 70–90% acetic acid produces nanofibers with appropriate quality and processing stability (Figure 16.8). They also reported that decreasing the acetic acid concentration in the solvent increases the mean diameter of the chitosan nanofibres.

In another study by Geng *et al.*, [36], chitosan nanofibres were electrospun from aqueous chitosan solution using concentrated acetic acid solution as a solvent. At acetic acid concentration of 30% or more, several thin nanofibers with about 40 nm average diameter were initially produced, together with large beads. At 90% acid concentration, the fibre diameter increased to 130 nm without beads formation (Figure 16.9). A uniform nanofibrous mat having average fibre diameter as 130 nm was obtained by electrospinning

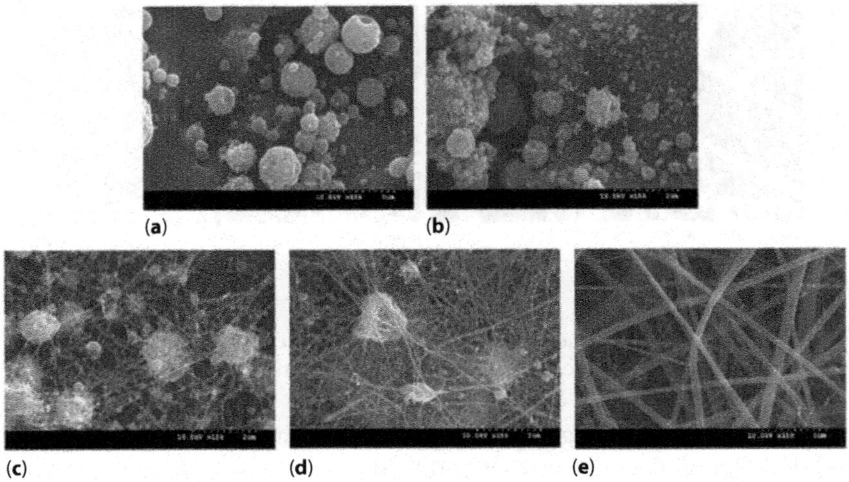

Figure 16.9 SEM images of electrospun fibers at different acid concentration (wt%): (a) 10, (b) 30, (c) 50, (d) 70, and (e) 90; 4 kV/cm, 7 wt% chitosan (106,000 g/mol) [36].

7% chitosan solution in aqueous 90% acetic acid in 4 kV/cm electric field. One of the most interesting results from this study was that the acetic acid concentration in water strongly influenced surface tension of chitosan solutions, which is remarkably important in chitosan electrospinning. As acetic acid concentration increased from 10% to 90%, surface tension decreased from 54.6 dyne/cm to 31.5 dyne/cm without significant change in viscosity. Net charge density of the chitosan solution also increased with increasing acetic acid concentration in water, resulting in more charged ions available for charge repulsion. Among three used chitosan samples with different molecular weights and concentrations, only about 7–7.5% of 1,06,000 g/mol chitosan (viscosity ranging from 484 cP to 590 cP) can produce a continuous and uniform fibre. The electrospun samples made from lower molecular weight chitosan solution (9.5–10.5%) usually contained large size beads and fragile fibres, while those made from higher molecular weight chitosan solution (2.5–3%) showed rougher and finer nanofibers with some bead defects.

16.4 Electrospun Chitosan Nanofibers Embedded with Silver Nanoparticles for Filtration of Water Contaminated with Bacteria

The addition of small amount of salt to the polymer solution improves its electrospinning ability, as well as results in the production of fibres

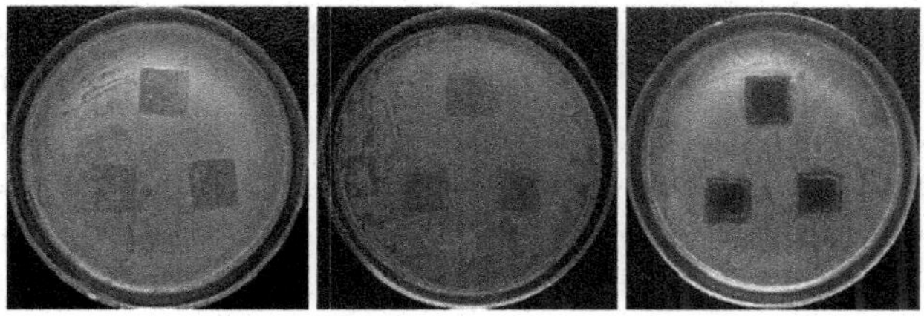

Figure 16.10 The bacterial growth inhibition halos against *Escherichia coli* (a) PVA/Chitosan nanofibers, (b) AgNPs/PVA/chitosan nanofibers with 5 mmol/L Ag, and (c) AgNPs/PVA/chitosan nanofibers with 10 mmol/L Ag [39].

with uniform structure [38]. When $AgNO_3$ is added to a chitosan/PVA blend solution, Ag^+ is reduced to Ag^0 during the electrospinning process. Moreover, chitosan is able to react with $AgNO_3$ by chelation, causing a decrease in the repulsive forces between ionic groups within polymer backbones. Therefore, continuous filaments can be electrospun from chitosan/PVA/Ag^+ solutions with high concentrations of chitosan [37].

Chitosan is a rigid and brittle natural polymer and has amino groups and hydroxyl groups in the backbone. Hang *et al.*, [37] found that as the concentration of chitosan is increased in the chitosan/PVA blends, it results in the production of an electrospun web with higher tensile strength at break. However, the elongation at break falls down for the same. Moreover, they found that addition of $AgNO_3$ to the chitosan/PVA blend solutions not only enhanced the electrospinning performance but also showed higher antibacterial activity [37, 38].

Zhao *et al.*, [39] manufactured an electrospun web from silver nanoparticles (AgNPs)/PVA/chitosan nanofibres taking $AgNO_3$ concentration as 5 mmol/L and 10 mmol/L. The silver nanoparticles were formed *in situ* in PVA/chitosan blend solution. This resulted in AgNPs/PVA/ chitosan nanofibres having smooth surface and uniform diameter. It was also observed that the average diameter of fibres slightly decreased with the decrease in Ag content. AgNPs were round in shape with diameter ranging from 4 nm to 14 nm observed on the surface of the nanofibers. With approximately 10 mg nanofibres in 50 mL bacteria nutrient solution with an initial concentration of ca. 10^4 CFU mL^{-1}, the AgNPs/PVA/chitosan nanofibres had an inhibition ratio of 98% against *Escherichia coli* up to a contact time of 12 h (Figure 16.10). These results suggest that silver deposited chitosan electrospun nanofibrous web can be explored as filters for purifying water contaminated with bacteria.

Adibzadeh *et al.*, [31] manufactured chitosan/PVA electrospun nanofibrous membrane, with and without sliver nanoparticles, and collected over polyester substrates. Their antibacterial properties were evaluated against phosphate buffer saline (PBS) samples containing *E. coli*, a gram-negative bacterium, using static and dynamic antibacterial filtration methods. For the static antibacterial test, 10 mg of nanofibers containing silver nitrate ($AgNO_3$) was added to 10 mL of PBS having pH 7.4 and inoculated with *E. coli* to prepare a bacterial suspension with a concentration of 1.5×10^5 colony forming unit (CFU)/mL. A suspension of 0.1 mL was removed and quickly poured into plates containing nutrient agar at different immersion times. The prepared plates were incubated at 37 °C for 24 h and the colonies were determined using the Pour Plate method. The reduction in bacterial colonies was measured using the Log Reduction method (Equation 16.2)

$$\text{Log Reduction} = (\text{Log initial bacteria concentration} - \text{Log final bacteria concentration}) \quad (16.2)$$

For dynamic antibacterial test which provides the practical approach for bacterial filtration testing of filter membrane, 100 mL of bacterial suspension of 1.5×10^5 CFU/mL concentration was prepared as described before, and filtered through the chitosan/PVA electrospun nanofibrous membrane (area of 4.7 cm^2) using a dead end filtration unit. The filtration unit was autoclaved at 121 °C for 15 min. The reduction in bacterial concentration after each filtration was measured using the Pour Plate method.

In case of static antibacterial testing, it was found that as the concentration of silver nitrate ($AgNO_3$) in chitosan/PVA electrospinning blend solution was increased from 2 wt% to 4 wt%, the time needed to kill the bacteria decreased from 60 min to 30 min. This was sufficient enough to completely eliminate all bacteria (Figure 16.11). The same trend is observed in case of both one and three filtration passages through electrospun nanofibrous membrane during dynamic antibacterial testing (Figure 16.12).

The leaching behavior of chitosan/PVA nanofibres containing 4 wt% $AgNO_3$ was evaluated using static and dynamic methods under same conditions. In case of static method, silver nanoparticles impregnated nanofibres were evaluated by immersing 10 mg of CS/PVA nanofibres containing 4 wt% $AgNO_3$ in 50 mL PBS. The amount of silver leached from the nanofibres at different exposure times was determined by Flame Atomic Absorption Spectrophotometry. The amount of silver released increased rapidly during the first 3 h of exposure time during static testing, whereas during dynamic testing it was found to decrease continuously from 6 h to 24 h immersion time. The total Ag^+ content released into the PBS from the CS/PVA nanofibers containing 4 wt% $AgNO_3$ during 24h time frame was just below 0.8 mg/L as showed in Figure 16.13.

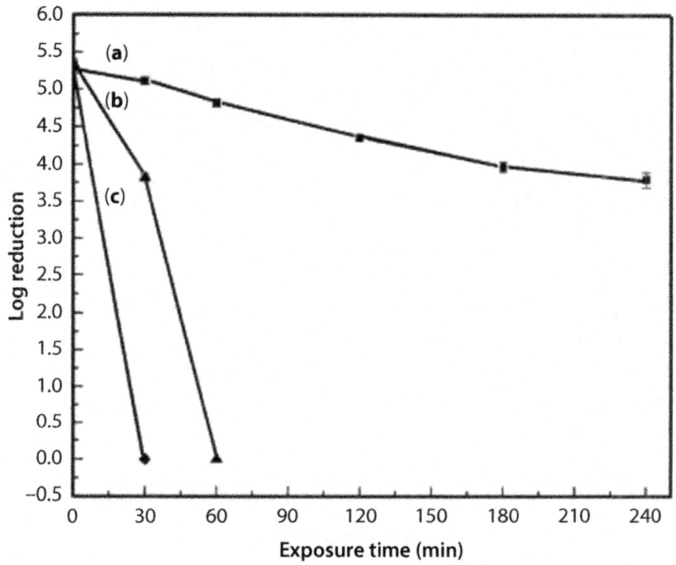

Figure 16.11 Log reduction of *E. coli* concentration in PBS samples after soaking 10 mg of nanofibers containing (a) 0, (b) 2, and (c) 4 wt% $AgNO_3$ [31].

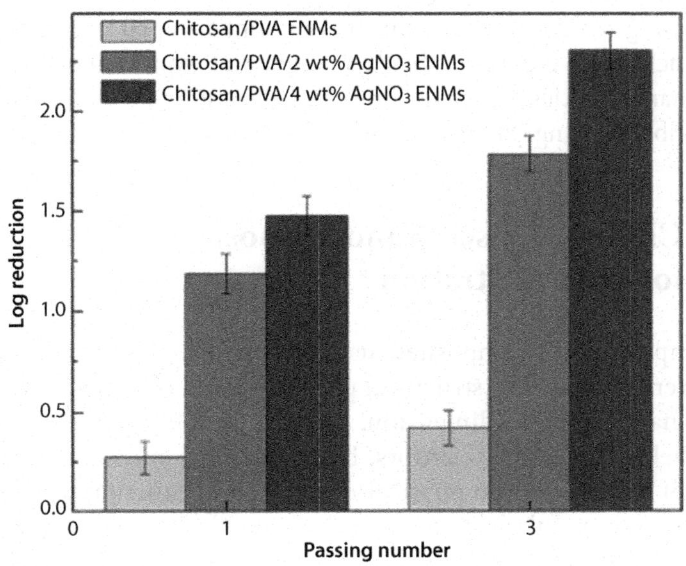

Figure 16.12 Log reduction of *E. coli* concentration after dynamic filtration of bacterial suspension through CS/PVA ENMs containing 0 wt%, 2 wt%, and 4 wt% $AgNO_3$ [31].

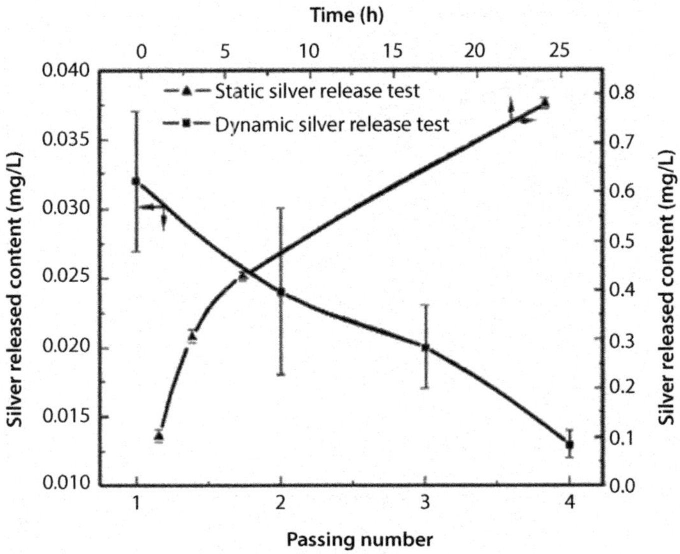

Figure 16.13 Static and dynamic silver release test [31].

In case of dynamic leaching test, four consecutive batches of 100 mL PBS were filtered through the CS/PVA ENMs containing 4 wt% $AgNO_3$ and the silver content was determined after each filtration. It was observed that the concentration of the Ag^+ lost during the filtration was rather high at the beginning of the filtration, while this amount dropped upon further filtration. Increase in the Ag^+ in the PBS sample was explained by the release of silver nanoparticles, which had lower stabilization rate on the surface of the nanofibers during the first and second filtrations (Figure 16.13).

16.5 Chitosan-Based Nanocomposites for Water Filtration

Nanocomposites are composites made up of two or more phases having different chemical or structural properties; out of which at least one phase is nanometric in dimension. They find their applications in the various fields like energy, sensors, biotechnology, and smart materials because of their improved physical/chemical and functional properties. Numerous successful attempts have been made for use of nanocomposites in water, air, and aerosol filters systems [40]. Using following approaches, it is possible to fabricate nanocomposites by utilizing electrospinning technique:

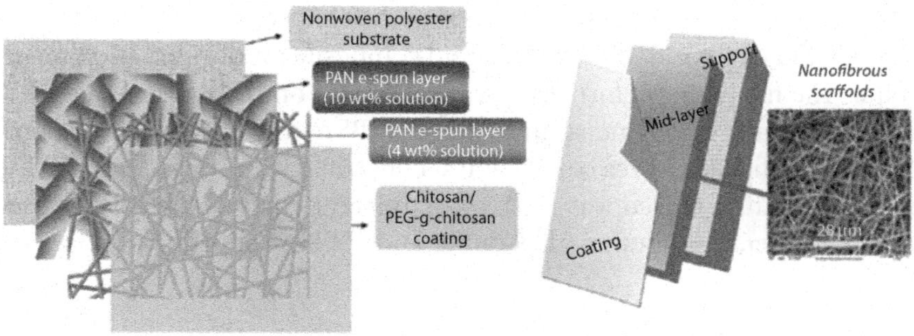

Figure 16.14 Three-tier composite structure [41].

- Using two or more precursor solutions for the electrospinning in order to achieve multicomponent nanofibers like bicomponent and core/shell nanofibers.
- Electrospinning of polymer solutions with dispersed inorganic nanoparticles such as carbon nanotubes (CNTs).
- Surface treatment of the electrospun nanofibres to achieve composite nanofibers.

Most of the ultrafiltration (UF) or nanofiltration (NF) filters for water treatment are based on porous membranes, usually manufactured by the phase immersion method. The torturous porosity in these membranes usually results in a low flux rate [41, 42]. In order to overcome this limitation, some researchers came with a new concept of high flux UF media. This high flux UF media is made up of porous electrospun nanofibrous composite [41, 42]. Yoon *et al.,* [41] demonstrated three-tier composite structure (Figure 16.14) for water filtration. The system had a "nonporous" hydrophilic chitosan coating as a top layer which restricted solid particles but allowed water to penetrate without losing too much flow rate. Moreover, the use of smooth coating at the top surface minimized the blockage problem. The middle layer was made up of polyacrylonitrile (PAN)-based electrospun scaffold with two different nanofibrous substrate in which top fibrous web was constructed with fibres of fine diameter which showed lower porosity and provided support for the chitosan coating and bottom fibrous web was constructed with fibre having relatively larger diameter which provided higher porosity to fibrous web. The third layer of conventional polyester (PET) nonwoven microfibrous structure provided support to all above layers. This new type of nanofiltration composite membranes exhibited flux rates comparable with commercial nanofiltration filter (NF 270 from Dow) after 24 h of operation.

Zhao *et al.*, [42] successfully fabricated the chitosan-coated nanofibrous polyvinylidene fluoride (PVDF) scaffold composite for ultrafiltration membrane. The filtration performance was calculated on the basis of the rejection percentage of bovine serum albumin (BSA) (Equation 16.3) against the effective filtration area of 13.4 cm^2 at constant temperature of 25 °C. The BSA concentration used was 1 g/L, with different types of solvents, including pure water, pH 7 buffer, pH 9 buffer and 1% HAc aqueous solution.

$$Rejection\ (\%) = \frac{C_f - C_p}{C_f} \times 100 \tag{16.3}$$

where, C_f and C_p represent the BSA concentration of the feed solution and that of the filtrate solution, respectively.

The chitosan was cross-linked with glutaraldehyde (GA) in order to form a bulk cross-linked chitosan, and then the surface of the chitosan was modified with terephthaloyl chloride (TPC) which increased the BSA rejection. The results showed that prepared chitosan ultrafiltration membrane gave 70.5 L/m^2 h of flux, and 98.9% of BSA rejection, at a pressure of 0.2 MPa compared with the commercial UF membrane (Sepro UF, PES10), which showed 57.1 L/m^2 h of flux and 98% of BSA rejection. These studies showed the practical possibility for the efficient use of chitosan-based electrospun nanofibrous composites for water filtration application.

16.6 Current Challenges and Future Perspectives

Chitosan is a promising alternative for the traditional chemical disinfectants that are disposed to create harmful disinfection by-products. However, more laboratory research and pilot scale testing will be needed to make commercial use of chitosan-based nanofibrous membrane in water purification systems. Different methods can be adopted to synthesize chitosan-based nanofibres, preparation of composites and nanohybrid materials with more advanced water purification capabilities. In upcoming days, nanomaterials will become major components of industrial as well as public water purification systems. At the same time, much more progress is required to make it cost effective. Future research should focus on the scalability, economics, and commercial viability of the chitosan-based nanofibrous membrane in water filtration system which could overcome many of the current limitations and provide new direction to reach a sustainable solution in this niche area.

References

1. Gomez, M., De la Rua, A., Garralon, G., Plaza, F., Hontoria, E., Gomez, M.A., Urban waste water disinfection by filtration technologies. *Desalination,* 190, 16–28, 2006.
2. Zhang, L., Luo, J., Menkhaus, T.J., Varadaraju, H., Sun, Y., Fong, H., Antimicrobial nano-fibrous membranes developed from electrospun polyacrylonitrile nanofibers.' *J. Membr. Sci.,* 369, 499–505, 2011.
3. Lalov, V.G., Guerginov, I.I., Krysteva, M.A., Fartso, K., Treatment of waste water from distilleries with chitosan.' *Wat. Res.* , 34(5), 1503–1506, 2001.
4. Santiago, M., Stuber, F., Fortuny, A., Fabregat, A., Font, J., Modified activated carbons for catalytic wet air oxidation of phenol. *Carbon,* 43, 2134–2145, 2005.
5. Savagel, N., Diallo, M.S., Nanomaterials and water purification: Opportunities and challenges. *J. Nanopart. Res.*, 7, 331–342, 2005.
6. Coopera, A., Oldinski, R.A., Hongyan Ma, Bryers, J.D., Zhanga, M., Chitosan-based nanofibrous membranes for antibacterial filter applications. *Carbohydr. Polym.*, 92(1), 254–259, 2013.
7. Rinaudo, M., Chitin and chitosan: properties and applications. *Prog. Polym. Sci.,* 31, 603–632, 2006.
8. Paul, W., Sharma, C.P., Chitosan, a drug carrier for the 21st century: a review. *S.T.P Pharma. Sci.*, 10, 5–22, 2000.
9. Muzzarelli, R.A.A., Muzzarelli, C. , Chitosan chemistry: relevance to the biomedical sciences. *Adv. Polym. Sci.,* 186, 151–209, 2005.
10. Chandy, T., Sharma, C.P., Chitosan-as a biomaterial. *Biomater. Artif. Cells Artif. Org.*, 18, 1–24, 1990.
11. Kaya, M., Baran, T., Karaarslan, M., A new method for fast chitin extraction from shells of crab, crayfish and shrimp. *Nat. Prod. Res.*, 29(15), 1477–1480, 2015.
12. Kaur, S., Dhillon, G.S., Recent trends in biological extraction of chitin from marine shell wastes: a review. *Crit. Rev. Biotechnol.*, 35(1), 44–61, 2015.
13. Ravi Kumar, M.N.V., A review of chitin and chitosan applications. *React. Func. Polym.,* 46, 1–27, 2000.
14. Crini, G., Recent developments in polysaccharide-based materials used as adsorbents in waste water treatment. *Prog. Polym. Sci.*, 30, 38–70, 2005.
15. Huang, Z.M., Zhang, Y.Z., Kotaki M., Ramakrishna S., A review on polymer nanofibers by electrospinning and their applications in nanocpomposites. *Comp. Sci. Technol.*, 63, 2223–2253, 2003.
16. Karakaş, H., Electrospinning of nanofibers and their applications, *MDT Electrospinning,* seventh framework program, 1-35 (http://www.2bfuntex.eu/, accessed on 29th Nov. 2016)
17. Gopal, R., Kaur, S., Maa, Z., Chanc, C., Ramakrishn, S., Matsuura, T., Electrospun nanofibrous filtration membrane. *J. Memb. Sci.*, 281, 581–586, 2006.

18. Decostere, B., Daels, N., Vrieze, S.D., Dejans, P., Camp, T.V., Audenaert, W., Hogie, J., Westbroek, P., Performance assessment of electrospun nanofibers for filter applications. *Desalination*, 249, 942–948, 2009.
19. Abebe, L.S., Chen, X., Sobsey, M.D. Chitosan coagulation to improve microbial and turbidity removal by ceramic water filtration for household drinking water treatment. *Int. J. Environ. Res. Pub. Health*, 13(3), 269–279, 2016.
20. Chopra, H., Ruhi, G. Eco friendly chitosan: An efficient material for water purification. *Pharma. Innovation J.*, 5(1), 92–95, 2016.
21. Larry, L.H., Biomaterials: a forecast for the future. *Biomaterials*, 19, 1419–1423, 1998.
22. Dutta, P.K., Dutta, J., Chitin and chitosan: chemistry, properties and applications, *J. Sci. Ind. Res.*, 63, 20–31, 2004.
23. Li, Q., Mahendra, S., Lyon, D.Y., Brunet, L., Liga, M.V., Li, D., Alvarez, P.J.J., Alvarez Antimicrobial nanomaterials for water disinfection and microbial control: Potential applications and implications. *Wat. Res.*, 42, 4591–4602, 2008.
24. Pan, J.R., Huang, C., Chen, S., Chung, Y.C., Evaluation of a modified chitosan biopolymer for coagulation of colloidal particles. *Colloids Surf. A: Physicochem. Eng. Asp.*, 147(3A), 359–364, 1999.
25. Bergamascoa R., Bouchardb, C., Vieira da Silvaa, F., Reisa, M.H.M, Fagundes-Klenc, M.R., An application of chitosan as a coagulant/flocculant in a microfiltration process of natural water. *Desalination*, 245, 205–213, 2009.
26. Zhang, L., Zeng, Y., Cheng, Z., Removal of heavy metal ions using chitosan and modified chitosan: A review. *J. Mol. Liq.*, 214, 175–191, 2016.
27. Mohanasrinivasan, V., Mishra, M., Paliwal, J.S., Singh, S.K., Selvarajan, E., Suganthi, V., Devi, S.C., Studies on heavy metal removal efficiency and antibacterial activity of chitosan prepared from shrimp shell waste. *3 Biotech.*, 4, 167–175, 2014.
28. Kyzas, G.Z, Bikiaris, D.N., Recent modifications of chitosan for adsorption applications: A critical and systematic review. *Mar. Drugs*, 13, 312–337, 2015.
29. Hunley, M.T., Long, T.E., Perspective electrospinning functional nanoscalefibers: a perspective for the future. *Polym. Int.*, 57, 385–389, 2008.
30. Doshi, J., Reneker, D.H., Electrospinning process and applications of electrospunfibers. *J. Electrostat.*, 35, 151–160, 1995.
31. Adibzadeh, S., Bazgir, S., Katbab, A.A., Fabrication and characterization of chitosan/poly(vinyl alcohol) electrospun nanofibrous membranes containing silver nanoparticles for antibacterial water filtration. *Iran Polym. J.*, 23, 645–654, 2014.
32. Elsabee, M.Z., Naguib, H.F., Morsi R.E., Chitosan based nanofibres-review. *Mater. Sci. Eng.*, 32C, 1711–1726, 2012.
33. Duan, B., Dong, C., Yuan, X., Yao, K., Electrospinning of chitosan solutions in acetic acid with poly(ethylene oxide). *J. Biomater. Sci., Polym. Ed.*, 15(6), 797–811, 2004.

34. Desai, K., Kit, K., Li, J., Davidson, P.M., Zivanovic, S., Meyer H., Nanofibrous chitosan non-wovens for filtration applications. *Polymer*, 50, 3661–3669, 2009.
35. Homayoni, H., Abdolkarim, S., Ravandi, H., Valizadeh, M., Electrospinning of chitosan nanofibers: Processing optimization. *Carbohydr. Polym.*, 77, 656–661, 2009.
36. Geng, X., Kwon, O.-H., Jang, J., Electrospinning of chitosan dissolved in concentrated acetic acid solution. *Biomaterials*, 26, 5427–5432, 2005.
37. Hang, A.T., Tae, B., Park, J.S., Non-woven mats of poly(vinyl alcohol)/chitosan blends containing silver nanoparticles: Fabrication and characterization. *Carbohydr. Polym.*, 82, 472–479, 2010.
38. Azizi, S., Ahmad, M.B., Hussein, M.Z., Ibrahim, N.A., Namvar, F., Preparation and properties of poly(vinyl alcohol)/chitosan blend bionanocomposites reinforced with cellulose nanocrystals/ZnO-Ag multifunctional nanosized filler. *Int. J. Nanomedicine.*, 9, 1909–1917, 2014.
39. Zhaoa, Y., Zhoua, Y., Wub, X., Wanga, L., Xua, L., Weib, S., A facile method for electrospinning of Ag nanoparticles/poly (vinyl alcohol)/carboxymethyl-chitosan nanofibers. *Appl. Surf. Sci.*, 285, 8867–8873, 2012.
40. Huanga, Z.-M., Zhangb, Y.-Z., Kotakic, M., Ramakrishn, S., A review on polymer nanofibers by electrospinning and their applications in nanocomposites. *Compos. Sci. Technol.*, 63, 2223–2253, 2003.
41. Yoon, K., Kim, K., Wang, X., Fang, D., Hsiao, B.S., Chu B., High flux ultrafiltration membranes based on electrospun nanofibrous PAN scaffolds and chitosan coating. *Polymer*, 47, 2434–2441, 2006.
42. Zhao, Z., Zheng, J., Wang, M., Zhang, H., Han, C.C., High performance ultrafiltration membrane based on modified chitosan coating and electrospun nanofibrous PVDF scaffolds, *J. Memb. Sci.*, 394–395, 209–217, 2012.
43. http://www.oxolutia.com/technology/electrospinning 2016.

17

Application of Chitin/Chitosan and its Derivatives as Adsorbents, Coagulants, and Flocculants

Parappurath Narayanan Sudha[1*], Soundararajan Aisverya[1], Thandapani Gomathi[1], Kumar Vijayalakshmi[1], Madhavan Saranya[1], Kirubanandam Sangeetha[1], Srinivasan Latha[2] and Sabu Thomas[3]

[1]*PG & Research Department of Chemistry, D.K.M College for Women, Vellore, India.*
[2]*Department of Chemistry, SRM University, Chennai, India*
[3]*School of chemical Sciences, Mahatma Gandhi University, Kottayyam, India*

Abstract

Wastewater is actually the water produced by different domestic and industrial activities which contains various inorganic, organic, and biological contaminants that are of environmental significance. These contaminants can produce health hazards if discharged without proper maintenance and treatment into streams or oceans. Especially the rapid growth in industries has led to the complexity of toxic effluents. The efficient removal of toxic metals from wastewater is an important matter and a number of technologies such as precipitation, reduction, and ion exchange were developed. Coagulation or flocculation process was conducted for the treatment of industrial wastewater to achieve maximum removal of chemical oxygen demand (COD), total dissolved solids (TDS), and total suspended solids (TSS). Chitosan, a biological cationic polymer, was used in treating dairy wastewater, agriculture, food processing, medicine, cosmetics, wastewater treatment, and biotechnology. As environmental protection is becoming an important global problem, industries are finding novel ways of developing technologies that do not cause environmental problems. Moreover, inappropriate water management has reduced the amount of available resources. The majority of commercial polymers and ion-exchange resins are derived from petroleum-based raw materials using processing chemistry that is not always safe and environment friendly. Today, there

**Corresponding author*: drparsu8@gmail.com

Shakeel Ahmed and Saiqa Ikram (eds.) Chitosan, (453–488) © 2017 Scrivener Publishing LLC

is growing interest in developing natural low-cost alternatives to synthetic polymers. Marine carbohydrates have great potential in the treatment of wastewater. Biosorption is an emerging technology, which uses natural materials as adsorbents for wastewater treatment. Low-cost adsorbents of polysaccharide-based materials obtained from marine, such as chitin, chitosan, alginate, agar, and carrageenan are acting as rescue for wastewater treatment. This chapter reviews the use of marine polysaccharides such as chitin and chitosan and its derivatives as adsorbent and coagulant for effective water treatment.

Keywords: Wastewater remediation, coagulation, flocculation, biosorption, adsorption, flocculants

17.1 Introduction

In recent years, severe water pollution is caused by the production of various wastewaters containing different types of dissolved and undissolved contaminants from the rapidly developing industries throughout the world [1]. Due to the dispersal of toxic substances from the wastewater, the environment faces tragic effects [2]. And hence the increasing environmental awareness along with more stringent regulation standards has triggered industries to challenge themselves in seeking appropriate wastewater treatment technologies [3]. Pollution has become one of the most important problems in the global context as a consequence of industrialization. Even though in this developed society, environmental requirements are becoming of great importance, because of over pollution. Therefore, much more attentions have been paid to sustainable, green and environmentally friendly materials for various applications [4, 5] and also to eradicate water pollution. In this regard, the application of renewable and biodegradable material such as chitin, chitosan, and its derivatives played a major role for all the requirements. Since there is an increased interest in the use of renewable resources, considerable efforts are now being made in the research and development of polysaccharide derivatives as the basic materials for new applications. In particular, the chitosan-based materials are one of the most attractive biosorbents for wastewater treatment. Recent and continuing interest in these macro molecules is evident from the number of papers that appear each year in the literature on this topic. Chitin and chitosan biopolymers have demonstrated outstanding removal capabilities for certain pollutants such as dyes and metal ions as compared to other low-cost sorbents and commercial-activated carbons. Biopolymer adsorbents are efficient and can be used for the decontamination of effluents (removal of pollutants) and for separation processes (recovery of valuable metals).

17.2 Chitin and Chitosan

Chitin, a homopolymer containing repeating units of *N*-acetylglucosamine (NAG), is a second abundant structural polysaccharide synthesized by an enormous number of living organisms. The deacetylation of chitin leads to the formation of chitosan which converts the acetamido ($NHCOCH_3$) group to amino group (NH_2) of most of the glucose rings to which it is attached [6]. The presence of free amino and hydroxyl groups in chitosan is responsible for the greater reactivity and stability when compared with cellulose. Chitin exists in three polymorphic forms designated as α-, β-, and γ-chitin which differ in their degree of hydration, the size of unit cell, and a number of chitin chains per unit cell [7].

Recently numerous approaches have been studied for the development of cheaper and more effective adsorbents containing natural polymers. The removal of metals, compounds, and particulates from solution by the biological material is recognized as an extension to adsorption and is named as biosorption [8]. Many biosorptions such as fungi [9], algae [10], seaweeds [11, 12], microorganisms [13, 14], and several biopolymers [15, 16] have been utilized in the removal of heavy metals from wastewater. But the polysaccharides from renewable resources are currently being explored intensively in water treatment applications [17]. Among the polysaccharide compounds such as chitin [18], starch [19], and their derivatives, chitosan [20] deserves particular attention. These polysaccharides are abundant, renewable, and biodegradable, low cost and are the best choice in water treatment and useful tool for protecting the environment [21]. Chitosan is a modified, natural, carbohydrate polymer derived from the chitin component of the exoskeleton of crustacean such as shrimp, crab, crawfish, etc. Chitosan's properties like nontoxicity, biodegradability, flocculating ability, polyelectrolisity, and the possibility for regeneration lead to a number of applications. Its largest use continues to be as a nontoxic cationic flocculent in the treatment of organically polluted wastewaters and as a chelator of toxic (heavy and radioactive) metals [22].

17.3 Properties of Chitin and Chitosan

17.3.1 Color and Appearance

Chitin is colorless to half white, hard, inelastic, nitrogeneous polysaccharide [23], whereas chitosan powder is quite flabby in nature and color varies from pale yellow to white.

17.3.2 Solubility

Chitin is highly hydrophobic and it is insoluble in aqueous media while chitosan is soluble in dilute acidic conditions due to the presence of free protonable amino groups present in the D-glucosamine units. The hydrogen bonding in chitin is responsible for its poor solubility nature. To overcome the problem of solubility the derivatives of chitin are generally prepared and used. Chitin is soluble in dimethylacetamide (DMAc) containing 5% lithium chloride (LiCl) [24], chloroalcohols in conjunction with aqueous solutions of mineral acids and fluorinated solvents such as hexafluoroisopropanol, hexafluoroacetone [25]. Chitosan is soluble in dilute organic acidic solutions where the pH is <6.5 (e.g., formic, acetic, pyruvic, 10% citric and lactic acid) [26, 27].

17.3.3 Molecular Weight and Degree of Polymerization

The molecular weight of native chitin is usually larger than 1 million Daltons [28]. The degree of polymerization (DP) of chitin ranging between 2000 and 4000. Depending on the source and preparation procedure, the average molecular weight of chitosan may range from 50 kDa to 1000 kDa [29], 3.8 kDa to 2000 kDa [30] or 50 kDa to 2000 kDa [31]. The functional properties of chitosan are mainly dependent on its molecular weight or viscosity. Viscometry, light scattering, or gel permeation chromatography are the general methods used to determine the molecular weight of chitin and chitosan.

17.4 Modification of Chitin and Chitosan

Modification of Chitin and Chitosan can introduce special properties which can enlarge its fields of potential applications [32], also it overcomes the drawbacks of pure chitin and chitosan. As chitosan is versatile, it can be manufactured into films, membranes, fibers, sponges, gels, beads, and nanoparticles, or supported on inert materials. Physical and chemical modifications have been performed for improving its unique properties and also to introduce special properties for specific applications. It can readily be modified under different conditioned polymeric forms such as powder, nanoparticles [33], gel beads [34], membranes [35], sponge [36], honeycomb [37], fibers [38], and hollow fibers [39]. For various fields of applications such as wastewater treatment, biomedical and textiles, the various forms of chitin and chitosan are discussed below.

17.4.1 Physical Modification

Generally, Chitosan has a very low specific area ranging between 2 and 30 m^2/g [40]. Glutaraldehyde cross-linked chitosan gel beads have a higher specific surface area around 180–250 m^2/g [41]. The cross-linked chitosan beads had very high adsorption capacities to remove the anionic dyes whose maximum monolayer adsorption capacity ranges from 1911 g/kg to 2498 g/kg at 300 °C. The material with large surface area and porosity has proven as the good biosorbent for treating the wastewater. Maruca *et al.* [42] used chitosan flakes of 0.4–4 mm for the removal of Cr (III) from wastewater. The sorption of arsenate on to chitosan flakes has been studied. The maximum adsorption capacity occurs at an initial pH 3.5 [43]. The chemically modified chitosan flakes were used as an adsorbent for the removal of Cd ions from electroplating waste effluent under laboratory conditions [44]. Cr (VI) removal ratio was also optimized by surface response methodology of chitosan flakes at pH 3 [45]. Compared with the beads, chitosan flakes and powders have minimal removal applications due to its low surface area.

On considering the chitin and chitosan beads, the physical characteristics are much essential estimates of shape, size, volume, specific gravity, surface area, and porosity. These characteristics will decide the efficiency of the polymer beads in various applications, especially in water treatment. For water treatment applications, surface area is an important parameter, which estimates the removal efficiency of the sorbent material. Besides the beads containing highly significant porosity may explain the conflicts in the sorption properties of the material [46].

Today, membranes are used on a large scale to produce potable water from the sea and brackish water, to clean industrial effluents and recover valuable constituents, purity and to separate gasses and vapor in the petrochemical process [47, 48]. Chitosan membranes as sorbents for trace elements determination in surface waters have been tried by Eliseveta K. Mladenova and the authors have successfully used their membranes as an efficient sorbents for the preconcentration and they have recommended their membranes for solid phase extraction of Cd(II), Eu(II), Ni(II), and Pd(II) from surface water [49]. The freeze-drying method was generally employed to prepare chitin sponges. Chitin sponges possess unique properties including very low density, high porosity, good mechanical properties, and excellent oil absorption performance. The swelling property makes chitin-based sponges a versatile template for biomineralization in both calcification and silification reactions. The biomimetic potential of chitin-based composite biomaterials of poriferan origin is analyzed in detail in a recently published monograph entitled "Biomimetic biomaterials: structure and applications" [50]. The preparation of chitin in nanoscale range and the removal of the amorphous phase of chitin result in the formation of chitin whisker. Coating with chitin whiskers improves the physical properties of paper sheets that could be used as food packing materials [51].

17.4.2 Chemical Modification

The chemical modification of chitosan with the different organic compounds and polymeric materials can lead to promising materials for heavy metal removal. Chethan and Vishalakshi attempted the selective modification of chitosan by incorporating ethylene-2-diamine molecule in a regioselective manner using *N*-phthaloylchitosan and chloro-6-deoxy *N*-phthaloylchitosan as precursors. The derivatives that resulted were ethylene-1,2-diamine-6-deoxy-*N*-phthaloylchitosan (PtCtsEn) and ethylene-1,2-diamine-6-deoxy-Chitosan (CtsEn) and were prepared for Cu(II), Pb(II) and Zn(II) removal [52].

Xiao-Yi Huang *et al.*, have reported TEPA–CS exhibited strong sorption capacity toward anionic dye eosin Y (the maximum adsorption capacity was 292.4 mg/g at 298 K), which was attributed to the presence of introduced amino groups. The Langmuir model represented the best fit of the experimental data, and the kinetic measurements showed that the sorption process was rapid and followed a pseudo-second-order model. The sorption mechanism was a multistep process, involving adsorption on the external surface, diffusion into the bulk and electrostatic interactions. It is concluded that TEPA–CS could be used as a low cost and efficient adsorbent for removal of anionic dye (acid dye) from wastewater and has good potential for further application in effluent treatment [53].

Mohammadtaghi Vakili *et al.*, (2017) have successfully applied a cationic surfactant, HDA, and Organosilane (APTES) to improve adsorption performance of chitosan beads for RB4 absorption. It revealed that removal of RB4 decreased with increasing pH (4–10) and Temperature (30–50 °C) of coloring solution. The maximum adsorption capacity CS balls after modification using ADH Y APTES increased by 317.1 mg/g of 468.8 mg/g at pH 4 and 30 °C. The adsorptivity CS/HDA/APTES was 1.48 times higher than that of Pearls CS. This increase in absorption capacity was attributed to increase in surface charge and the presence of multiple adsorption sites in Adapted modified surface caused by APTES HAD and equilibrium. The study suggested that RB4 multilayer adsorption was different distribution of energy levels to the adsorbent surface [54].

17.4.3 Multiple Modifications

There are some cases in which:

1. The reactivity of chitin/chitosan itself is insufficient to participate in the desired reaction,
2. The modified chitin/chitosan does not possess the desired properties,
3. Some sites of chitin/chitosan must be protected to sustain during the modification reactions. In such instants, there are two general approaches via chemical modification when the graft polymerization is a certain pathway to achieve desired characteristics:
   (a) *In-situ*-and/ or post-treatment of the graft copolymer,
   (b) Graft copolymerization onto a previously modified chitin/chitosan. So, the products may generally be referred to as multiple modified materials [55]. Chitosan can be modified as cross-linked chitosan, blends, composites and its derivatives.

For batch adsorption studies a variety of materials are successfully grafted with chitosan and possess a great potential as an adsorbent to remove toxic heavy metals. Okoya *et al.*, grafted chitosan with cocoa husk char, an eco-friendly and feasible material successfully remove chromium and lead from industrial wastewater [56]. Nithya *et al.*, grafted chitosan with poly(butyl acrylate)/silica gel(Cs-g-PBA/SG) nanocomposite and reported better adsorption of chromium was achieved by grafting [57].

Cross-linking of chitosan has been performed with many reagents like formaldehyde, glutaraldehyde, ethylene glycol, In Palladium sorption on glutaraldehyde cross-linked chitosan, the optimum pH was close to pH 2 [58]. Crown ethers bound chitosan had a strong complexing capacity and better selectivity for metal ions because of the synergistic effect of high molecular weight. Tang prepared the crown ether bound chitosan with Schiff's base type. It had not only good adsorption capacities for metal ions Pb^{2+}, Au^{3+}, Ag^{+} but also high selectivity for the adsorption of Pd^{2+} in the presence of $Cu^{2+}$and Hg^{2+} [59]. The spherical chitosan-tripolyphosphate chelating resin as sorbents for the removal of Cu (II) was achieved by [60].

17.4.3.1 Blends of Chitosan

A chitosan/cellulose acetate/polyethylene glycol ultra-filtration membrane was prepared with DMF as a solvent. It was focused to be efficient in removing chromium from artificial and tannery effluent wastewater. The highest rejection rate was responding [61]. Cross-linked chitosan/polyvinyl alcohol blend beads were prepared and studied for the adsorption capacity of Cd^{2+} from wastewater. The maximum adsorption of Cd (II) ions was found to be 73.75% at pH 6 [62].

17.4.3.2 *Composites of Chitosan*

Chitosan/magnetite nanocomposite beads have the maximum adsorption capacities for Pb (II), and Ni(II) occurred at pH 6 under room temperature were as high as 63.33 mg/g and 52.55 mg/g, respectively. Chitosan magnetite nanocomposite beads could serve a promising adsorbent not only for Pb (II) and Ni(II) but also for other heavy metal ions in wastewater treatment technology [63].

Chitosan/polyurethane porous composite based on chitosan, polyether polyols, and tolylene di-isocyanate were prepared. Adsorption of Cu^{2+} and Cd^{2+} on CS/PU was studied by atomic adsorption spectroscopy. The removing rate of Cu^{2+} reached 96.67% the removing rate of Cd^{2+} reached 95.67% the adsorption capacity of CS/PU for $Cu^{2+}$and Cd^{2+} was 28.78 mg/g and 25.32 mg/g, respectively. The selective adsorption of CS/PU for $Cu^{2+}$is higher than that for Cd^{2+} when $Cu^{2+}$and Cd^{2+} co-exist [64].

17.4.3.3 *Grafted Chitosan*

Among the various methods of modifications, graft copolymerization has been mostly used method. Grafting of chitosan allows the formation of functional derivatives by covalent binding of a molecule, the graft, on to the chitosan backbone. Chitosan has two types of reactive groups that can be grafted. First, the free amine groups on deacetylated units and secondly, the hydroxyl groups on the C_3 and C_6 carbons on acetylated (or) deacetylated units. Recently researches have shown that after primary derivation followed by graft modification, chitosan would obtain much improved water solubility, antibacterial and antioxidant properties and enhanced adsorption properties [65].

Grafting is a method wherein monomers are covalently bonded onto the polymer chain. Two major types of grafting may be considered as

- Grafting with a single monomer.
- Grafting with a mixture of two monomers [66].

Gopal Reddi *et al.*, have observed the study appends to the increasing interest in the exposure of removal of heavy metals from water bodies contaminated with toxic heavy metals. The use of low cost adsorbent material and the enhanced sorption process of chromium and copper ions by chitosan-g-maleic anhydride g-EDMA are described in this study. The removal process was found to be dependent on pH of the solution, initial metal ions concentration, contact time and adsorbent dosage. The process herein is cost effective and attractive which might help to reduce the disposal problems and thereby can also be used for heavy metal removal from aqueous systems as well as industrial effluent [67].

17.5 Application of Natural Polymers in Wastewater Treatment as Promising Adsorbents

Today there are many different ways of treating industrial effluent which is hazardous to the entire world when it is let into the water stream. Recently, numerous approaches have been studied for the development of cheaper and more effective technologies, both to decrease the amount of wastewater produced and to improve the quality of the treated effluent [68]. Among the various synthetic materials available, natural materials are being used to treat wastewaters. The marine polysaccharides' sorbents are low-cost materials obtained from natural raw resources, and these materials are versatile in properties with repetitive functional groups. Biopolymers provide excellent chelating and complexing materials for a wide variety of pollutants including dyes, heavy metals [69].

On seeing to the drawbacks of chitosan, chitosan has poor mechanical properties, pH sensitivity, and variability in the polymer characteristics; is nonporous, not useful to remove cationic dyes, and not much useful in the native form; and requires derivatization to improve its performance [70].

The adsorption properties of an adsorbent depend on the source of raw materials. The sorption capacity of chitin and chitosan materials depends on the origin of the polysaccharide, the degree of *N*-acetylation, molecular weight, and solution properties and varies with crystallinity, affinity for water, percent deacetylation, and amino group content [71].

17.5.1 For Heavy Metal Removal

Heavy metals are generally introduced into the environment through natural phenomena and human activities [72]. The contamination of the existing water resources is increasing day-by-day with industrialization. The disposal of wastewater containing heavy metal ions is always a challenging task for environmentalists [73].

Abdeen *et al.* (2014) prepared chitosan polyvinyl alcohol (PVA/CS) nanoparticles using ball milling methods and as an adsorbent to remove Mn (II) of an aqueous solution. It was found that batch parameters study, pH of solution, adsorption dose, adsorption concentration and contact time were effective on Mn (II) adsorption efficiency. The obtained experimental data and isothermal kinetic studies well inserted in the pseudo-second kinetic model and the isotopic model of Freundlich. Thermodynamic study indicated that adsorption of Mn (II) ion on PVA/CS crosslinked hydrogel nanoparticles is spontaneous and endothermic nature. The results revealed that the preparation of PVA/CS crosslinked hydrogel nanoparticles as adsorbent for contaminants without crosslinking and with the ball mill as a simple method is economical. In addition, because of their biodegradability, it is considered to be environmentally friendly adsorbent [74].

Mohamed Ali (2016) reported that the nanocomposite adsorbent of Cs/CDTA/GO was prepared and characterized .The obtained results showed that the optimum adsorbent dose was 2 g/l at pH of 3.5 and equilibrium time of 60 min. The adsorption kinetics of Cr6+onto Cs/CDTA/GO followed the pseudo-second order model and the adsorption isotherm was well fitted by the Langmuir model. The maximum adsorption capacity of the adsorbent was 166.98 mg/g, and the equilibrium parameter (RL) at different concentrations was less than unity indicating the adsorption of Cr6+ ions onto Cs/CDTA/GO is favorable [75].

Abbas Teimouri *et al.*, (2016) was prepared, a new low-cost biosorbent material chitosan/zeolite Y/ZrO_2 nanocomposite, using chitosan, zeolite and environmentally friendly zirconium oxide. The results showed that it could effectively remove nitrate in acidic conditions by chelation between No3 and the amine group of the complex. It has also been found that nitrate adsorption depends on the CTS ZY/Nano ZrO_2 molar ratios, the initial pH of the nitrate solution and the temperature. By checking these parameters, optimal conditions for nitrate adsorption have been obtained. It was observed that the pseudo-second-order kinetics adsorption model and the isotope followed the Langmuir monolayer model. Compared with the CTS, ZY, Nano ZrO_2, CTS/CTS Nano ZrO_2 and /ZY, CTS/ZY/ZrO_2 nanoparticles had a higher absorption capacity. Therefore, nanocomposite can be used effectively as an adsorbent for the removal of nitrates [76].

Chitosan–g–maleic anhydride–g–(acrylonitrile) copolymer was prepared via free radical polymerization using ceric ammonium nitrate as the initiator. This graft copolymer was used as an adsorbent in dye effluent treatment [77]. Katarzyna Jaros *et al.,* reported the application of chitosan in the form of beads to the removal of copper, zinc and chromium ions from water [78]. Donia Hitchu *et al.*, evaluated the composite chitosan magnetite microparticles for the removal of Co^{2+} and Ni^{2+} ions in aqueous solution [79].

17.5.2 For Dye Removal

Expelling of dye stuffs into water resource system causes major thread to the environment. Dyes are an important class of pollutants, and can even be identified by the human eye. Disposal of dyes in precious water resources must be avoided, however, and for that various treatment technologies are in use. Among various methods adsorption occupies a prominent place in dye removal. Modification of chitosan changes the original properties of this material so that it can be more suitable for adsorption of different types of dye. Adsorption is the cost effective and potential method to remove the dyes from the effluents.

Pranee Lertsutthiwong *et al.*, suggested chitosan–montmorillonite can be prepared by a cationic exchange reaction and its properties depend on the molecular weight and degree of deacetylation of chitosan, and on the chitosan/montmorillonite ratio. These results suggest that chitosan can be used as an enhancer for improvement of the dye adsorption [80]. Solairaj Dhananasekaran *et al.*, 2015. α-Chitin nanoparticles from the shells of *P. monodon* (Fabricius, 1798) were found to be a promising material for the purification of water dye stuff contamination. The prepared CNP have 49 nm average particle size with 96.8% DA and 83.73% crystalinity. The experiments done at various physical parameters has showed that CNP adsorbs dyes in a very short period of exposure in normal environmental conditions and does not need any specific conditions for the adsorption process. The experimental data were analyzed [81].

Abdollah Hajivandi *et al.*, introduced data acquired from new adsorbent, bimetalized chitosan particle that is successfully synthesized and applied to remove the orange II dye, an azo dye, from textile wastewater. The adsorbent was meso – and macro-porous material with BET surface area of 12.69 m^2/g and pH zpc 6.6. The simulated textile-wastewater can be significantly treated using a relatively low quantity of the adsorbent. Overall, the use of bi-metalized chitosan particle can be considered a promising method for eliminating the azo dye from wastewater effectively. Accordingly, these data will be useful for decolorizing of azo dyes from textile wastewater [82].

Javed Iqbal, 2012 Chitosan flakes, extracted from prawns and labeo rohita scales, with high adsorption capacity were prepared after chemical treatment and were used to remove acid yellow dye from water. The results showed that adsorption capacity is dependent on pH, initial concentration of dye, BET,

Langmuir surface area and pore volume of the adsorbent. In acidic conditions, the polymer amino groups were protonated (positively charged polymer chain), which showed attraction with negative ions of anionic dye [83]

Abdollah Hajivandi *et al.*, (2016) introduced data acquired from new a adsorbent, bimetalized chitosan particle that is successfully synthesized and applied to remove the orange II dye, an azo dye, from textile wastewater. The adsorbent was meso - and macro-porous material with BET surface area of 12.69 m^2/g and pH zpc 6.6. The simulated textile-wastewater can be significantly treated using a relatively low quantity of the adsorbent. Overall, the use of bi-metalized chitosan particle can be considered a promising method for eliminating the azo dye from wastewater effectively. Accordingly, these data will be useful for decolorizing of azo dyes from textile wastewater [82].

Javed Iqbal *et al.*, (2011) extracted Chitosan flakes from prawns and labeo rohita scales, with high adsorption capacity were prepared after chemical treatment and were used to remove acid yellow dye from water. The results showed that adsorption capacity is dependent on pH, initial concentration of dye, BET, Langmuir surface area and pore volume of the adsorbent. In acidic conditions, the polymer amino groups were protonated (positively charged polymer chain), which showed attraction with negative ions of anionic dye [83].

Mohammadtaghi Vakili *et al.*, (2014) has suggested that chitosan base adsorbents have received much attention for the absorption of dyes. They have studied various modifications of this polysaccharide to improve adsorption properties and mechanical and physical properties of chitosan. In this review article discusses the main research topics related to chitosan and its derivatives for use in the removal of water colorants. Chitosan modification modifies the original properties of this material to make it more suitable for adsorption of different types of dyestuff. Many have been derived from chitosan derivatives by chemical and physical modification including crude chitosan crosslinking, grafting and chitosan skeleton impregnation. A better understanding of these varieties and their affinity with different types of dyes can help future research properly orientate to bridge the gaps in knowledge in this area. This review provides researchers with the opportunity to better explore the potential of chitosan adsorbent derivatives for the removal of a variety of dyes [84].

17.5.3 For General Wastewater Treatment

A study on the removal of arsenic from groundwater using iron–chitosan composites was conducted. Removal of As (III) and arsenic (V) was studied through adsorption at pH 7.0 under equilibrium and dynamic conditions.

The adsorbent was also successfully applied for the removal of total inorganic arsenic down to <10 μg/L from arsenic contaminated groundwater samples [85]. Cross-linked chitosan/polyvinyl alcohol blend beads were prepared and studied for the adsorption capacity of Cd^{2+} from wastewater. The maximum adsorption of Cd (II) ions was found to be 73.75% at pH 6 [86]. Effluent waste from Electroplating industries considerably pollutes the environment. The study results indicate that the cost incurred toward extraction was less than that the cost incurred by the commercial treatment facilities. The –OH and –NH2 functional groups in chitosan facilitate the adsorbent function and make it an ideal adsorbent for the treatment of wastewater. This study brings to light that chitosan is function as an economically useful adsorbent for the treatment of electro plating effluent containing heavy metals [87].

Heavy metals discharge associated with industrial activities constitute serious environmental problems, most especially water pollution. Many materials have been identified for the adsorption of these heavy metals and chitosan, a derivative of natural polymer chitin, has been identified as one of such materials. This study was carried out to investigate the removal of chromium (VI) from chromate solution using chitosan from mussel shell, obtained from Gubi dam in Bauchi Nigeria.[88]. Chitosan-abrus precatorius (CS/Ab) blended beads were used as adsorbent for the removal of phenolic compounds from aqueous solution. The maximum monolayer adsorption capacity of phenol, 2-chlorophenol and 4-chlorophenol on to the (CS/Ab) beads was found to be 156 mg/g, 204 mg/g and 278 mg/g, respectively. A novel adsorbent namely magnesia/chitosan (MgOC) composite was prepared to remove fluoride from drinking water by Sundaram *et al.*, The maximum Langmuir monolayer capacity was found ca.11 mg/g. The mechanism offluoride removal of both MgO and MgOC composite was mainly governed by adsorption. The higher defluoridation capacity (DC) of MgOC composite might be attributed to the fact that chitosan also contributed in the enhancement of DC by removing fluoride through hydrogen bonding [89].

17.6 Chitin and Chitosan as a New Type of Polymer Coagulant/Flocculants

Coagulation or flocculation process was conducted for the treatment of industrial wastewater to achieve maximum removal of chemical oxygen demand (COD), total dissolved solids (TDS), and total suspended solids (TSS). Therefore, the effect of coagulant dose, polyelectrolyte dose, pH of solution, and addition of polyelectrolyte as coagulant aid and found to be important parameters for the effective treatment of beverage industrial wastewater was investigated by Amudha *et al.* Conventional treatment involves a process of coagulation/flocculation. This is a versatile process, which can be used alone or combined with biological treatments, as a way of removing suspended solids and organic material, as well as promoting the extensive removal of dyes from textile industry effluents [90–92]. The coagulation–flocculation of model sulphonated azo dyes, that is, Acid Black 1, Acid Violet and Reactive Black 5 with the use of chitosan. Experiments were performed at three initial pH values (pH 3, 5, and 7) and four concentrations 925, 50, 100, and 200 mg/L by Szygula *et al.*, [93]. The kinetics of the process were considered by varying the settling time, and the dye/polymer molar ratio was modified in order to determine the optimum relationship between the functional groups of the dye and the amine groups of chitosan for a maximum dye recovery. The results indicated that increasing the initial dye concentration from 25 mg/L to 600 mg L decreases the dye adsorption/mass ratio (mol/g) whereas the pressure of the surfactant DBS and a temperature increase of 25–35 °C increases it. The adsorption capacity of chitosan beads decreased in the presence of DBS as a result of the competitive adsorption for the same sites between the dye and surfactant molecules. The factorial experiments demonstrate the existence of a significant antagonistic interaction effect between the "concentration" and "surfactant" factors. The quantitative adsorption (Nf) modeling presents good correlations with the experimental data [94].

The main advantage of coagulation and flocculation is the decolorization of the waste stream by the removal of dye molecules from the dye bath effluents, and not by a partial decomposition of the dye, which can lead to even more harmful and toxic aromatic compounds [95]. It is dissolved in acetic acid, proved to be very efficient for the coagulation–flocculation of Acid Black 1, Acid Violet 5, and Reactive Black 5. In acidic solutions, the charge neutralization (associated to a "bridging-like" effect) is responsible for the dye removal in acidic solutions (pH 3–4.5). Several wastewater treatment technologies were developed by many researchers such as coagulation/flocculation [96], oxidation/reduction [97], membrane filtration [98], reverse osmosis [99], nanofiltration [100], biosorption [101], phytoremediation [102–104], etc.

But among the various technologies, due to its simplicity, low cost effectiveness and low energy consumption the coagulation/flocculation process is considered as the most commonly used techniques utilized in wastewater treatment to achieve efficient removal of turbidity arising mainly due to the soluble colloidal suspended particles, removal of coloring materials, particulates and natural organic matter (NOM). The efficient solid-liquid separation can be achieved through induced aggregation of both micro and macro particulates into larger-sized ones and followed by the sedimentation process [105–107]. Besides water treatment, the coagulation/flocculation technique is also widely employed in many industrial fields including oil extraction, mining operations and paper production. Coagulation process is usually a versatile process accomplished by the addition of coagulants having the charge opposite to that of the colloidal particles present in wastewater. During this coagulation, the destabilization process of small colloids suspended in wastewater [108], happens and followed by the coagulation, the flocculation process takes place in which the destabilized particles aggregate to form larger flocs that can be effectively removed by sedimentation finally. Different degrees of destabilization were achieved by different coagulants. Higher the valence of the counter-ion, greater the coagulant's destabilizing effect will be and lower the dose required [109].

Polymeric flocculants, especially cationic polymers possess the dual functions of coagulation and flocculation which involves the neutralization of the negative charges and bridging the aggregated destabilized particles in direct flocculation processes [110]. Both coagulants and flocculants based on synthetic polymers have been reported to pose a number of environmental problems as some of the intermediate products of their degradation are hazardous to human health as their monomer is neurotoxic and carcinogenic and the derivatives are nonbiodegradable [111]. The disadvantage of some of the synthetic organic polymer flocculants such as polyacrylamide (PAM) and its derivatives is the extremely severe neurotoxicity of residual unreacted monomers including acrylamide (AM) and ethyleneimine. Hence due to these neurotoxic and carcinogenic residual monomer productions, the application of synthetic polymers as coagulant has impeded its use in water treatment [112].

The primary function of flocculation is the removal of particles that can create turbidity, and/or retain bacterial organisms. The flocculants aggregate and agglutinate fine particles resulting in the settlement of fine particles from the liquid phase and it leads to the reduction in solution turbidity [113]. Certain parameters such as the turbidity, viscosity, total solids, total suspended solids, total polyphenols, and the zeta potential were mainly used to monitor the progress of the flocculation process. The floc properties play a vital role in the study of flocculation mechanism and in the practical operation of industrial processes [114, 115]. The floc size, compactness, and strength are some of the parameters mainly used to describe floc properties [116]. Since the purpose of flocculation is to make particles to form large aggregates and settle down, the larger flocs sizes are apparently favored. In addition, the compactness of flocs is even more important which indicate that the denser flocs lower sludge volumes and hence giving rise to lower total disposal cost [117].

Even though these conventional coagulants play an important role in the water treatment it can also produce adverse side effects on the environment and living organisms. The excessive use of coagulants might lead to corrosion problem and this will result in high concentrations of metal coagulants in treated water which is detrimental to the human health [118, 119]. In addition to this, the hydrolyzing metal coagulants also results in the production of the huge amount of sludge, which is nonbiodegradable and causing disposal problems to the environment [120]. Some other drawbacks of inorganic coagulants include high sensitivity to pH, inefficiency toward very fine particles and applicability only to a few disperse systems [121, 122]. Due to these limitations, the search for a better alternative for conventional coagulants and flocculant has become an important challenge in recent years and the current research goals focused on identifying more specific compounds with potent removal efficiencies in water treatment process. As an alternative, with the increasing awareness of sustainable development regarding the drawbacks of conventional coagulants and flocculants, the water sector seeks various natural coagulants for water treatment process. The main advantage of natural coagulants is the lesser sludge production and due to its biodegradability properties, the sludge treatment is also easier and so less harmful to the environment. Certain natural coagulants such as *Moringa oleifera, nirmali seeds, Jatropha curcas* seeds, chitosan and Tannin have been tested and proven as effective alternative coagulants for conventional coagulants in removing natural organic matter and turbidity [123, 124].

The increasing knowledge of structural analysis and chemical modifications enables the use of the natural coagulants chitin and chitosan in a newer way for the human welfare. The abundant free amino groups on the chitosan backbone protonated in acidic medium behave as a typical cationic polyelectrolyte and hence due to its properties, efficient charge neutralization and bridging effects, the chitosan acts as an outstanding anticoagulant with good coagulation and flocculation performance. Chitin and chitosan derivatives are mainly used in water purification plants throughout the world to remove heavy metals, oil, grease and fine particulate matter that cause turbidity in wastewater streams. Chitosan is natural water coagulant/flocculant having the ability to reduce turbidity, color and clay particles. From water and in addition by utilizing the natural coagulant chitosan, the chemicals can be saved and the sludge handling can be achieved [125]. The removal of turbidity from the synthetic raw waters by chitosan biopolymer mainly occurs by the neutralization mechanism for low alkalinity solution and for the solution with higher alkalinity and calcium ions the bridging mechanism played a role [126]. The chitin and chitosan applications in water treatment focus mainly on coagulation-flocculation processes to remove especially the organic residues, suspended solids, amino acids and dyes [127].

The chitosan biopolymer has been effectively used as a coagulant or flocculant for a wide variety of suspensions including silt in river water to microorganisms and it is also used as an effective agent for the coagulation of suspended solids from various food processing wastes. The chitosan coagulation produced flocs of better quality, namely larger flocs, and faster settling velocity. The effectiveness of chitosan for coagulating mineral suspensions has been improved at high pH, due to the addition of materials extracted from soils or due to the presence of inorganic solutes [128]. Madhukar *et al.*, investigated chitosan performance as a coagulant for flocculation of local clay suspensions of different turbidities by jar test [129]. Experimental results of this study proved that the turbidity removal efficiency has been increased from 5% to 28% at settling time of 30 min with readily biodegradable less voluminous sludges and the turbidity removal efficiency by this new technique was superior to that of the conventional process which meets WHO drinking water guidelines.

The comparison of three chitosans of different molecular weight and cationic charges with two traditional cationic polymers [poly-diallyldimethyl-ammonium chloride (PDADMAC) and poly-ethylene imine (PEI)] used in wet end chemistry as flocculation additives in paper making on two systems: calcium carbonate (GCC) and pulp/GCC suspension were investigated as by Nicu *et al.* [130]. By utilizing a focused beam reflectance measurement (FBRM) technique the flocculation efficiency was evaluated by flocculation parameters (mean floc size and number of counts) and by floc behavior under shear conditions. The obtained results indicated that the when compared to PDADMAC and PEI the chitosans were found to be more efficient for flocculating small particles of the GCC suspension, but less efficient for increasing floc sizes, regardless of their MW. The flocculation parameters for pulp/GCC suspensions suggested that the flocculation behavior of chitosan was close to that of PEI, but chitosan had higher efficiency and affinity toward cellulose fibers.

Fatma Gassara *et al.* reported that the chitin and chitosan can be used as natural flocculants for beer clarification and from the obtained results it was evident that of all the tested flocculants, chitosan (5 mg/L) was found to be the most effective flocculant [131]. Since the chitosan is produced from nonmammalian sources (typically arthropod shells), it can also be used as an ideal flocculant for biotechnology applications. The chitosan-based flocculants are firmly believed to have a wide application in near future owing to their numerous notable advantages. Zheng Yang *et al.* reported about the flocculation performance of carboxymethyl chitosan-graft-poly[(2-methacryloyloxyethyl) trimethyl ammonium chloride (CMC-g-PDMC) systematically evaluated using kaolin suspension, humic acid (HA) solution and kaolin-HA mixed suspension as synthetic wastewater under acidic, neutral and alkaline conditions, respectively [132]. The observed results indicate that the well-prepared novel chitosan-based flocculant, CMC-g-PDMC, exhibited high potential applicability from aspects of optimal dosage, contaminant removal efficiency, pH sensitivity, effluent toxicity, settling velocity, and floc properties.

17.6.1 For Heavy Metal Removal

All living beings are affected by the presence of heavy metals, which are frequently discarded in water flow near cities [133]. The majority of the mining operations, tanneries, and metal plating industries plays a vital role in contaminating the environment with toxic heavy metals [134].

The permanent pollutants present in water are the soluble heavy metal ions which can be adsorbed onto particles or become particles by hydrolysis and hence the heavy metal content of water can be reduced only to some extent by flocculation. Even after flocculation process, since there are still a large number of heavy metals that are soluble in water which is mainly metal complexing species with different ligands the flocculation unit must be followed by other treatment units, for example, chemical precipitation, adsorption, ion exchange and membrane filtration for the complete removal process. Macromolecule heavy metal flocculants is a new kind of flocculant [135–138] which has the ability to remove soluble heavy metal species from water by coordination and chelation. Culing Zhang *et al.* reported about the synthesis of macromolecule mercapto-acetyl chitosan (MAC) by amidization reaction with chitosan and l-cysteine and its application as flocculant for the Cu(II) removal also has been investigated [139]. Results indicated that the mercapto-acetyl chitosan (MAC) flocculant was used to chelate the heavy metals for the removal of Cu(II) ions with 100% removal rate.

17.6.2 For Dye Removal

Contamination of water resources with dyes is not desirable, since colored effluents may contain toxic, carcinogenic, and mutagenic chemicals. Approximately, 10,000 different dyes and pigments are used in industry. Chitin and chitosan both have efficiency in removing dyes from wastewater.

Chitosan chelation is the procedure of choice for dye removal from aqueous solution. Chitosan, a biological cationic polymer, was used in treating dairy wastewater, agriculture, food processing, medicine, cosmetics, wastewater treatment, and biotechnology [140]. Reactive dyes are widely used for textile dyeing, paper printing, leather dyeing, color photography, and as additives in petroleum products.

The amine groups on chitosan bind metal cations at pH close to neutral. At low pH, chitosan is more protonated and therefore it is able to bind anions by electrostatic attraction [141]. Chitosan's functional groups and natural chelating properties make chitosan useful in wastewater treatment by allowing for the binding and removal of metal ions such as copper, lead, mercury, and uranium from wastewater. It can also be utilized to remove dyes and other negatively charged solids from wastewater streams and processing outlets. Chitosan grafted with poly(acrylonitrile) has been further modified to yield amidoximated chitosan [142], a derivative having a higher adsorption for Cu^{2+}, Mn^{2+}, and Pb^{2+} compared to cross-linked chitosan. The adsorption capacity had a linear dependence on pH in cases of Cu^{2+} and Pb2+. However, a slight decrease in the adsorption capacity was observed in the case of Zn^{2+} and Cd^{2+} [143]. Chitosan has been modified with different mono as well as disaccharides.

Yang *et al.*, have also reported the metal uptake abilities of macrocyclic diamine derivative of chitosan. The polymer has high metal uptake abilities, and the selectivity property for the metal ions was improved by the incorporation of aza crown ether groups in the chitosan [144]. Chitosan was successfully coated on PET granules through a dip and phase inversion process and the coated granules were examined for the performance and mechanism of humic acid removal through a series of batch adsorption tests. With its positive charge, chitosan can be used for coagulation and recovery of proteinaceous materials present in food processing operations [145].

Chitin and chitosan are also good adsorbents for removal of phenol and other pollutants from industrial wastewaters. In addition to this, when chitosan was modified with some organic compounds, such as aldehydes and organic acids, it was found that the modified products with salicylaldehyde, cinnamaldehyde, benzaldehyde, and carbohydrate showed better Fe and Cu metal ions uptake. Comparatively, carbohydrate modified chitosan showed lower metal ion uptake, which they attributed to the higher ability of carbohydrate to be leached out of the chitosan matrix due to its higher solubility in water. In yet another study, chitosan was cross-linked using glutaraldehyde in the presence of magnetite [146].

17.6.3 For General Wastewater Treatment

The environmental disposal of untreated effluents to its surroundings often leads to the following consequences: coloration of water bodies with the creation of an aesthetic problem limitation of reoxygenation capacity of the receiving water, decrease in sunlight penetration which in turn disturbs photosynthetic activities in the aquatic system, and chronic and acute toxicities [147, 148]. This diverse spectrum of wastes requiring efficient treatment has focused the attention of researchers on membrane, ion-exchange, and biological technologies. The most effective and ecological technological systems developed during the past 20 years are as a rule based on a combination of the above methods.

In addition to other conventional methods, adsorption is another effective and widely used method for wastewater remediation. This method is cost effective and easy to adopt. Today, there is a growing interest in developing natural low-cost alternatives to synthetic polymers [149]. Various adsorption materials have been studied, such as activated carbons, clays, polymeric synthetic resins, metal oxides, and some natural materials. Among them, activated carbon is one of the oldest and most commonly used adsorbents. Generally, a suitable adsorbent for adsorption process of pollutants should meet several requirements such as (i) efficient for removal of a wide variety of target pollutants, (ii) high capacity and rate of adsorption, (iii) important selectivity for different concentrations, (iv) granular type with the good surface area, (v) high physical strength, (vi) Able to be regenerated if required, (vii) tolerant for a wide range of wastewater parameters, and (viii) low cost. The removal of metals, compounds, and particulates from solution by biological material is recognized as an extension to adsorption and is named as biosorption [150, 151]. Sharma *et al.*, reported the various natural and cheaper alternatives such as wood ash, brick powder, sugarcane bagasse, tea leaves, ash, and charcoal based on batch adsorption studies for decolorizing the effluent. Several authors presented review articles on the removal of dyes using low-cost adsorbents [152].

The release of colored substances from the industries into water bodies not only can aesthetically cause issues but it is also harmful to biological organisms and ecology [153]. The discharge of colored wastewaters to the ecosystem causes eutrophication and perturbations in the aquatic life. The reduction of photosynthetic activity happens by the interference of colored wastes in the receiving streams with the transmission of sunlight into streams [154]. The major industrial wastewater sources are the textile industries and other dyeing industries such as paper, printing, leather, food and plastic. Among all industrial sectors, based on both the discharged volumes and effluent composition, the wastewater generated by the textile industry is rated as the most polluting in the world [155]. Due to more water consumption, the textile dyeing industry produces large volumes of wastewater from different processes in dyeing and finishing processes. Generally, the discharged wastewater volume from each step of textile operation takes place approximately at a high rate of between 40 L/kg and 65 L/kg of the product [156]. Today around seven lakh tonnes of dye is produced annually worldwide, ten thousand different dyes and pigments available commercially and about 10–15% of these dyes are released as effluents during the dyeing processes [157].

The term "biosorption" is given to adsorption processes, which use biomaterials as adsorbents (or bio sorbents). Chitosan is a biopolymer and has an extremely high affinity for many classes of dyes, including disperse, direct, reactive, acid, vat, sulfur, and naphthol dyes. The rate of diffusion of dyes in chitosan is similar to that in cellulose. Only for basic dyes has chitosan a low affinity. Fouling is a problem encountered with most filtration processes as it reduces productivity and impairs the capabilities of the membrane due to the decline in flux because of concentration polarization or fouling of the membrane [158]. Le Roux *et al.,* demonstrated the feasibility of using chitosan as an antifouling agent in a bioreactor to help decrease fouling was successfully [159]. One potential application of chitosan complexes in seafood processing is the treatment of surimi wash water (SWW), which contains 0.1–2.3% protein to obtain not only clean wash water for reuse in the plant but also to recover suspended proteins for use in feed production [160]. Flocculation of SWW protein by using Chi–Alg at a concentration of 100 mg complex/L SWW for 1 h achieved high protein adsorption and turbidity reduction [161]. The Chi–Alg complex is an effective alternative in not only recovering soluble proteins that would otherwise be discarded into the environment but more so an economically viable downstream process over expensive, commercial membrane treatments and their limited use due to periodic fouling [162]. Insoluble proteins can be recovered in the same step used for SWW [163]. Fluoride normally enters the environment and human body through water, food, industrial exposure, drugs, cosmetics, etc. However, drinking water is the single major source of daily intake [164].

The dye removal from textile wastewater has been a big challenge over the last decades since the dyes discharged from the industries have high visibility, resistance [165]. Direct and indirect toxic effects on humans as they are associated with certain health disorders such as tumor, cancer, allergies, jaundice, skin irritation, heart defects, and mutations [166, 167]. In order to overcome these health disorders, the coagulation/flocculation, the conventional versatile treatment methodology has been either used alone or in combination with biological treatment since it helps in promoting the extensive removal of dyes from the textile industry effluents [168]. The highly selective nature attempts the researchers to focus their attention on studying the coagulant as an economic and viable alternative treatment for dye wastewater.

The two natural coagulating agents namely the chitosan alone and chitosan as coagulant aid in conjunction with polyaluminium chloride (PAC) have been selected for the removal of acid dyes (Acid Blue 292; AB292, and Acid Red 398; AR398) from dye-containing solutions by Zonoosi *et al.* [169]. Around 90% of excellent dye, removal results were achieved with relatively low dosages of chitosan (30–35 mg/L for AB292 and 50–60 mg/L for AR398). The investigated results show that chitosan, as a natural coagulant, efficiently removed the acid dyes and more efficiency was obtained for AB292 in comparison with AR398.

Easily available, low cost naturally prepared coagulants like Maize seed powder (MSP), Chitosan and Surjana seed powder (SSP) as an ideal alternative to recent expensive coagulant methods for Congo Red (CR) dye removal has been investigated by Himanshu Patel and Vashi in the year 2012. The optimization of various process parameters like pH, flocculation time, coagulant dose, the temperature was exploited and in addition the sludge volume index (SVI), turbidity were calculated. The calculated results of sludge volume index (SVI) and turbidity indicated that chitosan was more preferable than SSP and MSP respectively [170].

The performance of chitosan and extracted pandan leaves as a flocculant toward treatment of textile wastewater using jar test method was evaluated by Norzita Ngadi *et al.* The effect of dosage, pH, and settling time on reduction of COD, turbidity and color of textile wastewater was studied. The optimized condition for COD and turbidity removal was achieved at 0.2 g dosage, pH 4 and 60 minutes of settling time and under this condition about 58 and 99% of COD and turbidity was removed respectively. The obtained result shows that the chitosan was found to be effective for reduction of COD, turbidity, color but the results of extracted pandan were found to be opposite to that of chitosan. The extracted pandan was not able to remove both COD and turbidity of the waste [171].

The amphoteric grafted chitosan-based flocculant (carboxymethyl chitosan-graft-poly(2-methacryloyloxyethyl) trimethyl ammonium chloride) used for the removal of cationic and anionic dyes namely acid Green 25 (AG25) and Basic Bright Yellow (7GL), from water, was studied by Wu and his co-workers. By utilizing the flocculant dosage, initial solution pH and temperature as input variables, the flocculation conditions have been optimized on the basis of central composite design (CCD) by response surface methodology (RSM). More detailed evaluation indicates that the second-order regression model well described the process of AG25 removal whereas the cubic one is more suitable for that of 7GL.9 [172].

Polymeric nanocomposites and membrane technologies were widely explored and used to remove toxic dyes from water [173, 174]. The treatment of wastewater polluted with carcinogenic dyes using chitosan–starch-based polymer nanocomposites by coagulation-flocculation process was investigated by Amtul Jamil Sami *et al.* Reported results indicate that the starch/chitosan nanocomposite provide better dye removal effects due to electrostatic and hydrophobic interactions as compared to the use of chitosan alone [175].

17.7 Future Directions for Research

Marine carbohydrates have a great potential in the treatment of wastewater. In addition, marine carbohydrates' derivatives can be prepared for treating wastewater by modifying it with various other bioadsorbents, which explore its innovative properties against environmental pollution. Also, it can be used in various fields like food and agriculture, cosmetics, pharmaceutics, and tissue-engineering applications.

17.8 Conclusion

Industrial wastewater treatment covers the mechanisms and processes used to treat waters that have been contaminated in some way by anthropogenic industrial or commercial activities prior to its release into the environment or its reuse. Some form of pretreatment was done to the wastewater obtained from industrial process prior to discharge to a sewer. This form of pretreatment was mainly done to minimize corrosion and clogging of sewer lines and to prevent reductions in biological treatment process efficiency by toxic effects from the toxic concentration of organic and inorganic substances. The efficient removal of toxic metals from wastewater is an important matter and a number of technologies such as precipitation, reduction, and ion exchange were developed. This review summarizes the industrial applications of natural polysaccharides such as chitin – and chitosan-based materials in wastewater treatment. In addition, this review also opens up various factors affecting the adsorption and desorption processes. We expect that this chapter will provide insights on the use of these natural polysaccharides for researchers working to discover new materials with new properties for the valuable applications.

Acknowledgments

The authors are grateful to the authorities of DKM College for Women and Thiruvalluvar University, Vellore, Tamil Nadu, India for their support. Thanks are also due to the editor for the opportunity to review such an innovating field.

References

1. Shannon, M.A., Bohn, P.W., Elimelech, M., Georgiadis, J.G., Marinas, B.J., Mayes, A.M., Science and technology for water purification in the coming decades. *Nature,* 452, 301–310, 2008.
2. Muthumanickam D., Sengottuvelan., Synthesis and characterization of chitosan-g-N-methyl piperaziniumchloride: A hybrid flocculant., *Int. J. Biol. Macromol.,* 81, 778–784, 2015.
3. Teh, C.Y., Budiman, P.M., Shak, K.P.Y., and Wu T.Y., Recent advancement of coagulation–flocculation and its application in wastewater treatment. *Ind. Eng. Chem. Res.*, 55, 4363–4389, 2016.
4. Abdul Khalil, H.P.S., Bhat, A.H., and Ireana Yusra, A.F., Green composites from sustainable cellulose nanofibrils. A review. *Carbohydr. Polym.*, 87, 963–979, 2012.
5. Abdul Khalil, H.P.S., Sri Aprilia, N.A., Bhat, A.H., Jawaid, M., Paridah, M.T., and Rudi, D., A Jatropha biomass as renewable materials for biocomposites and its applications. *Renew. Sustainable Energy Rev.,* 22, 667–685, 2013.
6. Kim, C.H., Choi, J.W., Chun, H.J., Choi, K.S., Synthesis of chitosan derivatives with quaternary ammonium salt and their antibacterial activity. *Polym. Bull.*, 38, 387, 1997.
7. Kurita, K., Hirakawa, M., Aida, K., Yang, J., and Nishiyama, Y., Trimethylsilylated chitosan: a convenient precursor for chemical modifications. *Chem. Lett.*, 32, 11, 1074, 2003.
8. Boddu, V., Abburi, K., Talbott, J.L., and Smith, E.D., Removal of hexavalent chromium from wastewater using a new composite chitosan biosorbent. *Environ. Sci. Technol.,* 37, 4449–4456, 2003.
9. Acosta, R.I., Rodriguez, X., Guiterrez, C., Motctezuma, G., Bio sorption of Chromium (VI) from aqueous solutions on to fungal biomass. *Bio. Inorg. Chem. Appl. J.,* 2, 1–7, 2004.

10. Gupta, V.K., Shrivastava, A.K., and Jain, N., Bio sorption of chromium (VI) from aqueous solution by green algae spirogyra species, *Water Res.*, 35, 4079–4085, 2001.
11. Kratochvil, D., Pimentel, P., and Volesky, B, Trivalent and hexavalent chromium by seaweed bio sorbent, *Env. Sci. Technol.*, 32, 2693–2698, 1998.
12. Elangovan, R., Philip, L., and Chandraraj, K., Bio sorption of chromium species by aquatic weeds: Kinetics and mechanism studies, *J. Hazard. Mater.*, 152, 100–112, 2008.
13. Sahin, Y., and Ozturk, A., Bio sorption of chromium (VI) ions from aqueous solution by bacterium Bacillus thuringiensis, *Process Biochem.*, 40, 1895–190, 2005.
14. Fan, T., Liu, Y., Feng, B., Zeng, G., Yang, C., Zhou, M., Tan, Z., and Wang, X., Bio sorption of cadmium(II), zinc(II), and lead(II) by *Penicillium Simplicissimum*: Isotherm, kinetics, and thermodynamics. *J. Hazard. Mater.*, 160, 655–661, 2008.
15. Bailey, S.E., Olin, T.J., Bricka, R.M., Adrian, D.D., A review of potentially low-cost sorbents for heavy metals. *Water Resour.*, 33, 2469–2479, 1999.
16. Wu, F.C., Tseng, R.L., and Juang, R.S., Comparative adsorption of metal and dye on flake and bead types of chitosan prepared from fishy wastes. *J. Hazard. Mater.*, 73, 63–75, 2000.
17. Gupta, K.C., and Ravikumar, M.N.V., Preparation characterization and release profiles of pH sensitive chitosan beads. *Polym. Int.*, 49, 141–146, 2000.
18. RaviKumar, M.N.V., A review of chitin and chitosan applications. *React. Func. Polym.*, 46, 1–27, 2000.
19. Wurzburg, O.B., Ed., *Modified starches, properties and uses.* Boca Raton, FL: CRC Press, 1986
20. Varma, A.J., Deshpande, S.V., and Kennedy, J.F., Metal complexation by chitosan and its derivatives: a review, *Carbohydr. Polym.*, 55, 77–93, 2004.
21. Bolto, B.A., Soluble polymers in water purification. *Prog. Polym. Sci.*, 20, 987–1041, 1995.
22. Knorr, D., Use of chitinous polymers in food-A challenge for food research and development. *Food Technol.*, 38, 85–97, 1984.
23. Kurita, K., Controlled functionalization of polysaccharide chitin. *Prog. Polym. Sci.*, 26, 1921–1971, 2001.
24. Rutherford, F., Austin, P.R., Proceedings of the First International Conference Oil Chitin Chitosan; MIT: Cambridge, MA, 182–192, 1978.
25. Capozza, R.C., Enzymically decomposable biodegradable pharmaceutical carrier, German Patent 2505305, 1975.
26. Hayes, E.R., Davies, D.H., Munroe, V.G., Organic solvent systems for chitosan. Proceedings of 1st International Conference on Chitin and Chitosan. MIT Sea Grant Program. Massachusetts, 103, 1977.
27. Kumar, M.N.V.R., A review of chitin and chitosan applications. *React. Funct. Polym.*, 46, 1–27, 2000.

28. Li, Q., Dunn, E.T., Grandmaison, E.W., Goosen, M.F.A., Applications and properties of chitosan. *J. Bioact. Compat. Polym.*, 7, 370–397, 1992.
29. Francis Suh, J.K., Matthew, H.W.T., Application of chitosan based polysaccharide biomaterials in cartilage tissue engineering: a review. *Biomater.*, 21, 2589–2598, 2000.
30. Sinha, V.R, Singla, A.K., Wadhawan,S., Kaushik, R., Kumria, R., Bansal, K., Dhawan, S., Chitosan microspheres as a potential carrier for drugs. *Int. J. Pharm.*, 274, 1–33, 2004.
31. Chenite, A., Buschmann, M., Wang, D., Chaput, C., Rheological Characterization of thermogelling chitosan/glycerol phosphate solution. *Carbohydr. Polym.*, 46, 39–47, 2001.
32. Hong-Mei Kang., Yuan-Li Cai., Peng-Sheng Liu., Synthesis, characterization and thermal sensitivity of chitosan-based graft copolymers. *Carbohydr. Res.*, 341, 2851–2857, 2006.
33. Van der Lubben, I.M., Verhoef, J.C., Borchard, G., Junginger H.E., Muzzarelli R.A.A., Eds., *Chitosan per os from Dietary supplement to drug carrier*, ATEC, Grottammare, Italy, 115–126, 2000.
34. Guibal, E., Milot, C., Roussy J., Molybdate sorption by cross-linked chitosan beads; dynamic studies, *Water Env. Res.*, 71, 10–17, 1999.
35. Wang, X.P., Shen Z.Q., Studies on the effects of copper salts on the separation performance of chitosan membranes. *Polym Int.*, 49, 1426–1433, 2000.
36. Mi, F.L., Shyu, S.S., Wu, Y.B., Lee, S.T., Shyong, J.Y., Huang, R.N., Fabrication and characterization of a sponge-like asymmetric chitosan membrane as a wound dressing. *Biomater.*, 22, 165–173, 2001.
37. Amaike, M., Senoo, Y., Yamamoto H., Sphere, honeycomb, regularly spaced droplet and fiber structure of polyion complexes of chitosan and gellan. *Macromol. Rapid Comm.*, 19, 287, 1998.
38. Vincent, T., Guibal E., Non-dispersive liquid extraction of Cr (VI) by TBP/aliquat 336 using chitosan-made hollow fibers. *Solv. Ext. Ion Exchange,* 18, 1241–1260, 2000.
39. Agboh, O.C., Quin, Y., Chitin and chitosan fibers. *Adv. Polym. Technol.*, 8, 355, 1997.
40. Dzul Erosa, M.S., Saucedo Medina, T.I., Navarro Mendoza, R., Avila Rodriguez, M., Guibal., Cadmium sorption on chitosan derivatives. *Hydrometallurgy.*, 61, 157–167, 2001.
41. Milot, C., Adsorption de Molybdate VI sur Billes de Gel de Chitosan-Application au Traitement d'Efflyents Molydiferes, Ph.D., Unite doctorale "Mecanique des Materiaux, Structures et Genie des Procedes-Laboratoire d' accueil, Laboratoire Genie de I'Environment Industriel (Ecole des Mines d' Ales). 214, 1998.
42. Maruca, R., Suder, B.J., Wightman J.P., Interaction of heavy metals with chitin and chitosan. III. Chromium, *J. Appl. Polym. Sci.*, 27, 4827–4837, 1982.
43. Kwok, K., Lee, V., Gerente, C., McKay, G., Novel model development for sorption of arsenate on chitosan, *J. Chem.Eng.*, 151, 122–133 2009.

44. Chauhan, D., Sankararamakrishnan, N., Highly enhanced adsorption for decontamination of lead ions from battery wastewaters using chitosan functionalized with Xanthate. *Bioresour. Technol.*, 99, 9021–9024, 2008.
45. Aydın, Y., Aksoy, N., Adsorption of chromium on chitosan; Optimization, kinetics, and thermodynamics. *Chem. Eng. J.*, 151, 188–194, 2009.
46. Guibal, E., Jansson-Charrier, M., Saucedo, I., Le cloirec P., Enhanced of metal ion sorption performances of chitosan; effect of structure on the diffusion properties, *Langumuir II,* 591–598, 1995.
47. Bhattacharya, A., Misra B.N., Grafting-A versatile means to modify polymers techniques, factors and applications, *Prog. Polym. Sci.,* 29, 767–814, 2004.
48. Baker, R.W., *Membrane technology and applications,* J. Wiley & Sons, Chictester, U.K., 2004.
49. Elisaveta, K., Mladenova, I., Grigorova, D., Irina K.B., Chitosan membrances as sorbents for trace elements determination in surface waters, *Environ. Sci. Pollut. Res.* 18, 1633–1643, 2011.
50. Wysokowski, M., Bazhenov, V.V., Tsurka , M.V., Galli, R., Stelling, A.L., Stöcker, H., Kaiser, S., Niederschlag, E., Gartner, G., and Behm, T., Isolation and identification of chitin in three-dimensional skeleton of Aplysina. fistularis marine sponge. *Int. J. Biol. Macromol.*, 62, 94–100, 2013.
51. Li, H., Huang, G., An, C., Hu, J., Yang, S., Removal of tannin from aqueous solution by adsorption onto treated coal fly ash: kinetic, equilibrium, and thermodynamic studies. *Ind. Eeg. Chem. Res.,* 52, 15923–15931, 2013.
52. Chethan, P.D., Vishalakshi, B., Synthesis of ethylenediamine modified chitosan and evaluation for removal of divalent metal ions. *Carbohydr Polym.*, 97, 530–536, 2013.
53. Huang, X., Mao, X., Bu, H., Yu, X., Jiang, G., Zeng, M., Chemical modification of chitosan by tetraethylenepentamine and adsorption study for anionic dye removal. *Carbohydr. Res.*, 346, 1232–1240, 2011.
54. Vakili, M., Rafatullah, M., Salamatinia, B., Abdullah, A., Ibrahim, M., Tan, K., Gholami, Z., Amouzgar, P., Application Of chitosan and its Derivatives as adsorbents For dye removal From water and wastewater: A Review. *Carbohydr. Polym.*, 113, 115–130, 2014.
55. Jalal Zohuriaan-Mehr, M., Advances in Chitin and Chitosan Modification through Graft Copolymerization:A Comprehensive Review. *Iran Polym. J.*, 14, 235–265, 2005.
56. Aderonke., Okoya, A., Abimbola, B., Akinyele., Ifeanyi., Ofoezie, E., Omotayo., Amuda, S., Oluwagbemig., Alayande ., Oladotun, S., Makinde, W., Adsorption of heavy metal ions onto chitosan grafted cocoa husk char. *Afric J. Pure. Appl. Chem.,* 8, 147–161, 2014.
57. Nithya, R., Gomathi,T., Sudha, P.N., Jayachandran Venkatesan., Sukumaran Anil., Se-Kwon Kim., Removal of Cr(VI) from aqueous solution using chitosan-g – poly(butylacrylate)/silica gel nanocomposite. *Int. J. Biol. Macromol.*, 87, 545–554, 2016.

58. Ruiz, M., Sastre, A. M., Guibal, E., Palladium sorption on glutaraldehyde cross-linked chitosan. *React. Funct. Polym.*, 45, 155–173, 2000.
59. Tang, X.H., Tan, S.Y., Wang, Y.T., Study of the synthesis ofchitosan derivatives containing benzo-21-crown-7 and their adsorption properties for metal ions. *J. Appl. Polym. Sci.*, 83, 1886–91, 2002.
60. Lee, S., Mi, F., Shen, Y., Shyu S., Equilibrium and kinetic studies of copper(II) ion uptake by chitosan-tripolyphosphate chelatin resin. *Polymer.*,42, 1879–1892, 2001.
61. Sudha, P.N., Celine, S., and Jayapriya, S., Removal of heavy metal cadmium from industrial wastewater using chitosan coated coconut charcoal. *Nat. Environ. Pollut. Technol.*, 6, 421–424, 2008.
62. Kumar, N.S., Subbaiah, M.V., Reddy, A.S., *J. Chem. Technol. Biotechnol.*, 84, 972–981, 2009.
63. Tran, H., Tran, L., Nguyen, T., Preparation of chitosan/ magnetite composite beads and their application for removal of Pb(II) and Ni(II) from aqueous solution. *Mater. Sci. Eng. C.*, 30, 304–310, 2010.
64. Li, Q., Yue, Q. Y., Sun, H. J., Su, Y., and Gao, B. Y., A comparative study on the properties, mechanism and process designs for the adsorption of non-ionic or anionic dyes ontocationic-polymer/bentonite. *J. Env. Manag.*, 91, 1601–1611, 2010.
65. Thanou, M., Nihot M.T., Jansen, M., Verhoef, J.C., Junginger. H.E., Mono-N-Carboxymethyl chitosan (MCC), a polyampholytic chitosan derivative, enhances the intestinal absorption of low molecular weight heparin across intestinal epithelia *in vitro* and *in vivo*. *J. Pharm. Sci.*, 90, 38–46, 2001.
66. Bhattacharya, A., Misra B.N., Grafting: a versatile means to modify polymer-sTechniques, factors and applications. *Prog. Polym. Sci.*, 29, 767–814, 2004.
67. Gopal Reddi, M.R., Gomathi, T., Saranya, M., Sudha, P.N., Adsorption and kinetic studies on the removal of chromium and copper onto Chitosan-g-maliec anhydride-g-ethylene dimethacrylate. *Int. J. Biol. Macromol.*, 16, 32638, 2017.
68. Barakat, M.A., New trends in removing heavy metals from industrial wastewater, *Arab. J.Chem.*, 4, 361–377, 2011.
69. Abdel-Salam, O.E., Reiad, N.A., ElShafei, M. M., A study of the removal characteristics of heavy metals from wastewater by low-cost adsorbents. *J. Adv. Res.*, 2, 297–303, 2011.
70. Sudha, P.N., *Chitin, chitosan, oligosaccharides and their derivatives*, Se Kwon Kim, Ed., CRC Press, pp 561–585, 2010.
71. Kurita, K., Controlled functionalization of polysaccharide chitin. *Prog. Polym. Sci.*, 26, 1921–1971, 2001.
72. Abollino O., Aceto M., Malandrino M., Sarzanini C., Mentasti E., Adsorption of heavy metals on Na-montmorillonite. Effect of pH and organic substances, *Water Res.*, 37, 1619–1627, 2003.

73. Ngah W.S.W., Endud C.S., Mayanar R., Removal of copper (II) ions from aqueous solution onto chitosan and cross-linked chitosan beads, *React. Funct. Polym.*, 50, 181–190, 2002.
74. Abdeen, Z., Somaia G., Mohammad, M. S., Mahmoud Adsorption of Mn (II) Ion on Polyvinyl Alcohol /Chitosan Dry Blending from aqueous solution. *Environ. Nanotechnol. Monit. Manag.,* http://dx.doi.org/10.1016/j.enmm.2014.10.001., 2014.
75. Mohamed, Ali., E.A., Synthesis and adsorption properties of chitosan-CDTA-GO nanocomposite for removal of hexavalent chromium from aqueous solutions. *Arab. J. Chem.*, http://dx.doi.org/10.1016/j.arabjc.2016.09.010., 2016.
76. Teimouri, A., Nasab, S., Vahdatpoor, N., Habibollahi, S., Salavati, H., Chermahini, A., Chitosan/ Zeolite Y/Nano ZrO2 nanocomposite as an adsorbent for the removal of nitrate from the aqueous solution. *Int. J. Biol. Macromol.*, 93, 254–266, 2016.
77. Hemalatha, R., Chitra, R., Xavier Raja Rathinam., Sudha.P.N., Synthesizing and characterization of chitosan graft co polymer: adsorption studies for Cu (II) and Cr (VI). *Int. J. Env. Sci.*, 2, 805–827, 2011.
78. Jaros K., Kamiński W., Albińska J., Nowak U., Removal of heavy metal ions: copper, zinc and chromium from water on chitosan beads. *Env. Protec. Eng.*, 31, 3–4, 2005.
79. Hritcu, D., Dodi, G., Popa, M.I., Heavy Metal Ions Adsorption on Chitosan-Magnetite Microspheres. *Int. Rev. Chem. Eng.*, 4, 364–368, 2012.
80. Lertsutthiwong, P., Noomun, K., Khunthon, S., Limpanart, S., Influence of chitosan characteristics on the properties of biopolymeric chitosan–montmorillonite. *Progress in Natural Science: Mater. Int.*, 2, 502–508, 2012.
81. Dhananasekaran, S., Palanivel, R., Pappu, S., Adsorption of Methylene Blue, Bromophenol Blue and Coomassie Brilliant Blue by α-Chitin Nanoparticles. *J. Adv. Res.*, 7, 113–124.
82. Hajivandi, A., Farjadfard, S., Ramavandi, B., Akbarzadeh, S., Experimental data for synthesis of bi-metalized chitosan particle for attenuating of an azo dye from wastewater. *Data in Brief.*, 7, 71–76, 2016.
83. Iqbal, J., Wattoo, F., Wattoo, M., Malik, R., Tirmizi, S., Imran, M., Ghangro, A., Adsorption of acid yellow dye on flakes of chitosan prepared from fishery wastes. *Arab. J. Chem.*, 4, 389–39, 2011.
84. Vakili, M., Rafatullah, M., Salamatinia, B., Abdullah, A., Ibrahim, M., Tan, K., Gholami, Z., Amouzgar, P., Application of chitosan and its derivatives as adsorbents For dye removal from water and wastewater: a review. *Carbohydr. Polym.*, 113, 115–130, 2014.
85. Gupta, A., Chauhan, V.S., Sankararamakrishnan ,N., Preparation and evaluation of iron-chitosan composites for removal of As (III) and As (V) from arsenic contaminated real life groundwater. *Water Res.*, 43, 3862–3870, 2004.

86. Kumar, N.S., Subbaiah, M.V., Reddy, A.S. Biosorption of phenolic compounds from aqueous solutions onto chitosan-abrus precatorius blended beads. *J. Chem. Technol. Biotechnol.*, 84, 972–981, 2009.
87. Bhavani, K., Roshan Ara Begum, E., Selvakumar, S., Shenbagarathai, R., Chitosan a low cost adsorbent for electroplating waste water treatment. *J. Bioremed. Biodegrad.*, 7, 346, 2016.
88. Abdulkarim, A., Isa, M. T., Ameh., A. O., Muhammad, J.A., Chromium Ion Adsorption Using Chitosan from Mussel. *J. Appl. Chem.*, 7, 24–28, 2014.
89. Sundaram, C.S., Viswanathan, N., Meenakshi, S.,Defluoridation of water using magnesia/ chitosan composite. *J. Hazard. Mater.*, 163, 618, 2009.
90. Olcay Tünay., Isik Kabdasli., Gulen Eremektar., Derin Orhon., Color removal from textile wastewaters. *Water Sci. Technol.*, 34, 9–16, 1996.
91. Anjaneyulu, Y., Sreedhara Chary, N., Suman Raj, D.S., Decolourization of industrial effluents - available methods and emerging technologies - a review. *Rev. Env. Sci. Biotechnol.*, 4, 245–273, 2005.
92. Raghunadh Acharyulu, S., Prakash, N., Sudha, P.N., Chitosan blended polystyrene synthesis and its use to remove the toxic effects of chromium and copper from industrial wastewater. *Der Pharma. Chemica.*,6, 177–186, 2014.
93. Szyguła, A., Guibal, E., Ruiz, M., Sastre, A.M., The removal of sulphonated azo-dyes by coagulation with chitosan. *Colloids Surf. A: Physicochem. Eng. Aspects.*, 330, 219–226, 2008.
94. Cestari, A., Vieira, E., Tavares, A., Bruns, R., The removal of the indigo carmine dye from aqueous solutions using crosslinked chitosan—Evaluation of adsorption thermodynamics using a full factorial design. *J. Hazard. Mater.*, 153, 566–574, 2008.
95. Golob ,V., Vinder, A., Simoni, M., Efficiency of the coagulation/flocculation method for the treatment of dyebath effluents. *Dye Pigment.*, 67, 93–97, 2005.
96. Dao, V.H., Cameron, N.R., Saito, K., Synthesis, properties and performance of organic polymers employed in flocculation applications. *Polym. Chem.*, 7, 11–25, 2016.
97. Arena, F., Di Chio, R., Gumina, B., Spadaro, L., Trunfio, G., Recent advances on wet air oxidation catalysts for treatment of industrial wastewaters. *Inorg. Chim.Acta.*, 431, 101–109, 2015.
98. Fane, A.G., Wang, R., Hu, M.X., Synthetic membranes for water purification: status and future. *Angew. Chem. Int. Edit.*, 54, 3368–3386, 2015.
99. Pérez-González,A., Urtiaga,A.M., Ibáñez, R., Ortiz, I., State of the art and review on the treatment technologies of water reverse osmosis concentrates. *Water Res.*, 267–283, 2012.
100. Al-Rashdi, B.A.M., Johnson, D.J., Hilal, N., Removal of heavy metal ions by nanofiltration. *Desalin.*, 315, 2–17, 2013.
101. Mezaguer, M., Kamel, N., Lounici, H., Kamel, Z., Characterization and properties of Pleurotus mutilus fungal biomass as adsorbent of the removal of uranium(VI) from uranium leachate. *J. Radioanal. Nucl. Chem.*, 295, 393–403, 2013.

102. Ali, H., Khan, E., Sajad, M., Phytoremediation of heavy metals—Concepts and applications. *Chemosphere.*, 91, 869–881, 2013.
103. Demim, S., Drouiche, N., Aouabed, A., Benayad, T., Couderchet, M., Semsari, S., Study of heavy metal removal from heavy metal mixture using the CCD method. *J. Ind. Eng. Chem.*, 20, 512–520, 2014.
104. Belouchrani, A., Mameri, N., Abdi, N., Grib, H., Lounici, H., Drouiche, N., Phytoremediation of soil contaminated with Zn using Canola (Brassica napus L). *Ecolog. Eng.*, 95, 43–49, 2016.
105. Liu, T., Chen, Z.L., Yu, W.Z., Shen, J.M., Gregory, J., Effect oftwo-stage coagulant addition on coagulation–ultrafiltrationprocess for treatment of humic-rich water. *Water Res.*, 45, 4260–4268, 2011.
106. Nurdeen Abiola Oladoja., Headway on natural polymeric coagulants in water and wastewater treatment operations. *J. Water Proc. Eng.*, 6, 174–192, 2015.
107. Jiang, J.Q., The role of coagulation in water treatment. *Curr. Opin. Chem. Eng.*, 8, 36–44, 2015.
108. Yang, R., Li, H., Huang, M., Yang, H., Li, A., A review on chitosan based flocculants and their applications in water treatment. *Water Res.*, 95, 59–89, 2016.
109. Al-Malack, M.H., Abuzaid, N.S., El-Mubarak, A.H., Coagulation of polymeric wastewater discharged by a chemical factory. *Water Res.*,33, 521–529, 1999.
110. Chong, M., Direct flocculation process for wastewater treatment, In: *Advances in water treatment and pollution prevention*, S.K. Sharma, R. Sanghi, Eds., pp. 201–230, Springer, Netherlands, 2012.
111. Ruden, C., Acrylamide and cancer risk - expert risk assessments and the-public debate. *J. Food Chem. Toxicol.* 42, 335–349, 2004.
112. Sciban, M., Klasnja, M., Antov, M., Skrbic, B., Removal of water turbidity by natural coagulants obtained from chestnut and acorn, *Bioresour.Technol.*, 100, 6639–6643, 2009.
113. Riske, F., Schroeder, J., Belliveau, J., Kang, X., Kutzko, J., Menon, M.K., The use of chitosan as a flocculant in mammalian cell culture dramatically improves clarification throughput without adversely impacting monoclonal antibody recovery. *J. Biotechnol.*, 128, 813–23, 2007.
114. Yu, W.Z., Gregory, J., Campos, L., Breakage and regrowth of al-humic flocs-effect of additional coagulant dosage. *Environ. Sci. Technol.*, 44, 6371–6376, 2010.
115. Fabrizi, L., Jefferson, B., Parsons, S.A., Wetherill, A., Jarvis, P., The role of polymer in improving floc strength for filtration. *Environ. Sci. Technol.*, 44, 6443–6449, 2010.
116. Wei, J.C., Gao, B.Y., Yue, Q.Y., Wang, Y., Li, W.W., Zhu, X.B., Comparison of coagulation behavior and floc structure characteristic of different polyferric-cationic polymer dual-coagulants in humic acid solution, *Water Res.*, 43, 724–732, 2009.
117. Moghaddam, S.S., Moghaddam, M.R.A., Arami, M., Coagulation/flocculation process for dye removal using sludge from water treatment plant:

Optimization through response surface methodology. *J. Hazard. Mater.*,175, 651–657, 2010.

118. Renault, F., Sancey, B., Badot, P. M., Crini, G., Chitosan for coagulation/flocculation processes – An eco-friendly approach, *Eur. Polym. J.*, 45, 1337–1348, 2009.
119. Zhou, W., Zorn, M., Nawroth, P., Bütehorn, U., Perzborn, E., Heitmeier, S., Veltkamp, R., Hemostatic therapy in experimental intracerebral hemorrhage associated with rivaroxaban. *Stroke.*, 44, 771–778, 2013.
120. Choy, S.Y., Prasad, K.M.N., Wu, T.Y., Raghunandan, M.E., Ramanan, R.N., Utilization of plant-based natural coagulants as future alternatives towards sustainable water clarification. *J. Environ. Sci.*, 26, 2178–89, 2014.
121. Bratby, J., *Coagulation and flocculation in water and wastewater treatment*, 2nd ed. IWA Publishing, UK, 2006.
122. Sharma, B., Dhuldhoya, N., Merchant, U., Flocculants-an ecofriendly approach. *J. Polym. Environ.*, 14, 195–202, 2006.
123. Bergamasco, R., Bouchard, C., Da Silva, F.V., Reis, M.H.M., Fagundes-Klen, M.R., An application of chitosan as a coagulant/flocculant in a microfiltration process of natural water. *Desalin.*, 245, 205–213, 2009.
124. Chun-Yang Yin., Emerging usage of plant-based coagulants for water and wastewater treatment. *Proc.Biochem.*, 45, 1437–1444, 2010.
125. Diaz, A., Rincon, N., Escorihuela, A., Fernandez, N., Chacin, E., Forster, C.F., A preliminary evaluation of turbidity removal by natural coagulants indigenous to Venezuela. *Proc. Biochem.*, 35, 391–395, 1999.
126. Ang, W.L., Mohammad, A.W., Benamor, A., Hilal, N., chitosan as natural coagulant in hybrid coagulation-nanofiltration Membrane process for water treatment. *J. Environ. Chem. Eng.*, 4, 4857–4862, 2016.
127. Pastor, A., Quitina y Quitosano: obtención., caracterización y aplicaciones., Programa Cyted, Ciad, A.C., Fondo Editorial de la Pontificia Universidad Católica del Perú, Lima, Perú, 2004.
128. Guibal, E., Roussy, J., Coagulation and Flocculation of Dye Containing Solutions Using a Biopolymer (Chitosan). *React. Func. Polym.*, 67, 33–42, 2007.
129. Jadhav, M., Mahajan, Y., Investigation of the performance of chitosan as a coagulant for flocculation of local clay suspensions of different turbidities. *KSCE J. Civil. Eng.*, 17, 328–334, 2013.
130. Nicu,R., Bobu, E., Miranda, R., Blanco, A., Flocculation efficiency of chitosan for papermaking applications. *Bioresour.*, 8, 768–784, 2013.
131. Gassara, F., Antzak, C., Ajila, C., Sarma, S., Brar, S., Verma, M., Chitin and chitosan as natural flocculants for beer clarification. *J. Food Eng.*, 166, 80–85, 2015.
132. Yang, Z., Li, H., Yan, H., Wu, H., Yang, H., Wu, Q., Li, H., Li, A., Cheng, R., Evaluation of a novel chitosan-based flocculant with high flocculation performance, low toxicity and good floc properties. *J. Hazard. Mater.*, 276, 480–488, 2014.
133. Davydova, S., Heavy metals as toxicants in big cities, *Microchem. J.*, 79, 133–136, 2005.

134. Quek, S.Y., Wase, D.A.J., Forster, C.F., The use of sago waste for the sorption of lead and copper. *Water SA.*, 24, 251–256, 1998.
135. Chaudhari, S., Tare, V., Heavy metal-soluble starch xanthate interactions in aqueous environments. *Appl. Polym. Sci.*, 71, 1325–1332, 1999.
136. Ronald, R.N., Kenji, T., Enhancement of phosphonomethylate PEI–Cu complex flocculation by Ca ions: a new approach for heavy metal removal from aqueous solutions. *Sep. Sci. Technol.*, 37, 203–216, 2002.
137. Yu, M.Q., Chang, Q., Study on characteristics and action mechanisms of macromolecule heavy metal flocculant. *Acta Scientiae Circumstantiae.*, 25, 180–185, 2005.
138. Chang, Q., Preparation of flocculant PEX and treatment of wastewater containing copper. *Environ. Chem.*, 25, 176–179, 2006.
139. Zhang, C., Zhang, M., Chang, Q., Preparation of mercaptoacetyl chitosan and its removal performance of copper ion and turbidity. *Desalin. water treat.*, 1109–1116, 2015.
140. Tokura, S., Nishi, N., Specification and characterization of chitin and chitosan. In *Chitin and chitosan: The versatile environmentally friendly modern materials*, M.B. Zakaria, W.M.W. Muda, M.P. Abdullah, Eds., pp. 68–86, Bangi: Universiti Kebangsaan, Malaysia, 1995.
141. Guibal, E., Interactions of metal ions with chitosan-based sorbents: A review. *Sep.Purif.Technol.*, 38, 43, 2004.
142. Kang, D. W., Choi, H. R., Kweon, D. K.,Selective adsorption capacity for metal ions of amidoximated chitosan bead-g-PAN copolymer. *Polymer (Korea).*, 20, 989–995, 1996.
143. Kang, D.W., Choi, H.R., Kweon, D.K., Stability constants of amidoximated chitosan-g poly (acrylonitrile) copolymer for heavy metal ions. *J. Appl. Polym.Sci.*, 73, 469–476, 1999.
144. Yang, P.Y., Chen, H.J., Kim, S.J., Integrating entrapped mixed microbial cell (EMMC) process for biological removal of carbon and nitrogen from dilute swine wastewater. *Bioresour Technol.*, 86, 245–252, 2003.
145. Knorr, D., Recovery and utilization of chitin and chitosan in food processing waste management. *Food Technol.*, 45, 114–122, 1991.
146. Milhome, M., Keukeleire, D., Ribeiro, J., Nascimento, R., Carvalho, T., Queiroz, D., Removal of phenol and conventional pollutants from aqueous effluent by chitosan and chitin. *Quimica Nova.*, 32, 2122– 2127, 2009.
147. Kadirvelu, K., Karthika, C., Vennilamani, N., Pattabhi, S., Activated carbon from industrial solid waste as an adsorbent for the removal of rhodamine –B from aqueous solution: Kinetic and equilibrium studies. *Chemosphere.*, 60, 1009–1017, 2005.
148. Arami, M., Limaee, N., Mahmoodi, N., Tabrizi, N., Equilibrium and kinetics studies for the adsorption of direct and acid dyes from aqueous solution by soy meal hull. *J. Hazard. Mater.*, 135, 171–176, 2006.
149. Crini, G., Non-conventional low-cost adsorbents for dye removal: A review. *BioresourTechnol.*, 97, 1061–1085, 2006.

150. Boddu, V., Abburi, K., Talbott, J.L., Smith, E.D., Removal of hexavalent chromium from wastewater using a new composite chitosan biosorbent. *Environ. Sci. Technol.*, 37, 4449–4456, 2003.
151. Sharma, J. K., Kaushik, C. P., Kaushik, N., Low cost adsorbents in decolorization of effluents from dying of cotton fabric with congo red and chryosophenine-G. *Ind JEnviron Protect.*, 25, 61–65, 2005.
152. Singh, D.K., Srivastava, N., Basic dyes removal from wastewater by adsorption on rice husk carbon. *Ind J ChemTechnol.*, 8, 133–139, 2001.
153. Prasad, A. L., Santhi, T., Adsorption of hazardous cationic dyes from aqueousolution onto Acacia nilotica leaves as an eco friendly adsorbent. *Sust. Env. Res.*, 22, 113–122, 2012.
154. Gao, J., Zhang, Q., Su, K., Chen, R., Peng, Y., Biosorption of Acid Yellow 17 from aqueous solution by non-living aerobic granular sludge. *J. Hazard. Mater.*, 174, 215–225, 2010.
155. Rahman, F.B.A., Akter, M., Removal of dyes form textile wastewater by adsorption using shrimp shell. *Int J Waste Resour.*, 6, 1–5, 2016.
156. Mezohegyi, G., Van der Zee, F.P., Font, J., Fortuny, A., Fabregat, A., Towards advanced aqueous dye removal processes: A short review on the versatile role of activated carbon. *J. Environ. Manag.*, 102, 148–164, 2012.
157. Gomez, V., Larrechi, M.S., Callao, M.P., Kinetic and adsorption study of acid dye removal using activated carbon. *Chemosphere.*, 69, 1151–1158, 2007.
158. Fane, A.G., Fell, C.J.D., A review of fouling and fouling control in ultrafi ltration. *Desalin .*, 62, 117–136, 1987.
159. Le Roux, I., Krieg, H.M., Yeates, C.A., Breytenbach, J.C., Use of chitosan as an antifouling agent in a membrane bioreactor. *J. Membr. Sci.*, 248, 127–136, 2005.
160. Morrissey, M.T., Park, J.W., Huang, L., Surimi processing waste: Its control and utilization. In: *Surimi and Surimi Seafood*, J.W. Park, Ed., pp. 127–165, Marcel Dekker Inc, New York, 2000.
161. Wibowo, S., Savant, V., Cherian, G., Savage, T.F., Torres, J.A., Evaluation as a feed ingredient of surimi wash water protein recovered using a chitosan-alginate complex. *J. Aquat. Food Prod. Technol.*, 14, 55–72, 2005.
162. Savant, V.D., Protein adsorption on chitosan-Polyanion complexes: Application to aqueous food processing wastes, PhD Thesis, Oregon State University, Corvallis, OR, 2001.
163. Wibowo, S., Velazquez, G., Savant, V., Torres, J.A., Surimi wash water treatment for protein recovery: Effect of chitosan–alginate complex concentration and treatment time on protein adsorption. *Bioresour. Technol.*, 96, 665–671, 2007.
164. Sarala, K., Rao, P.R., Endemic fluorosis in the village Ralla, Anantapuram in Andhra Pradesh—an epidemiological study. *Fluoride.*, 26, 177–180, 1993.
165. Ali, N., Hameed, A., Ahmed, S., Physicochemical characterization and Bioremediation perspective of textile effluent, dyes and metals by indigenous bacteria. *J. Hazard. Mater.*, 164, 322–328, 2009.

166. Alver, E., Metin, A.U., Anionic dye removal from aqueous solutions using modified zeolite: Adsorption kinetics and isotherm studies. *Chem. Eng. J.*, 200, 59–67, 2012.
167. Hariharasuthan, R., Rao, A., Bhaskaran, A., Adsorption studies on reactive blue by varying the concentration of Mgo In Sorel's cement. *Adsorption*, 2, 287–292, 2013.
168. Anjaneyulu, Y., Sreedhara Chary, N., Suman Raj, D.S., Decolourization of industrial effluents - available methods and emerging technologies - a review. *Rev. Env. Sci. Biotechnol.*, 4, 245–273, 2005.
169. Zonoozi, M.H., Alavi Moghaddam, M.R., Arami, M., Study on the removal of acid dyes using chitosan as a natural coagulant/coagulant aid. *Water. Sci. Technol.*, 63, 403–409, 2011.
170. Himanshu Patel, R.T., Vashi., Removal of Congo Red dye from its aqueous solution using natural coagulants. *J. Saudi. Chem. Soc.*, 16, 131–136, 2012.
171. Norzita Ngadi., Nor Aida Yusoff., Treatment of Textile Wastewater using Biodegradable Flocculants of Chitosan and Extracted Pandan Leaves. *J. Eng. Sci. Technol.*, 64, 1–7, 2013.
172. Wu, J., Liu, J., Lin, L., Zhang, C., Li, A., Zhu, Y., Zhang,Y., Evaluation of several flocculants for flocculating microalgae. *Bioresour. Technol.*, 197, 495–501, 2015.
173. Riera-Torres, M., Gutiérrez-Bouzan, C., Crespi, M., Combination of coagulation–flocculation and nanofiltration techniques for dye removal and water reuse in textile effluents. *Desalin.*, 252, 53–59, 2010.
174. Zahrim, A.Y., Tizaoui, C., Hilal, N., Coagulation with polymers for nanofiltration pretreatment of highly concentrated dyes: a review. *Desalin.*, 266, 1–16, 2011.
175. Sami, A., Khalid, M., Iqbal, S., Afzal, M. and Shakoori, A. R., Synthesis and application of chitosan-starch based nanocomposite in wastewater treatment for the removal of anionic commercial dyes. *Pakistan. J. Zool.*, 49, 21–26, 2017.

Index

Also of Interest

Check out these published related titles from Scrivener Publishing

Journal of Renewable Materials
Editors: Alessandro Gandini and Ramaswamy Nagarajan
www.scrivenerpublishing.com

Handbook of Composites from Renewable Materials in 8 volumes
Edited by Vijay Kumar Thakur, Manju Kumari Thakur and Michael R. Kessler
Volume 1: Structure and Chemistry
ISBN: 978-1-119-22362-7
Volume 2: Design and Manufacturing
ISBN: 978-1-119-22365-8
Volume 3: Physico-Chemical and Mechanical Characterization
ISBN: 978-1-119-22366-5
Volume 4: Functionalization
ISBN: 978-1-119-22367-2
Volume 5: Biodegradable Materials
ISBN: 978-1-119-22379-5
Volume 6: Polymeric Composites
ISBN: 978-1-119-22380-1
Volume 7: Nanocomposites: Science and Fundamentals
ISBN: 978-1-119-22381-8
Volume 8: Nanocomposites: Advanced Applications
ISBN: 978-1-119-22383-2
8-volume set
ISBN 978-1-119-22436-5

Lignocellullosic Biomass Production and Industrial Applications
Edited by Arindam Kuila and Vinay Sharma
Published 2016. ISBN 978-1-119-32360-0

Advances in Biofeedstocks and Biofuels. Volume 2
Production and Technologies
Edited by Lalit Kumar Singh and Gaurav Chaudhary
Published 2016. ISBN 978-1-119-11752-0

Advances in Biofeedstocks and Biofuels. Volume 1
Biofeedstocks and Their Processsing
Edited by Lalit Kumar Singh and Gaurav Chaudhary
Published 2016. ISBN 978-1-119-11725.4

Biobased and Environmentally Benign Coatings
Edited by Atul Tiwari, Anthony Galanis, and Mark D. Soucek
Published 2016. ISBN: 978-1-119-18492-8

Lignocellulosic Fibers and Wood Handbook
Edited by Mohamed Naceur Belgacem and Antonio Pizzi
Published 2016. ISBN: 978-1-118-77352-9

Biodegradable and Biobased Polymers for Environmental and Biomedical Applications
Edited by Susheel Kalia and Luc Avérous
Published 2016. ISBN: 978-1-119-11733-9

Advanced Composites Materials
Edited by Ashutosh Tiwari, Mohammad Rabia Alenzi and Seong Chan Jun
Published 2016. ISBN: 9781119242536

Fundamentals of Conjugated Polymer Blends, Copolymers and Composites
Synthesis, Properties, and Applications
Edited by Parveen Saini
Published 2015. ISBN: 978-1-118-54949-0

Biofuels Production
Edited by Vikash Babu, Ashish Thapliyal, and Girijesh Kumar Patel
Published 2014. ISBN 978-1-118-63450-9

Lignocellulosic Polymer Composites
Processing, Characterization, and Properties
Edited by Vijay Kumar Thakur
Published 2014. ISBN 978-1-118-77357-4

Nanocellulose Polymer Nanocomposites
Fundamentals and Applications
Edited by Vijay Kumar Thakur
Published 2014. ISBN 978-1-118-87190-4

Handbook of Cellulosic Ethanol
By Ananda S. Amarasekara
Published 2014. ISBN 978-1-118-23300-9

The Chemistry of Bio-based Polymers
By Johannes Karl Fink
Published 2014. ISBN 978-1-118-83725-2

Handbook of Bioplastics and Biocomposites Engineering Applications
Edited by Srikanth Pilla
Published 2011. ISBN 978-0-470-62607-8

Renewable Polymers: Synthesis, Processing, and Technology
Edited by Vikas Mittal
Published 2011. ISBN 978-0-470-93877-5

Plastics Sustainability
Towards a Peaceful Coexistence between Bio-based and Fossil Fuel-Based Plastics
Michael Tolinski
Published 2011. ISBN 978-0-470-93878-2

Green Chemistry for Environmental Remediation
Edited by Rashmi Sanghi and Vandana Singh
Published 2011 ISBN 978-0-470-94308-3

Lightning Source UK Ltd.
Milton Keynes UK
UKOW01n1351150817
307352UK00001B/1/P